Rajkumar Roy (Ed.)

Industrial Knowledge Management

A Micro-level Approach

With 175 Figures

Springer

Rajkumar Roy, PhD
Department of Enterprise Integration, School of Industrial and Manufacturing
Science, Cranfield University, Cranfield, Bedford, MK43 0AL, UK

ISBN 1-85233-339-1 Springer-Verlag London Berlin Heidelberg

British Library Cataloguing in Publication Data
Industrial knowledge management : a micro-level approach
  1.Knowledge management 2.Knowledge management - Computer
  network resources 3.Knowledge management - Data processing
  I.Roy, R. (Rajkumar), 1966-
  ISBN 1852333391

Library of Congress Cataloging-in-Publication Data
Industrial knowledge management : a micro-level approach / Rajkumar Roy (ed.).
    p. cm.
  Includes bibliographical references and index.
  ISBN 1-85233-339-1 (alk. paper)
  1. Information resources management. 2. Knowledge management. I. Roy, R.
  (Rajkumar), 1966-
  T58.64.I454 2000
  658.5--dc21   00-044029

Typesetting: Camera ready by contributors
Printed and bound by Athenæum Press Ltd., Gateshead, Tyne & Wear
69/3830-543210 Printed on acid-free paper SPIN 10746129

Industrial Knowledge Management

Springer

London
Berlin
Heidelberg
New York
Barcelona
Hong Kong
Milan
Paris
Singapore
Tokyo

*The book is dedicated to the memory of
my beloved grandmother
Late Subhashini Roy*

Preface

The book presents state of the art practices and research in the area of Knowledge Capture and Reuse in industry. This book demonstrates some of the successful applications of industrial knowledge management at the micro level. The Micro Knowledge Management (MicroKM) is about capture and reuse of knowledge at the operational, shopfloor and designer level. The readers will benefit from different frameworks, concepts and industrial case studies on knowledge capture and reuse. The book contains a number of invited papers from leading practitioners in the field and a small number of selected papers from active researchers.

The book starts by providing the foundation for micro knowledge management through knowledge systematisation, analysing the nature of knowledge and by evaluating verification and validation technology for knowledge based system development. A number of frameworks for knowledge capture, reuse and integration are also provided. Web based framework for knowledge capture and delivery is becoming increasingly popular. Evolutionary computing is also used to automate design knowledge capture. The book demonstrates frameworks and techniques to capture knowledge from people, data and process and reuse the knowledge using an appropriate tool in the business. Therefore, the book bridges the gap between the theory and practice. The 'theory to practice' chapter discusses about virtual communities of practice, Web based approaches, case based reasoning and ontology driven systems for the knowledge management. Just-in-time knowledge delivery and support is becoming a very important tool for real-life applications. The tools chapter also discusses about STEP PC, WebGrid and datamining. Industrial case study chapter is a key chapter for the book. This chapter presents examples from a number of companies, and discusses about the challenges in handling real life problems, methodology used, results obtained and some pointers for future successful applications. The future of industrial knowledge management at the micro level is going in the direction of more standardisation in knowledge capture and reuse, developing specialist knowledge systems, and working as focused knowledge communities. Finally, the resource guide is a unique feature of the book. The chapter will help practitioners in getting access to leading research groups in the area, know about conferences, journals and magazine.

May I also take this opportunity to acknowledge the contribution made by Mr. Jesse Bailey, Mr. Benjamin Adesola and Mrs. Urbee Roy for the success of the book.

Rajkumar Roy

Cranfield University, UK
12th May 2000

Foreword

The term *Knowledge Management* has been defined (by management consultants), redefined (by computer scientists), and undefined (by marketers of software products). Rather than rendering the term meaningless, the adoption of the label by these groups reflects an evolution of the problem and opportunity it represents. The core idea is that knowledge is a critical asset for the modern enterprise, and therefore it should be used well --- that is, it should be captured, applied, and reused effectively. The ambiguity of the term reveals the different perspectives on how best to harness the value of corporate knowledge. The management consultant focuses on the capture and reuse of intellectual property, business processes, and organizational learning. The computer scientist will look for opportunities for formalizing the knowledge so that software can help apply it to tasks such as design and manufacturing. The vendors of knowledge management products have found a substantial market for software that helps knowledge workers collaborate in virtual teams while drawing from corporate knowledge sources.

This book brings together research that inhabits the blurry boundary between human knowledge processes (such as design, manufacturing, and employee learning) and computational support (design support tools, manufacturing automation, corporate Intranets). How to better capture knowledge? Design ontologies and representations of product and process, then use methods of knowledge acquisition to encode domain knowledge. How to better apply it? Create resources and tools that help people learn from a collective body of what they know and how they work. How to enable reuse? Integrate corporate knowledge, as captured and formalized, with the tools and processes of the workplace. The book invokes a vision of the enterprise in which humans and computers form a buzzing hive of synergistic activity, linked by a network of knowledge bases and tools.

Implicit throughout the collection is the assumption that the techniques of formalizing and operationalizing knowledge, which were developed for knowledge-based systems, can help with this human-computer-organization problem of knowledge management. Put another way, the assumption implies that the benefits of formalization --- mainly, the ability to get some software services out of it --- are useful to the human activities of knowledge workers in the modern enterprise. Sometimes formalization brings structure and organization to the chaos of documents, email, and web pages where knowledge work occurs. Sometimes the systematic mapping-out of design and manufacturing processes leads to better process design. Sometimes a knowledge management system can help coordinate the concurrent engineering activities of large virtual teams --- something not possible by collocation or hierarchical management techniques. But sometimes the well-intended effort to formalize and systematize runs into the messy reality of human activity: people don't have time to become knowledge engineers, and knowledge-based systems can't automate much of what we call knowledge work. The reader of this volume should keep a critical eye on treatment of the integration of knowledge acquisition and knowledge use. Where there is synergy among the methods for capturing knowledge and the ways in which it is applied, there is insight. By bringing together research on both sides of this chasm, and a few gems that span the waters, this collection offers a fresh perspective on the potential of technology to support knowledge management in industry.

Tom Gruber

Founder and CTO, Intraspect Software
Silicon Valley, California
24th March, 2000

Table of Contents

Chapter 1

Foundation

Micro-Scale Knowledge Management: a necessary condition for getting corporate knowledge properly implemented
E. Vergison

Foundation of Knowledge Systematization: Role of Ontological Engineering
R. Mizoguchi and Y. Kitamura

A Conceptual Model for Capturing and Reusing Knowledge in Business-Oriented Domains
G. P. Zarri

Domain Knowledge in Engineering Design: Nature, Representation, and Use
F. Mili and K. Narayanan

Selecting and Generating Concept Structures
D. Janetzko

Evaluating Verification and Validation Methods in Knowledge Engineering
A. Preece

Micro-scale Knowledge Management: A Necessary Condition for Getting Corporate Knowledge Properly Implemented

Emmanuel VERGISON
SOLVAY Research & Technology
Rue de Ransbeek, 310
B – 1120 Brussels
E-mail: emmanuel.vergison@solvay.com

Abstract: The role of Knowledge Management (KM) as a discipline that promotes an integrated, inter-disciplinary and multi-lingual approach to identifying, eliciting, structuring, diffusing and sharing a company's knowledge assets is now widely recognised.

Although several management initiatives to deploy KM at corporate level have been taken, sometimes including intellectual capital valuation and accounting, in many, if not most enterprises, Knowledge Management has definitely been initiated from the floor.

Addressing problems with reasonable size and complexity at Business Unit level, *but thinking company-wide,* allows, amongst other things, a better understanding of the rift that we perceive between the present working practices and those which we expect in the knowledge-enabled organisation of the future.

Examples of local KM initiatives that can be easily embedded in a corporate scheme will be discussed.

1. Introduction

For many decades, most industrial companies and organisations have been seeking methods and tools capable of helping them in making appropriate management decisions and it would be rather unfair to assert that they have only recently discovered the need for and the virtues of Knowledge Management. However, the deep structural changes and the globalisation of the economy which have taken place in the past fifteen years, have radically affected their business schemes as well as their working practices.

Knowledge Management (in short KM), has emerged in this very fierce competitive context as an integrated, inter-disciplinary and multi-lingual discipline providing methodologies and tools for identifying, eliciting, validating, structuring and deploying knowledge within the Enterprise.

KM is now recognised as a core competency in its own right and a way of helping industries to gain competitive advantage and increase profits, rather than just increasing revenues.

Two major strands have developed and spread out almost independently:

- *Micro–scale Knowledge Management* which focuses on the capture, structuring and use of Knowledge at local level, which does not necessarily need strong top management support and is not very sensitive to strategic plan variations.
- *Macro-scale Knowledge Management* which is very sensitive to company strategic plans, addresses corporate and transverse inter-Business Unit concerns and definitely does require strong senior management commitment as well as pro-active top management support.

Micro-scale KM that deals with problems with reasonable scale and complexity is, by far, the most popular and the most widespread. It allows, in particular, to learn about, seize and grasp properly the inevitable behavioural barriers such as resistance to change at working level and the reservations of a management not yet very familiar with knowledge processes.

Whilst a company-wide integration of knowledge assets including human, structural, customer and supplier capital, is the ultimate objective, it was pointed out in the EIRMA[1] Working Group 54 Report on Corporate Knowledge Management [1], that:

" ... *a top-down approach (of KM) is usually too ambitious at the start, ... when starting to use dedicated KM tools, act locally but think company-wide.*"

In the next section, we will briefly discuss the philosophy that has underpinned our methodology in transposing these principles into action. So doing, we wanted to integrate properly and from the outset, the most determining factor of Knowledge Management: *people.*

We will detail how we coupled a concept developed by the Hungarian philosopher Arthur Koestler for modelling social systems to software engineering techniques, in order to get a user-friendly way to simulate organisations involving human activities.

Thereafter, in Chapters 3, 4 and 5, we will describe and discuss three rather different micro-scale KM initiatives which are now in operation and are to be considered as a hard core for further development at Corporate and large scale Business Unit level:
- Competitive (Market) Intelligence Services
- Process Modelling Tools for Administration and Research
- Training and Education in KM

---

[1] EIRMA stands for European Industrial Research Management Association, 34, Rue de Bassano, 75008 PARIS, France

2. An Underpinning Model for KM

Since the eighties, industry has moved quickly from a protected and rather secure market, with medium and long term perspectives, towards an open and fiercely competitive one, where the short term dominates.

The customer and the supplier relationships have been and are still being impacted not only by structural changes like outsourcing, mergers and downsizing, but also by human factors such as retirement, increased mobility and not the least, defection.

Knowledge Management issues today, whether they are, micro or macro, are much more strongly coupled to organisational and human factors than they used to be in the past and we felt it imperative to built our KM system on a strong and reliable philosophy.

2.1 Koestler's holons and holarchies

More than thirty years ago, Arthur Koestler, the Hungarian philosopher, in a famous trilogy, analysed the greatness and poverty of human condition. In the last volume entitled "The Ghost in the Machine" [2], he elaborated the concept of Open Hierarchical Systems and proposed necessary conditions to be fulfilled by social organisations in order to survive.

As M. Höpf pointed out [3]: "Unlike the title may suggest, this (book) is not a science fiction story on anthropoid robots but a brilliant polemic analysis of human evolution."

In a holistic perspective of living organisms and social organisations, Koestler developed the concept of Holon which originates from the Greek *holos*, meaning "whole" coupled to the suffix "on", suggesting a part. A holon is to be looked at as "an identifiable part of a system that has a unique identity, yet is made up of subordinates parts and, in turn, is part of a larger whole".

The characteristics of holons, which can be viewed as so many necessary conditions for an organisation to perpetuate, are threefold:

- *Autonomy* that refers to the self-management ability of a holon to react against disturbances
- *Co-operation* which means that holons are standing together with other similar entities or holon layers and are capable of working on common projects and goals
- *Stability*, a crucial concept, that means that a holon can face and properly react against severe requests and/or perturbations.

A single holon exists only in the context of its membership of a larger system where intelligence is distributed where needed and appropriate, through processes between, and knowledge within holons.

Koestler did not restrict himself to just characterising holons and holonic systems, so called holarchies; he suggested more than sixty rules that specify the duality *co-operation-autonomy* of holons, the concepts of communication between them and between holonic architectures.

In a holarchy, the bottom layers are basically composed of reactive entities, whilst in the successive layers towards the top, holons become more and more endowed with cognitive and learning abilities.

Intensive studies based on the holonic concepts have been reported in various areas, especially in Intelligent Manufacturing Systems [4], [5], [6] and Human Systems Modelling [7], [8]. In both areas, holarchies are considered as one paradigm of the future, either for the factory or for the work organisation.

Achieving an industrial and operational way to model co-operative work in administrative, technical and R&D activities was, and still is, one of our main objectives [9], [10].

2.2 Lessons from experience.

Firstly, modelling processes involving humans is such a complex matter, that it is almost an illusion to try doing it with just a single technique. For example, when modelling administrative workflow processes in the patent department, four different points of view have been implemented: three for modelling static aspects of the processes and one for modelling the system's dynamics.

In increasing order of complexity:
- The Object Modelling Technique (OMT) [11] was used to design the information system in this case restricted to documents.
- The Structured Analysis and Design Technique (SADT) [12] was found, after several comparative tests with other models, to be quite attractive and best suited to represent activities, whilst
- The Office Support and Analysis Design Method (OSSAD) [13] is very powerful in modelling operations on documents as well as co-operation between actors.
- A Petri Net where the concepts of events, occurrence of events, event (a)synchronisation and/or parallelism play a key role, is the cornerstone of a rule-based system developed to model the dynamics [14], [15].

The second lesson learnt from experience in modelling is the imperative necessity to set up structured benchmarks based on sound measurement indicators [16], in order to compare existing methods and tools. Benchmarking's role is indeed twofold:
- To avoid re-inventing the wheel and preferably to re-use sound existing techniques
- To check the relevance of a particular method or tool to the problem at hand and to compensate for the fact that, in many cases, its domain of applicability and usability is not clearly stated or, at least, blurred around the edges.

3. Competitive Intelligence Services for New Business

In contrast to technological watches which mostly focus on very technical issues, Competitive or Market Intelligence relates to market and product information concerns [17], [18].

Information is to be taken here in a broad sense and will refer either to documents in the usual sense (paper or electronic), to WEB sites, dedicated pages within WEB-Sites or even to pointers towards experts.

3.1 The concept

Generally, if about 80 up to 90% of the information, called the memory [17], is in written form and can be found or, at least accessed internally, in notes, reports, internal data bases, etc., or in various open sources like Internet, fairs, conferences, exhibitions, external data bases or magazines, a small part with significant added value, is located in oral form within the network of marketing people and services (Fig. 1).

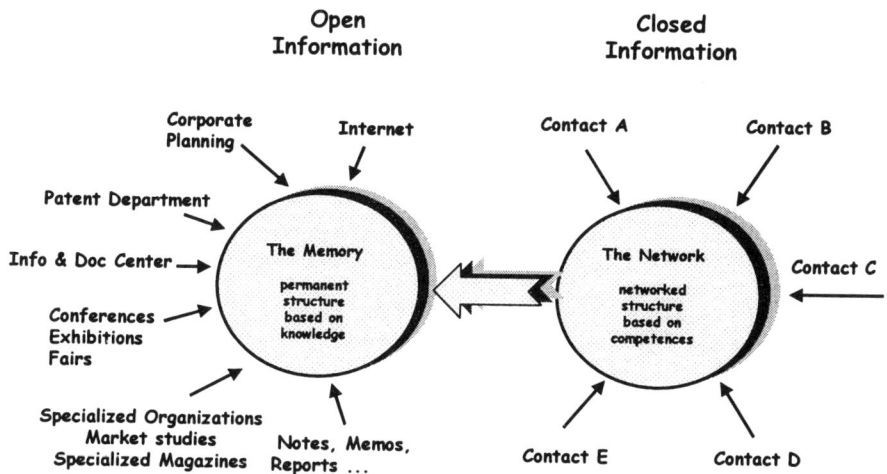

Figure 1: The structure of a competitive intelligence service[2]

Competitive intelligence services are to be viewed as highly responsive, reactive and almost real-time systems operated by experts and capable of providing decision makers with very reliable and value adding information They should detect, locate, capture, structure and push information, as automatically as possible, to the end-users.

They can also be looked at from both the macro and the micro level. We focused on systems dedicated to New Business activities at Business Unit level, thinking

---

[2] Derived from "Du renseignement à l'intelligence économique" de B. Besson et J-C. Possin

in the long term about a network of more or less coupled entities. Instead of a tall man of war, we definitely opted for a fleet of scout ships [19], [20].

This approach has the advantage of better management of human aspects (i.e. end-user's priorities and level of acceptance, the needs and requirements of the watch team) and of financial constraints (small investment and maintenance costs), at the expense of setting up, from the outset, a sound supervision system.

The word "Service", in turn, emphasises the fact that if technology undoubtedly plays a key role in the development of competitive intelligence systems and is probably close to the first step, it still remains an enabler that will never supersede human brain.

3.2 The market watch process

For the sake of simplicity, we will describe (Fig. 2) a market watch process where three types of information are referred to:
- Information coming from an R&D team in charge of New Business
- Information accessible on Internet
- Information coming from other sources

The Market Watch process consists of five steps:
- Information capture
- A validation step
- Information structuring and embedding within the system
- System consulting modes
- Information publishing, profiling and pushing

The information capture is essentially made from electronic sources where Internet holds a key position [21]; we must not however forget that a critical part of the information comes from other sources such as oral ones [17],[18]. After a benchmark on "intelligent" search engines had been completed, the reconciliation and filing operations were significantly automated.

Data validation, as one would expect, requires substantial and decisive human contributions.

The computerised system that provides the framework for structuring, embedding and publishing the information, runs on an off-the shelf software called CI-MASTER[3] It allows and carries out, amongst other things:
- The creation of individual profiles
- The automatic push of individual profiles through E-mail
- Reports on demand
- Alert triggering

---

[3] CI-MASTER is a software developed and delivered by ALCOR Sprl, a Belgian software house that specialised in industrial R&D software.

The consulting mode provided by CI-MASTER was conceived to avoid as far as possible, the use of a separate search engine.

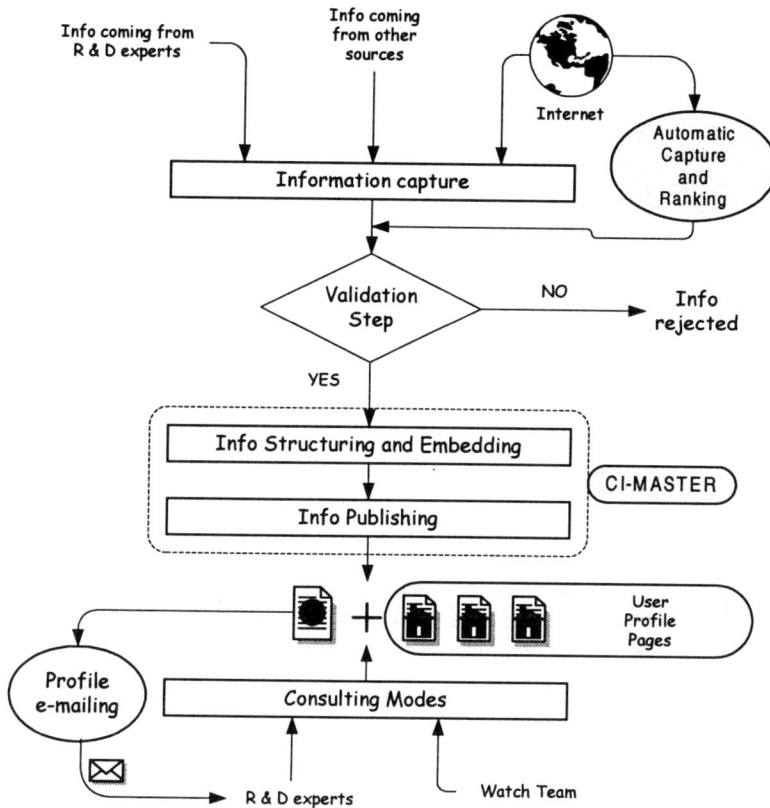

Figure 2: The market watch process

Two competitive intelligence systems are now in regular operation within the company, one in an R&D department and the second in a business unit; a third one has been launched recently.

4. Process Modelling Tools for Administration and R&D

The detailed analysis of knowledge-enabled organisations and of the underpinning processes is a prerequisite for optimising their working mechanisms; it is the first step in achieving efficient knowledge-based information systems.

The idea of building up a computerised workshop to help people analyse their working environment, emerged from the following observations:

- It has been recognised that the evolution of the working practices is so radical nowadays, that we cannot avoid the need to have a good and deep understanding of the underpinning working processes.
- The multi-disciplinary character of the now widespread project management paradigm requires user-friendly platforms to make the exchange of ideas easier.
- The globalisation of the economy has amplified the problems related to and raised by the use of multiple languages. Resorting to a "broken-English like" solution however appealing it seems, is definitely not a panacea.

Modelling an organisation means laying down the underlying processes, static and dynamic. It depends upon:
- Using concepts that are familiar to *all the actors* who are involved in the different processes
- Relying on technologies that are suitable for simulation exercises.

The analysis of processes is a well established technique to people working in Quality Assurance and Knowledge Management.

4.1 The concept of a modelling workshop

Generally speaking, a Modelling Workshop is a computerised integrated system devised as a toolbox and capable of addressing efficiently a targeted range of applications. In our case, human organisations and activities are to be taken into account.

The workshop helps to put the concepts and the philosophy we discussed earlier into operation and is considered to be a decisive factor of the whole methodology.

4.2 The computerised system

The Workshop we describe here, called MAMOSACO, has been developed in co-operation with the LAMIH[*], a research institute depending on the French University of Valenciennes [7], [22].

It consists of 6 views or pages:
- The first page aims at referencing the process under consideration and at giving information on its structure (list of sub-processes).
- The second, called the Activity – Role Matrix (Fig. 3) contains:
  - a detailed list of the different players within the process and sub-processes.
  - the explicit connections between activities and roles.
- The other four views each correspond to one mode of representation (Fig. 4); three of them focus on static views of the process and the fourth one, more complex, allows for dynamic simulations.

---

[*] LAMIH stands for Laboratoire d'Automatique et de Mécanique Industrielles et Humaines

The methods that are used to construct the last four views were introduced and referenced earlier in 2.2 whilst the VISIO[©]^5 toolbox was used to design the graphical end-user interface.

	head of department	assistant	head of sector	patent consultant	inventor	administrative staff
Patent demand formulation				X	X	X
Patent demand management		X	X	X	X	X
Patent demand filing	X	X				X

Figure 3: The *activity–role* matrix

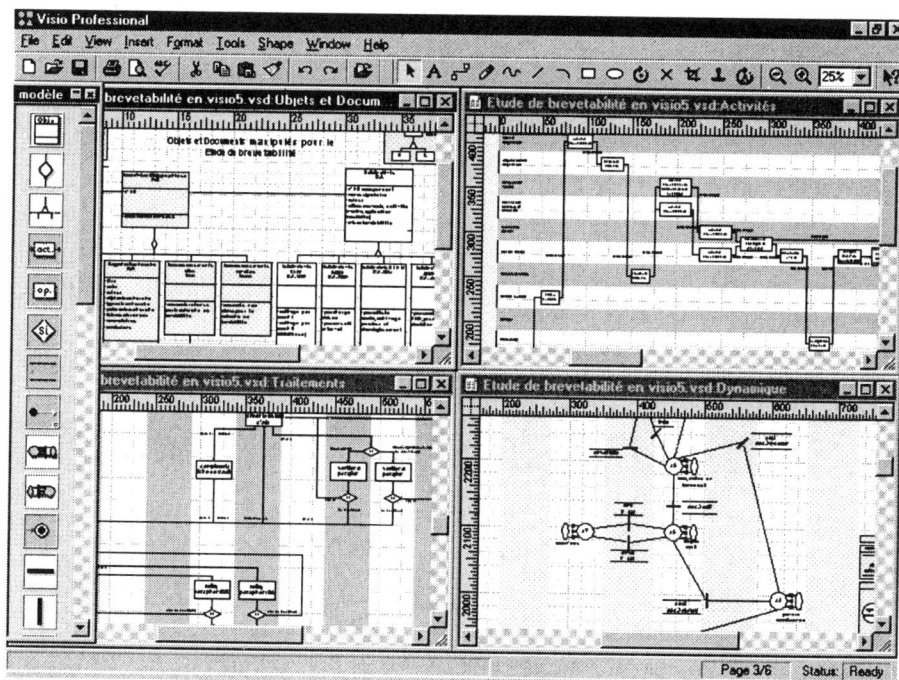

Figure 4: Screen dump of the MAMOSACO workshop

---

^{5} VISIO is a Visio Corporation Trademark

5. Training and Education in KM

Very quickly, we realised that it would take rather a long time to familiarise people with KM concepts, methods and tools that substantially change their working practices.

For this reason, we launched two initiatives:
- teaching people the foundations and basic principles of KM and keeping them informed about internal realisations
- improving written communications and adapting them to the requirements of the future knowledge-enabled organisation

5.1 Basic principles of KM and operational applications

Nowadays, it is not a problem to find excellent references that can help in structuring a general presentation of the foundations, the rationale and the major facets of Knowledge Management [23]. Attention should however be paid to giving as fair and as exhaustive as possible, a review of the state of the art within the enterprise.

Success stories undoubtedly constitute an essential part of the message to win people over. Just as an example, the first time we gave the seminar, the following success stories were presented:
- An example of process modelling exercise within the Patent Department
- How to share knowledge and improve the efficiency of working practices within a group of businessman
- A group-ware facility for managing R&D projects
- Capitalising knowledge in corrosion
- Benchmarking on plant performances

5.2 KM oriented document writing

In conjunction with the development of specific methodologies and tools, a strong emphasis was placed on the refocusing and revitalising of the written culture, reshaping it towards KM-enabled activities.

The Information Mapping®* Method developed by Professor Robert E. Horn while at Harvard and Columbia Universities, has been adopted for structuring both business and technical information.

This method breaks down a document into self-contained and reusable chunks of knowledge and structures it according to the reader's cognitive behaviours: the way he actually reads, processes, browses and remembers information.

---

* Information Mapping is an Information Mapping, Inc. Trademark

Information Mapping® revealed itself particularly well suited to KM exercises and, as a by-product written material is now much more under control and better focused on the reader's needs and expectations.

Many different categories of documents benefited from Information Mapping©:
- technical & scientific reports
- notes & memos
- regulatory information
- ISO documentation
- procedure manuals
- user guides and manuals
- policy reports

The method that started being taught, on a voluntary basis, in R&D departments and in the Patent Department is now used and applied in technical and administrative units as well as in plants.

Notwithstanding the contribution of the method itself, we emphasise the fact that it is just one component in the whole communication process. It is, for example, well known that structuring a paper document is different from dealing with the same information online. Structuring big documents and organising remote training courses are examples of matters requiring further study.

Many problems are still open and a lot of time is spent both internally and in co-operation with the Business Graduate School of Chambéry, in order to build up what we will consider as a knowledge-enabled information system.

6. Perspectives and Conclusions

If, as a starting point, Micro-Scale Knowledge Management enabled the concepts and the ideas of the KM-paradigm to be tested and experimented, we may assert that today, it has reached such a degree of maturity that it is no longer a risk to launch applications at local level.

KM is a sustainable activity in its own right that is undoubtedly winning the trust of the business. Many lessons have been and are still being drawn from experience, which allow us today to prepare much more ambitious projects at corporate level. This will permit Corporate Knowledge Management and methodologies like Technology Roadmapping [24] and Technology Monitoring [25] to be launched in optimal conditions.

Micro-Scale Knowledge Management has also confirmed both the central role of people and the determining role of information technology. However the primary role of people cannot be disputed: *technology is close to the first step, but is never first.*

Deployment of Micro-Scale KM also revealed how deeply the normal working practices need to be reassessed and how much the mentalities will have to adapt; many problems are still open and need to be solved.

From a personal point of view, I believe that it is much more at the conceptual and methodological level that the main part of the effort should be spend. Especially we need to obtain a better understanding of the network concept (networks of processes, of people, communities of practice, activities, etc.) and on how to link properly and efficiently the ongoing KM initiatives.

Acknowledgements

KM being in essence a collaborative initiative, nothing would have happened without the contribution of colleagues and academic partners. I would like to particularly thank my colleague F. Gallez, one of my first fellow travellers in KM and who initiated the Competitive Intelligence project. On the academic side, my thanks go jointly to Prof. C. Kolski, Dr. R. Mandiau and E. Adam[7] from the University of Valenciennes (F), to Prof. J-P. Barthès from the Technology University of Compiègne (F) and to Mrs. J. Gerbier from the Business Graduate School of Chambéry (F).

References

[1] EIRMA Working Group 54 Report, 1999, *The Management of Corporate Knowledge*, EIRMA, Paris.
[2] Koestler A., 1969, *The Ghost in the Machine*, Arkana Books, London
[3] Höpf E., 1996, *Intelligent Manufacturing Systems – The Basic Concept*, Internet: http://www.ncms.org/hms/public/holo2.html
[4] Valckenaers P. and Van Brussel H., 1996, *Intelligent Manufacturing Systems – Technical Overview*, Internet: http://www.ncms.org/hms/public/hms_tech.html
[5] Van Brussel H., Valckenaers P., Wyns J. et al, 1996, *Holonic Manufacturing Systems and Iim*, IT and Manufacturing Partnerships, J. Browne et al. (Eds.), IOS Press.
[6] Gerber C., Siekman J., Vierke G., 1999, *Holonic Multi-Agent Systems*, Research Report RR-99-03, Deutches Forschungzentrum für Künstliche Intelligenz GmbH, FRG.
[7] Adam E., Vergison E., Kolski C., et al., 1997, *Holonic User Driven Methodologies and Tools for Simulating Human Organisations*, ESS97, 9th European Simulation Symposium, W. Hahn & A. Lehman (Eds.), Passau, FRG, pp. 57-61.
[8] Adam E., Kolski C.,Vergison E., 1998, *Méthode adaptable basée sur la modélisation de processus pour l'analyse et l'optimisation de processus dans l'Entreprise*, ERGO IA 98, Actes du 6ème colloque, M-F. Barthet (Scientific Dir.), Biarritz, France.
[9] Adam E., 1999 , « *Specifications of human machine interfaces for helping co-operation in human organisations* », 8th international conference on human-machine interaction, HCI'99, Munich.

---

[7] Emmanuel ADAM is a PhD Student whose thesis was jointly sponsored by the "Nord – Pas de Calais" Region (F) and SOLVAY S.A.

[10] Adam E., Kolski C., Mandiau R., 1999, *Approche holonique de modélisation d'une organisation orientée workflow : SOHTCO*, 7ème Journées Francophone pour l'Intelligence Artificielle Distribuée et les Systèmes Multi-Agents, JFIADSMA'99, Saint-Gilles, Ile de la Réunion, France.

[11] Rumbaugh J. Blaha M., Premerlani W. et al., 1991, *Object-oriented modelling and design*, Prentice-Hall.

[12] I.G.L. Technology., 1989, *SADT, un langage pour communiquer*. Eyrolles, Paris.

[13] Dumas P., Charbonnel G., 1990, *La méthode OSSAD, pour maîtriser les technologies de l'information. Tome 1 : principes*, Les éditions d'organisation, Paris.

[14] Adam E., Mandiau R., Vergison E., 1998, *Parameterised Petri nets for modelling and simulating human organisations in a workflow context*, Workshop on Workflow Management, 19th International Conference on Application and Theory of Petri Nets, Lisbon.

[15] Gransac J., 1997, "*Construction of an Object Oriented Petri Net Simulator – Contribution to the study of Complex Business/Administrative Systems*" (in French), DESS Technical Report, University of Valenciennes .

[16] Adam E., Kolski C., 1999 ,*Etude comparative de méthodes du Génie Logiciel en vue du développement de processus administratifs complexes* », to appear in "Revue de Génie Logiciel".

[17] Besson B. et Possin J-C., 1996, *Du renseignement à l'intelligence économique*, DUNOD, Paris.

[18] Collins B., 1997, *Better Business Intelligence*, Management Book 2000 Ltd, Chalford, Gloucestershire, UK.

[19] Pereira A., 1998, *Conception et mise en oeuvre d'un système informatisé pour la consultation, la publication et la gestion d'informations relatives à la veille marché stratégique*, Student Report, IUP – G.E.I.I., University of Valenciennes, France

[20] Broudoux L., 1999, *Conception et mise en oeuvre d'un système intégré de veille marché stratégique*, Student Report, IUP – G.E.I.I., University of Valenciennes, France

[21] Revelli C., 1998, *Intelligence Stratégique sur Internet: comment développer efficacement des activités de veille et de recherche sur les réseaux*, DUNOD, Paris

[22] Adam E., Doctoral Thesis, University of Valenciennes, France (to appear)

[23] Liebowicz J., 1999, *Knowledge Management Handbook*, CRC Press, London

[24] EIRMA Working Group 52 Report, 1997, *Technology Roadmapping*, EIRMA, Paris.

[25] EIRMA Working Group 55 Report, 1999, *Technology Monitoring*, EIRMA, Paris.

Foundation of Knowledge Systematization: Role of Ontological Engineering

Riichiro Mizoguchi and Yoshinobu Kitamura

The Institute of Scientific and Industrial Research, Osaka University, 8-1, Mihogaoka, Ibaraki, Osaka, 567-0047, Japan, {miz,kita}@ei.sanken.osaka-u.ac.jp

Abstract: The objective of the research described in this article is to give a firm foundation for knowledge systematization from AI point of view. Our knowledge systematization is based on ontological engineering to enable people to build knowledge bases as a result of knowledge systematization. The basic philosophy behind our enterprise is that ontological engineering provides us with the basis on which we can build knowledge and with computer-interpretable vocabulary in terms of which we can describe knowledge systematically. The ontology we developed includes device ontology and functional ontology both of which play fundamental roles in understanding artifacts by a computer. Redesign of a manufacturing process through its functional understanding in industrial engineering is taken as an example task to show how to use the ontologies as well as the way of knowledge systematization of manufacturing processes. A sample ontology of a manufacturing process is also built in terms of functional ontology from device ontology point of view.

1. Introduction

In AI research history, we can identify two types of research. One is "Form-oriented research" and the other is "Content-oriented research". The former investigates formal topics like logic, knowledge representation, search, etc. and the latter content of knowledge. Apparently, the former has dominated AI research to date. Recently, however, "Content-oriented research" has attracted considerable attention because a lot of real-world problems to solve such as knowledge reuse, facilitation of agent communication, media integration through understanding, large-scale knowledge bases, etc. require not only advanced formalisms but also sophisticated treatment of the content of knowledge before it is put into the formalisms.

Formal theories such as predicate logic provide us with a powerful tool to guarantee sound reasoning and thinking. It even enables us to discuss the limit of our reasoning in a principled way. However, it cannot answer any of the questions such as what knowledge we should prepare for solving the problems given, how to scale up the knowledge bases, how to reuse and share the knowledge, how to manage knowledge and so on. In other words, we cannot say it has provided us with something valuable to solve real-world problems.

In expert system community, the knowledge principle proposed by Feigenbaum has been accepted and a lot of development has been carried out with a deep appreciation of the principle, since it is to the point in the sense that he stressed the importance of accumulation of knowledge rather than formal reasoning or logic. This has been proved by the success of the expert system development and a lot of research activities has been done under the flag of "knowledge engineering". The authors are not claiming the so-called rule-base technology is what we need for future knowledge processing. Rather, treatment of knowledge should be in-depth analyzed further to make it sharable and reusable among computers and human agents. Advanced knowledge processing technology should cope with various knowledge sources and elicit, transform, organize, and translate knowledge to enable the agents to utilize it.

Although importance of such "Content-oriented research" has been gradually recognized these days, we do not have sophisticated methodologies for content-oriented research yet. In spite of much effort devoted to such research, major results were only development of KBs. We could identify the reasons for this as follows:

1) It tends to be ad-hoc, and

2) It does not have a methodology which enables knowledge to accumulate.

It is necessary to overcome these difficulties in order to establish the content-oriented research.

Ontological Engineering has been proposed for that purpose [1]. It is a research methodology which gives us design rationale of a knowledge base, kernel conceptualization of the world of interest, semantic constraints of concepts together with sophisticated theories and technologies enabling accumulation of knowledge which is dispensable for knowledge processing in the real world. The authors believe knowledge management essentially needs content-oriented research. It should be more than information retrieval with powerful retrieval functions. We should go deeper to obtain the true knowledge management.

The major topic of this paper is the knowledge systematization necessary for an intelligent design support system. This is indeed a topic of content-oriented research and is not that of a knowledge representation such as production rule, frame or semantic network. Although knowledge representation tells us how to represent knowledge, it is not enough for our purpose, since what is necessary is something we need before the stage of knowledge representation, that is, knowledge organized in an appropriate structure with appropriate vocabulary. This is what the next generation knowledge base building needs, since it should be principled in the sense that it is based on well-structured vocabulary with an explicit conceptualization of the assumptions. This nicely suggests ontological engineering is promising for the purpose of our enterprise.

While every scientific activity which has been done to date is, of course, a kind of knowledge systematization, it has been mainly done in terms of analytical formulae with analytical/quantitative treatment. As a default, the systematization is intended for human interpretation. Our knowledge systematization adopts another way, that is, ontological engineering to enable people to build knowledge bases on the computer as a result of knowledge systematization. The philosophy behind our enterprise is that ontological engineering provides us with the basis on which we can build knowledge and with computer-interpretable vocabulary in terms of which we can describe knowledge systematically in a computer-understandable manner.

This paper summarizes the results of the research the authors have conducted to date on this topic and is organized as follows: The next section briefly describes ontological engineering. Section 3 presents the framework of knowledge systematization and function in design. Basic ontologies used as the foundation of knowledge systematization are discussed in Section 4. Section 5 discusses the main topic, functional ontology, followed by its applications to knowledge of manufacturing process in industrial engineering.

2. Ontological Engineering

An ontology is
an explicit specification of objects and relations in the target world intended to share in a community and to use for building a model of the target world.
Amongst its many utilities, it works as:

- **A common vocabulary.** The description of the target world needs a vocabulary agreed by people involved. Terms contained in an ontology contribute to it.
- **Explication of what has been left implicit.** In all of the human activities, we find presuppositions/assumptions that are left implicit. Typical examples include definitions of common and basic terms, relations and constraints among them, and viewpoints for interpreting the phenomena and target structure common to the tasks they are usually engaged in. Any knowledge base built is based on a conceptualization possessed by the builder and is usually implicit. An ontology is an explication of the very implicit knowledge. Such an explicit representation of assumptions and conceptualization is more than a simple explication. Although it might be hard to be properly appreciated by people who have no experience in such representation, its contribution to knowledge reuse and sharing is more than expectation considering that the implicitness has been one of the crucial causes of preventing knowledge sharing and reuse.
- **Systematization of knowledge.** Knowledge systematization requires well-established vocabulary/concepts in terms of which people describe phenomena, theories and target things under consideration. An ontology thus contributes to providing backbone of systematization of knowledge. Actually, the authors have been involved in a project on Knowledge systematization of production knowledge under the umbrella of IMS: Intelligent Manufacturing Systems project [2].
- **Design rationale.** Typical examples to be explicated include intention of the designers of artifacts, that is, part of design rationale. An ontology contributes to explication of assumptions, implicit preconditions required by the problems to solve as well as the conceptualization of the target object that reflects those assumptions. In the case of diagnostic systems, fault classes diagnosed and range of the diagnostic inference, in the case of qualitative reasoning systems, classes of causal relations derived, and so on.
- **Meta-model function.** A model is usually built in the computer as an abstraction of the real target. And, an ontology provides us with concepts and relations among them that are used as building blocks of the model. Thus, an ontology specifies the models to build by giving guidelines and constraints that should be satisfied in such a model. This function is viewed as that at the meta-level.

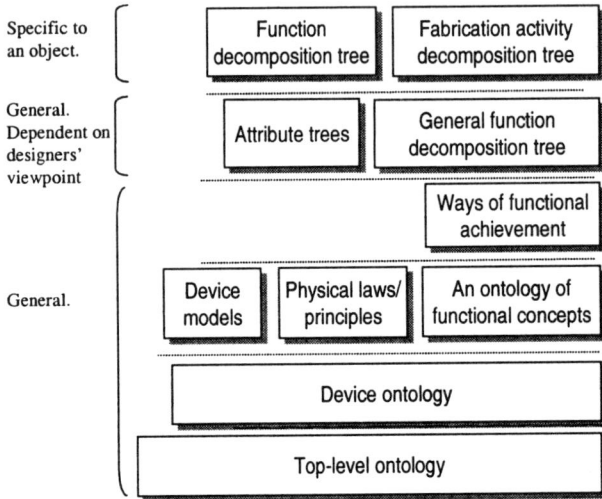

Figure 1: Hierarchical organization of ontologies.

In summary, an ontology provides us with "a theory of content" to enable research results to accumulate like form-oriented research avoiding ad-hoc methodologies which the conventional content-oriented activities have been suffering from.

3. The Framework of Knowledge Systematization

3.1 Framework

Figure 1 shows the sketch of multi-layered ontologies for ontological knowledge systematization together with some kinds of functional knowledge. We utilize the two major functionalities of an ontology, that is, common vocabulary function and meta-model function as well as knowledge systematization functionality. The former is essential for describing knowledge. The latter is for formulating ontologies in a multi-layered structure to keep it in a moduler structure. Ontologies are arranged in up-side-down to stress the analogy of physical construction. An ontology in a layer plays the role of a meta-ontology of those in an upper layer. Top-level ontology stays at the bottom layer to govern all the ontologies in the upper layers. Next is the device ontology which is effectively the basis of our knowledge systematization. On top of device ontology, functional ontology is built together with ontology of ways of function achievement discussed in section 5.2 as above knowledge are organised. Further, several kinds of knowledge is organized in terms of the concepts in the ontologies. It is important to note here that the first four layers from the bottom contain **definitions** of concepts and other higher layers are for **description** of various kinds of knowledge in terms of concepts defined in the lower layers.

3.2 Function in design

No one disagrees that the concept of function should be treated as a first class category in design knowledge organization. That is, function is an important member of a top-level ontology of design world. A top-level ontology is like one discussed in philosophy and usually consists of a few categories such as entity, relation, role, time, space, etc.

One of the key claims of our knowledge systematization is that the concept of function should be defined independently of an object which can possess it and of its realization method. Contrary to the possible strong disagreement from the people working in the engineering design, it has a strong justification that the concept of a function originally came from the user requirements which is totally object- and behavior-independent, since common people have no knowledge about how to realize their requirements and are interested only in satisfaction of their requirement. Another justification is reusability of functional knowledge. If functions are defined depending on objects and their realization, few function are reused in and transferred to different domains. As is well-understood, innovative design can be facilitated by flexible application of knowledge or ideas across domains.

In the abstraction hierarchy of function, there are various functional concepts. At the higher levels, we find very general ones such as to transmit, to achieve, to drive, et al., at the lower levels, many domain-specific functions appear. Our interest includes to identify how many categories of function there are, in what relations they are one another, how those categories facilitate use of functional knowledge in design process, etc. Furthermore, we investigate the relationship between function and behavior. That is, a functional ontology

Function has been a major research topic in intelligent design community[3,4,5,6]. In many of the workshops held in IJCAI, AAAI and ECAI, workshops on functional modelling have been held almost every time. The following is a list of reasons why the concept of function is critical in design:

1. Most of the requirements are given in terms of functional concepts as is seen in the statement "Design is a mapping from functional description to structural one".
2. Understanding of functional terms is necessary for computers to interpret requirements specification.
3. In order to realize functions, computers have to know the relations among functions and behaviors.
4. A rich set of functional concepts enable a computer to make inference on functional space to find a solution.
5. Major portion of design rationale of design is left as functional description.
6. Philosophically, the concept of the "whole" of any artifact largely depends on its functional unity. Without functional concepts, we cannot explain why an artifact or a module is meaningful and complete as it is.

4. Basic Ontologies

Although our major objective in the current research is to establish a functional ontology for design knowledge systematization, we cannot avoid to mention other ontologies which give foundations of it. It is device ontology and top-level ontology discussed below.

4.1 Top-level ontology

This is the most fundamental ontology and provides necessary set of categories to talk about every thing in the world. It may include the following concepts:

-Substrate
-Space
-Time
-Substance
-Entity
  -Concrete things
   -Objects
  -Process
   -Action
   -Phenomena
  -Abstract things

-Relation
-Role
-Attribute
-Matter
 -Event
 -Fact
 -State
-Representation
 -p/o form: Representation form
 -p/o content: Matter
-Representation form

In spite of its importance, we do not go into details of the top level ontology here.

4.2 Device ontology

Concerning modelling of artifacts, there exist two major viewpoints: Device-centered and Process-centered views. The device ontology specifies the former and the process ontology the latter. Device-centered view regards any artifact as composition of devices which process input to produce output which is what the users need. Process-centered view concentrates on phenomena occurring in each place(device) to obtain the output result with paying little attention to the devices existing there.

The major difference between the two is that while device ontology has an *Agent* which is considered as something which plays the main actor role in obtaining the output, process ontology does not have such an Agent but has *participants* which only participate in the phenomena being occurring. Needless to say, such an agent coincides with the device in device ontology. It is natural to consider process ontology is more basic than device ontology. Before humans had started to make an artifact, only natural phenomena had been out there in the world. Any device and/or artifact is something built by limiting the environment where phenomena of interest occur to what the device provides and by connecting them to achieve the goal set in advance. Every device utilizes several natural phenomena to realize its functions.

In spite of its less basic characteristics, device ontology has been dominant to date in modelling artifacts for many reasons as follows:

1. Device ontology is straightforward because every artifact is a device.
2. Every artifact is easily and nicely modelled as a part-whole hierarchy of devices because it can be considered as being composed of sub-devices recursively.

3. The concept of a function has to be attributed to an agent(a device) which is viewed as a main actor to achieve the function.
4. Theoretically, process ontology cannot have function associated with it because it has no *Agent*.
5. Device-centred view saves a load of reasoning about design, it hides internal details of devices.
6. One can design almost all artifacts in the device ontology world by configuring devices, except cases where one needs innovative devices based on new combination of phenomena or new phenomena themselves.

It is important to note, however, device ontology is not almighty. Essentially, device ontology views any device as a black box. Of course, in the device ontology world, one can go into smaller grain world in which the parent device is seen as configuration of several sub-devices of smaller grain size. But, the smaller devices themselves are still devices whose internal behavior is also hidden. Process ontology is different. It is useful for explaining why a device works, since it directly explains what phenomena are occurring in the device to achieve its main function. Furthermore, process ontology is better than device ontology at modelling, say, chemical plants where chemical reactions have to be modelled as phenomena.

4.3 Constituents of device ontology

In this article, we basis all of our discussion on device ontology. Top-level categories of device ontology include *Entity, Role, Structure, Behavior* and *Function*. *Entity* represents the real and basic things existing in the world such as human, device, fluid, solid, etc. *Role* represents roles the entity plays during the process where the device works. It has functional role and structural role. The former includes *Agent* which is an actor who does the action to cause the effect required, *Object* which is processed by the agent, *Medium* which carries something used by the device, *material* which is the main source of the results output from the device and *product* which is the regular output of the device and is made of/from the material. The structural role has sub-roles such as *input, output* and *component* which is the role played by any device to form the structure. Not all the *input* has material role, though all *material* have input role. Similarly, not all the *output* have product role, though all *product* have output role. *Structure* has sub-categories such as *inlet, outlet* and *connection* which is a relational concept connecting devices. A component has inlets and outlets whose role is to connect it with others. Input (output) is anything coming into (going out of) a device through an inlet (outlet).

Behavior of a device (*DO-Behavior* in short) is defined as "situation-independent conceptualization of the change between input and output of the device" and it has two subcategories such as *transitive behavior* and *intransitive behavior* which roughly correspond to transitive and intransitive verbs, respectively. Roughly speaking, the former is concerned with the change of the *object* and the latter with that of the *agent*. Note here that we do not define behavior in general, that is, we only discuss behavior of a device. *Function(of a device)* is defined as "teleological interpretation of a *DO-behavior*" [5].

We now would like to defend our definition of *DO-Behavior* from a possible critic that it seems not compliant to what a normal simulation does. The counter argument would look like as follows:

In numerical simulation, which is the typical way of simulation, by simulation of the behavior of a system(device), we mean trace of the change of a parameter value over time. In other words, the behavior is an interpretation of change of each parameter value. It is straightforward and convincing, since it is compatible to the basis of how a thing behaves which is not explained by DO-Behavior.

This view is parallel to the naïve view of behavior like the idea based on "motion" such as walking and running. It is true that the above idea of numerical simulation is basic because it depends on neither device nor process ontologies. There is, however, another kind of simulation, that is, causal simulation in qualitative reasoning. If we interpret it in the device ontology world, causal simulation is different from the numerical simulation in that it tries to simulate the change of an object between that when it exists at the inlet of the device and that when it exists at the outlet of the device, that is, causality of the change made by the device. This is what AI research needs about behavior of a system from device-centered point of view.

In design community, what is necessary are the ideas of function and behavior as well as structure. In order to relate the idea of function to that of behavior, both should be based on the same ontology. As discussed above, functional concepts inherently need device ontology because we always associate function with a device. This suggests us that behavior should be defined in terms of device ontology, that is, DO-behavior which is consistent with not the idea of numerical simulation but that of causal simulation. In the rest of this article, we use the term "**behavior**" instead of "*BO-behavior*" for simplicity.

5. An Ontology of Functional Concepts

Typical examples of function are seen in the phrases of "this machine has speaking and speech recognition functions" or "he temporarily lost walking function". This is one of the source of confusions in understanding what a function is. This view tells us every action itself can be a function. In design research, however, what we want is not such a view of function but one which contributes to distinction between behavior and function because functional concepts are usually found in the requirement specification and behavior is something to realize it. In the design context, speaking is a behavior and its function would be transmitting information in such a manner that a human can hear; walking is a behavior and its function would be carrying itself from one place to another or consuming energy.

It is very important to note that by the above phrase "behavior is something to realize it(function)", we do not mean we go inside of a device to explain the mechanism of how it realizes the function but mean a different way of interpretation of the same event at the same level of grain size as is seen in our definitions of behavior and function. Going into finer-grained explanation (understanding) is done by "functional decomposition" or "behavioral decomposition" when necessary. Note also that functional decomposition still produces another set of functions based on device ontology rather than explaining internal behavior of the device.

We already defined *Function* of a device as "teleological interpretation of a DO-behavior". This tells us function of a device is determined contextdependently, though DO-behavior is constant independently of the context. Considering that, in most cases of design, the context of a device is determined by the goal to be achieved by the device, function of a device is determined goal- or purpose-dependently. This reflects the reality that definition of function tends to be context-dependent and hence, in many cases, functional knowledge about design is hardly reusable.

The major goals of our research include to give a framework for organizing functional concepts in a reusable manner and to define them operationally as an essential step towards knowledge systematization. By operationally, we mean computers can make use of functional knowledge in the reasoning tasks of the functional modelling, understanding of the functional structure of a device and revising it. What we have to do for these goals are as follows:

1. To define functional concepts independently of its realization so as to make the definition reusable.
2. To devise a functional modelling method to enable a modeller to relate such reusable functional definitions to specific application problems, that is, to get functional concepts **grounded** onto the behavior and hence structure.
3. To formulate a functional decomposition scheme to obtain efficient functional knowledge for design.
4. To identify categories of functional concepts for systematization of functional knowledge.
5. To provide rich vocabulary for reasoning in the functional space.

The following is an overview of our work on ontologies of function.

5.1 Several categories of functional concepts

The ontology of the functional concepts is designed to provide a rich and comprehensive vocabulary for both human designers and design supporting systems. It consists of the four spaces as shown in Figure 2. We identified three categories of function: (1) Base-function, (2) function type and (3) meta-function.

5.1.1 Base-function

We call the function defined above a *base-function* of a device, since it is something to do with the entities in the real world. A base-function can be interpreted from object-related aspect (called *object functions*), energy-related aspect (called *energy functions*) or information-related aspect (called *information functions*). For example, the object function and the energy function of a turbine are "to rotate the shaft" and "to generate kinetic energy", respectively.

Figure 2a shows the energy functions organized in an is-a hierarchy with clues of classification. A base-function is defined by minimal conditions about the behavior and the information for its interpretation called Functional Toppings (FTs) of the functional modeling language FBRL (abbreviation of a Function and Behavior Representation Language) [5]. There are three types of the functional toppings; (1) O-Focus representing focus on attributes of objects, (2) P-Focus representing focus on ports (interaction to neighboring devices), and (3) Necessity of objects. For example, a base-function "to take energy" is defined as "an energy flow between two mediums" (a behavioral condition), and "focus on the source medium of the transfer" (functional toppings). The definition of "to remove" is that of "to take" plus "the heat is unnecessary". Thus, "to take" is a super concept

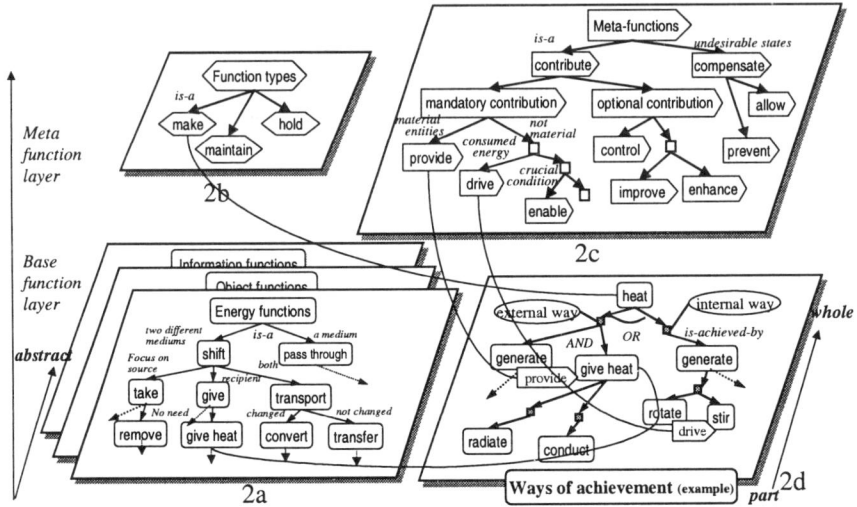

Figure 2: Four kinds of functional planes.

of "to remove" as shown in Figure 2a. The values of O-Focus and P-Focus represent intentional focus of the function. Such a parameter that is a kind of O-Focus and is an attribute of the entity in the focused port indicated by P-Focus is called as the focused parameter. The entity (object or energy) having the focused parameter is called as the focused entity.

The object functions are categorized into the following two categories according to the kinds of the focused parameter. The one is called the *amount function* that changes the parameter values (target objects) or changes the type of the objects, say, from liquid to gas. The other is called the *attribute function* that changes the other attributes of the objects. These categories are used in the definitions of the meta-functions.

Note that definition of base-function using FTs is highly independent of its *realization*, that is, the details of behavior and internal structure of the device. For example, P-Focus specifies not concrete location but abstract interaction with the neighboring devices. The realization is represented by the ways of achievement shown in Figure 2d.

5.1.2 Function types

The function types represent the types of goal achieved by the function [7]. We have redefined the function type as "ToMake", "ToMaintain", and "ToHold" as shown in Figure 2b. For example, consider two devices, an air-conditioner (as a heating device) and a heater, having the same function "to give heat". The former keeps the room temperature at the goal temperature. The latter does not. These are said to be "ToMaintain" and "ToMake", respectively.

5.1.3 Meta-function

The meta-functions (denoted by *mf*) represent a role of a base-function called an *agent function* (f_a) for another base-function called a *target function* (f_t). A meta-

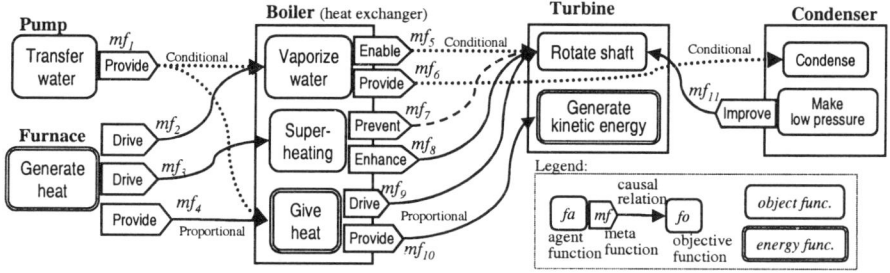

Figure 3: Meta-functions in a power plant (part)

function is not concerned with changes of objects of these devices but with interdependence between base-functions, while other three kinds of functional concepts are concerned with existence or changes of objects.

We have defined the eight types of meta-functions as shown in Figure2c (an is-a hierarchy). Figure 3 shows examples of the meta-functions $\{mf_1 ...mf_{11}\}$ in a simple model of a power plant. Note that furnace which is a sub-component of the boiler is separated from the boiler as a heat exchanger for explanation. The details of definitions and examples in the different domains are shown in [8].

We begin definition of meta-functions with the condition where there is a causal relation from the focused parameter of f_a to that of f_t. If the goal of f_t is not satisfied when f_a is not achieved, the f_a is said to have a *mandatory contribution* for the f_t. Although we can intuitively say that f_a has a ToEnable meta-function for f_t, the authors define ToEnable to have a narrower meaning by excluding the cases of ToProvide and ToDrive as follows:

ToProvide: When a function f_a generates (or transfers) the *materials* which another function f_t intentionally processes, the function f_a is said to perform a meta-function "to provide (material)" for f_t. The material of f_t can be basically defined as input objects (or energy) that will be a part of the output objects (*products*) on which the function f_t focuses. For example, the "to transfer water" function of the pump has a ToProvide meta-function for the "to vaporize" function of the boiler (see mf_1 in Figure 3). If f_t is an attribute function, the function changing the same attribute of the focused material object is said to have a ToProvide meta-function.

ToDrive: When a function generates or transfers such an energy (called *driving energy*) that is *intentionally* consumed by another function f_t, the function is said to have the meta-function "to drive f_t". For example, the "to give heat" function of the boiler has a ToDrive meta-function for the "to rotate" function of the turbine (see mf_9), because the heat energy is not material of the shaft and is consumed by the rotation. A part of the heat energy is transformed into the kinetic energy that is carried by the focused object (the shaft). On the other hand, for "to generate kinetic energy", the same function performs not ToDrive but ToProvide (see mf_{10}), because the heat energy is material of the kinetic energy.

ToEnable: This meta-function is used for representing a mandatory condition playing a crucial role in f_t except ToProvide and ToDrive. What we mean by this weak definition is that the conditions such as the existence of the material and the existence of the driving energy are too obvious to be said to enable the function. For example, because the steam of which phase is gas plays a crucial role in occurrence of the heat-expansion process in the turbine and the phase is neither material of rotation nor the consumed energy, the "to vaporize" function of the boiler is said to have a meta-function ToEnable (see mf_5).

ToAllow and ToPrevent: These two meta-functions are concerned with the undesirable side effects of functions. A function f_a having positive effects on the side effect of a function f_{t1} is said to have a meta-function "to allow the side-effects of f_{t1}". If a serious trouble (e.g., a fault) is caused in a function f_{t2} when a function f_a is not achieved, the function f_a is said to have a meta-function "to prevent malfunction of f_{t2}". For example, the "super-heat" function of the boiler prevents malfunction of the turbine (mf_7), because the steam of low temperature would damage the turbine blade.

ToImprove and ToEnhance: These meta-functions represent *optional* contribution for f_t. The discrimination between ToImprove and ToEnhance is made by increment of the amount of the input energy. For example, the "to keep low pressure" function of the condenser contributes to the efficiency of the "to rotate" function (see mf_{11}) without increment of input energy. That is, it improves the efficiency of "to rotate" function. As to "ToEnhance" meta-function, on the other hand, the "to super-heat" function of the boiler optionally increases the amount of the input energy (mf_8) to increase the output energy, which is enhancement.

ToControl: When a function f_a regulates the behavior of f_t, its meta-function is said to be "to control f_t". For example, consider a control valve that changes the amount of flow of the combustion gas for the boiler in order to maintain the amount of flow of the steam. It is said to have a meta-function ToControl for the "to vaporize" function of the boiler (not shown in Figure 3).

5.2 Ways of functional achievement

A base-function f_u can be achieved by different groups of sub-functions. We call a group of sub-functions $\{f_t, ... f_n\}$ constrained by the relations among them (such as meta-functions) *a functional method* of an achievement of f_u. On the other hand, we call the basis of the method *a functional way*. The way is the result of conceptualization of the physical law, the intended phenomena, the feature of physical structure, or components used which explain how the functional methods make sense.

Figure 2d shows some ways of achievement of "to heat an object", which are described in terms of concepts in other three spaces. There are two ways, that is, the external and internal ways. According to the external way, it is decomposed into two sub-functions, that is, "to generate heat" and "to give heat". The former should perform a ToProvide meta-function for the latter. In the figure, the latter function can be decomposed according to "radiation way" or "conduction way".

Note that Figure 2d shows *is-achieved-by* (whole-part) relations among the functional concepts in OR relationship, while Figure 2a shows *is-a* relations as the definitions of them, which are independent of "how to realize them".

6. Aplications to Industrial Engineering

We now can go into the detailed discussion on systematization of manufacturing process knowledge on the basis of the ontological consideration on function made thus far. Redesign of a manufacturing process through its functional understanding

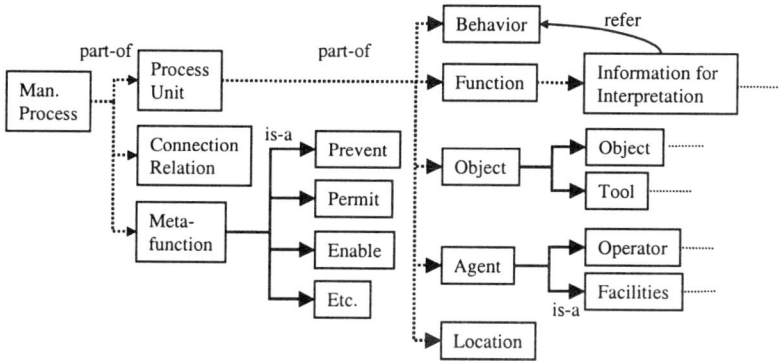

Figure 4: An ontology of manufacturing process (portion).

in industrial engineering is taken as an example task to show how to use the ontologies as well as the way of knowledge systematization of manufacturing processes. A sample ontology of a manufacturing process is also built in terms of functional ontology from device ontology point of view.

6.1 Knowledge model for manufacturing process in industrial engineering

In spite of the long history of the research of industrial engineering, knowledge is not satisfactorily systematized yet and well-established common vocabulary is not available. Furthermore, know-how of redesign of the manufacturing process is kept as heuristics that can be transferred only by apprenticeship. We first tried to establish a system of basic concepts as shown in Figure 4.

Solid lines represent *is-a* relations and dotted lines *part-of* relations. Process unit is a key concept in a manufacturing process and is composed of five sub-concepts such as *Behavior, Function, Agent, Object* and *Location*. An *Agent* processes the *Object* at *Location* by showing *Behavior* that is interpreted as *Function*.

A manufacturing process can be viewed as a sequence of process units (P). The model of a process unit consists of *Behavior, Function, Objects* (O), *Agents* engaged in the process (A) and *Location* where the processing is done. Following our definition of behavior and function, the *Behavior* of a process unit represents the change of the object input to it and is does not change according to the context in any manufacturing processes. Contrary to this, *Function* of the *Behavior* is an interpretation of the *Behavior* under the given context. For example, let us consider a part of a manufacturing process shown in Figure 5. The behavior of the process unit P_2 is "to *assort* input objects O_1 into two categories O_2 and O_3 according to the criteria". Because the next process unit P_4 discharges the assorted objects O_3, we can interpret the behavior of P_2 as the function that "to *remove* the needless objects O_3". Moreover, if the criteria are the quality of the product, it can be called as "*inspect*". This functional interpretation of P_2 depends on the function of P_4. Therefore, under other contexts, P_2 could achieve different functions such as "*select*". Such definition of function is different from others such as that in [9]. Using the functional representation language FBRL [5], we can describe such a functional model as behavior plus functional toppings (FTs). FTs represent the

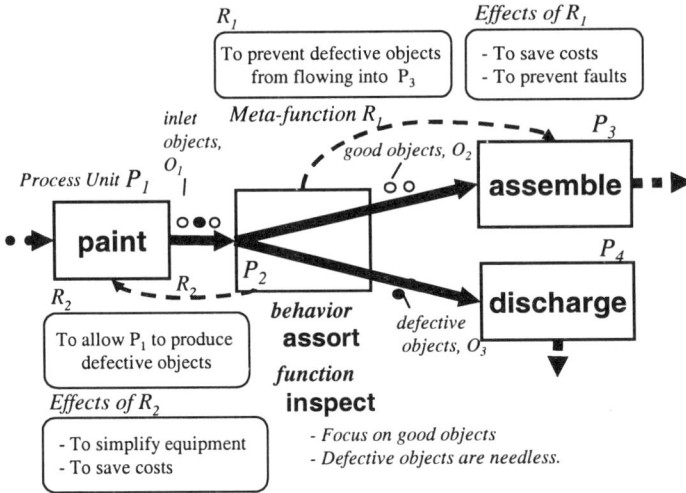

Figure 5: Example of behavior, function, and meta-function of a process

ways of interpretation as discussed in 5.1.1. A model of the "inspect" function mentioned above consists of a behavioral model (constraints over attributes of objects) and two FTs "focus on good objects O_2" and "defective objects O_3 are needless". More detail of modeling based on FBRL is shown in [5].

Connection relation represents the structure of the manufacturing process and meta-function represents how each process unit contributes to each other. A process unit contributes to other process units with specific goals.

Figure 5 also shows the meta-function of the inspection process P_2 mentioned above. It *allows* the paint process P_1 to produce the defective objects, and *prevents* them from flowing into the assembling process P_3. The former contributes to simplifying the equipment of P_1 and hence to saving of costs. The later prevents faults of P_3 and contributes to saving costs.

Such meta-functions represent the justification of the existence of the process units. Justification of the inspection process P_2 is production of the defective objects by the paint process. Such justification is a part of "design rationale" of the manufacturing process.

Figure 6 shows is-a hierarchy of "Process unit". A process unit has sub-concepts such as *Operation, Delay, Transportation and Inspection*. One of the contributions of our systematization of manufacturing process knowledge is the explication of attributes on which these four activities operate. We identified four categories of attributes such as *physical attributes, information attribute, spatial attribute and temporal attribute*. On the basis of these categories of attributes, we define the above four activities together with *Storage* activity as follows:

- *Operation*: change of physical attributes
- *Transportation*: change of spatial (location) attributes
- *Inspection*: change of information attributes
- *Delay*: change of temporal attributes (the change can be done without permission)
- *Storage*: change of temporal attributes (the change cannot be done without permission)

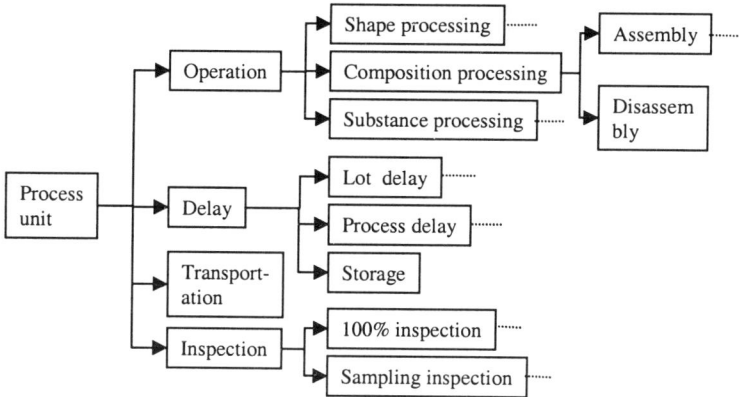

Figure 6: Is-a hierarchy of process unit.

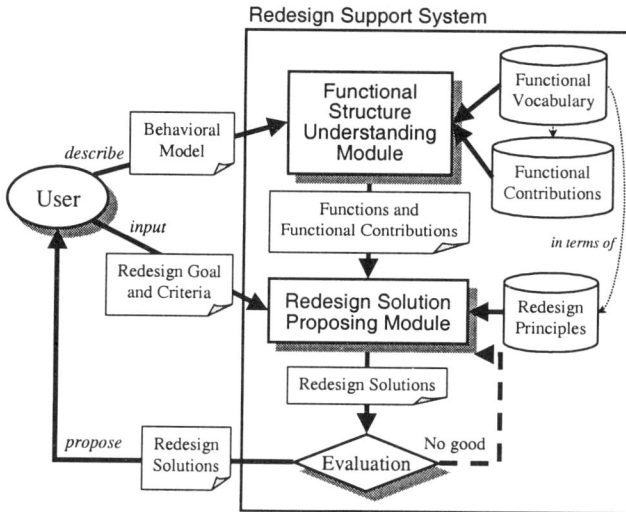

Figure 7: Framework of the redesign support system

6.2 Redesign of a manufacturing process

Figure 7 shows the framework of the redesign support system we are developing. It consists of the following two major functional modules:
   (1) Functional structure understanding module and
   (2) Redesign solution proposing module.

6.2.1 Functional understanding

The former module can generate plausible functional structures from the given behavior models of the target process. Firstly, the module generates all possible functional interpretations of the given behavior according to a functional vocabulary. It contains definitions of a number of functional concepts as constraints over FTs. Since the search space over FTs is limited, the functional

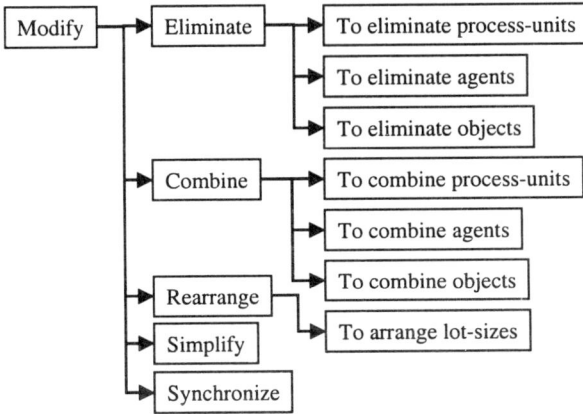

Figure 8: Is-a hierarchy of Modify operation.

Table 1: Examples of the redesign principles

	Omission Principle	**Agent Sharing Principle**
Goals	Reduction of processing time and/or costs	Reduction of costs for agents
Targets	A process unit P_1	Process units P_1,P_2, Agents A_1,A_2
Conditions	P_1 has no functional contribution	The performance of A_1 is enough to do P_1,P_2
Changes	To omit P_1	To remove A_2 and to engage A_1 in P_2
Effects	Reduction of processes	Reduction of agents

understanding module generates all tuples of values of FTs and then matches them with definitions in the functional vocabulary.

Next, meta-functions among the generated functions are identified according to the definitions of meta-functions shown in Section 5.1.3. The algorithm is shown in [8]. It enables the module to identify plausible functional interpretations of P from the generated possible functional interpretations, because a functional interpretation that does not contribute to any processes is not plausible. Such contributions cannot be captured by causal relations as pointed out in [9].

6.2.2 Redesign

The later module proposes how to change the manufacturing process (redesign solutions). Then, they are evaluated according to the given criteria. The module uses the redesign principles knowledge representing general improving principles in Industrial Engineering (IE), called ECRS principles shown in Figure 8. They are highly abstracted in terms of the functional vocabulary and hence reusable. Table 1 shows examples of them.

6.3 Example of redesign

Figure 9 shows an example of redesign of a manufacturing system. Firstly, the redesign system identifies the meta-functions (R_1 and R_2) shown in Figure 9a from

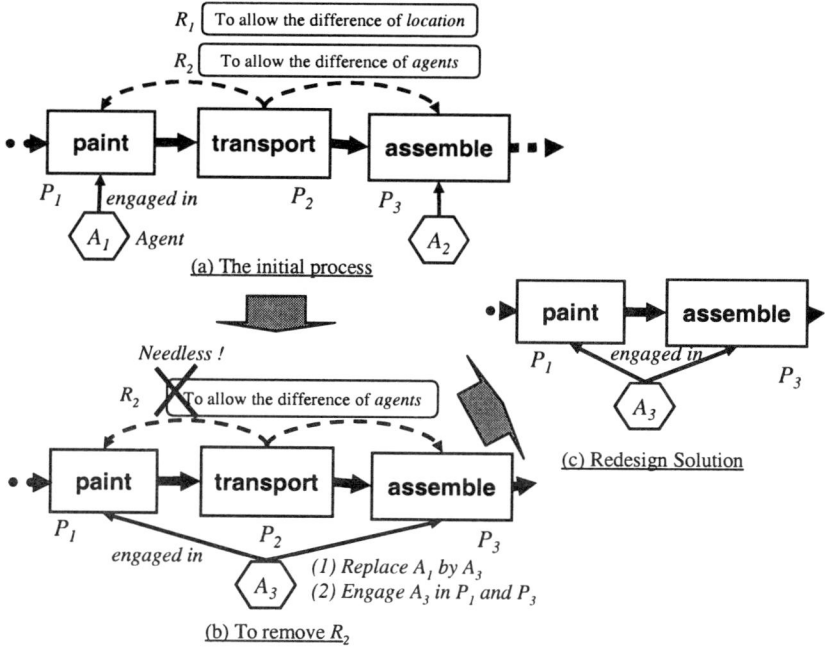

Figure 9: Example of redesign

the given behavioral model. The transportation process unit P_2 allows the two differences between P_1 and P_3, that is, the difference of location (R_1) and the difference of agents (R_2). In other words, the two differences are the justification for the existence of P_2.

Next, the system tries to generate redesign solutions. The goal (criteria) of redesign here is reduction of the processing time. Thus, the system tries to omit P_2 according to the "omission principle" shown in Table 1. In order to satisfy its condition, that is, no meta-function, the redesign system has to remove the meta-functions by changing P_1 and/or P_3.

The meta-function R_1, that is, to allow the difference of location, can be removed by changing the layout. On the other hand, for the meta-function R_2, that is, to allow the difference of agents, the system tries to apply the "agent sharing principle" shown in Table 1. In the cases where the performances of the agent A_1, A_2 are not enough to do so, the system also applies the "agent replacement principle". Then the two processes share an agent A_3 (Figure 9b). At this point, there is no meta-function of P_2. Then, the process unit can be omitted (Figure 9c).

Lastly, the redesign solution is evaluated. Its effects include saving of the processing time by omitting the transportation and reduction of costs for an agent. The side effect is costs for replacing the agents.

Another example is shown in Fig. 10. The story is as follows: The quality of painting was turned out not to be satisfactory. Diagnosis revealed the fact that the power of compressor used for painting is not enough. The compressor was shared by shot blast that gets rid of the stains on the painting objects, so the possible solutions were

(1) to replace the compressor with that of more power
(2) to install another compressor for the painting unit.

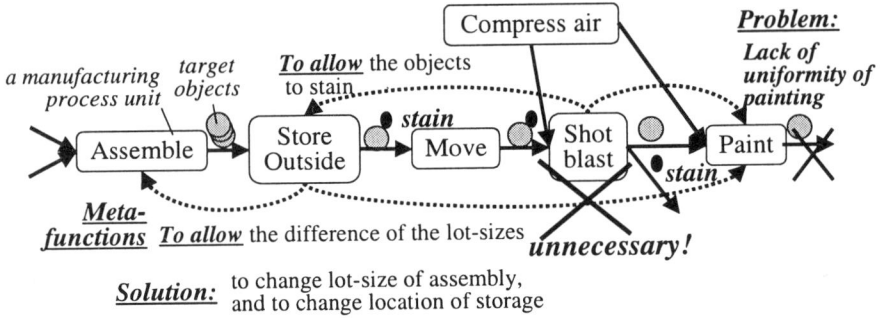

Figure 10: Another example (Shot blast elimination case).

Needless to say, these solutions are not good because it costs a lot. A better solution has been obtained as follows: Investigation on meta-functions of the shot blast uncovered that it allows the objects to stain and to prevent from low quality of painting and that of storage of objects outside are to allow difference of the lot-sizes of Assembling unit and Painting unit. Considering the fact that getting objects stained is caused by the storage outside, if we find a space for storing objects inside, we can prevent objects from being stained, and hence we can eliminate the shot blasting unit, and hence the painting unit can enjoy necessary power of compressor by using the current one to attain the required quality of painting.

7. Related Work

7.1 Ontology of functional concepts:

The functions in [3,5,6,10,11] represent abstract behavior of components and are defined as base-functions in our framework. The meta-function represents a role for another function without mention of changes of incoming or outgoing objects of components and hence is totally different from such base functions. In [12,13], function is defined as a kind of hierarchical abstraction of behavior. It corresponds to the is-achieved-by relations among functions in our framework.

The CPD in CFRL [13] represents causal relations among functions. Lind categorizes such relations into Connection, Condition and Achieve [10]. Rieger identifies "enablement" as a type of the causal relation between states and action [14]. The meta-functions are results of interpretation of such causal relations between functions under the role of the agent function for the target functions.

In [15], the sets of "primitives of behavior" are proposed. Lind identifies a few general functions which are categorized into multiple levels [11]. We added more intention-rich concepts and organized in is-a and part-of hierarchy. Furthermore, we identify some types of the meta-functions.

In Value Engineering [16], standard sets of verbs (i.e., functional concepts) for value analysis of artifacts are proposed [17]. It enables the human designers to share descriptions of functions of the target artifacts. However, they are designed only for humans, and there is no machine understandable definition of concepts.

In a literature on design, many general "patterns" of synthesis are proposed (e.g., [18]). Our ways of achievement, however, explicitly represent the feature of achievement such as theory and phenomena. The importance of such key concepts in design is pointed out in [19]. They enable the redesign system to facilitate the smooth interaction between models at the structural and functional levels.

7.2 Ontology-based approach for knowledge modelling

Borst et al. [20] discuss very similar idea of layered ontologies for engineering composed of Mereology, Topology, System theory, Component ontology, Process ontology and EngMath ontology. While their ontologies are well-formalized, they do not discuss the function or behavior which are our main topic. Although their Component ontology should correspond to our device ontology, it mainly is concerned about structural aspect and leaves content part for Process ontology. Our device ontology is self-contained for modeling artifacts.

8. Conclusions

We have discussed ontology-based knowledge systematization. Functional knowledge is set to the main target for investigation, since **function**, which has left untouched to date, plays the key role in AI-based knowledge modelling. Ontological engineering contributes to our enterprise in some respects:
1. It bridges the gap between humans and computers by providing well-structured vocabulary (concepts) in terms of which we can model functional knowledge,
2. it gives a system of concepts used as foundation of knowledge modeling,
3. and hence, it enables us to avoid ad hoc knowledge representation.

We applied the results of ontological engineering of functional knowledge to modelling of manufacturing processes and demonstrated how the redesign system works using systematized knowledge. Future plans include incorporating Value Engineering (VE) results [16] into the redesign module to realize creative redesign as well as implementation of the system.

Acknowledgements

This research has been partly supported by the Japan Society for the Promotion of Science (JSPS-RFTF97P00701) and GNOSIS, IMS. The authors are grateful to Dr. Fuse, Sumitomo Electric Industries for his help on manufacturing process modelling.

References

[1] Mizoguchi, R., and Ikeda, M. 1997, Towards ontology engineering. In *Proc. of PACES/SPICIS '97*, 259-266.
[2] Mizoguchi, R. et al., 1999, A Methodology of Collaborative Synthesis by Artificial Intelligence, 2nd International Workshop on Strategic Knowledge and Concept Formation, pp.221-232.

[3] Chandrasekaran, B., and Josephson, J. R., 1996, Representing function as effect: assigning functions to objects in context and out. In *Proc. of AAAI-96 Workshop on Modeling and Reasoning with Function.*

[4] Iwasaki, Y., et al., 1993, How things are intended to work: Capturing functional knowledge in device design. *Proc. of IJCAI-93*, pp.1516-1522.

[5] Sasajima, M.; Kitamura, Y.; Ikeda, M.; and Mizoguchi, R., 1995, FBRL: A Function and Behavior Representation Language. In *Proc. of IJCAI-95*, 1830-1836

[6] Umeda, Y. *et al.*, 1990. Function, behavior, and structure. *AI in Engineering*, 177-193, 1990.

[7] Keuneke, A. M., 1991, Device representation: the significance of functional knowledge. *IEEE Expert*, Vol.24, pp.22-25.

[8] Kitamura, Y., and Mizoguchi, R. 1999. Meta-functions of artifacts, Papers of *13th International Workshop on Qualitative Reasoning (QR-99)*, 136-145

[9] Chandrasekaran, B., Goel, A. K., and Iwasaki, Y., 1993, Functional representation as design rationale. *COMPUTER*, pp.48-56.

[10] de Kleer, J. 1984, How circuits work, *Artificial Intelligence* 24:205-280.

[11] Lind, M., 1994, Modeling goals and functions of complex industrial plants. *Applied Artificial Intelligence*, Vol.8, pp.259-283.

[12] Sembugamoorthy, V., and Chandrasekaran, B. 1986, Functional representation of devices and compilation of diagnostic problem-solving systems. In *Experience, memory and Reasoning*, 47-73.

[13] Vescovi, M., et al., 1993, CFRL: A language for specifying the causal functionality of engineered devices. In *Proc. of AAAI-93*, 626-633.

[14] Rieger, C., and Grinberg, M., 1977, Declarative representation and procedural simulation of causality in physical mechanisms. In *Proc. of IJCAI-77*, 250-256.

[15] Hodges, J., 1992, Naive mechanics - a computational model of device use and function in design improvisation. *IEEE Expert* 7(1):14-27

[16] Miles, L. D., 1961, *Techniques of value analysis and engineering.* McGraw-hill.

[17] Tejima, N. *et al.* eds, 1981, Selection of functional terms and categorization. Report 49, Soc. of Japanese Value Engineering (In Japanese).

[18] Bradshaw, J. A., and Young, R. M., 1991, Evaluating design using knowledge of purpose and knowledge of structure. *IEEE Expert* 6(2):33-40.

[19] Takeda, H., et al., 1990, Modeling design processes. *AI Magazine*, 11(4), 37-48.

[20] Borst P., Akkermans, H., and Top, J., 1997, Engineering ontologies, Int. J. Human-Computer Studies, 46, pp.365-406.

A Conceptual Model for Capturing and Reusing Knowledge in Business-Oriented Domains

Gian Piero Zarri

CNRS, 44 rue de l'Amiral Mouchez, 75014 Paris, France, zarri@ivry.cnrs.fr

Abstract: A considerable amount of important, 'economically relevant' information is buried into natural language 'narrative' documents. In these, the information content (the 'meaning') consists mainly in the description of facts or events relating the real or intended behaviour of some (not necessarily human) actors. In this paper, we describe the methodology used in NKRL (acronym of Narrative Knowledge Representation Language) to accurately represent this meaning. NKRL is a conceptual modelling language that has a long history of successful, concrete applications.

1. Introduction

A considerable amount of important, 'economically relevant' information, is buried into natural language (NL) documents: this is true, e.g., for most of the corporate knowledge documents (memos, policy statements, reports, minutes etc.), as well as for the vast majority of information stored on the Web. A fundamental (and quantitatively relevant) subset of this useful NL information is formed by documents that are structured in the form of 'narratives' of an economic import. In these 'narrative' documents, the main part of the information content consists in the description of facts or events that relate the real or intended behaviour of some 'actors' (characters, personages, etc.): these try to attain a specific result, experience particular situations, manipulate some (concrete or abstract) materials, send or receive messages, buy, sell, deliver etc. Note that, in this sort of documents, the actors or personages are not necessarily human beings; we can have narrative documents concerning, e.g., the vicissitudes in the journey of a nuclear submarine (the 'actor', 'subject' or 'personage') or the various avatars in the life of a commercial product. Moreover, narrative information is not necessarily associated with the 'material' existence of NL documents. We can make use of narratives, e.g., to describe (to 'annotate', see below) the situations reproduced by a set of images on the Web, or those described in a video or a digital audio document.

Being able to represent in a general, accurate, and effective way the semantic content of this business-relevant, narrative information — i.e., its key 'meaning' — is then both conceptually relevant and economically important. In this paper, we introduce NKRL (Narrative Knowledge Representation Language), see [1, 2], a language expressly designed for representing, in a standardised way, the 'meaning' of economically-relevant narratives. NKRL has already been used as modelling knowledge representation language in European projects like Nomos (Esprit P5330), Cobalt (LRE P61011) and WebLearning (GALILEO Actions). It is currently employed in the CONCERTO project (Esprit P29159) to encode the 'conceptual annotations' (formal description of the main semantic content of a document) that will be added to digital documents to facilitate their 'intelligent' retrieval, processing, displaying, etc., see [3]. Two pilot applications are being developed in CONCERTO. The first is in a 'digital libraries and electronic publishing' style, and concerns the creation of NKRL conceptual annotations for some of the abstracts stored in the NL databases of the Publishing and New Media Group of Pira International (UK). The second is more in the style of 'intelligent data mining on the textual component of the Web', and concerns the annotation of Internet NL documents used by a Greek company, Biovista, to compile and commercialise written reports about emerging sectors of biotechnology.

In the following, we present firstly (Section 2) a brief 'state of the art' in the domain of ontological and conceptual languages. After a general description of the NKRL's architecture (Section 3), we will illustrate by means of examples (Section 4) some important NKRL features. Section 5 will supply some details about the new implementation of NKRL according to an RDF/XML format that has been achieved in a CONCERTO framework; Section 6 is the "Conclusion".

2. Short State of the Art in the Ontological Languages Domain

The term "ontology" — which should normally denote the branch of philosophy that concerns the essence of "Being", see [4] — is increasingly used to indicate the explicit specification of a given 'conceptualisation'. A conceptualisation refers to a collection of entities in a domain of interest, that must be described with respect to the knowledge these entities convey and to the relationships that hold among them.

In practice, an ontology consists i) of terms or concepts — general terms expressing the main categories that organise the world, like *thing_*, *entity_*, *substance_*, *human_being*, *physical_object* etc., or the particular terms that describe a very specific domain of application (domain ontologies) — ii) of their definitions, and iii) of the axioms that associate them or impose particular constraints. Terms are organised into a taxonomy, normally a tangled one; from an operative point of view, ontologies are not so different from the hierarchies of computational data structures ("concepts") currently used in the AI domain to represent the properties of some 'interesting entities' to be modelled.

The level of complexity of the implemented ontology systems varies considerably from a system to another. For example, WORDNET [5] is a particularly sophisticated electronic dictionary, which describes lexical knowledge more than concepts in the traditional meaning of this word. It is, however, assimilated to an ontology given that the lexical objects — even if classified basically into nouns, verbs, adjectives and adverbs — are then organised according to some specific semantic criteria. The central semantic notion used in WORDNET is that of "synset", i.e., a set of synonyms; a simple synset is, e.g., {person, human being}. If a word has more than one sense, it will appear in more than one of the 70,000 synsets presents in the systems. The synset data structures are, like the traditional concepts, organised in a hierarchy where the hypernymy-hyponymy relationships have the traditional role of superclass-subclasses relationships. Another general ontology that, like WORDNET, is linguistically motivated, but which represents a level of abstraction in between lexical knowledge and conceptual knowledge is the GENERALIZED UPPER MODEL [6]. This model includes, among other things, a hierarchy of about 250 general concepts (like ELEMENT, PROCESS, CIRCUMSTANCES, PARTICIPANT, SIMPLE-QUALITY etc.) as well as a separate hierarchy of relationships.

At the opposite end of the complexity scale there is CYC [7], one.of the most controversial endeavour of the Artificial Intelligence history. Launched in late 1984, CYC has spent more than fifteen years in building up an extremely large knowledge base (and inference engine, plus interface tools, application modules etc.) which deals with a huge amount of 'common sense knowledge' including time, space, substances, intention, causality, contradiction, uncertainty, belief, emotions, planning etc. CYC's KB presently contains approximately 400,000 logical assertions, which include simple statements of facts, rules about what conclusions to draw if certain statements of facts are satisfied (i.e., they are 'true'), and rules about how to reason in the presence of certain facts and rules. New conclusions are derived by the inference engine using deductive reasoning. The coherence of this huge KB is assured by the "Upper CYC Ontology", a set of approximately 3,000 terms intended to capture the most general concepts of human reality that can be freely consulted on the Web.

A popular approach to the construction of ontologies consists in making use of description (terminological) logics, see [8] — this techniques has been chosen by, e.g., several recent projects dealing with the 'intelligent' indexing of documents on the Web (see, e.g., Ontobrocker [9], or SHOE [10, 11] and its derivative OML/CKML). A well-known project that tries to build up ontologies corresponding to full first-order theories is Ontolingua [12].

Returning now to the main topic of this paper, we can remark that the representation of the 'meaning' of narratives cannot be realised making use only of the traditional ontological tools, even though the most sophisticated implementations of the ontological principles, like the description logics systems, are used. In all the ontological languages, the representation of the concepts is always based, in substance, on the traditional, 'binary' model built up in the 'property-value' style. For the narratives, the use of this sort of concepts is not enough, and must be integrated by the description of the mutual relationships between concepts — or, in other terms, by the description of the 'role' the different concepts and their instances have in the framework of the global actions, facts, events etc. If, making use of a concrete example, we want to represent a narrative fragment like "NMTV (an European media company) ... will develop a lap top computer system...", asserting that NMTV is an instance of the concept company_ and that we must introduce an instance of a concept like lap_top_pc will not be sufficient. We must, in reality, create a sort of 'threefold' relationship, see next Section. This will include a 'predicate' (like DEVELOP or PRODUCE), the two instances, and a third element, the 'roles' (like SUBJECT or AGENT for NMTV and OBJECT or PATIENT for the new lap top system) that will allow us to specify the exact function of these two instances within the formal description of the event. Note that some sort of 'horizontal' relations among concepts are also present in the 'traditional' ontologies. But, given their intrinsic 'taxonomic' nature, the introduction of horizontal links among concepts is, in a way, a sort of unnatural operation, and the relationships used are always particularly simple, see, e.g., "the e-mail address of X is Y" (Ontobrocker) or "the infected species V has symptoms W" (SHOE).

3.   The Architecture of NKRL

NKRL is organised according to a two-layer approach. The lower layer consists of a set of general representation tools that are structured into several integrated components, four in our case.

The 'definitional' component of NKRL supplies the tools for representing the important notions (concepts) of a given domain; in NKRL, a concept is, therefore, a definitional data structure associated with a symbolic label like *physical_entity*, *human_being*, *city_*, etc. These definitional data structures are, substantially, frame-like structures; moreover, all the NKRL concepts are inserted into a generalisation/specialisation (tangled) hierarchy that, for historical reasons, is called H_CLASS(es), and which corresponds well to the usual 'ontologies' of terms.

A fundamental assumption about the organisation of H_CLASS concerns the differentiation between 'notions which can be instantiated directly into enumerable specimens', like "chair" (a physical object) and 'notions which cannot be instantiated directly into specimens', like "gold" (a substance). The two high-level branches of H_CLASS stem, therefore, from two concepts labelled as *sortal_concepts* and *non_sortal_concepts*, see [1]. The specialisations of the former, like *chair_*, *city_* or *european_city*, can have direct instances (chair_27, paris_), whereas the specialisations of the latter, like *gold_*, or *colour_*, can admit further specialisations, see *white_gold* or *red_*, but do not have direct instances.

The enumerative component of NKRL concerns then the formal representation of the instances (lucy_, wardrobe_1, taxi_53, paris_) of the sortal concepts of H_CLASS. In NKRL, their formal representations take the name of 'individuals'. Throughout this paper, we will use the italic type style to represent a *concept_*, the roman style to represent an individual_.

The 'events' proper to a given domain — i.e., the dynamic processes describing the interactions among the concepts and individuals that play a 'role' in the contest of these events — are represented by making use of the 'descriptive' and 'factual' tools.

The descriptive component concerns the tools used to produce the formal representations (predicative templates) of general classes of narrative events, like "moving a generic object", "formulate a need", "be present somewhere". In contrast to the dyadic structures used for concepts and individuals, both in NKRL and in the 'traditional' ontological languages, templates are characterised by a complex threefold format connecting together the *symbolic name* of the template, a *predicate* and its *arguments*. These, in turn, are linked with the predicate through the use of a set of named relations, the *roles*. If we denote then with L_i the generic symbolic label identifying a given template, with P_j the predicate used in the template, with R_k the generic role and with $a_{k\,the}$ corresponding argument, the NKRL data structures for the templates will have the following general format:

$$(L_i \, (P_j \, (R_1 \, a_1) \, (R_2 \, a_2) \ldots (R_n \, a_n))) \,.$$

see the examples in the following Sections. Presently, the predicates pertain to the set {BEHAVE, EXIST, EXPERIENCE, MOVE, OWN, PRODUCE, RECEIVE}, and the roles to the set {SUBJ(ect), OBJ(ect), SOURCE, BEN(e)F(iciary),

MODAL(ity), TOPIC, CONTEXT}. Templates are structured into an inheritance hierarchy, H_TEMP(lates), which corresponds to a taxonomy (ontology) of events.

The instances (predicative occurrences) of the predicative templates, i.e., the representation of single, specific events like "Tomorrow, I will move the wardrobe" or "Lucy was looking for a taxi" are, eventually, in the domain of the last component, the factual one.

The upper layer of NKRL consists of two parts. The first is a catalogue describing the formal characteristics and the modalities of use of the well formed, 'basic templates' of the H_TEMP hierarchy (like "moving a generic object" mentioned above), built up using the lower layer tools and permanently associated with the language. Presently, the basic templates are about 200, pertaining mainly to a (very broad) socio-economico-political context where the main characters are human beings or social bodies. By means of proper specialisation operations it is then possible to obtain from the basic templates the 'derived' templates that could be concretely needed to implement a particular application. The second part of the layer is given by the general concepts that belong to the upper levels of H_CLASS, such as *sortal_concepts*, *non_sortal_concepts*, *event_* (reified events), *physical_entity*, *modality_*, etc., see Figure 6. They are, as the basic templates, invariable.

4. A Brief Survey of Some Important NKRL Features

4.1 Descriptive and factual components

Figure 1 reproduces the NKRL (simplified) 'external' representation of a simple narrative like "Three nice girls are lying on the beach", that could represent, e.g., the NL 'annotation' associated with a WWW image.

```
c1)     EXIST SUBJ     (SPECIF girl_1 nice_ (SPECIF cardinality_
                                            3)): (beach_1)
              MODAL    lying_position

        [ girl_1
           InstanceOf :   girl_
           HasMember :  3 ]
```

Figure 1: Annotation of a WWW image represented according to the NKRL syntax.

The 'predicative occurrence' c1, instance of a basic NKRL template, brings along the main features of the event to be represented. EXIST is a predicate, SUBJ(ect) and MODAL(ity) are roles. In the complex argument ('expansion') introduced by the SUBJ role, girl_1 is an individual (an instance of an NKRL concept); *nice_* and *cardinality_* are concepts, like the argument, *lying_position*, introduced by MODAL. A 'location attribute' (a list that contains here the individual beach_1) is linked with the SUBJ argument by using the colon code, ':'. The 'attributive operator', SPECIF(ication), is one of the NKRL operators used to build up structured arguments or 'expansions', see [1] (and Section 5 below). The SPECIF lists, with syntax (SPECIF e_1 p_1 ... p_n), are used to represent some of the properties which can be asserted about the first element e_1, concept or individual, of the list — i.e., in c1, the properties associated with girl_1 are *nice_* and (*cardinality...*), the property associated with *cardinality_* is '3'.

The non-empty HasMember slot in the data structure explicitly associated with the individual girl_1, instance of an NKRL concept (*girl_*), makes it clear that this individual is referring in reality to several instances of *girl_* ('plural situation'). In Figure 1, we have supposed, in fact, that the three girls were, *a priori*, not sufficiently important *per se* in the context of the caption to justify their explicit representation as specific individuals, e.g., girl_1, girl_2, girl_3 ; note that, if not expressly required by the characteristics of the application, a basic NKRL principle suggests that we should try to avoid any unnecessary proliferation of individuals. Individuals like girl_1 in Figure 1 (or girl_99, the numerical suffix being, of course, totally irrelevant) are 'collections' rather then 'sets', given that the extensionality axiom (two sets are equal iff they have the same elements) does not hold here. In our framework, two collections, say girl_1 and girl_99, can be co-extensional, i.e., they can include exactly the same elements, without being necessarily considered as identical if created at different moments in time in the context of totally different events, see [13].

A conceptual annotation like that of Figure 1 can be used for posing queries in the style of: "Find all pictures of multiple, recumbent girls", with all the possible, even very different, variants; the queries must be expressed in NKRL terms giving then rise to data structures called 'search patterns'. Search patterns are NKRL data structures that represent the general framework of information to be searched for, by filtering or unification, within a repository of NKRL 'metadocuments' (e.g., conceptual annotations in the CONCERTO style), see also Figure 2 below and, for the technical details, [13].

We reproduce now in Figure 2 the NKRL coding of a narrative fragment like: "On June 12, 1997, John was admitted to hospital" (upper part of the Figure). This occurrence can be successfully unified with a search pattern (lower part of the Figure) in the style of: "Was John at the hospital in July/August 1997?" (in the absence of explicit, negative evidence, a given situation is assumed to persist within the immediate temporal environment of the originating event, see [2]).

From Figure 2 (upper part), we see that temporal information in NKRL is represented through two 'temporal attributes', date-1 and date-2. They define the time interval in which a predicative occurrence (the 'meaning' represented by the occurrence) 'holds'. In c2, this interval is reduced to a point on the time axis, as indicated by the single value, the timestamp 2-june-1997, associated with the temporal attribute date-1; this point represents the 'beginning of an event' because of the presence of 'begin' (a 'temporal modulator'). The temporal attribute date-1 is then represented 'in subsequence' (category of dating); see [2] for the full details.

```
c2)      EXIST  SUBJ  john_: (hospital_1)
                     [ begin ]
                     date-1: (2-july-1997)
                     date-2:

(?w   IS-PRED-OCCURRENCE
                     :predicate EXIST
                     :SUBJ      john_
                     (1-july-1997, 31-august-1997))
```

Figure 2: Coding of temporal information, and an example of search pattern.

The two timestamps of the search pattern in the lower part of Figure 2 constitute the 'search interval' linked with this pattern, to be used to limit the search for unification to the slice of time that it is considered appropriate to explore. Examples of complex, high-level querying (and inferencing) procedures that are characteristic of the NKRL approach are the so-called 'transformation rules', see [14].

As a further example of descriptive/factual structures, we give now, Figure 3, an NKRL interpretation of the sentence: "We have to make orange juice" that, according to Hwang and Schubert [15], exemplifies several interesting semantic phenomena. A narrative fragment like this could be included in an audio document.

```
c3) BEHAVE   SUBJ  (SPECIF human_being (SPECIF cardinality_
                                                 several_))
          [oblig, ment]
          date1:  observed date
          date2:

c4)   *PRODUCE SUBJ  (SPECIF human_being (SPECIF cardinality_
                                                 several_))
                OBJ  (SPECIF orange_juice (SPECIF amount_))
                date1: observed date + i
                date2:

c5)   (GOAL  c3  c4)
```

Figure 3: Representation in NKRL of 'wishes and intentions'.

To translate then the general idea of "acting to obtain a given result", we use:

- A predicative occurrence (c3 in Figure 3), instance of a basic template pertaining to the BEHAVE branch of the template hierarchy (H_TEMP), and corresponding to the general meaning of "focusing on a result'". This occurrence is used to express the 'acting' component, i.e., it allows us to identify the SUBJ of the action, the temporal co-ordinates, possibly the MODAL(ity) or the instigator (SOURCE), etc.
- A second predicative occurrence, c4 in Figure 3, with a different NKRL predicate and which is used to express the 'intended result' component. This second occurrence, which happens 'in the future', is marked as hypothetical, i.e., it is always characterised by the presence of a 'uncertainty validity attribute', code '*'.
- A 'binding occurrence', c5, linking together the previous predicative occurrences and labelled with GOAL, an operator pertaining to the 'taxonomy of causality' of NKRL. Binding structures — i.e., lists where the elements are symbolic labels, c3 and c4 in Figure 3 — are second-order structures used to represent the logico-semantic links that can exist between (predicative) templates or occurrences.

The general schema for representing the 'focusing on an intended result' domain in NKRL is then:

c_α)   BEHAVE   SUBJ   <human_being_or_social_body>

c_β)   *<predicative_occurrence>, with any syntax

c_γ)   (GOAL c_α c_β)

In Figure 3, oblig and ment are 'modulators', see [1]. ment(al) pertains to the 'modality' modulators. ment(al) pertains to the 'modality' modulators. oblig(atory) suggests that 'someone is obliged to do or to endure something, e.g., by authority', and pertains to the 'deontic modulators' series. Other modulators are the 'temporal modulators', begin, end, obs(erve), see Figure 2 and, again, [2]. In the constructions for expressing 'focus on...', the absence of the ment(al) modulator in the BEHAVE occurrence means that the SUBJ(ect) of BEHAVE takes some concrete initiative (acts explicitly) in order to fulfil the result; if ment(al) is present, as in Figure 3, no concrete action is undertaken, and the 'result' reflects only the wishes and desires of the SUBJ(ect).

We reproduce now, in Figure 4, one of the basic templates (descriptive component) which are part of the NKRL upper level catalogue, in particular, the template that gives rise to the occurrence c1 of Figure 1. Optional elements are in round brackets — e.g., the role SOURCE introduces the possible origin of the situation; note that, if the role OBJ(ect) is present, the syntactic construction of Figure 4 takes the specific meaning of 'stay with someone', i.e., with the OBJ's filler. In the corresponding occurrences, the variables (x, y, u, etc.) are replaced by concepts (definitional component) or individuals (enumerative component) according to the associated constraints; constraints are expressed mainly under the form of associations of high-level concepts of H_CLASS.

```
{ EXIST2.1 ;  'being physically present somewhere'

IsA: EXIST2 ;  'actual presence of a character or social body'
EXIST  SUBJ      x  :  [<location_>]
       (OBJ      y  :  [<location_>])
       (SOURCE   u  :  [<location_>])
       (MODAL    v)
       (CONTEXT  w)
       ({'modulators'})
       [date-1: <date_>, 'initial date' | 'observed date']
       [date-2: <date_>, 'final date']

       x = <human_being_or_social_body>
       y = <human_being_or_social_body>
       u = <human_being_or_social_body>
       v = <modality_> | 'label of a predicative occurrence'
       w = <situation_framework> }
```

Figure 4: An example of NKRL template.

Figure 5 is a (very abridged) representation of the H_TEMP hierarchy, where only some BEHAVE, MOVE and PRODUCE templates has been introduced. In this Figure, the syntactic descriptions are particularly sketchy: e.g., all the symbolic labels have been eliminated, and the templates are discriminated only through the associated natural language comments. For a more realistic picture of a template, see Figure 4 above. The codes '!' and '•' mean, respectively, 'mandatory' and 'forbidden'.

4.2 Definitional and enumerative components

Figure 6 gives a simplified representation of the upper level of H_CLASS (hierarchy of concepts, definitional component). From this Figure, we can note that substance_ and colour_ are regarded in NKRL as examples of non-sortal concepts. For their generic terms, pseudo_sortal_concepts and characterising_concepts, we have adopted the terminology of [16]. For a discussion about substance_ see, e.g., [1].

. **H_TEMP** ("hierarchy of predicative templates")

.. **BEHAVE templates** ; predicate : BEHAVE

... "external manifestation of the subject" ; •OBJ
....   "acting in a particular role" ; MODAL < *role_* [ex: *rugby_player*] >

... "focus on a result"; • OBJ, BENF, TOPIC; •mod. against, for; !GOAL bind. struct.
....   "act explicitly to obtain the result" ; •modulator ment
....   "wishes and intentions" ; ! modulator ment

... "concrete attitude toward someone/s.thing"; !OBJ, MODAL; •BENF, GOAL; •ment

.. **EXIST templates** ; predicate : EXIST ; !location of the SUBJ ; •BENF

.. **EXPERIENCE templates** ; predicate : EXPERIENCE ; !OBJ ; •BENF

.. **MOVE templates** ; predicate : MOVE

... "moving a generic entity" ; OBJ < *entity_*>
....   "move a material thing" ; OBJ < *physical_entity* >
.....   "change the position of something" ; •BENF (ex: "move the wardrobe")
.....   "transfer something to someone" ; !BENF (ex: "send a letter to Lucy")

... "generic person displacement"; SUBJ = OBJ = <*human_being*> ; !loc. of SUBJ, OBJ

... "transmit an information to someone" ;
....   "transmit a generic information" ; OBJ < *type_of_information* [ex: *message_*] >
....   "transmit a structured information" ; OBJ "label of binding/predicat. occurrence"

.. **OWN templates** ; predicate : OWN ; !OBJ ; •BENF

.. **PRODUCE templates** ; predicate : PRODUCE ; !OBJ

... "conceive a plan or idea" ; !modulator "ment"

... "creation of material things" ; OBJ < *physical_entity* > ; •modulator "ment"

... "perform a task or action" ; OBJ < *action_name* >
....   "acquire, buy" ; OBJ < *purchase_* >
....   "sell" ; OBJ < *sale_* >

... "relation involvem." ; SUBJ (COORD); OBJ *mutual_agreem.* MODAL <*relation_*>

.. **RECEIVE templates** ; predicate : RECEIVE ; !OBJ ; •BENF

Figure 5: Schematic H_TEMP hierarchy (ontology of events).

h_class

non_sortal_concepts

sortal_concepts

characterising_concepts

pseudo_sortal_concepts

...

location_

attribute_value
• logical_quantifier
•• all_
•• all_except
•• few_
•• many_
•• several_
•• some_
• relational_quantifier
•• equal_to
•• less_than
•• more_than
•• percent_

property_
• magnitude_
•• amount
•• numerical_property
•• cardinality_
• ownership_
• quality_
• relational_property
•• concordance_with
•• corresponding_to
•• has_member
•• has_part
•• member_of
•• part_of

colour_

substance_

human_being_or_soc.body

physical_entity

abstract_entity
• beliefs_
• relationships_

modality_

event_

process_
• industrial_process
• intellectual_process

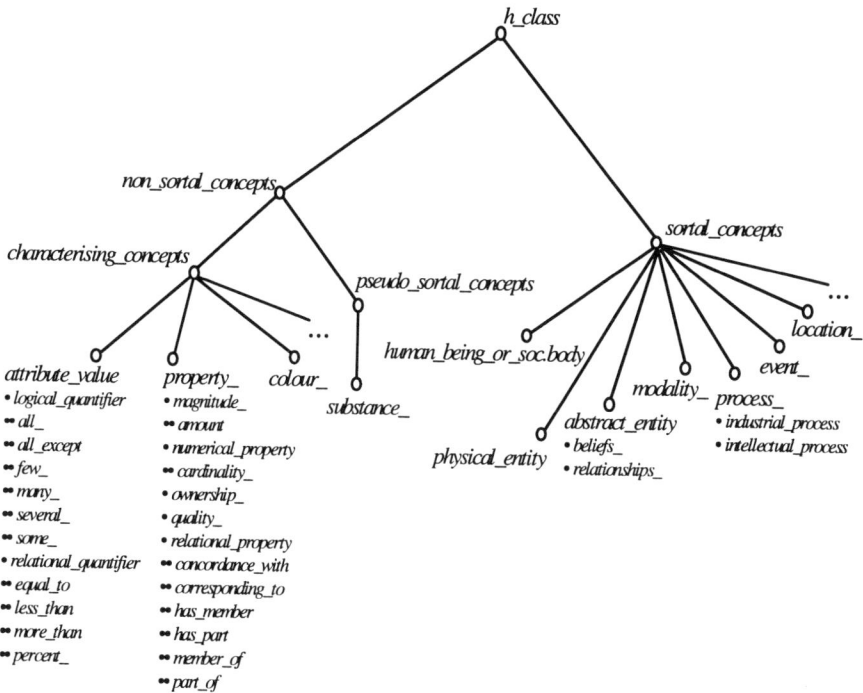

Figure 6: An abridged view of the 'upper level' of H_CLASS.

Coming now to the data structures used for concepts and individuals, they are essentially frame-based structures and their design is relatively traditional. These structures are composed of an OID (object identifier), and of a set of characteristic features (slots). Slots are distinguished from their fillers. Three different types of slots are used, 'relations', 'attributes', and 'procedures'; see again [1] for more information on this subject.

5. Implementation Notes : NKRL and RDF

The usual way of implementing NKRL has been, until recently, that of making use of a three-layered approach: Common Lisp + a frame/object oriented (commercial) environment + NKRL. In the framework of the new CONCERTO project, the decision has been taken to realise a new version of NKRL, implemented in Java and RDF-compliant (RDF = Resource Description Format). Given the importance that all the 'metadata' issues (RDF included) have recently acquired thanks to the impressive development of the Internet, it may be of some interest to examine quickly some of the problems brought forth by this decision of the CONCERTO consortium. Metadata is now a synonymous of "machine understandable knowledge describing Web 'resources'", see, e.g., [17], where a resource may concern all sorts of multimedia documents, texts, videos, images, and sounds.

RDF [18] is a proposal for defining and processing WWW metadata that is developed by a specific W3C Working Group (W3C = World Wide Web Consortium). The model, implemented in XML (eXtensible Markup Language), makes use of Directed Labelled Graphs (DLGs) where the nodes, that represent any possible Web resource (documents, parts of documents, collections of documents etc.) are described basically by using attributes that give the named properties of the resources. No predefined 'vocabulary' (ontologies, keywords, etc.) is in itself a part of the proposal. The values of the attributes may be text strings, numbers, or other resources.

The first, general problem we had to solve for the implementation of the XML/RDF-compliant version of NKRL has concerned the very different nature of the RDF and NKRL data structures. The first — modelled on the 'traditional' ontologies — are *dyadic*, i.e., the main RDF data structure can be assimilated to a triple where two resources are linked by a binary conceptual relation under the form of a property. We have seen on the contrary, in Section 3, that the basic building block of the NKRL descriptive and factual structures is a complex *threefold* relationship associating a symbolic label, a predicate, one or more roles and the corresponding fillers (arguments of the predicate). To assure then the conversion into RDF format, the NKRL data structures have been represented as intertwined binary 'tables', see Figure 7 that describes the RDF-compliant, general structure of an NKRL template. For comprehensibility's sake, this figure has been considerably simplified and, e.g., it does not take into account the hierarchical relations between templates or the relationships between templates and occurrences.

More specific problems have concerned the (still limited) choice of knowledge representation tools that are presently associated with RDF. To give only an example, let us consider the solutions that, making use of the *containers* — RDF tools for describing collections of resources, see [18] — we have adopted in order to reproduce the semantics of the so-called *AECS sublanguage* [1] of NKRL. The AECS operators are used to build up expansions (structured arguments) like those included in Figures 1 and 3 above.

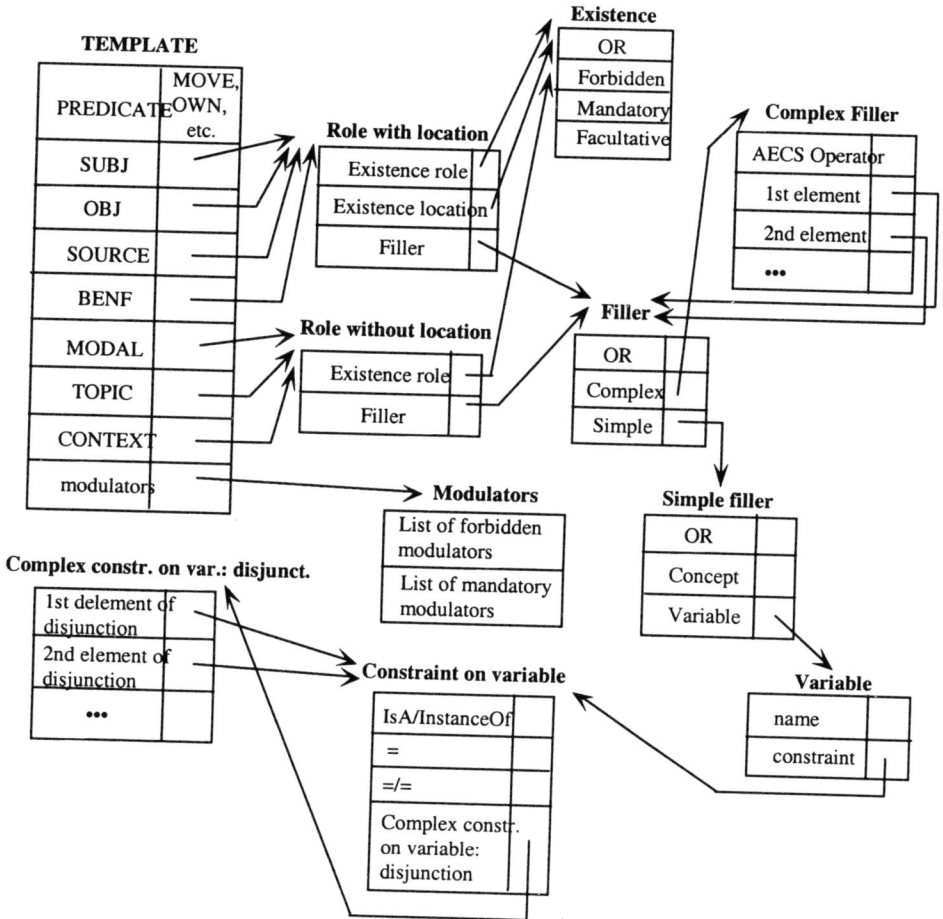

Figure 7: General structure of an NKRL template according to a 'binary' view.

AECS includes four operators, the disjunctive operator (ALTERNative = A), the distributive operator (ENUMeration = E), the collective operator (COORDination = C), and the attributive operator (SPECIFication = S). The semantics of SPECIF has already been explained in Section 4.1 above; the semantics of ALTERN is self-evident. The difference between COORD and ENUM consists in the fact that, in a COORD expansion list, all the elements of the expansion take part (necessarily) together in the particular relationship with the predicate defined by the associated role. As an example, we can imagine a situation similar to that described in Figure 2 above, where two persons, John and Peter, have been admitted *together* to hospital: the SUBJ of EXIST would be, in this case: (COORD john_ peter_). In an ENUM list, each element satisfies the same relationship with the predicate, but they do this separately. RDF defines on the contrary three types of containers:

- 'Bag', an unordered list of resources or literals, used to declare that a property has multiple values and that no significance is associated with the order in which the values are given;
- 'Sequence', Seq, an ordered list of resources or literals, used to declare that a property has multiple values and that the order of these values is significant;
- 'Alternative', Alt, a list of resources or literals that represent alternative for the (single) value of a property.

Of these, only Alt presents a very precise coincidence with an AECS operator, obviously, ALTERN; moreover, we have at our disposal only *three* container constructions to represent *four* NKRL operators. We can note, however, that the use of a Bag construction is an acceptable approximation in order to represent the semantics of COORD. For example, in [18], the use of Bag is associated with the representation of the sentence: "The committee of Fred, Wilma and Dino approved the relation". As the editors of this report say, "...the three committee members *as a whole* voted in a certain manner...". This situation corresponds certainly in NKRL to a COORD situation. As a first conclusion, we can then state that:

- The ALTERN constructions (a disjunction between the arguments of the constructed complex filler) are described making use of the RDF container Alt.
- The COORD constructions (an unordered list of elements of the conceptual hierarchy) are represented making use of the RDF container Bag.

With respect now to the RDF representation of the ENUM and SPECIF constructions, we make use of a 'liberal' interpretation of the semantics of Seq containers. Seq introduces an order relation between the elements of a list. In Section 4.1, we have seen that the operator SPECIF introduces a 'partial order' relationship of the type (SPECIF e_i p_1 ... p_n), where the element e_i necessarily appears in the first position of the list, while the order of the residual elements p_1 ... p_n (the 'properties' or 'attributes' associated with e_i) is, on the contrary, indifferent. We can then include these properties in an (implicit) Bag, and insert e_i and this Bag in a Seq list. The same solution has been adopted for the ENUM operator, with the (only) difference that an explicit enum identifier appears in the first position of the Seq list. Note that a previous solution, see [19], where ENUM was represented making use of the RDF 'repeated properties' has been abandoned because that it could lead to an unacceptable proliferation of final RDF structures. We can than say that:

- For the SPECIF constructions, we use the RDF operator Seq followed directly by the elements of the SPECIF list. Of these, the elements representing the 'properties' are to be considered as inserted into a Bag; to simplify the RDF code, this Bag construction is conceived as an 'implicit' one, and the Bag operator is not expressly mentioned.
- The ENUM constructions are coded by using a Seq container where the first element is the identifier enum.

As an example, let us consider, Figure 8, the RDF representation of occurrence c2 of Figure 2 where, for completeness' sake, we have assumed again that John *and* Peter have been admitted *together* (COORD) to hospital. In general, the RDF text associated with each predicative occurrence is composed by several tags, all nested inside the <CONCEPTUAL_ANNOTATION> tag and belonging to two different namespaces, rdf and ca. The first namespace describes the standard environment under which RDF tags have to be interpreted. The second namespace describes specific tags defined in the context of an application, in this case, the CONCERTO project. For example, the tag <ca:Template$_i$> is used to denote that the predicative occurrence is an instance of the template identified by Template$_i$. The other tags specify the various roles of the predicative occurrence, together with the associated value. The tag <ca:subject> specifies, e.g., the role SUBJ of c2. The code li means 'list item', and it was chosen in RDF to be mnemonic with respect to the corresponding HTML term.

6. Conclusion

In this paper, we have described some of the main data structures proper to NKRL, a language expressly designed for representing, in a standardised way ('metadata'), the semantic content of narratives. Even if we attribute the most general meaning to the word "narratives", we can agree on the fact that NKRL is *not* a universal language, capable of representing any sort of knowledge in the world. We would like, however, to emphasise at least two very important features of NKRL:

- Making use of a relatively simple, flexible and easily manageable formalism NKRL offers some interesting solutions to very hard problems concerning the 'practical' aspects of the knowledge representation endeavour. These include, e.g., the representation of implicit and explicit enunciative situations, of wishes, desires, and intentions (modals), of plural situations, of causality, and of other second order, intertwined constructions.
- A second, important characteristics of NKRL is represented by the fact that the catalogue of basic templates, see Section 3.1 above, can be considered as part and parcel of the definition of the language. This approach is particularly important for practical applications, and it implies, in particular, that: i) a system-builder does not have to create himself the structural knowledge needed to describe the events proper to a (sufficiently) large class of narrative texts and documents; ii) it becomes easier to secure the reproduction or the sharing of previous results.

```
<?xml version="1.0" ?>
<!DOCTYPE DOCUMENTS SYSTEM "CA_RDF.dtd">
<CONCEPTUAL_ANNOTATION>
  <rdf:RDF xmlns:rdf="http://www.w3.org/1999/02/22-rdf-syntax-ns#"
  xmlns:rdf="http://www.w3.org/TR/1999/PR-rdf-schema-19990303#"
  xmlns:ca="http://projects.pira.co.uk/concerto#">
    <rdf:Description about="occ3">
    <rdf:type resource="ca:Occurrence"/>
      <ca:instanceOf>Template2.31</ca:instanceOf>
      <ca:predicateName>Exist</ca:predicateName>
      <ca:subject rdf:ID="Subj2.31" rdf:parseType="Resource">
        <concerto:filler>
          <rdf:Bag>
```

```
        <rdf :li rdf:resource="#john_"/>
        <rdf :li rdf:resource="#peter_"/>
      </rdf:Bag>
    </concerto:filler>
    <concerto:location>hospital_1</concerto:location>
  </ca:subject>
  <ca:listOfModulators>
    <rdf:Seq><rdf:li>begin</rdf:li></rdf:Seq>
  </ca:listOfModulators>
  <ca:date1>02/06/1997</ca:date1>
</rdf:Description>
  </rdf:RDF>
</CONCEPTUAL_ANNOTATION>
```

Figure 8: The RDF format of a predicative occurrence.

References

[1] Zarri, G.P. (1997) NKRL, a Knowledge Representation Tool for Encoding the 'Meaning' of Complex Narrative Texts. Natural Language Engineering **3**, 231-253.

[2] Zarri, G.P. (1998) Representation of Temporal Knowledge in Events : The Formalism, and Its Potential for Legal Narratives. Information & Communications Technology Law **7**, 213-241.

[3] Zarri, G.P., Bertino, E., Black, B., Brasher, A., Catania, B., Deavin, D., Di Pace, L., Esposito, F., Leo, P., McNaught, J., Persidis, A., Rinaldi, F. and Semeraro, G. (1999) CONCERTO, An Environment for the 'Intelligent' Indexing, Querying and Retrieval of Digital Documents. In : *Foundations of Intelligent Systems – Proc. of 11th International Symposium on Methodologies for Intelligent Systems, ISMIS'99*. Berlin: Springer-Verlag.

[4] Guarino, N., and Giaretta, P. (1995) Ontologies and Knowledge Bases : Towards a Terminological Clarification. In : *Towards Very Large Knowledge Bases - Knowledge Building and Knowledge Sharing*, Mars, J.I., ed. Amsterdam: IOS Press.

[5] Fellbaum, C., ed. (1998) *WordNet, An Electronic Lexical Database*. Cambridge (MA): The MIT Press.

[6] Bateman, J. A., Magnini, B., and Fabris, G. (1995) The Generalized Upper Model Knowledge Base : Organization and Use. In : *Towards Very Large Knowledge Bases - Knowledge Building and Knowledge Sharing*, Mars, J.I., ed. Amsterdam: IOS Press.

[7] Lenat, D.B., and Guha, R.V. (1990) *Building Large Knowledge Based Systems*. Reading (MA): Addison-Wesley.

[8] Heinsohn, J., Kudenko, D., Nebel, B., and Profitlich, H-J.(1994) An Empirical Analysis of Terminological Representation Systems. Artificial Intelligence **68**, 367-397.

[9] Fensel, D., Decker, S., Erdmann, M., Studer, R. (1998) Ontobroker : Or How to Enable Intelligent Access to the WWW. In: *Proc. of the 11th Banff Knowledge Acquisition for KBSs Workshop, KAW'98*. Calgary: Dept. of CS of the University.

[10] Heflin, J., Hendler, J., and Luke, S. (1999) *SHOE : A Knowledge Representation Language for Internet Applications* (Tech. Rep. CS-TR-4078). College Park (MA): Dept. Of CS of the Univ. of Maryland.

[11] Heflin, J., Hendler, J., and Luke, S. (1999) Coping with Changing Ontologies in a Distributed Environment. In: *Proc. of the AAAI-99 Workshop on Ontology Management*. Menlo Park (CA): AAAI.

[12] Gruber, T.R. (1993) A Translation Approach to Portable Ontology Specifications. Knowledge Acquisition **5**, 199-220.

[13] Franconi, E. (1993) A Treatment of Plurals and Plural Quantifications Based on a Theory of Collections. Minds and Machines **3**, 453-474.

[14] Zarri, G.P., Azzam, S (1997) Building up and Making Use of Corporate Knowledge Repositories. In : *Knowledge Acquisition, Modeling and Management - Proc. of EKAW'97*, Plaza, E., Benjamins, R., eds. Berlin: Springer-Verlag.

[15] Hwang, C.H., Schubert, L.K. (1993) Meeting the Interlocking Needs of LF-Computation, Deindexing and Inference : An Organic Approach to General NLU. In: *Proc. of the 13th Int. Joint Conf. on Artificial Intelligence*. San Francisco: Morgan Kaufmann.

[16] Guarino, N., Carrara, M., Giaretta, P. (1994) An Ontology of Meta-Level Categories. In: *Proc. of the 4th Int. Conference on Principles of Knowledge Representation and Reasoning*. San Francisco: Morgan-Kaufmann.

[17] Boll, S., Klas, W., Sheth, A. (1998) Overview on Using Metadata to Manage Multimedia Data. In: Sheth, A., Klas, W. (eds.): *Multimedia Data Management - Using Metadata to Integrate and Apply Digital Media*. New York: McGraw Hill.

[18] Lassila, O., Swick, R.R. (eds.) (1999) *Resource Description Framework (RDF) Model and Syntax Specification*. W3C.

[19] Jacqmin, S., Zarri, G.P. (1999) *Preliminary Specifications of the Template Manager* (Concerto NRC-TR-4). Paris: CNRS.

Domain Knowledge in Engineering Design: Nature, Representation, and Use

Fatma Mili[1], Krish Narayanan[2], Dan VanDenBossche[3]

1 Oakland University, Rochester MI 48309-4478, mili@oakland.edu
2 Ford Motor Company, Dearborn, Michigan
3 DaimlerChrysler Corporation, Auburn Hills, Michigan

Abstract: Engineering design is a highly regulated activity. Design decisions need to be negotiated in light of engineering laws, professional standards, and other federal regulations. In this paper we are interested in the integrated and flexible support of engineering design. Integration requires that domain knowledge be an integral part of the design support system rather than an add-on. Flexible support requires that laws and regulations remain accessible and updatable to reflect technological advances and changes in regulations as they occur. We present a knowledge and object model that satisfies these two criteria. Designing a compliant engineering artifact is a lengthy and non-monotonic refinement process. We define a set of design milestones. We define the semantics of the object model and of the associated design milestones.

1. Introduction

Engineering design is the process of transforming a set of requirements into a product description in compliance with a set of predetermined laws and procedures. Design is a labor intensive, knowledge rich, and creative activity. Traditionally, engineering design environments (CAD, CAM, CASE) have focussed on the mechanistic and book-keeping aspects of design. The creative nature of design has typically been thought of as a purely human activity. As such, creativity does not need to be directly supported, but the support of other aspects must not inhibit creativity. There is a shift in thinking; recent research efforts have been directly targeting the support of creativity [11, 15, 16]. The knowledge related aspects of design on the other hand, had benefited from some attention, although their support tends to be localized to specific subtasks, and supported in a non integrated way through add-on layers of specialized libraries. A number of research efforts have since targeted an integrated and flexible support of knowledge related issues of quality assurance, change propagation, and error detection and correction [8, 16, 19, 29].

In this paper, we are interested in the support of the verification and validation aspects of design. Early error detection and quality assurance in design have often been identified as areas of critical need and high return potential [9, 19, 23, 27]. For the support of validation to have a notable and durable impact on the quality of products and processes, it needs to be tightly coupled with design support rather than designed as an after-thought add-on to the system. Furthermore, the relevant knowledge needs to be easy to maintain and adapt to advances in technology and changes in laws and regulations. The approach presented in this paper is characterized by a tight coupling between domain knowledge and design decisions provided by a common object model. It is also characterized by an underlying concern for knowledge independence and knowledge evolutivity. The domain knowledge is represented in a way that facilitates its access, analysis, review, and update. This work shares its motivations and goals of enforcing domain constraints during design with [19, 23, 26, 30, 33]. The organization of the knowledge builds on the modularity and encapsulation provided by object oriented databases and active databases [2, 5, 12, 24, 18]. We build on the principle of knowledge independence underlying active database [32] whereby business rules are represented centrally within the database model rather than distributed across application programs. Because the knowledge of interest here is more akin to integrity constraints than to business rules, we use the same approach used in the Ode active database environment [1, 12, 13] by representing the constraints declaratively, independently of their enforcement processes, thus increasing the level of knowledge independence.

This paper is organized as follows. In Section 2, we present an example and use it to motivate the choices made. In Section 3, we present an integrated object model encompassing objects' domain constraints, design requirements, and design decisions. In Section 4, we define semantics for the object model proposed. In Section 5, we discuss the use of the knowledge during design decision making. We summarize and conclude in Section 6.

2. Example, Approach

2.1 Domain constraints; design decisions

We consider the design of the interior of a vehicle. This activity consists of selecting (or designing) the various components and placing them in the vehicle's interior. The typical components include the driver's seat, the passengers' seats, the control panel, the mirrors, and the foot pedals. This design activity is typically performed after the overall geometry as well as other global attributes of the vehicle have been set. The decisions are materialized in a mostly geometric world but are motivated by high level considerations of function, style, safety, cost, and so forth. We show here a sample of functional considerations. They are naturally focused on the driver's space. The design and placement of the driver's seat, the control panel, the mirrors and other indicators must all be coordinated to provide the driver with a functional space. In particular, the following conditions must be met:

- The drivers can be sat comfortably to sustain the position for hours.

- The drivers have an adequate reach to the controls. Primary controls must be at immediate reach and secondary controls must be reachable with minimal effort.

- The drivers have an unobstructed view of the road, the rear and side mirrors, and the primary indicators.

The above constraints are part of the standards and recommendations developed and maintained by the Society of Automotive Engineers (SAE). SAE maintains a handbook [25] with thousands of standards and recommended practices. These standards are documented, formalized, and cross referenced. More importantly, they are updated yearly reflecting scientific findings, technological advances, and new regulations. The standards included in the SAE handbook are sufficiently detailed and unambiguous to allow their automatic verification. For example, the functional constraints on the interior are formalized using the following data and processes:

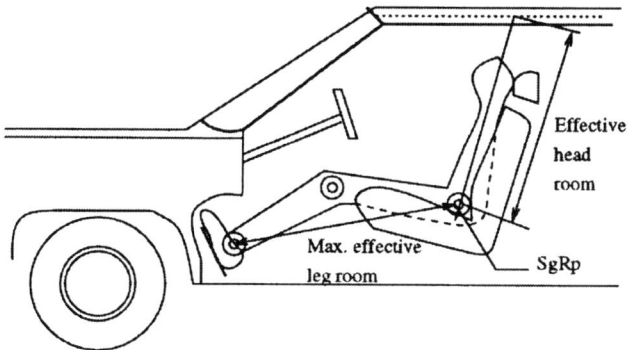

Figure 1: Driver Space

- The SAE Handbook contains census tables listing drivers' anthropometry by gender and by percentile.

- The comfortable seating constraint sets minimal values for the distance between the hip point of the driver (SgRp in Figure 1) and the foot pedal (leg room) and the distance between the hip point and the car ceiling (effective head room). These minimal values are computed for each pair (driver percentile, seat position).

- Given a seat position and a driver percentile, the SAE handbook provides a procedure for computing hand reach areas. The adequate reach constraint restricts the placement of standard primary and secondary controls within designated hand reach areas.

- Similarly, given a seat position and a driver percentile, the SAE handbook provides a procedure for computing various fields of view. The adequate visibility constraint translates into a set of constraints on intersections between fields of view and other surfaces and volumes in the interior.

- Because the constraints need to be met for every driver, the handbook provides with an optimized procedure that identifies critical "points" for which the constraint needs to be verified.

The SAE handbook is mentioned here only to illustrate the availability and accessibility of design knowledge information in engineering domains. The SAE is only one such source. In a typical engineering setting, there are multiple, autonomous sources of domain constraints. Scientific communities, professional societies, and regulatory bodies (EPA, OSHA, etc.) maintain and distribute extensive documentation of domain knowledge and constraints. The trend of knowledge management identifying corporate know-how as a defining competitive resource [3, 6, 17, 28] has led many organizations to cultivate and document their in-house standards and best practices. The availability of these resources is an additional incentive for making use of the knowledge during design.

2.2 Issues, motivations

As knowledge becomes more widely available and the need for improving the process and product more pressing, the integration of the knowledge with design tools emerges as an inevitable next step. A number of considerations come into play in this integration. We briefly discuss them here

Differing levels of abstraction: One of the major hurdles in the active knowledge-based support of designers is the gap between the world of low level geometric artifacts and the world of abstract constraints that motivate and constrain the decisions defining them. A constraint on the size and placement of the odometer for example, is immaterial if the CAD/CAM system has no conception of what and where the odometer is and what design decisions do or do not relate to it. There is an increasing awareness of the shortcoming of existing geometry-bound CAD/CAM systems [7, 14, 27, 30, 31]. The trend in research and in commercial products is moving towards systems designed to handle concepts and features that are more abstract and more tangible to the users. A premise of this work is that the design support system is not a purely geometric system, and that it is possible to define a complete set of concepts that can underlay both domain constraints and design decisions.

Nature of the knowledge: Knowledge based systems traditionally assumed an operational representation of the knowledge. Production rules or active database rules for example are sets of "reactions" to conditions (violations of constraints). The difficulty of extracting and maintaining such operational knowledge has been acknowledged in research in knowledge acquisition and problem solving [22]. In the current context, the standards and recommendations documented and published are typically normative statements of facts rather than prescriptive rules. The hand reach constraint for example simply states that primary controls must be within reach of the driver without excessive leaning and bending. It does not prescribe what to do when the constraint is not met. Generally, declarative knowledge is more readily available from sources, more modular, and is easier to understand and to update than prescriptive knowledge.

Knowledge Independence: The knowledge of interest is changing in light of technological innovations and changes in laws and regulations. It is critical that changes can be made in a modular localized manner. The constraints on the interior of a vehicle for instance are not set in stone. As new controls and "indicators" such as navigation systems and wireless communications become standard, new constraints are added and old constraints are changed or withdrawn. As controls become voice-activated, they no longer need to be at hand-reach, the constraints related to their size and location need to be revised and withdrawn or updated. For the knowledge based validation support to be viable, it is essential that the knowledge be devised to enable and support the continuous update of the knowledge.

3. Knowledge and Data Model

We define an object model that captures the domain knowledge and design requirements and defines a context for design decisions. The model presented is devised in view of the considerations listed in 2.2.

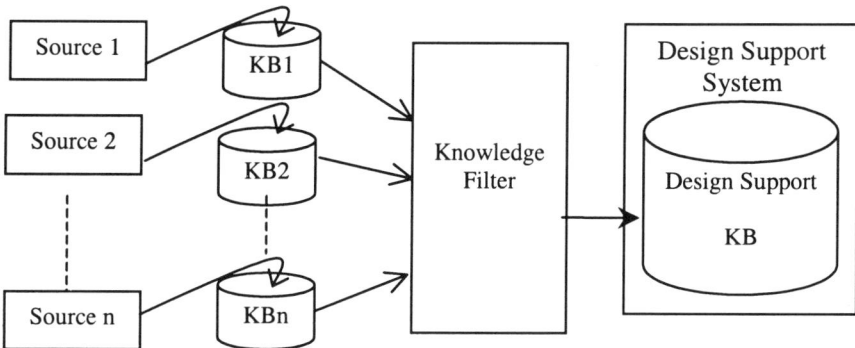

Figure 2: Reconciling integration and independence

3.1 Overview

The model proposed integrates three levels of information: the generic domain constraints, the design requirements, and the design decisions. In order to reconcile the two competing concerns of seamless integration of domain knowledge with design decision making on the one hand and the need for knowledge independence on the other hand, a level of indirection is introduced between the autonomous sources and the design support. This is shown in Figure 2.

The filter component can be thought of as a knowledge extractor, synthesizer, and translator. In a closely coordinated environment, the different knowledge sources would use a common ontology and a common representation, the role of the filter is then to select and extract the knowledge of interest and to monitor and propagate changes. In a setting with greater autonomy, the filter will be responsible for semantic conflict identification and resolution. These issues are outside the scope of this paper. We will solely focus here on the contents of the design support knowledge base.

3.2 Class constraints

Our goal is to capture the normative knowledge about categories of objects in such a way that it can be used when instances of these categories are being designed. The object model provides with a modular representation of classes. We follow on the steps of other proposals augmenting the definition of classes with the set of constraints on the instances of the class. A skeleton of the class *VehicleInterior* for example, is shown below using the graphic UML [4] notation.

VehicleInterior
target.:   PopulationRange
...
SelectDS(S:Seat)
...
Constraints
SeatPlacement

The constraint seatPlacement captures the functional requirement on comfort, reach, and view. For modularity and flexibility purposes, we have opted to have a separate class **Constraint** in which constraints are described in their full detail. The essence of a constraint is its formula, but other attributes such as source, illustration, and rationale, are also relevant to its use, analysis and update. The specific attributes for constraints can be chosen to fit the needs of the domain. We show below a possible description of the seatPlacement constraint.

seatPlacement:Constraint	
name:	seatPlacement
text:	For every driver in the targeted population, there exists a seat position in which the driver can reach adequately all of the primary and secondary controls, the driver can view the road ahead, the sides, the side and rear mirrors, and the driver has adequate leg and head room to be comfortable
source:	SAExyz
rationale:	functional (more details can be given here).
assumptions:	controls are hand controls. driver is right-handed....
OCLformula:	targetPop→ forAll(driver | driverSeat.type.positions→ exists (position | AdequateReach(driver,position) and AdequetView (driver,seat-position) and AdequateRoom (driver,seat-position)))
illustration:	Figure 1
Evaluation:	SAEProcedure xyz

The attributes used illustrate the multiple uses of the constraints. The free text reflects the fact that it is authored by regulation agents and read by designers. The graphical form used in the illustration gives it a "visual" form directly relevant to the engineer's work. The formal language used in the formula makes it accessible for automatic verification. In addition to the formula, an evaluation procedure is given here to validate this constraint efficiently. With all of their advantages, declarative formulas are not always efficient to evaluate. Additional information (axioms) can be provided to optimize the evaluation process. For example, consider the axioms "if the tallest person has enough room, then so does everyone in the targeted population", and "if the shortest person can reach and see then so can every one else in the targeted population." These axioms can be used to work around the universal quantifiers in the evaluation of the seatPlacement constraint. Evaluation information can come in the form of axioms, pre-packaged procedures, or any other form.

3.3 Design requirements specification

The constraints captured in the classes are constraints common to all instances of a given class. In addition to these general constraints, designers need to account for specific *design requirements*, i.e. requirements specific to the product at hand. Whereas domain constraints are shared by *all* the objects from the associated class, design requirements are constraints that are specific to design instances that we will call assignments. Design requirements are shared by all the competing alternatives generated for the same assignment. We capture design requirements by an attribute associated with all classes. We define a root class *DesignObject* in which this attribute is defined, and from which it is inherited by all classes.

DesignObject
Requirement: set(Constraint} = {True} {frozen}

The frozen qualifier of the Requirement attribute protects it from being changed accidentally during design. Changes in requirements can still occur, but they are handled with operators specifically designed for this purpose. Design requirements are used in conjunction with domain constraints. They may strengthen domain constraints (e.g. extra legroom) or involve different attributes (e.g. seats must be leather).

4. Semantics of the Object Model

4.1 Design milestones

With the expression of domain constraints and design requirements set, we turn our attention to their effect on design decisions. In programming settings, constraints are generally interpreted as invariants that are meant to be met at every state of the instance. Programmers are responsible for implementing the class methods in a way that preserves the invariants. In database settings where operations are of a smaller grain (not closed operations), more flexibility is introduced by making transactions rather than operations the scope of enforcement. Transactions encapsulate intermediate states in which the constraints may not be met. Transactions cannot commit unless all constraints are met. In design, the difference in gain size between design steps and objects is even wider. Designers aim to globally converge towards a final design that meets all of the constraints, but this convergence is typically non-monotonic, and relatively slow. Design transactions are long-lived [10] and can last for days or weeks; waiting for the end of a transaction to provide users with validity feedback may make them waste valuable time and effort. It is desirable to provide more timely feedback to the user. In order to balance between the need for feedback and the need for flexibility, we define design milestones in the design process. They are checkpoints in the design process where the user can get feedback and directions without requiring that decisions be final or that all constraints be met. The four milestones proposed are:

Assignment: An assignment is defined by the initial object creation and setting of the Requirement attribute.

Solution: A (partial) solution is an assignment for which some decisions have been made and for which no constraints are violated.

Design: A design is a complete solution (no free variables).

Draft: Any set of decisions made on an assignment produces a draft.

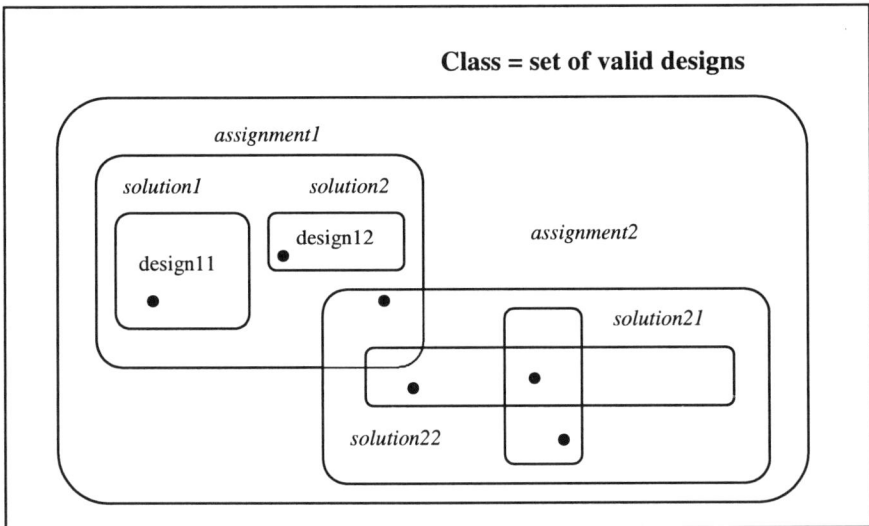

Figure 3: Illustration of assignments, solution, designs, and drafts

Additional milestones can be defined and added as needed. The Venn diagram in Figure 3 provides a first approximation of these milestones. The two outer sets represent respectively the space of the class, and the feasible space. The space represents all possible states whether they are compliant or not, complete or not. In other words, it is the set of all possible drafts. The feasible space represents the set of all instances in the space that meet the domain constraints. It is the set of all possible designs (correct with respect to some requirement). An assignment defines a subset of the class. Solutions are subsets of the associated assignment. Designs are singleton elements within their associated assignment.

4.2 Semantics of non monotonic refinement

At some level, engineering design is a progressive refinement process whereby an initial assignment is refined into solutions and then a final design. At the level of individual decisions, the process is not necessarily a refinement process (decisions are withdrawn and changed), and by the same token, it is non-monotonic. We define here semantics that capture this search process. Classes are typically thought of as a symbiosis between a state --- that we try to abstract away --- and a behavior that is directly visible to the user. In a database setting, and more so, in a design setting where the user is constructing the state of designed objects, the state is fully visible and the state rather than the behavior is what defines the class [12]. We define here a state-based semantics characterized as follows:
1. It captures refinement, by focusing on sets being refined and reshaped rather than on individual instances,

2. It builds on the concept of milestones by using the domain constraints as yardsticks used to assess progress rather than as invariant that must be met al all times, and

3. It accounts for design decisions even when they conflict with domain and design constraints.

Given a class *Cl*, we distinguish between two types of sets: The implicit sets specified by the class definition and the design requirements, and the explicit sets shaped by the users via design decisions. We start with the sets specified to the user.

Definition 1 Given a class *Cl* defined by:
*   A, a set of attributes $\{a_1:T_1, \ldots ,a_n:T_n\}$,
*   C, a set of domain constraints in the form of predicates on the attributes,

We define:
*   The **space** *S* **of** *Cl* by $\quad S = T_1 \times T_2 \times \ldots \times T_n;$

*   The **feasible space** *F* of *Cl* $\quad F = Sc = \{s | s \in S \wedge C(s)\};$

- Given a requirement Q, we define the **Requested space** R_Q, denoted simply by R when Q is implied by the context, by :

$$R_Q = F \cap S_Q = S_{C \cup Q} = \{s | s \in S \wedge C(s) \wedge Q(s)\}$$

On the other hand, as designers make decisions, they define subsets of the class space. Because decisions may or may not be compliant, these sets need to be defined independently of the feasible and requested spaces. We define design decisions as predicates d on the space. For example, x=5, x+y≠5, and DriverSeat.width=70cm are all decisions. A decision made for one task (e.g. defining seat width as part of interior design,) may become an assignment for a subordinate task (e.g. design a seat). Decisions represented by a set d of predicates define a subset D of the space $D = S_d = \{s | d(s)\}$.
We define now the semantics of classes.

Definition 2.
Given a class *Cl* defined by:
- A, a set of attributes $\{a_1 : T_1, \ldots.. a_1 : T_n\}$,
- C, a set of domain constraints in the form of predicates on attributes,

The instances of the class are pairs of sets <R,D>, subsets of S the space of the class, where R captures the domain constraints and design requirements, and D captures the decisions made on the instance.

Using this definition, we formalize the notions of assignment, solution, and design:

Definition 3.

Assignment: A pair < R,D > is an assignment iff D = S.

Solution : A pair < R,D > is a solution iff R ∩ D ≠ φ.

Design: A pair < R, D > is a design iff |R ∩ D| = 1.

Draft: A pair < R,D > is a draft iff true.

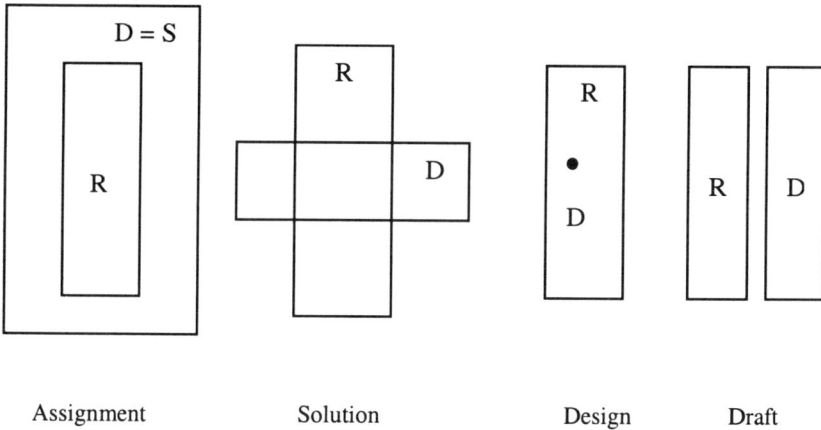

| Assignment | Solution | Design | Draft |

Figure 4: Decisions and Requirements

The four milestones are illustrated in Figure 4. Note that the categories are not exclusive. All four configurations shown qualify as drafts for example.

For the milestones to have any significance, we need to provide mechanisms for creating them and updating them. We define a set of such methods. As a first step, we assume the existence of stored sets AssignmentSet, solutionSet, designSet, and draftSet that are initialized upon initialization of the database, and that have their own object identification mechanism (two stages of the same object are two different solutions). The methods for a class Cl are defined as follows:

Create():Cl
pre: True
post: self.Req={true}
          self.R=F
          self.D=S
CreateAssignment(Q:set(Constraint)):Cl
pre: True
post: self.Req=Q
          self.R=F$\rightarrow$ intersection(S_Q)
          self.D=S
          AssignmentSet = AssignmentSet@pre$\rightarrow$including(self)
AddDecision(d:decision)
pre: True

post: self.R=self.R@pre
        self.D=self.D@pre$\rightarrow$ intersection(S_d)

CreateSolution()

pre: AssessCompliance(self)$\rightarrow$ isempty
post: self = self@pre
        SolutionSet= SolutionSet@pre$\rightarrow$ including(self)

CreateDesign()

pre: AssessCompleteness(self)$\rightarrow$ isempty and
                 AssessCompliance(self)$\rightarrow$isempty
post: self= self@pre
        DesignSet=DesignSet@pre$\rightarrow$ including(self)

CreateDraft()

pre: True
post: self=self@pre
        DraftSet=DraftSet@pre$\rightarrow$including (self)

AccessCompliance(self): set(Constraint)

pre: True
post: self-self@pre
        result=(self.Constraints$\rightarrow$union(self.Req))$\rightarrow$
                select$\{c|(S_c\rightarrow$intersectio(seld.D))$\rightarrow$ isempty$\}$

AccessCompleteness(self):set(attributes)

pre: True
post: self=self@pre
        result=$\{a|$(self.D$\rightarrow$ Select(a)) $\rightarrow$ size>1 and not
             (self.D$\rightarrow$ intersection(self.R))$\rightarrow$select(a)$\rightarrow$size=1$\}$

The create methods can be thought of as checkpoints storing a version of the design that can be backtracked to.

The AccessCompliance method identifies variables that still represent a degree of freedom. A variable a represents a degree of freedom if explicit decisions have not narrowed its values to 1, and the constraints and requirements are either not sufficient to restrict it to 1 ((self.D.a$\rightarrow$intersection(self.R.a))$\rightarrow$ size>1), or are conflicting with the decisions made (((self.D.a)$\rightarrow$ intersection (self.R.a))$\rightarrow$ size=0).

5. Monitoring Progress Through Active Rules

The domain constraints and design requirements set the conditions that a final design must meet. The four categories of objects define major stages in design progress. Constraint enforcement protocols can be added as needed to reflect the domain specific policies and procedures. Protocols can be of a sufficiently general nature that they are defined for all objects in a design system. The following protocol is an example of such a general policy: "whenever a federally mandated constraint is violated, notify the designer and tag the object with the constraint." More specific protocols can be associated with individual classes as needed. The following protocol is an example of local policies "For luxury vehicles, cost constraints take a lower priority." The separation between the constraints and the protocols enforcing them is a departure from the "traditional" active databases where the constraints have no independent existence; they are embodied within the active rules. With the approach proposed here, constraints are specifically tailored to classes, they are inherited and specialized. Protocols on the other hand, tend to be of a more general purpose nature and express a general design philosophy rather than specific rules of action. To illustrate this point, we provide a sample of active rules.

5.1 Active rules for passive feedback

The co-existence of the two sets D and R in the object semantics is necessary to allow the flexibility to temporarily violate requirements. The set D represents the direct actions taken by the user, whereas the set R represents the set where the final designs must fall. There is considerable room for support for the users in terms of relating the sets D and R.

One of the difficulties of design is to see all of the implications of the decisions that one makes. For example, if there is a requirement such as $x^2+y^2 \geq 1$, and the user makes a decision to set x to 1, the user has automatically satisfied the requirement and can make further decisions without needing to be concerned with the requirement. This situation is a situation where $D \subseteq R$. It is a very desirable situation as it ensures that the current state is a solution, and that all of its further refinements will also be solutions. The following rule dictates informing the user.

All requirements met

Whenever   AddDecision

If   self.D→*includes* self.R

Then   Notify user that all requirements are all met.

More frequently, users make decisions consistent with the requirements without fully satisfying them. For example, if there is a requirement $x+y \geq 1$ and the user makes the decision $x=1$, this decision is consistent with the requirement ($D \cap R \neq \Phi$) but further refinements may still violate the requirement. We may use a rule of the form:

Solution detected

Whenever   AddDecision

If   not self.D → includes (self.R) and

                           not (self.D → intersection(self.R) → isempty.

Then   Notify user that current state is a solution.

The contents of the set $D \cap R$ represent the range of compliant decisions. The system may play a pro-active role by using constraint propagation and approximating for computing that range. For example, given the requirement $x+y \geq 1$, and the decision $x=1$, the system may deduce $y \geq 0$.

Along the same lines, when the user makes decisions that violate a constraint, the system can provide support by notifying the user of the violation, and by suggesting corrective actions or backtracking points.

5.2 Active checkpoints

The active checkpoints are those points in the design process when the user "checks in" works in progress claiming a specific status for it. The Create methods are used to checkpoint work at the respective milestones. The methods assessCompleteness and assessCompliance areused to support the user in identifying the milestones reached. These methods take no reactive or corrective action. They only return an assessment of the situation. Domain specific rules can be added to provide a more pro-active support. When assessing compliance, the system may suggest corrective actions, alternate decisions, or retrieve similar cases that can be used. In order to be able to suggest corrective actions, the system may just backtrack in the decision tree, or it may resort to domain specific or condition specific heuristics. In particular, corrective actions can be attached to constraints, to objects, or to general configurations. We provide below a sample of active rules illustrating, notably, the difference between constraints and rules, the wide range of support that can be provided with the same set of constraints, and the general purpose-nature of the constraint-enforcement and design-support protocols.

Assignment Consistency

Whenever	CreateAssignment(Req)
If	(self.D $\rightarrow$ *intersection*(self.R)) $\rightarrow$ *isempty*
Then	Notify user that assignment has no solutions.

This rule checks that the assignment admits solutions at all. Because we have defined CreateAssignment with a precondition True, it is conceivable to create an assignment whose design requirements conflict with the domain constraints. We have opted not to put a consistency condition as a pre-condition to the method because of the potential computational cost and theoretical difficulty in systematically detecting inconsistencies. We felt it more appropriate to delegate this and similar tasks to active rules that can be based on heuristics, fine tuned with usage, and turned on and off as desired.

Another type of support that the system can provide is in promoting reuse by checking if existing solutions and designs meet the new requirements. For example, we can have a rule:

Solution Reuse:

Whenever	CreateAssignment(Req)
If	not Solutions-set$\rightarrow$ *select*(S_1|S_i.R$\rightarrow$ *include* (self.R))$\rightarrow$ *isempty*
Then	Notify user of existing solutions to the assignment.

A particular case of the above situation is when a stronger requirement exists in the database. Any solution to the stronger requirement will also be solution to the weaker requirement. The system may in fact maintain a lattice of requirements based on strength (difinedness [20]).

Note that the two active rules suggested above are independent of the classes. They can be applied to any class. Also the rules shown have a non-intrusive action. They do not cancel operations or automatically alter user's decisions. We are intentionally showing rules that play a supportive rule while leaving total control to the user. The gateways represent a sufficient control in most cases. In other words, the users are allowed to create all kinds of non-compliant objects for as long as they want. They are only accountable for the quality of the objects that they want to check in as solutions and designs.

6. Summary, Conclusion

Design is becoming a new bottleneck in manufacturing. Improving design requires that validity knowledge be brought to affect decisions at an opportune time. Such endeavor can only be accomplished effectively if design support systems are developed in light of and around concepts and knowledge from the domain [11]. At the same time, for design support systems to be viable and adaptable, domain knowledge must not be hard-coded and buried within the system, it must remain transparent, easy to access, understand, review, and revise. In this paper we have presented a database model capturing domain knowledge, design requirements, and design decisions in an integrated and flexible way. The domain knowledge in the form of constraints is encapsulated within the categories of objects to which it is relevant. At the same time, the knowledge is represented in a normative and modular format, promoting its accessibility and updatability. The normative rather than prescriptive representation of the knowledge increases knowledge independence. The timing of constraint checking and the reaction to their violation can be expressed independently of the constraints themselves. Instead these prescriptive details are a reflection of design strategies, level of interference desired, and priorities. We propose to formulate these prescriptive directives in a generic form, thus allowing their update and that of the constraints to take place independently.

The price paid for the transparency and ease of update of the knowledge comes in terms of efficiency. The same knowledge distributed across normative constraints and design strategies can be efficiently represented in customized active rules. Optimized algorithms can be used to compile the constraints into efficient rules, thus the importance of supplementing the constraints formulae with adequate information such as source, priority, and computational hints and pointers.

References:

[1] R. Agrawal and N.H. Gehani. Ode (Object database and environment); the language and the data model. In *Proc. ACM-SIGMOD 1989 Int'l Conf. Management of Data*, pages 36-45, May 1989.

[2] Francois Bancilhon, Claude Deloble, and Paris Kanellakis, editors. *Building an Object-Oriented Database System*: The Story of O2. Morgan Kaufmann, 1992.

[3] Frank Blackler, Michael Reed, and Alan Whitaker. Editorial introduction: Knowledge workers and contemporary organizations. *Journal of Management Studies*, 30(6):852-862, November 1993.

[4] Grady Booch, James Rumbaugh, and Ivar Jacobson. *The Unified Modeling Language User Guide*. Addison Wesley, 1999.

[5] A.F. Cardena and D. McLeod, editors. *Research foundations in object-oriented and semantic database system*. Prentice Hall, 1990.

[6] Chun Wei Choo. *The Knowing Organization*. Oxford University Press, 1998.

[7]   Ch-Ch. P Chu and R. Gadh. Feature-based approach for set-up minimization of process design from product design. *Computer-Aided Design,* 28(5):321-332, 1996.

[8]   Prekumar Devanbu and Mark A. Jones. The use of descriptive logics in kbase systems. *ACM Transactions on Software Engineering and Methodology,* 6(2):141-172, April 1997.

[9]   Clive Dym. *Engineering design- A Synthesis of views.* Cambridge University Press, 1995.

[10] Ahmed K. Elmagarmid, editor. *Database Transaction Models for Advance Applications.* Morgan-Kaufmann Publishers, Inc., 1996.

[11] G. Fisher, et al. Seeding evolutionary growth and reseeding: supporting incremental development of design environments. In *Human factors in computing systems (CHI'94),* Pages 292-298, 1994.

[12] N. Gehani and H.G. Agadish. Ode as an active database: Constraints and triggers. *In Proceedings of the Seventh International Conference on Vary Large data Bases,* Pages 327-336, 1991.

[13] N. Gehani and H.V. Jagadish. Ode as an active database: Constraints and triggers. *In Proceedings of the 17th International Conference on Vary Large Data Bases,* Pages 327-336, 1991.

[14] J. Gero and M.L. Maher. A framework for research in design computing. In *Proceedings of ECAADE,* 1997.

[15] John S, Gero. Creativity, emergence and evolution in design. *Knowledge-Based Systems,* 9:453-448, 1996.

[16] J.S. Gero and F. Sudweeks, editors. *Artificial Intelligence in design.* Kluwer Academic Publishers, 1996.

[17] William E. Halal, editor. *The infinite resource: Creating and leading the knowledge enterprise.* Jossey-Bass Publishers, 1998.

[18] Mansour, Val Collins and Dale Caviness. A survey of current object-oriented databases. *DATABASE Advances,* 26(1), February 1995.

[19] C.A. McMahon, J.H. Sims Williams, and J. Devlukia. Knowledge-based systems for automotive engineering design. In *Autotech 1991,* pages 1-7, 1991.

[20] Ali Mili, Jules Desharnais, and Fatma Mili. *Computer Program Construction.* Oxford University Press, 1993.

[21] Fatma Mili. Knowledge architecture for engineering design support. In *Proceedings of AAAI spring symposium,* March 2000.

[22] M. A. Musen, et al. Protégé-II:computer support for development of intelligent systems from libraries of components. *In Proceedings of MEDINFO'95, Eighth World Congress on Medical Informatics,* pages 766-770, 1995.

[23] K. Nichols. Getting engineering changes under control. *Journal of Engineering Design,* 1(1), 1990.

[24] Shridhar Ramaswamy M. H. Nodine and S.B. Zdonik. A cooperative transaction mode for design databases. In Ahmed K. Elmagarmid, editor, *Database Transaction Models for Advanced Applications.* Morgan-Kaufmann Publishers, Inc., 1996

[25] Society of Automotive Engineers, Inc., editor. *SAE Handbook.* Society of Automotive Engineers. Inc., 1996.

[26] K.A. Reinschmidt and G.A. Finn. Smarter computer-aided design. *IEEE Expert,* pages 50-55, 1994.

[27] Research opportunities in engineering design (nsf strategic planning work-shop, final report). Technical report, Directorate of Engineering, National Science Foundation, 1996.

[28] Gerald H. Taylor. Knowledge companies. In William E. Halal, editor, *The infinite resource,* pages 97-110. Jossey-Bass Publishers, 1998.

[29] C. Tong and D. Sriram, editors. *Artificial Intelligence in Engineering Design:* Volume *I.* Academic Press, 1991.

[30] Y. Umeda and T. Tomiyama. Functional reasoning in design. *IEEE Expert,* pages 42-48,March-April, 1997.

[31] Unigraphics. Addressing the CAD/CAM/CAE interoperability issue. *White paper,* Unigraphics, 1999.

[32] J. Widmon and S. Ceri, editors. *Active database systems: trigger and rules for advances database processing.* Morgan-Kaufman Publishers, Inc., 1996.

[33] A. Wong and D. Sriram. Shared: An information model for cooperative product development. *Research in engineering design,* (5):21-39, 1993.

Selecting and Generating Concept Structures

Dietmar Janetzko

University of Freiburg, D-79098 Freiburg i. Br.

Abstract: Knowledge can be used to build applications that carry out a number of complex activities (e.g., diagnosis, monitoring, decision-making). However, knowledge is difficult to elicit and not easy to represent. Moreover, the way knowledge may be related to other forms of data (e.g., collected in human-computer interaction) such that computer-based problem-solving becomes possible, is often unclear. In this paper, an approach to knowledge management is described that uses sequences of symbols. This sort of data is a by-product of numerous human activities, but it is also the output of many process control devices. We present some examples that support the claim that this method lessens the burden of knowledge acquisition and opens new perspectives for knowledge management.

1. Introduction

Knowledge has become a crucial asset that no organization can afford to disregard. In an industrial setting, management of knowledge comes in many forms [10]. We take the view that the best way when starting to analyze and to model processing of knowledge is to take a look at humans who do this job: They learn, collect experience, solve problems, invent, produce, control, monitor, plan, negotiate and decide. In modern industry, however, the central position of humans in all of these forms of knowledge management is no longer unchallenged since information technology, in particular knowledge-based systems or data mining software carry out some of these tasks more efficiently. A promising way to reshape knowledge management in organizations is to use information technologies without neglecting human information processing. By using a divide-and-conquer approach many cognitive activities can be precisely analyzed and then modeled by information technology. Expert systems and robots used in factories for industry production are two cases in point.

In this paper, we will also pursue a cognitive approach: we will delineate a cognitive model of human information processing that can be deployed in many areas of industrial knowledge management like analysis, diagnosis or explanation generation. What has a cognitive model to offer for industrial knowledge management? First, there is no better working model for modelling complex tasks in knowledge management (e.g., decision making, problem solving, planning) than human information processing. Second, a cognitive model that really captures principles of human information processing may support tasks at workplaces where the work of a human being is interleaved with that of a machine, i.e., in human-computer interaction. Knowledge tracking is a cognitive model that brings these advantages to bear. As we will see later, knowledge tracking owes much to three more general approaches to knowledge or data analysis. First, many ideas of knowledge acquisition fed into knowledge tracking. Second, knowledge tracking is relying on probability theory in particular on Marcov chains [3]. Third, knowledge tracking makes use of statistical methods. Knowledge tracking is a powerful model, and care must be taken that the reader will not be bogged down by all the nitty-gritty details that have to be mastered before understanding knowledge tracking. To introduce into knowledge tracking in a structured manner, this paper is organized as follows. We will first provide a bird's eye view on knowledge tracking. This will give the reader a more general idea of the way data processing is done via knowledge tracking. Second, we will then go into the formal details of this method. Third, a Web site is described that can be used to carry out remote calculations via knowledge tracking. Fourth, we introduce the usage of knowledge tracking by presenting simplified real-world examples. Finally, we discuss the perspectives for applying knowledge tracking for industrial knowledge management.

2. A Primer on Knowledge Tracking

What is the *objective* of knowledge tracking? Though knowledge tracking can be deployed for many purposes the most general way to use this model can be stated as follows: knowledge tracking is used to detect patterns in streams of symbolic data. Thus, the *input* of knowledge tracking is simply a sequence of symbols. This sort of data come in many forms, e.g., letters, words, URLs, graphics, products. Click-streams of users visiting Web sites, eye-movements, sequences of words, data collected on-line in air-control or industrial production are examples for streams of symbolic data. Usually, recordings of sounds or scenes do not qualify as an input for knowledge tracking. If, however, the stream of low-level signals is analyzed such that an interpretation with respect to symbols (e.g., objects seen or words uttered) is achieved, knowledge tracking can also be applied. Knowledge tracking is a knowledge-based method. This means, before using knowledge tracking to analyze sequences of symbolic data we have to specify in advance a concept structure, i.e., a kind of semantic network. Concept structures are taken to analyze the data. In fact, in knowledge tracking we have to specify several concept structures. Thus, we may also say, knowledge tracking combines bottom-up processing (streams of symbolic data) with top-down processing (concept structures).

There are two versions of knowledge tracking: Quantitative knowledge tracking provides a score that indicates which of several alternative and pre-specified concept structures gives the best fit to the symbolic data. Qualitative knowledge tracking, on the other hand, produces a different output. It synthesizes a new concept structure on the basis of concept structures that have been specified initially. Quantitative knowledge tracking is an example of an analytic task or analytical problem solving [2]. It is confirmative, since it specifies how well pre-specified concept structures fit to sequences of symbolic data. Qualitative knowledge tracking is an instance of a synthetic task or synthetic problem solving [2]. It is generative, as it produces a new concept structure on the basis of one or many concept structures so that a good fit of the new structure to the data is achieved.

3. Knowledge Tracking - A 5-Step Scheme

Analyzing cognitive structures via knowledge tracking involves five steps. Note that the first four steps are common to both the quantitative and the qualitative mode of knowledge tracking. Step five, however, differs between these two modes. We will first briefly mention all five steps and will then present a more detailed account of the 5-step scheme underlying knowledge tracking:

Step 1: We elicit concepts and relations in the domain under study and set up concept structures.
Step 2: We record empirical data (sequences of concepts).
Step 3: We express the concept structures by transition probabilities.
Step 4: We predict data by using concept structures and calculating goodness of fit scores.
Step 5/quantitative: On the basis of the goodness of fit scores we select the best-fitting structure.
Step 5/qualitative: We collect all bridging inferences activated in each prediction of a subsequent concept; on this basis, we form a new concept structure.

Local area networks are chosen as an example domain to introduce the theoretical concepts of knowledge tracking. Later, we will present an example also based on this domain.

3.1 Concepts, relations, and structures

We begin by specifying "concept", "relation", and "structure", which are taken to represent the background knowledge in a domain. Eliciting concepts, relations, and structures is a knowledge engineering task required to carry out knowledge tracking. Concepts and relations should give a good account of the domain under study.

3.1.1 Concepts

Let $C = \{c_1, \ldots, c_m\}$ be a non-empty finite set of concepts. Examples of concepts that may be elicited in the example field chosen (constructing local area networks) are adapter_card, requirements, physical_object etc. The *reference set of concepts* C_0 is meant to be the set of concepts actually employed by a sample of subjects taking part in a study.

3.1.2 Relations

We take the relations to be binary and symmetrical. Let $\overline{R} = \{R_1, \ldots, R_n\}$ be a non-empty finite set of relations. An examples of a relations is x is_physically_connected_to y.

3.1.3 Structures

In knowledge tracking, the knowledge or the theories that can possibly explain transitions in a trace of symbolic concepts are represented by structures. Every network (e.g., hierarchies ontologies, partonomies, semantic networks) of concepts, be it a cyclic or an acyclic graph can be called a structure. Viewed from a formal point of view, a structure consists of the following ingredients: (i) a non-empty set called the universe or domain of the structure, (ii) various operations on the universe and (iii) various relations on the universe. The operations are optional [1].[1] To analyze data via knowledge tracking we have to construct up several concept structures. Data processing basically proceeds by setting up a procedure that decides which of the concept structures explains the data best. Concept structures may differ both with respect to the size and the density of relations. However, each concept structure should form a coherent graph. Extreme differences between concept structures ought to be avoided. Seen from a formal vantage point, a concept structure is described as $(C, R_1, \ldots, R_n)$. Having elicited a set of concepts C and a set of relations $\overline{R}$ that describe the domain under study, we may take one instance from the set of relations and will then have to find arguments for the relation chosen among the set of concepts. If, for example, we instantiate the relation x is-a y with concepts, a concept structure (in this case a taxonomy) emerges.

Then, we take another relation (e.g., computer_unit_ethernet) (computer including all cards and cables as required in an ethernet) and a different relational structure emerges.[2] This procedure is repeated as long as each relation is instantiated by arguments taken from the set of concepts. In this way, we transform the concepts and relations into concept structures.

If not stated otherwise, we use the name of the relation as an abbreviation to refer to the concept structure. A concept structure may have more than one relation. For ease of interpretation, however, we mainly employ concept structures built on the basis of one relation. Each concept structure is used to test the hypothesis that a particular concept structure explains behavioral data best. Then, however, we also need the *random structure*. This structure is required to test the hypothesis that a concept structure provides results on a par with a structure based on the principle of chance. Only by comparing the goodness of fit score of a concept structure with that of the random structure, we may say whether a concept structure significantly explains the data. The random structure is set up by taking the relation y randomly_follows x and by instantiating it with all concepts.

---

[1]In mathematics and psychometrics a structure made up only of a universe and various relations is usually called a *relational structure* [10]. Thus, the concept structures used in knowledge tracking may also be called relational structures.

[2]While the first example of a concept structure is usually labeled a semantic network, the second one is generally called a frame.

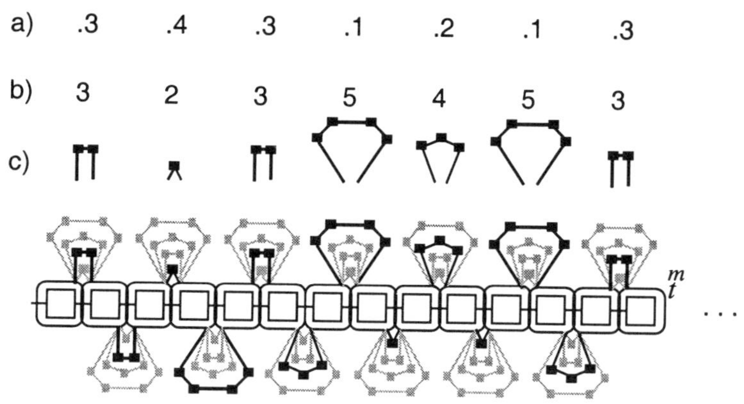

Figure 1: Trace $\overset{m}{t}$ and analysis of transitions

This figure serves as an illustration of the formula used to calculate Γ-scores: repeatedly pushing the two moving windows of prediction object and prediction base (rounded squares) over the concepts of a trace (white squares) results in intersecting pairs of prediction object and prediction base. The diamond-shaped figure refers to one of many alternative concept structures, viz., the background knowledge. For each transition of prediction base and prediction object in a trace, the shortest bridging inference (small black squares) among a number of possible concepts (small grey squares) is identified. Traversal is from w via the concept structure to c. In the quantitative mode of knowledge tracking we take the shortest path length. Deriving the transition probabilities allows us to calculate Γ-scores. This in turn permits the selection of the concept structure that fits the data best (level a). In the qualitative mode of knowledge tracking, we collect the bridging infereces for each transition (level c). On this basis, we form a new concept structure. The number of path lengths traversed is shown on level b. To prevent overload, in this figure only the numerical scores for every second transition are shown.

3.2   Empirical data / trace

The empirical data recorded and analyzed by knowledge tracking are traces, i.e., sequential symbolic data that reflect problem solving. This type of empirical data may be recorded in investigations of HCI, protocols of thinking aloud studies, or any other investigation where a sequence of concepts is recorded [12]. Note that in some domains the set of concepts is fixed, while in other domains it is not. An example of the former are click-stream analyses. Here the concepts refer to HTML-documents which are usually fixed. An example of the latter are all kinds of analyses of unconstrained verbal data (e.g., raw analyses of texts that are based on nouns). Here the number of concepts is usually not fixed. Hence, the data are noisy. The same is true for the type of sequences: In domains where the set of concepts is fixed the possible transitions between concepts is also fixed. Knowledge tracking is a robust technique for knowledge management that can be applied in fields that do not come up with a fixed number of concepts or transitions between concepts. Still, the performance of knowledge tracking is better the more concepts that can possibly occur in the data are known beforehand. Knowledge tracking is restricted to symbols. Hence, there is currently no possibility to analyze numerical data (e.g., two-dimensional data like symbol + time in click-stream analyses or symbol + parametrization in analyses of design activities) via knowledge tracking.

$$\Gamma = \frac{.3 + .4 + .3 + .1 + .2 + .1 + .3}{7}$$

Figure 2: Calculation of Γ-scores

The upper part of his figure serves as an illustration of the formula used to calculate Γ-scores: Deriving the transition probabilities for each transition allows us to calculate Γ-score, which is the average of all transitions. The lower part of this figure shows in a more intuitive way that the knowledge associated with all bridging inferences can be taken to synthesize a concept structure.

From a formal point of view, a trace of length m can be viewed as a mapping t : $\{1, \ldots, m\} \to C$. We assign to the number k the k-th concept in the trace. The abbreviation $\overset{m}{t}$ refers to $t_1, \ldots, t_m$. If the length of the trace is of no relevance, the abbreviation $\bar{t} := \overset{m}{t}$ is used. The formalization $(c_i, c_j) \in \overset{m}{t}$ refers to a transition of two concepts subsequently following one after the other. The trace of empirical data is taken to set up predictions for each concept of the trace. This proceeds by using a moving-window technique: two adjacent windows – the prediction base (w) and the prediction object (c) – are successively pushed over the trace. The concept of the prediction base is used to derive a prediction for the concept of the prediction object (Fig. 1). Quality of prediction is taken to calculate the goodness of fit between the empirical data under study and the concept structure used to derive a prediction. To rule out the possible bias that some concept structures will produce high goodness of fit scores no matter which kind of trace is used, we have to set up a *random trace*. This trace is a randomized sequence of concepts of C_0.[3]

3.3 Expressing structures by transition probabilities

In order to calculate the goodness of fit, we have to extend the symbolic network representation, i.e., the random structure and the concept structures, by a numerical description. This can be achieved by transforming the random structure and all concept structures into probability distributions. The approach taken relies on a Markov process [3]. We begin with a general definition for probability distributions. In the sections following, this definition will be adjusted so that all concept structures and the random structure may be expressed by using transition probabilities. As our starting point, we use the *reference set* C_0, viz., the set of concepts actually employed by

---

[3]The *random structure* represents the principle of chance on the side of the theory, the *random trace* does the same thing on the side of the data. The former is required for testing hypotheses, the latter is needed to detect possible biases via Monte-Carlo studies.

a sample of subjects under study. Generally speaking, in knowledge tracking we conceive of a distribution function ϑ as a mapping from the Cartesian product of the set of concepts C_0, i.e., $C_0 \times C_0$ into the interval $[0, 1]$. Hence, we obtain the definition $\vartheta : C_0 \times C_0 \to [0, 1]$.[4] An important property of this distribution is that the transition probabilities of a particular instance of the prediction base w to all concepts of C_0 sum up to 1. We end with a very simple definition for a probability distribution that explicitly refers to the notions of prediction base (w) and prediction object (c):

$$\vartheta : (w, c) \mapsto [0, 1].$$

3.3.1 Random distribution function

The random distribution function corresponds to the random structure. It assigns a fixed score to each transition from one concept to another concept. The fixed score expresses the probability of each transition (unconditional or a priori probability) and is simply the reciprocal of $|C_0 \setminus w|$. Stated more formally, we take $\vartheta^R(w, c)$ to denote the distribution function corresponding to the central assumption of the random structure. We can write the equation for the random-choice distribution function as

$$\vartheta^R(w, c) := \frac{1}{\mid C_0 \setminus w \mid}.$$

3.3.2 Concept structure distribution function

There are three basic tasks to be addressed when setting up the concept structure distribution functions:

I. Calculating Minimal Path Lengths. The probability distribution of a concept structure is calculated by using the minimal path lengths between concepts in w and the concept c along with a weighting of path lengths according to the branching of paths. From the minimal path lengths we choose the shortest one. We take $|p(w, c)|$ to symbolize the minimal path lengths between concepts in w and in c.

II. Calculating the Decay Rate Attached to the Minimal Path Lengths. Given that there is a path between a concept c_i and c_j of a structure, a low score will be assigned to this transition, if the path is relatively long and vice versa. Hence, we assume a monotonic decay of activation among concepts in a concept structure. Using a more formal lingo, this assumption can be specified by introducing a function δ with $\delta : \mathbb{N} \to \mathbb{R}$. In its simplest instantiation, the decay rate may also be calculated by using the reciprocal of the path length m.

III. Deriving a Distribution that Sums up to 1. The transition probabilities of a particular instance of the prediction base w to all concepts of C_0 should sum up to 1. What is needed then is a constant to accomplish this standardization. This constant is taken to multiply each reciprocal of a path length so that the resulting sum will be 1. To compute this constant, we add up all scores for the reciprocal of path lengths and calculate the reciprocal of the resulting sum. We take ξ to denote this standardization

---

[4]To alleviate readability, this definition is restricted to $|w| = k = 1$. However, it may be relaxed easily so that it also applies to $k \geq 1$.

constant. Intuitively, we may say that as a result of the ξ-standardization concept structures with a high number of transitions are "punished", i.e., the mean of their transition probabilities is low. By contrast, concept structures with a low number of transitions are "rewarded", i.e., the mean of their transition probabilities is high. Putting all three building blocks for the definition of a concept structure distribution function together, we are now in a position to state it as

$$\vartheta_R^{CN}(w,c) := \xi \cdot \delta(|p(w,c)|).$$

3.4 Prediction of adjacent concepts in the trace

The random structure and the concept structure distribution function put us in a position to carry out predictions of empirical data that form a trace. The basis for prediction are competing structures taken to explain the transitions between adjacent concepts in a trace.

3.5 The quantitative mode: setting up goodness of fit score and testing hypotheses

When investigating cognitive structures in a sample of subjects $\{s_1, \ldots, s_n\}$, a trace $\overset{m}{t_i}$ is obtained from each subject s_i. For each pair (w,c) in the trace $\overset{m}{t_i}$, we estimate the transition probability assigned to the occurrence of c, if w is preceding c. By analyzing a trace of length m, we get a probability score for $m-k$ transitions.[5] Notice that the $m-k$ transition probabilities obtained strongly depend on the structure (and its corresponding probability distribution) applied. The goodness of fit between a particular structure and the empirical data recorded is expressed via a goodness of fit measure called Γ-score. This measure is calculated by summing up the probability scores and by dividing the resulting sum by $m-k$. In the following,

$$\Gamma(\overset{m}{t_i}) = \frac{1}{m-k} \sum_{j=k+1}^{m} \vartheta(\{t_{i_{j-k}}, \ldots, t_{i_{j-1}}\}, t_{i_j})$$

figures as the formula of the averaged prediction scores, viz., the goodness of fit of the structure (C, R) to empirical data $\overline{t}_i$ of one subject s_i. The simple rationale behind this formula is illustrated by Fig. 1. Due to a standardization the range of possible values for this score is $0 \leq \Gamma(\overset{m}{t_i}) \leq 1$. We still have to calculate the mean $\overline{x}$ that expresses the average goodness of fit with respect to a *sample* of subjects $\{s_1, \ldots, s_n\}$. This is done by summing over the scores for goodness of fit for each single person and dividing the score by n:

---

[5] By definition, we do not have the prediction base for the first concept of the trace. The number of concepts in the trace for which there is no full prediction base is $m-k$ since it varies according to the size of k.

$$\overline{x}_R = \sum_{i=1}^{n} \frac{\Gamma_R(\overline{t}_i)}{n} \quad .$$

A more intuitive impression of how scores for goodness of fit are calculated is conveyed by Fig. 1. In knowledge tracking, testing hypotheses proceeds by structure-structure comparisons. First, the concept structures to be analyzed have to be free of biases. Analyses that make use of the random trace will reveal whether or not any structure produces very high Γ-scores. Second, for a concept structure to be meaningful it has to produce Γ-scores higher than that of the random structure. Third, among all concept structures, which score higher than the random structure, pairwise comparisons are carried out. Visual inspection is used for the first, and non-parametric statistical tests (Wilcoxon matched-pairs tests based on pre-post differences for overall pre-post differences, Mann-Whitney U test for group-specific differences) are used to carry out the second and the third steps.

3.6 The qualitative mode: setting up a concept structure formed by bridging inferences

The qualitative mode is an alternative way of data processing via knowledge tracking. In this mode, we do not calculate a goodness of fit score by computing the mean of the transition probabilities associated with bridging inferences that connect adjacent concepts in sequential data. Instead, in the qualitative mode we simply collect the knowledge associated with the bridging inferences and build a concept structures on the basis of this knowledge. Note that the concept structure that is built in this way is always a subset of the knowledge structure(s) hypothesized. In fact, this subset is nothing but the part of the knowledge that actually contributes to an explantion of the empirical data under study (cf. Table 1).

4. Knowledge Tracking via the WWW

We have put a device onto the Web that allows everyone to analyze data via knowledge tracking freely: `http://www.knowledge-tracking.com`. This Web site provides a tutorial and information on knowledge tracking.

5. Fitting the Model

A crucial question regarding the usage of knowledge tracking relates to the comparison of concept structures: To carry out a justified comparison, care must be taken concerning the maximum path length in structures. Here, the concept of the decay rate comes in. We have to select a decay rate that is adjusted according to the structure where the traversal of path lengths is shortest. In short, this path length sets the limit for the decay rate. If, for instance, in our set of competing structures there is a small structure where the longest path length is 5, we should fix the decay rate as follows: (7 4 2 1 0). This means, the user should make sure that the decay of spreading activation of each concept roughly equals an exponential decay. In our example $p= 7/14$ is attached to the transition to concepts that are 1 path length far apart (direct neighbours in a concept structure), $p=4/14$ is attached to the transition to concepts with path length of 2 etc. If we do not adjust the decay rate, the results may be distorted. If, for instance, the decay rate has a very long slope and there are some very small structures, these structures will be overestimated as they unjustifiably produce high Γ-scores. In this case a relatively high transition probability is attached to each concept, while the reverse is true for the concept structures that come up with longer path lengths [7].

6. Examples

We will now present and discuss some examples that show the usage of knowledge tracking. The examples are meant to clarify the theoretical concepts introduced in the last sections. As evidenced by the examples, knowledge tracking may be used in variety of domains in tasks of either analytical problem solving or synthetic problem solving. The former is concerned with data-driven decision making among pre-established concept structures (e.g., diagnosis, segmentation). The latter is realized by data-driven generation of new concept structures (e.g., formation of design patterns, theories, networks, explanations). The newly generated concept structures are always based on one or many concept structures that have to be fed into the system initially. In all the examples described, we will make use of the Web site where remote calculations of knowledge tracking can be carried out to answer this question. We kept the examples as simple as possible. But it is no principal problem to scale up the applications.

6.1 Analysis of eye-movements

In the first example, a psychologist wants to assess the quality of a picture that has been designed to be used in an advertisement of a car. The picture shows a car and also a girl sitting on the car (cf. Fig. 3). Let us assume that the picture really works in advertisement if two conditions are met: first, the eye movements are attracted by either the car and the lady or the car alone. If, however, the test persons are only looking at the lady, the advertisement will most probably not work. Second, we make the additional assumption that really concentrating on either the car, the lady or both will result in a sequence of eye movements tracking the car. Correspondingly, concentrating on the lady will result in a sequence of eye movements following her. For this reason, simply counting whether a spectator has focused on points of the picture of either the new car or the lady will not answer the question. In a word: the advertisement will really work, if the eye movements track the car or the lady and the car.

We need a method that detects whether or not the eyes of the spectators follow a particular pattern. Knowledge tracking is a method that can be taken to fulfill this

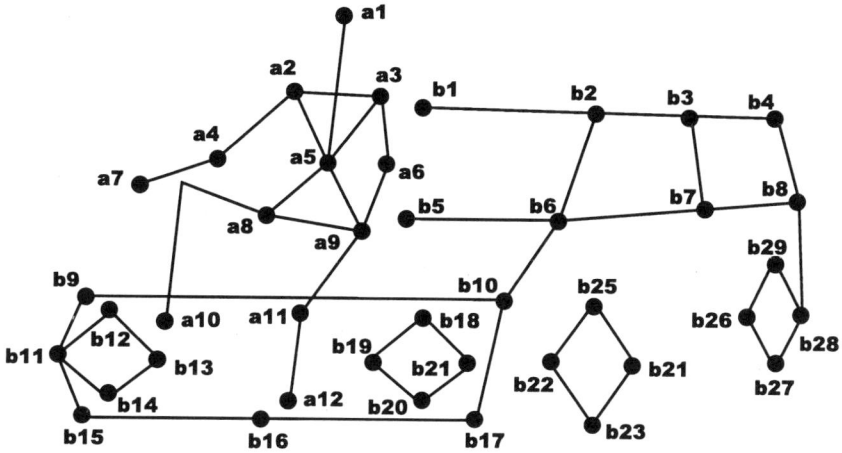

Figure 3: Network of nodes representing a graphics

objective. More precisely, we will use quantitative knowledge tracking since we have to decide between clear-cut possible patterns. The input to knowledge tracking is the sequence of eye-movements recorded by some test-persons. The background knowledge is provided by the concept structures made up by the outline of this figure. More precisely, we need concept structures that codify the network of nodes describing the car, the girl and conjointly the car and the girl. We can easily represent all concept structures by using a more formal notation. (Note that this representation covers more links than the figure!):

```
(girl
    (a1 a5) (a1 a2) (a1 a3) (a2 a3) (a2 a4) (a2 a5) (a3 a5) (a3 a6) (a4 a7) (a5 a8) (a5 a9) (a6 a9) (a7 a10)
    (a8 a9) (a8 a10) (a8 a11) (a9 a11) (a10 a11) (a10 a12) (a11 a12))
(car
    (b1 b2) (b1 b5) (b1 b6) (b2 b3) (b2 b5) (b2 b6) (b2 b7) (b2 b8) (b3 b4) (b3 b7) (b4 b6) (b5 b6) (b6 b25)
    (b7 b8) (b7 b25) (b7 b26) (b7 b29) (b8 b29) (b9 b10) (b9 b11) (b9 b16) (b10 b16) (b10 b21) (b10 b22)
    (b11 b12) (b11 b13) (b12 b13) (b12 b14) (b13 b14) (b11 b15) (b13 b16) (b13 b19) (b15 b16) (b16 b17)
    (b16 b19) (b17 b21) (b17 b22) (b18 b19) (b18 b21) (b20 b21) (b22 b23) (b23 b24) (b24 b25) (b24 b26)
    (b26 b27) (b27 b28) (b28 b29))
(girl_and_car
    (a1 a5) (a1 a2) (a1 a3) (a2 a3) (a2 a4) (a2 a5) (a3 a5) (a3 a6) (a4 a7) (a5 a8) (a5 a9) (a6 a9) (a7 a10)
    (a8 a9) (a8 a10) (a8 a11) (a9 a11) (a10 a11) (a10 a 12) (a11 a12) (b1 b2) (b1 b5) (b1 b6) (b2 b3) (b2 b5)
    (b2 b6) (b2 b7) (b2 b8) (b3 b4) (b3 b7) (b4 b6) (b5 b6) (b6 b25) (b7 b8) (b7 b25) (b7 b26) (b7 b29) (b8 b29)
    (b9 b10) (b9 b11) (b9 b16) (b10 b16) (b10 b21) (b10 b22) (b11 b12) (b11 b13) (b12 b13) (b12 b14) (b13 b14)
    (b11 b15) (b11 b13) (b13 b16) (b13 b19) (b15 b16) (b16 b17) (b16 b19) (b17 b21) (b17 b22) (b18 b19)
    (b18 b21) (b20 b21) (b22 b23) (b23 b24) (b24 b25) (b24 b26) (b26 b27) (b27 b28) (b28 b29) (a1 b1) (a3 b1)
    (a3 b5) (a9 b5) (a11 b5) (a11 b6) (a11 b10) (a12 b19) (a12 b13) (a12 b16)))
```

Below we show data of a test person who has watched the graphics thereby producing a sequence of symbolic data. These data are the input to knowledge tracking:

a1 a5 a9 b5 b6 b10 a11 a12 b16

In which way do we have to proceed, if we want to analyze the data of several subjects? In this case we can either consider each data set on its own or treat all data sets together. While the first approach gives us an analysis of the single subjects separately, the second one carries out an overall analysis.

For the sake of simplicity, we will only analyze the dummy data stated above. The background theory taken to analyze the data is given by the relational structures that describe the car and the girl, respectively (cf. Fig 3). The question underlying our analysis is whether the attention (and thus the eye-movements) have been attracted by the girl, by the car or the composite figure of the car and the girl. Obviously, we obtain high Γ-scores when analyzing the composite figure (the car and the girl, $\Gamma = 0.075$). In contrast, the goodness of fit scores of the two other structures are quite low (girl, $\Gamma = 0.059$; car, $\Gamma = 0.041$). We may thus conclude that the figure works well since it directs the attention to the car and the girl.

6.2 Analysis of click-streams

Our second example is concerned with a click-stream analysis of people visiting a big commercial Web site (e.g., purchasing statistical software). A click-stream is a recording of the HTML-documents that a user has requested when navigating through one or more Web sites. As it is well-known in human-computer interaction, navigation patterns of users provide a wealth of data that can be exploited by such analyses. The central question pursued is whether there are unique patterns of users who buy a product. Clearly, we can cut down the overall group of visitors in two groups: buyers and non-buyers. But it is an open question whether or not these two groups take different paths through a Web-site. Hence, we are dealing with the task of segmenting the groups of users. How could a possible result of this analysis look like? Let us assume, for instance, the outcome of such an analysis would be that visitors who pass more often informational documents buy a product via the Web site more often than vistors who do not or seldomly consider informational documents. This would be a strong – though not unequivocal – argument supporting the claim that extending the area of product information can foster the rate of sellings. Note, however, that this kind of analysis does not answer the question whether visitors had decided before or after visiting this area of the Web site to buy a product.

Though on the surface this example and that introduced in the last section seem to be completely different, there are important parallels. Again, we need a method that helps to answer the question of whether a sequence of symbols follows a particular pattern. In this example, we will also use the quantitative mode of knowledge tracking to fulfill this objective since we deal with distinct classes of patterns (i.e., groups of HTML-documents). The input to knowledge tracking is a sequence of HTML-documents that reflect the navigational path of a user visiting a Web-site. We use relations beween single HTML-documents as background knowledge. Note, however, that these relations are not simply hyperlinks. Instead, we use a relational schema that reflects the contents of the HTML-documents involved. For the sake of simplicity, we assume that the overall set of HTML-documents of the Web-Site under

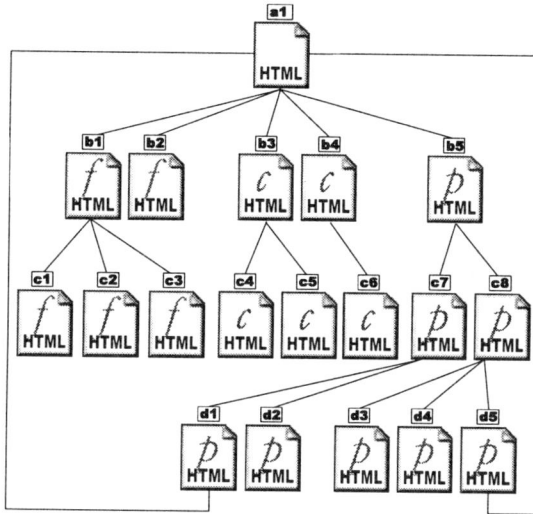

Figure 4: HTML-Documents of a Web Site

study can be subdivided in three classes: The first group of documents shares information on foundations of the products (e.g., white papers), the second group of documents covers informations of the company (e.g., a map showing all offices of the company around the world), finally all documents of the third group of documents provide information on products (e.g., press releases on new products). On the basis of these classes of HTML-documents we can establish a simple relational schema. The hyperlink that relates HTML-documents dealing with foundations is called f_relation. Correspondingly, we speak of c_relation (relation between HTML-documents on the company) and of the p_relation (relation between HTML-documents on products). In this way, we give a simple semantic interpretation to the hyperlinks. We end up with three concept structures that reflect the semantics of the HTML-documents. Again, we may easily represent these concept structures by using a more formal notation:

```
(f_relation
  (b1 c1)(b1 c2)(b1 c3))
(c_relation
  (b3 c4)(b3 c5)(b4 c6))
(p_relation
  (b5 c7)(b5 c8)(c7 d1)(c7 d2)(c8 d3)(c8 d4)(c8 d5))
```

Let us assume we run an analysis of click-streams that have been processed by persons who bought a product via the Web-site. As matter of fact, we have to pre-process data recorded by standard log-files [9] such that we obtain a sequence of symbols. We may then collect a stream of data that covers for instance the following sequence of concepts:

a1 b3 a1 b5 c7 d1 c7 b5 c8 d5 a1

To keep matters simple, we have taken a very short click-stream that could be analyzed easily by visual inspection. Note, however, that in real world click-streams we
have thousands or more clicks where visual inspection is doomed to failure but where
knowledge tracking is still applicable. We find supporting evidence for the p_relation
($\Gamma = 0.2188$), while the c_relation ($\Gamma = 0.063$) and the f_relation ($\Gamma = 0.0$) clearly produce lower Gamma-scores. The results strongly indicate that there is a tendency of
buyers to consider HTML-pages that present information on products before deciding
to buy a product.

6.3 Explanation of design activities

Hitherto, we have seen two examples that showed the application of quantitative
knowledge tracking. The third example will highlight the qualitative version of
knowledge tracking. Hence, it belongs to the category of synthetic problem solving. We will address the question in which way we can explain process data. These
data may be elicited when designing artifacts or when monitoring processes (e.g., air
control data). In which way can we derive automatic explanations of sequences of
symbolic data? Clearly, our explanations have to be restricted to sequences of symbols, since knowledge tracking can not process numerical data (e.g., parametrization
in design). By the term explanation we mean a deduction of a fact from a more general theory. Applied to knowledge tracking this means that we predict a symbol in
a sequence of symbols by a concept structure. The domain used by this example is
again design of computer networks. Let us assume that we have a tool that records
all activities when designing a computer network. Clearly, tools like these can only
be applied under very specific circumstances. In particular, all crucial design steps
have to be captured. In many fields of design (e.g., network design or web design)
tools are used that offer a limited number of design activities. The fact that such tools
allow only a fixed set of design activities limits the design space [5]. Note that this
is an advantage both for the beginner in the field and the recording of the process of
design activities.

In our third example, the input to knowledge tracking is a sequence of concepts
recorded by a design tool [6]. This tool is a window-based device that can be taken to
specify the layout of a computer network. It presents the user with a plan of a room
that should be equipped with a computer network. Additionally, it comes up with a
palette of icons that refer to standard components of a computer network. Whenever
the user picks up an icon and places it on the map, this action is recorded. The result
is a trace that consists of a recording of all design actions carried out by the user.
The background knowledge used to derive explanations should be a set of concept
structures that represents the body of theoretical knowledge necessary to describe the
domain under study. In the domain of network design we have identified 24 relations
that describe this domain [6]. However, in our simple example, we will consider only
three of them. The first one (is_a) represents knowledge about the ontology in the

domain. The second (is_physically_connected_to) refers to the conjunction of elements in a network. The third one (is_consistent_with_ethernet) specifies the compatibility of elements of a particular network protocol:

(is_a
    (twisted_pair_cable cable) (coaxial_cable cable) (bus_topology topology) (computer hardware)
    (transceiver conjunction_device) (t_connector connector) (bnc_connector connector) (connector hardware)
    (bridge conjunction_device) (conjunction_device hardware) (ethernet protocol) (bus_topology topology))
(is_physically_related_to
    (terminator t_connector) (computer adapter_card) (adapter_card cable) (t_connector cable)
    (bridge t_connector))
(is_consistent_with_ethernet
    (ethernet bus_topology (ethernet terminator) (ethernet transceiver) (ethernet ethernet_card)
    (ethernet t_connector (terminator bus_topology) (terminator transceiver) (terminator ethernet_card)
    (terminator t_connector))

An example of a sequence of symbols recorded in design that should be explained may be simply stated as

t_connector computer t_connector coaxial_cable t_connector computer t_connector terminator

In which way do we employ the trace and the concept network to derive explanations? Since this trace is a sequence of symbols it can easily be analyzed via knowledge tracking: for each transition of concepts in the trace we obtain a proposition made up by the concepts involved in the transition together with the relation selected by knowledge tracking to explain this transition (cf. Table 1). In fact, knowledge tracking comes up with two kinds of explantions to the sequence of symbolic data: First, each transition of concepts in the trace is explained by identifying one or possibly many relations that link these concepts in a concept structure. The figure below shows the explantion of each transition by referring to the concept structure that can be used to deduce this transition. Clearly, knowledge about the physical linkages between network components is dominating. This is not surprising since the design tool we used expects the user to connect elements of a computer network. Apart from this the system recognizes that the trace reflects the design of an ethernet. The structure that specifies ontological relations, however, is not used by the procedure. When analyzing the data (cf. Table 1), we obtain an explanation for each pair of two adjacent concepts in the trace (left column). The explanation is in terms of the concept structure that fits the transition (middle column), finally the system calcculates incrementally Γ-scores (right column). Second, while the first explanation closely follows the sequence of symbolic data, the second one is more condensed. It consists of all relations that fit the transition of two adjacent concepts in a trace. Again there are structures referring to the physical linkage between network components along with the structure that relates to ethernet as the network type.

(is_physically_connected_to (bridge t_connector) (t_connector cable) (terminator t_connector)) (is_a (coaxial_cable cable)) (is_consistent_with_ethernet (terminator t_connector))

Table 1: Explanation generation via knowledge tracking

Transition	Concept Structure	Gamma-Score
t_connector computer	is_physically_connected_to	0.142
computer t_connector	is_physically_connected_to	0.136
t_connector coaxial_cable	is_physically_connected_to	0.138
coaxial_cable t_connector	is_physically_connected_to	0.141
t_connector computer	is_physically_connected_to	0.141
computer t_connector	is_physically_connected_to	0.139
t_connector terminator	is_consistent_with_ethernet	0.142

7. Discussion and Conclusions

Clearly, each method has its strengths and weaknesses. There are some areas where we can expect quite satisfactory results and others where the performance of knowledge tracking diminishes. First, knowledge tracking is a knowledge-based method. Without specifying candidate concept structures on theoretical grounds, no analysis is possible via knowledge tracking. Second, knowledge tracking is more appropriate for analyzing behavioral data than verbal data, e.g., obtained by thinking aloud protocols. This type of data may also be expressed by sequences of symbolic concepts. Some features of thinking aloud protocols, e.g., quantification, qualification, negation etc., cannot be expressed by concept structures. Third, knowledge tracking is a competence model, but not a performance model of cognition. Thus, it cannot be used to generate a particular behavior. On the sunny side, however, it should be emphasized that knowledge tracking brings together insights from cognitive science and psychometrics to analyze cognitive representation, their development, and change. It can be employed in different modes: Used in a quantitative way, it calculates goodness of fit scores between sequential empirical data and concept structures. Used in a qualitative way, it selects from a bigger, hypothesized concept structure a subset that actually contributes to an explanation of the data under study. Both modes can be very well combined to form a powerful suite of methods: When comparing different samples, we may use knowledge tracking in a qualitative mode to pick out a sub-structure from a larger structure. Then we take this substructure to perform a cross-validation of this structure by using it to analyze a different sample. When comparing different concept structures, we will use knowledge tracking in the quantitative mode, run different calculations by using different concept structures. Since knowledge tracking is suitable for automatic on-line analyses of huge amounts of data, it can be applied to a wide range of applications of human-computer interaction.

References

[1] Bell, J., & Machover, M., 1986, *A course in mathematical logic*. North-Holland Amsterdam.

[2] Breuker, J. & Wielinga, B., 1989, Models of expertise in knowledge acquisition. In Guida G. and Tasso C. (Eds.), *Topics in Expert System Design - Methodologies and Tools*, 265-295, North-Holland Amsterdam.

[3] Chung, K., 1968, *Markov chains with stationary transition probabilities*. Springer-Verlag Berlin.

[4] Duffy, V. G. & Salvendy, G., 1999, *Problem solving in an AMT environment: difference in the knowledge requirements for an interdisciplinary team*. International Journal of Cognitive Ergonomics, 3, 23-35.

[5] Goel, V. & Pirolli, P., 1992, The structure of design problem spaces. *Cognitive Science*, 16, 119-140.

[6] Janetzko, D., 1996, *Knowledge Tracking: A method to Analyze Cognitive Structures*. University of Freiburg/Germany. IIG-Berichte, 2.

[7] Janetzko, D. (1998a). *Analyzing sequential data via knowledge tracking - An animated tutorial and a remote calculation device for knowledge tracking*. http://www.knowledge-tracking.com

[8] Janetzko, D. (1998b). Tracking Cognitive Representations. In M. A. Gernsbacher & S. J. Derry (Eds.), *Proceedings of the Twentieth Annual Conference of the Cognitive Science Society*. 1229.

[9] Janetzko, D. (1999). *Statistische Anwendungen im Internet - In Netzumgebungen Daten erheben auswerten und präsentieren (Collecting, Analyzing and Presenting Data via Internet*. Addison-Wesley Munich.

[10] Krantz, D., Luce, R., Suppes, P, & Tversky, A. (1971). *Foundations of measurement Vol. 1*. Academic Press New York

[11] Sobol, M. G. & Lei, D., 1994, *Environment, manufacturing, technology, and embedded knowledge*. International Journal of Human Factors in Manufactoring, 4, 810-829.

Evaluating Verification and Validation Methods in Knowledge Engineering

Alun Preece

University of Aberdeen, Computing Science Department, Aberdeen AB24 3UE, Scotland
Email: apreece@csd.abdn.ac.uk

Abstract: Verification and validation (V&V) techniques have always been an essential part of the knowledge engineering process, because they offer the only way to judge the success (or otherwise) of a knowledge base development project. This remains true in the context of knowledge management: V&V techniques provide ways to measure the quality of knowledge in a knowledge base, and to indicate where work needs to be done to rectify anomalous knowledge. This paper provides a critical assessment of the state of the practice in knowledge base V&V, including a survey of available evidence as to the effectiveness of various V&V techniques in real-world knowledge base development projects. For the knowledge management practitioner, this paper offers guidance and recommendations for the use of V&V techniques; for researchers in knowledge management, the paper offers pointers to areas where further work needs to be done on developing more effective V&V techniques.

1. The Art of Knowledge Engineering

Knowledge-based systems (KBS) have proven to be an effective technology for solving many kinds of problem in business and industry. KBS succeed in solving problems where solutions are derived from the application of a substantial body of knowledge, rather than by the application of an imperative algorithm. In the 1980s, KBS technology was widely applied to solve stand-alone problems. Classic examples of the successful use of the technology were in diagnostic problem-solving (for example, in medicine or engineering), provision of advice (for example, in "help-desk" applications), and construction/configuration (for example, product manufacturing and transportation loading).

In the 1990s, many organisations identified their collective knowledge as their most important resource, and are applying KBS technology in *knowledge management*: to capture and exploit these "knowledge assets" in a systematic manner [1]. The primary advantage of using KBS technology in this context is that the knowledge is then queryable in a rich way, so that users can pose complex questions to the organisation's knowledge bases, and directly use the knowledge in decision support and problem-solving.

The characteristic feature of problem domains where KBS technology is suitable is that the problems are *ill-defined*: they are not amenable to solution by algorithmic means; instead, the knowledge in the knowledge base of the KBS is used in some way to search for a solution. Often, the domain is such that there can be no guarantee that a solution will be found, or that found solutions will be optimal. Many KBS offer a "best effort" solution, which is good enough when the application requirements permit this (that is, the system is not safety or mission-critical). This is true of knowledge management applications: the knowledge is only as good as the human experts that supplied it, but this is typically considered good enough.

The literature on knowledge engineering recommends that the requirements for a KBS be divided into *minimum* and *desired* functionality [2]: *minimum requirements* will often dictate what a system must never do (for example, a vehicle loading application must never produce a configuration that is unbalanced to the point of being dangerous to vehicle operators), while *desired requirements* will attempt to specify the quality of solutions (for example, that at least 90% of the configurations produced by the vehicle loading application should be within 15% of optimal). In practice, desired requirements will be difficult to specify, due to the ill-defined nature of the problem to be solved (for example, in the vehicle loading application, it may be very difficult to determine what constitutes an "optimal solution" for the desired requirements) [3]. This is unsurprising; from a software engineering point-of-view, given the fact that the problem is ill-defined, it follows that the user requirements will be ill-defined also.

Knowledge engineering can be viewed as a special instance of software engineering, where the overall development strategy typically must employ exploratory prototyping: the requirements will typically be ill-defined at the outset, and it will take some effort in acquiring knowledge, and building prototype models, before the requirements can become more clearly defined. The knowledge engineer will have the hardest task when the domain knowledge itself is not well-understood; for example, when the knowledge is locked up in the heads of human experts who are not able to articulate it clearly. It is not unusual for a knowledge engineer to face a situation in which the users will be unable to say what they really want, experts will be unable to say what they really know, and somehow a KBS must be built! Building KBS is something of an art.

2. The Importance of Validation and Verification

Validation and verification (V&V) comprise a set of techniques used in software engineering (and, therefore, in knowledge engineering) to evaluate the quality of software systems (including KBS). There is often confusion about the distinction between validation and verification, but the conventional view is that *verification* is the process of checking whether the software system meets the *specified* requirements of the users, while *validation* is the process of checking whether the software system meets the *actual* requirements of the users. Boehm memorably characterised the difference as follows [4]:

Verification is building the system right.

Validation is building the right system.

Verification can be viewed as a part of validation: it is unlikely that a system that is not "built right" to be the "right system". However, verification is unlikely to be the whole of validation, due to the difficulty of capturing specifying user requirements. As noted above, this is a particularly important distinction in knowledge engineering. Of course, the goal in software/knowledge engineering is to try to ensure that the system is both "built right" and the "right system"; that is, the goal is to build "the right system, right".

This is no less important where KBS technology is used in knowledge management: V&V techniques provide ways to measure the quality of knowledge in a knowledge base, and to indicate where work needs to be done to rectify anomalous knowledge. In this context, verification tells us whether or not the knowledge bases are flawed as software artifacts, while validation tells us whether or not the content of the knowledge base accurately represents the knowledge of the human experts that supplied it. Both are clearly important. It is worth noting in this context that verification is essentially an objective test: there are absolute measures of the correctness of a piece of software. However, validation is typically subjective to a certain extent, where we must compare formally-represented knowledge to informal statements.

In software engineering, efforts have been made to formalise the development process so that user requirements may be stated as a fully-formal specification, from which it can be proven that the implemented software system meets the requirements. While formal methods are desirable - even essential - in some cases (notably safety and mission-critical systems), these methods are unsuitable in large classes of software applications:

- Where requirements are amenable to formal specification, it may be too difficult to create the specification within project time and budgetary constraints.
- There are many kinds of requirement that are not amenable to formal specification (for example, the "usability" of a graphical user interface).

The extent to which formal methods can be applied in knowledge engineering is debatable [5], but it is certainly unrealistic to expect formal verification to serve as the only V&V technique in a KBS development project, because it will rarely be possible to ensure that the formal specification is a complete and correct statement of the users' requirements. Therefore, KBS V&V will typically need to involve multiple techniques, including formal verification against formal specifications (where possible), and empirical validation (including running test cases and evaluating the system in the operational environment) [6]. This is especially important in the context of knowledge management, where a large part of validation will be fundamentally subjective: checking that the represented knowledge accurately captures what's going on in an expert's head.

Given that knowledge engineering is an inexact art, the most fundamental measures of the success of a KBS project would seem to be:

Did we get it right? That is, does it meet the users' actual requirements.

Can we keep it right? That is, is it sufficiently maintainable for anticipated future changes.

Can we do it again? That is, is the process repeatable to ensure success with future projects.

The final point refers to the *capability* of the knowledge engineers, and reflects the modern view of software quality being determined primarily by the quality of the development process [7].

While verification and validation are only part of the overall development process, they are extremely important because they are the only way to produce an answer to the first of the three questions above ("Did we get it right?"), and provide partial answers to the other two questions: V&V techniques assist in measuring maintainability, and a repeatable V&V capability is a prerequisite for success in knowledge engineering. Maintainability is also of enormous importance in knowledge management, where an organisation's knowledge bases will typically evolve over the organisation's lifetime.

Consideration of the importance of V&V to successful knowledge engineering and knowledge management raises another question: how effective are the KBS V&V techniques in current use? Obviously, if the techniques are incomplete or unsound, then they cannot be trusted to provide measurement of software quality and project success. The goal of this paper is to reflect upon studies which have been done to assess the effectiveness of current KBS V&V techniques, and to:

- summarise what the studies tell us about the current state-of-the-practice in KBS V&V;
- identify ways to improve the state of knowledge engineers' own knowledge about available KBS V&V techniques.

In doing this, the objective is not to propose *new* V&V techniques, but to determine what can be done with the *existing* techniques, and propose further ways of measuring the effectiveness of current (and future) V&V techniques.

3. Knowledge Engineering = Method + Measurement

The previous section emphasised the importance of V&V as measurement techniques for the knowledge engineering process. Knowledge engineering (and software engineering) can be seen as a combination of *methods* and *measurement*: the methods used in requirements specification, knowledge acquisition, system design, and system implementation result in the production of a series of *artifacts* [7] (Preece, 1995), each of which is amenable to some form of measurement (either individually or in combination). V&V techniques provide the means of obtaining the measurements. The following artifacts are of particular importance in the KBS development process:

Requirements Specification The requirements specification document states the minimum and desired user requirements (as described in Section 1), typically in natural language (or, less usually, in some restricted or semi-structured natural language subset). A framework for KBS requirements specification is given by Batarekh et al [3]. As a natural language document, the requirements specification is not amenable to analysis by V&V techniques - instead, it is used to establish the needs for V&V.

Conceptual Model The conceptual model describes the knowledge content of the KBS in terms of real-world entities and relations. This description is entirely independent of the ways in which the KBS may be designed or implemented: the idea is to allow the knowledge engineer to perform a knowledge-level (epistemological) analysis of the required system before making any design or implementation choices. The best-known framework for defining KBS conceptual models is KADS [8] (Wielinga, Schreiber and Breuker, 1992), in which models may be initially defined using a semi-formal, largely diagrammatic representation, from which a refined, formal model can be derived. The conceptual model forms the basis of the design model.

Design Model The design model serves to "operationalise" the conceptual model into an executable KBS; it describes the required system in terms of computational entities: data structures, processes, and so forth. For example, the design model may specify that a particular conceptual task is to be performed by a backward-chaining search, or that a concept taxonomy is to be represented using a frame hierarchy. The KBS specification language DESIRE is particularly well-suited to the representation of design models [9]. The design model dictates the form of the implemented system.

Implemented System This is the final product of the development process: the KBS itself. Once the design issues have been explored in the design model, the system may be implemented in any programming language, although typically a special-purpose KBS language is used.

In the context of knowledge management, it is worth noting that all of these stages are still necessary if a knowledge base is to be created that can be processed by an inference engine in response to users' queries, for example in decision support or problem-solving. Alternatively, if the knowledge is only to be captured for interpretation by humans, for example using a browser, then it may be sufficient to develop only a conceptual model, as described in [10].

There are many V&V techniques that have been developed for use on KBS - Gupta [11] and Ayel and Laurent [12] provide good entry-points to the KBS V&V literature. Five of the most common approaches are listed below.

Inspection According to a survey of developers of KBS in business applications, inspection is the most commonly-employed V&V technique [13]. Arguably, it is also the least reliable, as it essentially involves nothing more than human proof-reading the text of the various artifacts. Typically, a domain expert is asked to check the statements in the knowledge base; since the formal languages used in the design model and implemented system will be unfamiliar to domain experts, this technique is better-suited to use with the semi-formal conceptual model (which will typically use a more "reader-friendly" graphical representation).

Inspection is a highly-relevant technique to use in the context of knowledge management, where human experts need to review ("proofread") knowledge aquired from them. It is also the minimal form of validation that should always be applied in any knowledge management project.

Static Verification Static verification consists of checking the knowledge base of the KBS for logical *anomalies*. Frameworks for anomalies in rule-based KBS have been well-explored, and software tools exist to detect them [14]. The most commonly-identified anomalies - and the ones detected by most of the available tools - are *redundancy* and *conflict*. Redundancy occurs when a knowledge base contains logical statements that play no purpose in the problem-solving behaviour of the system; this typically indicates that the system is incomplete in some way. Conflict occurs when there are logical statements that are mutually inconsistent, and would therefore cause the system to exhibit erroneous behaviour. Anomalies may exist in any of the formal artifacts: the implemented system, the design model, and - if it is defined formally - the conceptual model.

One novel application of this kind of checking in the context of knowledge management lies in checking for conflicts between statements made by different experts. While it may not be necessary or even desirable to remove such conflicts (different opinions may be tolerable and often are useful) it is likely that they will reveal insights to the way an organisation applies its knowledge [15].

Formal Proof Formal proof is a more thorough form of logical analysis of the (formal) artifacts in the development process than that provided by static verification. As described in Section 1, where requirements are amenable to formal specification, proof techniques can be employed to verify that the formal artifact meets the specified requirements. A review of opportunities to use formal methods in knowledge engineering is provided by Meseguer and Preece [5]. In practice, however, while there are many formal specification languages for KBS, there are few documented examples of the use of proof techniques to very user requirements.

Formal proof is only likely to be applicable in knowledge management applications where organisational knowledge will be applied in safety-critical or costly mission-critical situations for decision-making or decision support.

Cross-Reference Verification When there exists descriptions of the KBS at different "levels", it is desirable to perform cross-checking between these, to ensure consistency and completeness. For example, we would expect the concepts that are specified as being required at the conceptual level to be realised in terms of concrete entities at the design level, and in terms of concrete data structures in the implemented system. Therefore, the most appropriate uses of cross-reference verification are to check correspondence between:
- conceptual model and design model;
- design model and implemented system.

A useful product of cross-reference verification in knowledge management lies in the linking of knowledge described at different levels of formality. Once a correspondence has been established between, for example, semi-formal statements in the conceptual model and formal statements in the implemented system, then users can use hyperlinking tools to move from one to the other. In one direction, users can move from formal to semi-formal in order to obtain a more understandable statement of the same knowledge; in the other direction, users can move from semi-formal to formal in order to apply some knowledge in automated decision support.

Empirical Testing All software testing involves running the system with *test cases*, and analysing the results. The software testing literature distinguishes between function-based testing and structure-based testing. *Function-based testing* bases the selection of test cases upon the functional requirements of the system, without regard for how the system is implemented. The success of function-based testing is dependent upon the existence of a "representative" set of test cases. In *structure-based testing*, test cases are selected on the basis of which structural components of the system they are expected to exercise; the objective is to show that the system produces acceptable results for a set of test cases that exercise all structural components of the system. Testing can be applied only to the executable artifacts: typically only the implemented system.

In knowledge management, testing often takes the form of systematically asking questions of an implemented KBS, the goal being to assess the acceptability of the responses in terms of both completeness and correctness.

Table 1 summarises the applicability of the various types of V&V technique to the KBS development artifacts. The table shows only potential applicability of techniques to artifacts. The really important questions go beyond this, to ask:
- How effective is each of the techniques listed in Table 1, to provide some measurement of the quality of the appropriate artifact(s)?
- What combination of V&V techniques work best to provide the most cost-effective assurance of high quality for each artifact?
- What V&V techniques work best with which *method* for creating the artifacts?

The last question acknowledges the fact that not all V&V techniques can be used with all methods. For example, static verification to detect logical anomalies can be applied to the implemented system only if the implementation is created using a suitable programming language (one for which logical anomalies can be defined). Similarly, formal proofs can be applied to the design model only if an appropriate proof theory exists for the modelling language.

The following section examines the available data to discover to what extent the above questions can be answered now.

4. KBS V&V: How Well Are We Doing?

Surprisingly few studies are known to have been performed to evaluate the effectiveness of KBS V&V techniques. This section examines the results of five studies:

1. A comparative evaluation of several KBS verification and testing techniques, conducted at the University of Minnesota, USA (referred to here as "the Minnesota study").
2. A study comparing the effectiveness of an automatic rule base verification tool with manual testing techniques, conducted at SRI, USA ("the SRI study").
3. An examination of the utility of an anomaly detection tool, conducted at Concordia University, Canada ("the Concordia study").
4. A comparative evaluation of several KBS verification tools/techniques, conducted by SAIC, USA ("the SAIC study").
5. A comparative evaluation of KBS verification and testing techniques, conducted at the University of Savoie, France ("the Savoie study").

Table 1: Applicability of V&V techniques to KBS development artifacts.

Artifact	V&V techniques
Conceptual model	Inspection, Static verification (if formalised), Cross-ref verification (against Design model)
Design model	Inspection, Static verification, Formal proof, Cross-ref verification (against Conceptual model, Implemented system)
Implemented system	Inspection, Static verification, Testing, Cross-ref verification (against Design model)

4.1 The Minnesota study

Kirani, Zualkernan and Tsai [16] at the University of Minnesota, USA, report on the application of several V&V techniques to a sample KBS in the domain of VLSI manufacturing. With the exception of a simple static verification (anomaly detection) tool, all of the methods used were manual testing techniques. The KBS itself was a 41-rule production system based upon well-understood physical properties of semiconductors, into which a variety of plausible faults were seeded. Interestingly, efforts were made to introduce faults at several different phases in the development process: at specification time, at design time, and at implementation time. A summary of the results is presented in Table 2.

The results of the study showed that the manual testing techniques, though labour-intensive, were highly effective, while the static verification tool performed poorly in detecting the seeded faults. Unfortunately, the success of the manual testing techniques could be attributed to the fact that this KBS application was exhaustively testable - which is rarely the case for industrial-scale KBS applications. Furthermore, given that the anomaly detection tool employed was of only the most basic type (able to compare pairs of rules only for conflict and redundancy), it is unsurprising that it performed poorly. Therefore, this study does not provide clear evidence - positive or negative - for the utility of modern KBS verification tools. Moreover, the study did not consider the complementary effects of the tools: no data was provided on which faults were detected by more than one V&V technique.

Table 2: Summary of results of the Minnesota study: percentage of faults found for each phase by each V&V method.

V&V method	Development phase		
	Specification	*Design*	*Implementation*
Static verification	38%	27%	19%
Structure-based testing	54%	68%	74%
Function-based testing	75%	92%	62%

4.2 The SRI study

Rushby and Crow [17] at SRI, USA, like the Minnesota study, compared manual testing techniques with a simple static verification tool. The application used was a 100-rule forward-chaining production system in an aerospace domain, but the structure of the system was largely "flat" and very simple. Faults were not seeded in this study - instead, actual faults were discovered in the real application! - so there was no way to control the results. While interesting, this study does not yield reliable evidence as to the effectiveness of the V&V techniques employed.

4.3 The Concordia study

Preece and Shinghal [18] at Concordia University, Canada, examined the use of a particular static verification tool, COVER, on a variety of KBS in different domains. The anomalies detected by COVER are as follows:

Redundancy Redundancy occurs when a KBS contains components which can be removed without effecting any of the behaviour of the system. This includes logically subsumed rules (if p and q then r, if p then r) rules which cannot be fired in any real situation, and rules which do not infer any usable conclusion.

Conflict Conflict occurs when it is possible to derive incompatible information from valid input. Conflicting rules (`if p then q, if p then not q`) are the most typical case of conflict.

Circularity Circularity occurs when a chain of inference in a KB forms a cycle (`if p then q, if q then p`).

Deficiency Deficiency occurs when there are valid inputs to the KB for which no rules apply (`p` is a valid input but there is no rule with `p` in the antecedent).

COVER was applied to the following KBS (all of these were independently-developed, real KBS applications, not "toy" systems):
- MMU FDIR: a fault diagnosis/repair KBS developed by NASA/Lockheed);
- TAPES: a "help desk" product recommendation system developed by an adhesive tape manufacturer;
- DISPLAN: a health care planning system developed by the UK Health Service;
- DMS1: a fault diagnosis/repair KBS developed by Bell Canada).

A summary of the anomalies found by COVER appears in Table 3; the table also gives a measure of the complexity of each application, in terms of the number of objects in each knowledge base (rules, frames, or the equivalent, depending on the actual implementation language employed).

COVER was shown to detect genuine and potentially-serious faults in each system to which it was applied (in contradiction to the negative results on the use of this technique in the Minnesota study). Unfortunately, the Concordia study did not compare the effectiveness of COVER with other kinds of V&V technique.

Table 3: Summary of results of the Concordia study: number of anomalies of each type found in each KBS.

Anomaly type	KBS			
	MMU	*TAPES*	*DISPLAN*	*DMS1*
Redundancy	10	5	5	7
Conflict	-	4	40	10
Circularity	-	-	4	-
Deficiency	-	16	17	-
KB size (objects)	170	230	405	1060

4.4 The SAIC study

Miller, Hayes and Mirsky [19] at SAIC, USA, performed a controlled experiment on two KBS built in the nuclear power domain. Faults were seeded in each system, and groups of KBS developers and domain experts attempted to locate the faults using three different V&V techniques: manual inspection, static verification using the VERITE tool (an enhanced version of COVER [18], and static verification using MetaCheck, a simulated tool based on a conceptual enhancement of VERITE. The VERITE tool and the MetaCheck pseudo-tool were shown to provide significant assistance to both the groups of KBS developers and domain experts in locating faults:

- Groups using a tool (either VERITE or MetaCheck) found almost twice as many faults as the groups who did not have a tool, in 18% less time, with half as many falsely-identified faults.
- Groups using VERITE found 59% of seeded faults correctly.
- Groups using MetaCheck found 69% of seeded faults correctly.

While providing good evidence for the utility of static verification tools, and confirming the unreliability of manual inspection, the SAIC study did not compare static verification with empirical testing techniques.

4.5 The Savoie study

Preece, Talbot and Vignollet [20] at the University of Savoie, France, performed a comparative study of three V&V tools:
- SACCO: a static verification tool performing redundancy and conflict detection;
- COCTO: a static verification tool performing deficiency detection;
- SYCOJET: a structure-based testing tool capable of generating test cases to provide a specified level of knowledge base test coverage.

SACCO and SYCOJET are described in detail by Ayel and Vignollet [21].
Independently-created sets of plausible faults were seeded into three different "mutated" versions of a real (207 rule) KBS application in an aerospace fault diagnosis domain. Each of the three tools was run on each of the three mutated KBS, and the results were aggregated; in summary:
- In each mutated system, at least 61% of faults were found by the combined effect of the three tools.
- SACCO always found at least 35% of the seeded faults.
- COCTO always found at least 27% of the seeded faults.
- SYCOJET always lead to the discovery of at least 27% of the seeded faults (with a test coverage of up to 46% of the rules - a level chosen for reasons of computational efficiency).
- The three tools were shown to be complementary in effect: less than 29% of faults detected were found by more than one tool.

Arguably, this study provides the best evidence yet that a combination of V&V techniques should be employed in any KBS development project. It also provides some useful evidence on the sensitivity of the different KBS techniques to different sets of seeded faults; however, three mutated KBS is not sufficient to provide any statistical confidence.

4.6 Conclusions from the studies

The overall conclusion from the studies is that the collective knowledge on the effectiveness of KBS V&V techniques is very limited. There is some evidence that different techniques have complementary effectiveness, and no technique has been shown to be so weak as to be not worth employing. However, the data that is available is sparse, being limited to a few instances of KBS and specific applications of tools or techniques. It is almost impossible to combine the results of the different studies, because they were run with different types of KBS (for example, the Minnesota study used a "toy" KBS that was exhaustively testable, while the Savoie study used a genuine KBS application that was computationally too costly to attempt exhaustive testing), different instances of V&V techniques (the static verifiers used in each of the five studies *all* have different capabilities!), and different assumptions (for example, while the types of errors seeded in the Minnesota, SAIC and Savoie studies are similar, there are subtle differences which make cross-comparison hard).

The sparse nature of the available data is also evidenced by the fact that there is no known data for the effectiveness of formal proofs or cross-reference verification. Moreover, none of the studies apply V&V techniques directly to any artifact except the implemented system, and the implemented systems are almost exclusively rule-based.

The following section considers what can be done to improve this situation.

5. KBS V&V: What Do We Need To Do?

Clearly, in order to improve the collective state of knowledge on the effectiveness of KBS V&V techniques, it is necessary to perform a considerably larger set of studies. In order to gather a sufficiently complete data set, the following process would need to be followed:

1. Create a sufficiently complete enumeration of the types of KBS requiring V&V. For each type of KBS, create instance artifacts at each stage of development (conceptual model, design model, and implementation), and *for each development method*. For example, instances of KBS with various distinct problem-solving methods would be required, and artifact instances would need to be created using different methods (and representation languages).
2. Define reference implementations for each V&V technique, either in the form of well-defined manual procedures, software tool specifications/implementations, or a combination of the two. Where necessary, variations on the V&V techniques will need to be defined for different representations used in the reference KBS artifacts produced in Step 1.
3. Define good fault models, based on observed error phenomena from actual experience in KBS projects.
4. Mutate the KBS artifacts from Step 1 using the fault models from Step 3 (ideally, this would be done automatically); then apply each of the V&V techniques defined in Step 2 to each mutated artifact; repeat for a statistically-significant set of mutated artifacts.

Such a study would be very ambitious but extremely valuable: it would provide conclusive evidence as to the effectiveness of each V&V technique for each type of KBS and development method, individually and in combination. Furthermore, it would support further research and development of KBS V&V techniques. Of course, such a study would be very difficult: Step 1 and Step 3 in particular are made hard by the fact that KBS technology is moving constantly forward: new kinds of KBS are always emerging - for example, witness the current interest in multiple-agent KBS [22] - and reliable information on actual error phenomena is had to come by (partly because knowledge engineers do not wish to advertise failures). It is worth noting, however, that the artifacts created in Step 1 would be of wider use that merely in a study of V&V techniques - they could facilitate complementary studies on the effectiveness of knowledge acquisition and design methods.

6. Conclusion and Perspective

This paper has argued that V&V techniques are an essential part of the knowledge engineering process, because they offer the only way to judge the success (or otherwise) of a KBS development project. This is equally true in the context of knowledge management, where V&V techniques tell us whether or not the KBS can be relied upon to accurately embody the knowledge of the human experts that supplied it.

However, examination of known studies on the effectiveness of existing KBS V&V techniques has shown, that the state of knowledge in this area is sparse. The way to improve this situation would be by systematically gathering data from a representative set of KBS projects and V&V techniques. Without such a study, knowledge engineering will remain very much an art and, by extension, so will the use of KBS technology in knowledge management.

In conclusion, however, it should be noted that the state of knowledge in software engineering is hardly much better! In particular, little is known about the relative effectiveness of V&V techniques in object-oriented software development. Despite this lack of knowledge, a huge number of successful, robust software systems have been created; similarly, a huge number of successful, robust KBS have been developed without perfect knowledge of the effectiveness of the methods employed. Clearly, software engineers, knowledge engineers, and knowledge managers have considerable artistic ability.

References

[1] Liebowitz, J., and Wilcox, L., 1997, *Knowledge Management and Its Integrative Elements*. CRC Press, New York.
[2] Rushby, J., 1990, Validation and testing of Knowledge-Based Systems: how bad can it get? In (Gupta, 1990).

[3] Batarekh, A., Preece, A., Bennett. A., and Grogono, P., 1996, Specifying an expert system. *Expert Systems with Applications*, Vol. 2(4).

[4] Boehm, B., 1984, Verifying and validating software requirements and design specifications. *IEEE Software*, Vol. 1(1).

[5] Meseguer, P. and Preece, A., 1995, Verification and Validation of Knowledge-Based Systems with Formal Specifications. *Knowledge Engineering Review*, Vol. 10(4).

[6] Preece, A., 1990, Towards a methodology for evaluating expert systems. *Expert Systems*, Vol. 7(4).

[7] Preece, A., 1995, Towards a Quality Assessment Framework for Knowledge-Based Systems. *Journal of Systems and Software*, Vol. 29(3).

[8] Wielinga, B.J., Schreiber, A.Th., and Breuker, J.A., 1992, KADS: a modelling approach to knowledge engineering. *Knowledge Acquisition*, Vol. 4(1).

[9] Brazier, F., Keplics, B., Jennings, N. and Treur, J., 1997, DESIRE: Modelling multi-agent systems in a compositional formal framework. *International Journal of Cooperative Information Systems*, Vol. 6(1).

[10] Kingston, J., and Macintosh, A., 1999, Knowledge Management through Multi-Perspective Modelling, In Research and Development in Intelligent Systems XVI, pages 221-239, Springer-Verlag, Berlin.

[11] Gupta, U.G., 1990, *Validating and Verifying Knowledge-based Systems*. IEEE Press, Los Alamitos, CA.

[12] Ayel, M. and Laurent, J-P., Eds., 1991, *Verification, Validation and Test of Knowledge-based Systems*. John Wiley & Sons, New York.

[13] O'Leary, D.E., 1991, Design, development and validation of expert systems: a survey of developers. In (Ayel and Laurent, 1991), pages 3-20.

[14] Preece, A., Shinghal, R. and Batarekh, A., 1992, Principles and practice in verifying rule-based systems. *Knowledge Engineering Review*, Vol. 7(2).

[15] O'Leary, D.E., 1998, On a common language for a "best practices" knowledge base. In *Using AI for Knowledge Management and Business Process Reengineering: Papers from the AAAI-98 Workshop*, AAAI Press, 1998..

[16] Kirani, S., Zualkernan, I.A., and Tsai, W.T., 1992, *Comparative Evaluation of Expert System Testing Methods*. Technical report TR 92-30, Computer Science Department, University of Minnesota, Minneapolis.

[17] Rushby, J. and J Crow, J., 1990, *Evaluation of an Expert System for Fault Detection, Isolation, and Recovery in the Manned Maneuvering Unit*. NASA Contractor Report CR-187466, SRI International, Menlo Park CA.

[18] Preece, A. and Shinghal, R., 1994, Foundation and application of knowledge base verification. *International Journal of Intelligent Systems*, Vol. 9(8).

[19] Miller, L., Hayes, J., and Mirsky, S., 1993, *Evaluation of Knowledge Base Certification Methods*. SAIC Report for U.S. Nuclear Regulatory Commission and Electrical Power Research Institute NUREG/CR-6316 SAIC-95/1028 Vol. 4.

[20] Preece, A., Talbot, S. and Vignollet, L., 1997, Evaluation of Verification Tools for Knowledge-Based Systems. *International Journal of Human-Computer Studies*, Vol. 47.

[21] Ayel, M. and Vignollet, L., 1993, SYCOJET and SACCO, two tools for verifying expert systems. *International Journal of Expert Systems: Research and Applications*, Vol. 6(2).

[22] Fisher, M. and Wooldridge, M., 1993, Specifying and verifying distributed artificial intelligent systems. In *Progress in Artificial Intelligence-Sixth Portugese Conference on Artificial Intelligence (LNAI Volume 727)*, pages 13-28, Springer-Verlag, Berlin.

Chapter 2

Frameworks

Design Knowledge Acquisition and Re-Use Using Genetic Engineering-Based Genetic Algorithms
J. Gero and V. Kazakov

A Web-Based Framework for Integrating Knowledge
J. T. Fernandez-Breis and R. Mertinez-Bejar

Re-Design Knowledge Analysis, Representation and Re-Use
I. Arana, H. Ahriz and P. Fothergill

Project Memory in Design
N. Matta, M. Ribiere, O. Corby, M. Lewkowicz and M. Zacklad

Micro Knowledge Management: A Job Design Framework
M. J. Leseure and N. Brookes

Building the KDD Roadmap: A Methodology for Knowledge Discovery
J.C.W Debuse, B. de la Iglesia, C.M. Howard, and V.J. Rayward-Smith

Design Knowledge Acquisition and Re-use Using Genetic Engineering-based Genetic Algorithms

John S. Gero and Vladimir Kazakov

Key Centre of Design Computing and Cognition, Department of Architectural and Design Science
The University of Sydney, NSW 2006 Australia. e-mail:{john,kaz}@arch.usyd.edu.au

Abstract: This chapter describes an application of genetic engineering-based genetic algorithms as a tool for knowledge acquisition and re-use. This version of genetic algorithms is based on a model of neo-Darwinian evolution enhanced by an analysis of genetic changes, which occur during evolution, and by application of various operations that genetically engineer new organisms using the results of this analysis. The genetic analysis is carried out using various machine learning methods. This analysis yields domain-specific knowledge in a form of two hierarchies of beneficial and detrimental genetic features. These features can then be re-used when similar problems are solved using genetic algorithms. Layout planning problem is used to demonstrate the process and the results obtainable.

1. Introduction

The use of machine learning techniques in design processes has been hampered by a number of problems. The first problem arises because the problem solving process and the learning process are usually considered as two different activities [1], which are conducted separately, analyzed separately, and which employ different processes and tools.

The second problem is that learning is considered a more general task than problem solving and thus a more difficult one. Therefore computational and other costs associated with the conduct of machine learning is normally significantly higher than the corresponding costs for problem solving only. This limits the scope of applications of learning methods in the industry. Also the more efficient types of learning require considerable domain-specific knowledge and thus additional cost/time for their development. The effect of this is that it has to be done separately for each domain.

The third problem relates to how much knowledge a problem-solving tool developer is willing to incorporate into the tool. Clearly, the higher is the knowledge-content of the tool the more effort its development will require and less general the domain where this tool can be used. The ease of development and wide applicability of such knowledge-lean problem-solving tools such as genetic algorithms [2], [3], simulated annealing [4], and neural networks have made them extremely popular.

The fourth problem is that the representations used by the learning tools/methods and the problem-solving tools/methods are usually incompatible. Therefore, an additional transformation of the knowledge that has been acquired during learning, has to be performed before it can be used in problem solving. Attempts to use this knowledge for further advanced processing (combining knowledge from different domains/problems, generalization, etc.) also leads to need for the development of some homogeneous knowledge representation.

In this chapter we present an approach that attempts addresses all these problems. Firstly, it is a problem-solving tool and also a learning tool. It includes a problem-solving component and a learning component that operate simultaneously and provide feedback to each other. Secondly, instead of reducing the efficiency of problem-solving process, the learning here enhances it. This makes the overall process more efficient than the problem-solving component of this algorithm alone. The learning approach that is employed by the algorithm is general and requires very little or no domain knowledge. Therefore the development of this method for new domains requires very little effort. Thirdly, the algorithm results in a knowledge-lean tool that can be initiated without any knowledge, but it acquires knowledge when it is being used and the longer it is used the richer in knowledge it becomes. Fourthly, because this algorithm acquires and utilizes the knowledge using a single homogeneous representation, no knowledge transformation is needed when it is re-used, re-distributed, generalized, etc

The problem-solving module of the proposed algorithm uses the standard genetic algorithm machinery (selection/crossover/mutation) [2] followed by an additional stage of genetic engineering. In this additional stage the population's genetic material is subjected to direct manipulation that simulates various genetic engineering techniques. The learning module can employ any one of the standard machine learning algorithms. At the end of each evolutionary cycle two training sets to be used by the learning module are formed from the best and the worst performing solutions within the current population. The learning module identifies which attribute (genetic) features are beneficial and which are detrimental for the current population. The genetic features that have been identified before from the previous populations are re-tested against the current population. The ones that fail are deleted from the list of beneficial/detrimental features. Next the standard GA processes produce an intermediate population from the current population. This population is subjected to genetic engineering processing, which promotes the presence of beneficial features and discourages the presence of detrimental features. This produces a new population and concludes the current evolution cycle and the next evolutionary cycle is initiated.

2. Genetic Engineering Genetic Algorithms

Genetic algorithms (GAs) [2], [3] are search algorithms that simulate Darwinian evolutionary theory. GAs attempt to solve problem (e.g., finding the maximum of an objective function) by randomly generating a population of potential solutions to the problem and then manipulating those solutions using genetic operations. The solutions are typically represented as finite sequences (genotypes) drawn from a finite alphabet of characters. Through selection, crossover and mutation operations, better solutions are generated out of current population of potential solutions. This process continues until an acceptable solution is found, Figure 1. GAs have many advantages over other problem-solving methods in complex domains. They are very knowledge-lean tools, which operate with great success even without virtually any domain knowledge.

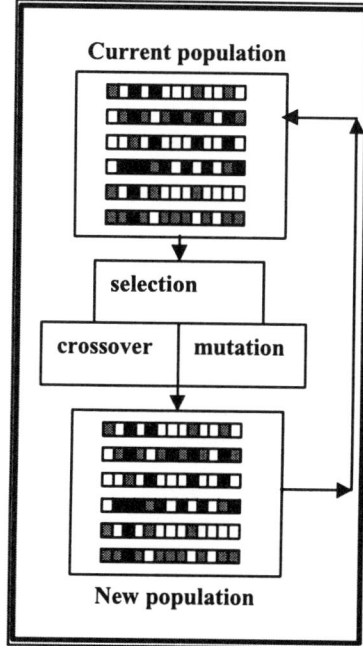

Figure 1: The structure of genetic algorithms evolutionary cycle.

Genetic-engineering based GA is an enhanced version of GAs that simulates Darwinian theory of the current stage of the natural evolutionary process into which genetic engineering technology has been introduced. This technology is based on the assumption that the genetic changes that occur during an evolution can be studied and that the genetic features that lead to wanted and unwanted consequences can be identified and then used to genetically engineer an improved organism. Thus, the model assumes that this technology includes two components:

(a) methods for genetic analysis, which yield the knowledge about the beneficial and the detrimental genetic features; and

(b) methods for genetically engineering an improved population from the new population generated by normal selection/crossover/mutation process by using the knowledge derived from genetic analysis.

The genetic-engineering GA uses a stronger assumption than (a), that standard machine learning methods can identify the dynamic changes in the genotypic structure of the population that take place during the evolution. The type of machine learning depends on the type of genetic encoding employed in the problem. For the order-based genetic encoding sequence-based methods [5] give the best results and for attribute-value based genetic encoding decision trees [6], [7] or neural networks [8] should be tried first. The two training sets for machine learning are formed by singling out the subpopulations of the best and worst performing solutions from the current population. The structure of the learning module is shown in Figure 2. Note that since new features are built by the system as derivatives of the already identified features (which are the elementary genes/attributes when the system is initiated), the approach readily leads to a hierarchy of genetic features. For example, if genetic features that occur are position-independent genetic "words" (as it is the case in the system in Figure 2) then the features that evolve later would by genetic "words" that include previously found "words" as additional letters of genetic alphabet.

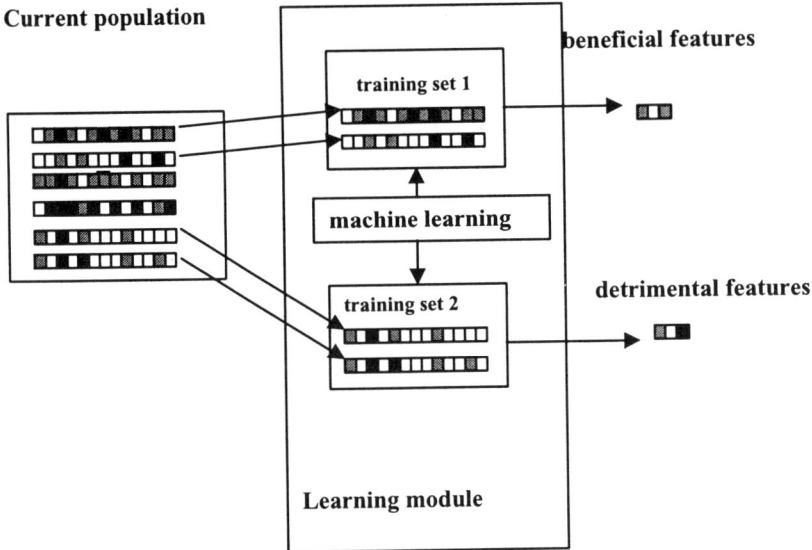

Figure 2: The structure of the learning module of genetic engineering GA. The genotypes in the current population are vertically ordered by their performances, so the higher the genotype is in the box, the higher is its performance.

The genetic engineering of the new population applies the following four operations to each genotype:

(a) screening for the presence of the detrimental genetic features;

(b) screening for the presence of incomplete beneficial genetic features;

(c) replacing the minimal number of genes to eliminate the detrimental features found during screening (a);

(d) replacing the minimal number of genes to complete the incomplete beneficial genetic features found during (b).

If some beneficial/detrimental genetic features are known when the genetic engineering GA is initiated, then even the initial population (which normally is generated randomly) is subjected to the genetic engineering operations (a)-(d). These are the principles, the implementation details of genetic engineering GA can be found in [9].

3. Learning in Genetic Engineering Genetic Algorithms

Let us discuss the major characteristics of learning process is genetic engineering GA. Firstly, note that if one employs any problem-solving method that is population-based (like GA), then the solution process will generate a large amount of information (the optimal solutions, the search trajectories, etc.) that can be used to carry out some form of learning. The possible recipes for the formation of the training sets for genetic learning here can vary between two extremes of:

(a) using only the optimal solutions [10]; here only the final population is used for learning and thus the cost of learning is minimal and the possibilities for knowledge acquisition are also minimal; and

(b) using the complete set of search trajectories.

Genetic engineering GA corresponds to some middle choice by using only the current population for learning. Nevertheless, it can be easily generalized to include a more comprehensive analysis of larger parts of the evolution history that lead to the current population. Obviously some tradeoff between the comprehensiveness of genetic analysis and its cost should be achieved to balance the advantages of more comprehensive genetic learning against its computational and developmental cost.

Secondly, note that the space of genetic sequences (genotypes) and its derivatives form the only domain that is used for learning in genetic engineering GA. Thus, whatever knowledge can be learned by the algorithm, it is always expressed in terms of some characteristics of these genetic sequences, that is, in terms of genetic representation. This is important since the learned knowledge is in the same representation as the source from which it was learned and hence it can be used directly. We shall call these learned characteristics genetic features. Many features types can emerge in a genetic sequence and a genetic engineering GA can be tailored to handle all of them [9] by including the machine learning methods that are capable of learning these regularities and by creating the specific versions of genetic engineering operators to target these genetic regularities. Nevertheless, the two most frequent and important types of features are:

(a) the set of fixed substrings in fixed positions (called building blocks in GA theory [2]); and

(b) the fixed substring that is position-independent (like words in natural languages).

The specific versions of genetic engineering GA that efficiently handle these feature types can be found in [9]. They should be used when it is known a priori that it is likely that the problem under consideration has these types of genetic regularities. When this is not known then a general learning method that is capable of identifying a wide range of features and the corresponding non-specialized genetic engineering operators should be used.

Thus, the learning process in genetic engineering GA leads to a step-wise recursive enlargement of its feature space, layer by layer. This recursion builds a hierarchical structure of this feature space, where each new layer is a derivative of all the previous layers of features. For the problem where the characteristic genetic features are position-independent genetic substrings (genetic "words", Figure 2), this hierarchy can be represented as a tree, Figure 3. Here the new features are derivatives of the already evolved features because they use them as building components (sub-substrings).

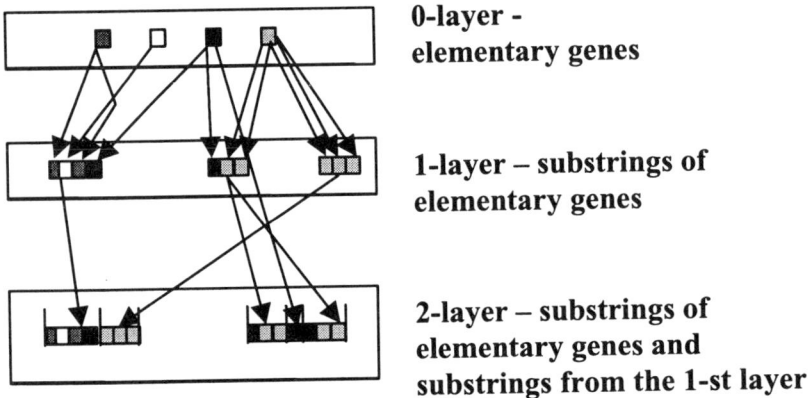

0-layer - elementary genes

1-layer – substrings of elementary genes

2-layer – substrings of elementary genes and substrings from the 1-st layer

Figure 3: The hierarchical knowledge structure (feature space) is built by genetic engineering GA for the problem where evolved genes are genetic "words". The arrows denote the component from the lower lever that is used to assemble new feature on the next level.

The problem's knowledge in the genetic engineering GA is its feature space, which is derived from its genetic representation. Therefore, if two problems have identical or even partially overlapping genetic encodings then the knowledge that is acquired when one problem is solved can be translated into complete or partial initial knowledge (feature space) for the second problem. If a number of instances of the problems from the same class are solved then the generalization of their knowledge into a class of knowledge (class feature space) is the straightforward procedure of finding the overlapping part of their feature space. This concept is important as it allows the problem-solving tool to improve its performance as it acquires more problem class-specific knowledge.

4. Layout Planning Problem

The space layout planning problem is fundamentally important in many domains from architectural design to VLSI floor-planning, process layouts and facilities layout problems. It can be formalized as a particular case of a combinatorial optimization problem – the quadratic assignment problem. As such it is NP-complete and presents all the difficulties associated with this class of problems.

Thus, we formulated the layout-planning problem as a problem where such one-to-one mapping ρ

$$\rho: \{M \rightarrow N\}, j = \rho(I), i \in M, j \in N$$

of the discrete set M with m elements (set of activities, for example office facilities) onto another discrete set N of n elements (set of locations, for example floors of the building where these facilities should be placed), $m \leq n$ is sought that the overall cost of the layout I is minimal, i.e.

$$I = \sum_i f_{i\rho(i)} + \sum_i \sum_j q_{ij} c_{\rho(i)\rho(j)} \rightarrow \min$$

where f_{ij} is the given cost of assigning element $i \in M$ to the element $j \in N$, q_{ij} is the measure of interaction of elements $i,j \in M$, and c_{ij} is the measure of distance between elements $i,j \in N$.

Usually the space layout-planning problem contains some additional constraints, which prohibit some placements and/or impose some extra requirements on feasible placements.

5. Example

As a test example we use Liggett's problem of the placement of a set of office departments into a four-storey building [11]. The areas of the 19 activities to be placed (office departments, numbered 0,..,18) are defined in Table 1 in terms of elementary square modules. There is one further activity (number 19) whose location is fixed. The objective of the problem does not have a non-interactive cost term, i.e. $f_{ij} = 0$. The interaction matrix q_{ij}, $i, j = 0, ..., 19$ is given in Table 2. The set of feasible placements is divided into 18 zones numbered from 0 to 17, Figure 4 (a). The areas of these zones are defined in Table 3. The activity number 19 is an access area which has a fixed location - zones numbers 16 and 17, Figure 5.

Since the layout cost does not depend on the areas of the zones numbered 16 and 17 or on the area of the activity number 19 they are not shown in Table 3. The matrix showing travel distances between all zones is defined in Table 4. We use the same genetic representation as was used in [12] and [13] – the genotype of the problem is a sequence that determines the order in which zones are filled with activity modules. Each zone is filled line by line starting from its highest one and from the leftmost position within it. For example, the genetic sequence

$x = \{13, 14, 0, 2, 6, 9, 16, 11, 3, 4, 18, 15, 17, 1, 12, 10, 8, 7, 5\}$

generates a layout plan in the following manner. All 18 elementary modules of activity 13, i.e. department 0800 are placed, followed by the 31 modules of activity 14, i.e. department 0900, followed by the 2 modules of activity 0, i.e. department 0210, etc.

Table 1: The definition of activities [11].

Activity	0	1	2	3	4	5	6	7	8
Dept	0210	0211	0220	0230	0240	6815	0300	0400	0500
Num. Of modules	2	2	8	15	15	13	15	7	6

Act.	9	10	11	12	13	14	15	16	17	18
Dept	0600	0700	6300	6881	0800	0900	1000	Extra	extra	Extra
Mod	12	53	10	16	18	31	61	1	1	1

Table 2: Activity interactions matrix [11].

	0	1	2	3	4	5	6	7	8	9	10	11	12	13	14	15	16	17	18	19
0		3	3	2	2	2			3							2				3
1			3	2	2	2			3							2				
2				2	2		3								2					
3					3	2			3							2				3
4						2			3							2				
5									3							2				
6														3		2				
7														3		2				3
8															2					
9															3	2				
10															3	2				
11															3	2				
12																2				
13																2				
14																2				
15																				3
16																				
17																				
18																				
19																				

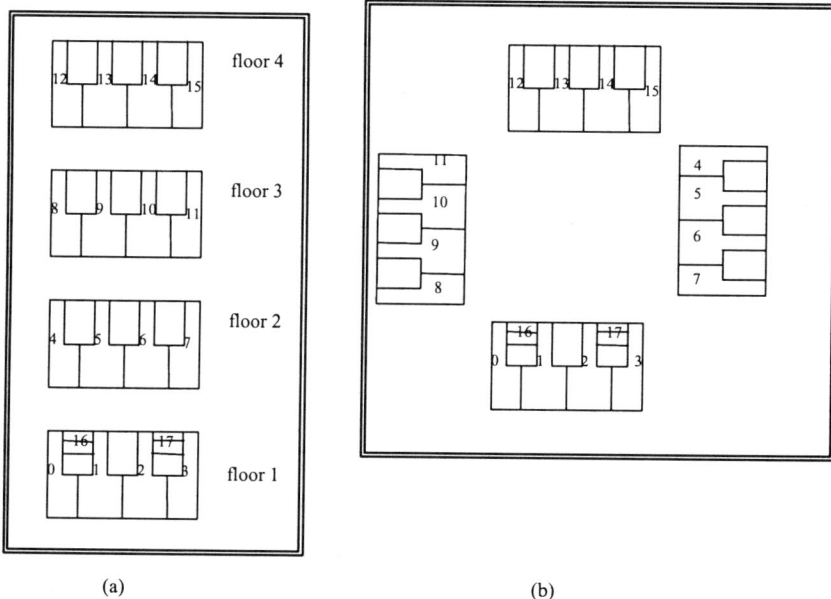

(a)   (b)

Figure 4: Zone definitions for Liggett's example [11] (a) and its modification (b).

Table 3: Zone definitions, measured in number of modules [11].

zone	0	1	2	3	4	5	6	7	8	9	10	11	12	13	14	15
Modules	20	22	20	20	18	20	18	18	16	18	16	16	14	16	14	14

The genetic engineering GA will be first used to solve this layout planning problem. During the process of solving this problem knowledge about the problem will be acquired. The acquired knowledge will be then re-used during the solution of a similar problem where the same set of activities is to be placed into a modified zone location, Figure 4(b). This is a common situation where alternate possible receptor locations are tried. In building layouts this can be the result of alternate existing buildings. The same applies to the layout of processes and department stores. In this case the change is from a four-storey building to a courtyard-type single storey building. This new zone structure yields the modified distance matrix shown in Table 5.

Table 4: Distance matrix [11].

	0	1	2	3	4	5	6	7	8	9	10	11	12	13	14	15	16	17	18
0		4	9	21	27	9	10	23	28	10	11	24	29	11	12	25	30	6	24
1			5	17	23	5	6	19	24	6	7	20	25	7	8	21	26	2	20
2				13	18	6	5	18	23	7	6	19	24	8	7	20	25	2	15
3					5	23	18	5	6	24	19	6	7	25	20	7	8	15	2
4						24	19	6	5	25	20	7	6	26	21	8	7	20	2
5							5	18	23	5	6	19	24	6	7	20	25	3	21
6								13	18	6	5	18	23	7	6	19	24	3	16
7									9	23	18	5	6	24	19	6	7	16	3
8										24	19	6	5	25	20	7	6	21	3
9											5	18	23	5	6	19	24	4	22
10												13	18	6	5	18	23	4	17
11													5	23	18	5	6	22	4
12														24	19	6	5	17	4
13															5	18	23	5	23
14																13	18	5	18
15																	5	18	5
16																		23	5
17																			18
18																			

Table 5: The modified distance matrix for new zone location, Figure 4(b).

	0	1	2	3	4	5	6	7	8	9	10	11	12	13	14	15	16	17	18
0		4	9	21	27	9	10	23	28	10	11	24	29	11	12	25	30	6	24
1			5	17	22	29	34	46	51	58	53	41	36	29	24	12	7	2	20
24				12	17	24	29	41	46	51	56	49	44	37	30	25	13	2	15
3					5	12	17	29	34	41	46	58	53	46	41	29	24	15	2
4						7	12	24	29	36	41	53	58	51	46	34	29	23	10
5							5	17	22	29	34	46	51	58	53	41	36	28	15
6								12	17	24	29	41	46	53	58	46	41	40	27
7									5	17	22	34	39	46	51	58	48	45	32
8										7	12	24	29	36	39	51	56	52	39
9											5	17	22	29	34	46	51	57	44
10												12	17	24	29	41	46	59	56
11													5	12	17	29	34	45	65
12														7	12	24	29	35	55
13															5	17	23	30	51
14																12	17	18	38
15																	5	15	33
16																		6	25
17																			18
18																			

6. Results of Simulations

In the initial problem the genetic engineering GA was not able to find any detrimental genetic regularities but found four instances of beneficial genetic regularities. They were of a non-standard type of genetic regularity that is characteristic not only for this particular example but also for many other layout planning problems. This feature is a gene cluster – a compact group of genes. The actual order of genes within each group is less significant for the layout performance than the presence of such groups in compact forms in genotypes. The genetic engineering GA was able to find first two clusters {0, 1, 2, 8, 3, 4} and {12, 14} after the 5-th generation. Then it found two more clusters {6, 13}and {15, 12, 14} after the 10-th generation. Note that the last cluster was derived from the previously identified cluster {12, 14}. This is an example of how the genetic engineering GA builds a multilayer hierarchy of evolved genetic structures, where each new layer is derived from the components below it in the hierarchy (from the features, which have been already identified).

The comparison of the standard GA evolution process and the genetic engineering GA evolution processes shown in Figure 5, illustrates the significant computational saving genetic engineering brings in terms of the number of generations that are needed for convergence. The overall computational saving here was 60%. In the general case, this saving typically varies between 10% and 70%, depending on the computational cost of the evaluation of the genotype relative to the computational cost of the machine learning employed.

If these clusters are given to genetic engineering GA when it is initiated, then the number of generations needed for convergence drops from 21 to 15.

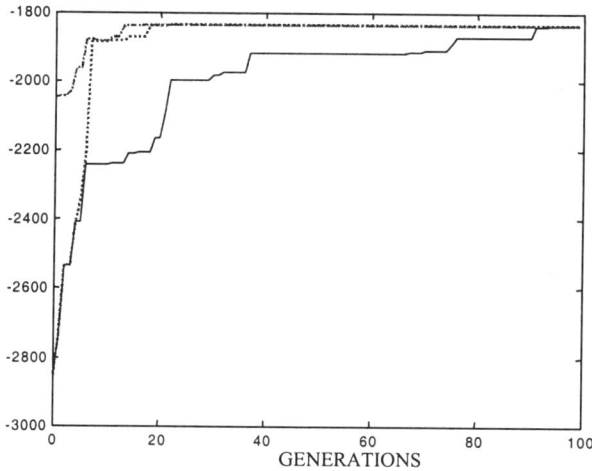

Figure 5: The best layout cost vs generation number for the standard GA (solid line), genetic engineering GA without prior knowledge (dotted line) and genetic engineering GA with prior knowledge (dashed line). The results are averaged over 10 runs with different initial random seeds.

The simulation was then carried out for the modified example, Figure 4(b). The results are shown in Figure 6 for genetic engineering GA's runs under the following conditions:

(a) without any prior knowledge;
(b) with partially incorrect prior knowledge (using the four clusters that have been acquired by solving the non-modified example); and
(c) with correct prior knowledge for the modified example.

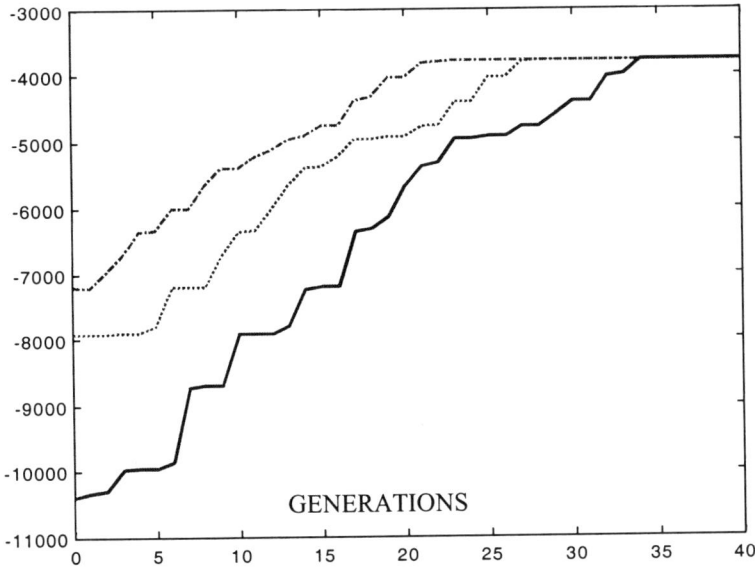

Figure 6: The best layout cost vs. generation number in modified example genetic engineering GA without prier knowledge (solid line), with non-perfect prior knowledge (dotted line) and with perfect prior knowledge (dashed line). The results are averaged over 10 runs with different initial random seeds, which converge to the best solution found.

Only one of the four beneficial clusters from the initial problem survived the problem's modifications and was retained by the genetic engineering GA in simulation (b). This is the first cluster $\{0, 1, 2, 8, 3, 4\}$ of a group of very small but strongly attracted to each other activities. After modification it even grew into a larger cluster $\{0, 1, 2, 8, 3, 4, 5, 17\}$. The genetic engineering GA also found one new cluster $\{6, 7, 13, 15\}$.

To demonstrate the benefits of these clusters for the layout cost we calculated the average cost of layouts with and without these clusters for the modified test problem, Table 6. The average layout costs were calculated using the Monte-Carlo method with averaging over 500 trial points. For the first cluster, which is beneficial in both problems, we also present the decomposition of the average layout cost into two components: the cost of the inter-cluster activity interactions I_I and the $I_c = I - I_i$.

Table 6: The average layout cost in the modified test problem with and without evolved beneficial genetic clusters.

Gene clusters	No clusters	{0, 1, 2, 8, 3, 4, 5, 17}	{1, 5, 6, 7, 13}
Average layout cost	10880	5541	8924

This cost decomposition is presented in Table 7 for a sample of gene sequences within the cluster for both original and modified problems. For the original problem the reduction in I_i. was 55% (with standard deviation of 2%) and the reduction in I_c was 2%. For the modified problem the reduction in I_i. was 66% (with standard deviation of 5%) and I_c was actually increased by 2.5%. This cost decomposition provides some insight as to why this cluster is beneficial - the placement of its genetic components in a tight group leads to the large reduction of I_i. but very small variation of I_c. Thus, the genetic engineering GA in layout planning problem "discovers" the gravitational-type model that is hidden in the problem statement.

Table 7: The average layout cost in the original and modified test problem with and without cluster of genes {0, 1, 2, 8, 3, 4}

Gene cluster	Original problem		Modified problem	
	I_i	I_c	I_i	I_c
Scattered	1539	2539	4025	6854
{8, 0, 1, 2, 3, 4}	607	2540	4025	6832
{0, 1, 2, 8, 3, 4}	637	2536	972	6827
{0, 1, 2, 3, 8, 4}	671	2536	985	6847
{4, 1, 2, 8, 0, 3}	729	2510	1015	6790
{8, 2, 1, 0, 3, 4}	628	2535	1013	6801
{4, 2, 1, 8, 0, 3}	687	2509	984	6718

7. Discussion

In this chapter we have presented an approach to automated knowledge acquisition and re-use in design through the deployment of a genetic engineering-based genetic algorithm that operates in conjunction with the problem-solving process. This approach obviates a number of problems that occur when automated knowledge acquisition systems are used separately from the problem-solving process. Genetic engineering-based genetic algorithms make use of the notions of genetic engineering: that structural features of the genotype influence the fitness or behavior of the resulting designs. If these features can be isolated they can be manipulated to the benefit of the resulting design process.

The genetic features in the genotypic representation of designs are problem specific knowledge that has been acquired using a knowledge lean process. The knowledge acquired in this manner is in the same representational form as the genotypic design representation and can therefore be used without the need for any additional interpretive knowledge. These genetic features can also be subjected to genetic engineering style operations. Operations such as gene surgery and gene therapy have the potential to improve the performance of the resulting system.

Since the genetic features are in the form of genes, they can be replaced by a single "evolved" gene and thus extend the range of symbols used in the genetic representation. This extension allows for any genetic algorithm that uses it to search for design solutions in a more focussed way since it it using knowledge that was previously not available in the formulation of the problem. The genetic engineering approach to knowledge acquisition provides opportunities to acquire different kinds of knowledge. It can be used to acquire knowledge about solutions as was demonstrated in the example above, or it can be used to acquire knowledge about design processes [14]. In acquiring knowledge about solutions it is possible to caquire design knowledge from two different design problems that are linked only by a common representation and then to combine this knowledge in the production of a novel design that draws features from both sources [15]. This provides a computational basis for the concept of combination.

This form of knowledge acquisition has applications beyond the design domain. Any domain where there is a uniform representation in the form of genes in a genotype may benefit from this approach.

Acknowledgments

This work is directly supported by a grant from Australian Research Council.

References

[1]   Weiss, S. and Kulikowski, C., 1991, *Computer Systems that Learn*, Morgan Kaufmann, Palo Alto, CA.

[2]   Goldberg, D.,E., 1989, *Genetic Algorithms in Search, Optimization and Machine Learning*, Addison-Wesley, Reading, Mass.

[3]   Holland, J., 1975, *Adaptation in Natural and Artificial Systems*, University of Michigan, Ann Arbor.

[4]   Kirkpatrick, S., Gelatt, C. G. and Vecchi, M., 1983, Optimization by simulated annealing, *Science*, **220** (4598): 671-680.

[5]   Crochemore, M., 1994, *Text Algorithms*, Oxford University Press, New York.

[6]   Quinlan, J. R., 1986, Induction of decision trees, *Machine Learning*, **1**(1) 81-106.

[7] Salzberg, S, 1995, Locating protein coding regions in human DNA using decision tree algorithm, *Journal of Computational Biology*, **2** (3): 473-485.

[8] Rumelhart, D. E. and McClelland, J. L., 1986, *Parallel Distributed Processing*, Vol. 1, MIT Press, Cambridge, Mass.

[9] Gero. J. S. and Kazakov, V., 1995, Evolving building blocks for genetic algorithms using genetic engineering, *1995 IEEE International Conference on Evolutionary Computing*, Perth, 340-345.

10] McLaughlin, S. and Gero, J. S. 1987, Acquiring expert knowledge from characterized designs, *AI EDAM*, **1** (2): 73-87.

[11] Liggett, R.S., 1985, Optimal spatial arrangement as a quadratic assignment problem, *in*: Gero, J. S. (ed.), *Design Optimization*, Academic Press, New York: 1-40.

[12] Gero, J. S. and Kazakov, V.,1997, Learning and reusing information in space layout problems using genetic engineering, *Artificial Intelligence in Engineering*, **11** (3): 329-334.

[13] Jo, J.H. and Gero, J.S., 1995, Space layout planning using an evolutionary approach, *Architectural Science Review*, **36** (1): 37-46.

[14] Ding, L. and Gero, J. S., 1998, Emerging Chinese traditional architectural style using genetic engineering, *in* X. Huang, S. Yang and H. Wu (eds), *International Conference on Artificial Intelligence for Engineering*, HUST Press, Wuhan, China, 493-498.

[15] Schnier, T. and Gero, J. S., 1998, From Frank Lloyd Wright to Mondrian: transforming evolving representations, *in* I. Parmee (ed.), *Adaptive Computing in Design and Manufacture*, Springer, London, 207-219.

A Web-based Framework for Integrating Knowledge

Jesualdo T. Fernández-Breis[1], Rodrigo Martínez-Béjar[2]

[1] Grupo de Inteligencia Artificial e Ingeniería del Conocimiento, Facultad de Informática, Universidad de Murcia, 30071- Espinardo (Murcia), Spain. Phone: +34 968 367 345, Fax: +34 968 364 151, email: jfernand@perseo.dif.um.es

[2] Grupo de Inteligencia Artificial e Ingeniería del Conocimiento, Facultad de Informática, Universidad de Murcia, 30071- Espinardo (Murcia), Spain. Phone: +34 968 364 634, Fax: +34 968 364 151, email: rodrigo@dif.um.es

Abstract: Knowledge Integration is a key point in Knowledge Management and it can be viewed from two perspectives: integration of different knowledge bases or integration of different representations (ontologies) of the same knowledge at different formalisation levels. The use of WWW is considered to be an important factor for co-operative knowledge development provided that the WWW can be said to be the most important knowledge source in the world, while its underlying technology allows for the co-operation in the context of ontology construction processes. In addition to this, while a (global) ontology can be centralised, its instances may be distributed over the WWW. This article introduces a co-operative philosophy-based framework that allows for these processes.

1. Introduction

For the last ten years, considerable advances have been devoted to the development of the foundations for constructing technologies that make it possible to reuse and share knowledge components. For it, Ontologies and Problem-Solving Methods (PSMs) have been proposed. Ontologies deal with static knowledge, while PSMs are concerned with dynamic knowledge. In Knowledge Management, a key point is that of knowledge integration, which can be viewed from two perspectives: integration of different knowledge bases and integration of different representations (ontologies) of the same knowledge at different formalisation levels ([14]). This work is focused on the second point of view, that is, we will deal with the problem of ontology integration.

In the approach introduced here, the basic idea is that several experts on the same topic are encouraged to work on a knowledge construction process in a co-operative way. However, working co-operatively can give rise to several problems. Some of these are: redundant information, synonym terms and inconsistent knowledge.

As [2] have put forward, the use of the WWW is an important factor for the co-operative ontology development. On the one hand, the WWW can be said to be the most important knowledge source in the world. On the other hand, the technology that supports the WWW allows for the co-operation in the context of ontology construction processes and a (global) ontology can be centralised while its instances can be distributed over the WWW. The purpose of this work was the development of a tool for accomplishing ontology integration processes through the WWW. In this article, we present a co-operative philosophy-based framework that allows for these processes. Moreover, the ontologies must be supplied by a set of users, who must be interconnected through the Internet.

In co-operative work-based tools it is important that these tools possess a component that allows for dialogue between the agents who co-operate. One of the most famous examples of current technologies that support this dialogue is KARAT[1]. In this system, dialogue is promoted through knowledge sharing, since knowledge is accessible to every user. This system makes use of dialogue in order to facilitate refinement-oriented discussion between different users over a pre-existent (i.e., pre-created) ontology. However, in our approach the dialogue is different because agents play the role of both ontology suppliers and global ontology constructors. Another quite famous system that supports dialogue of the characteristics quoted above is the Ontolingua Server [7]. This system deals with the dialogue issue by means of the concept of shared session, through which users can modify the same ontology, while communication via e-mail is facilitated.

In the system outlined here, the selection of the ontologies that will take part of the integration-derived ontology (i.e., the one constructed co-operatively) is based on two criteria. Firstly, each of these ontologies must be consistent with the last version that the corresponding agent possesses. Secondly, that ontology must be maximal in terms of the amount of knowledge contained with respect to the rest of (equivalent) ontologies. Another fundamental feature of the approach presented here is that agents can consult the contents of an ontology built co-operatively at a given instant by means of the (co-operative) dialogue supplied by the framework. This consult can, in turn, be done with the purpose of obtaining as much information as possible in constructing their respective ontologies. Moreover, the system facilitates a synonym concept management that makes it possible for agents to use always their own terminology.

The structure of the paper is as follows. Section 2 provides the philosophical foundations of this work. In Section 3, the framework we use for integrating knowledge is explained. Section 4 gives a brief description of the web-based system implemented. In Section 5, we present an example for illustrating the functionality of the framework. Finally, some remarks are made in Section 6 standing as the conclusions of our work.

2. Ontological Foundations

2.1 Ontologies and PSMs

The development of technologies for facilitating shareable and reusable knowledge components has been a hot research topic in the last decade. Ontologies and PSMs have been pointed out as the most adequate tools for enabling sharing and reuse of knowledge and reasoning behaviour across domains and tasks. PSMs and ontologies can be considered to be complementary solutions due to the fact that PSMs deal with dynamic knowledge and ontologies are concerned with static knowledge. Ontologies typically contain modelling primitives such as classes, relations, functions, axioms and instances. As stated in [10], ontologies provide a common vocabulary of a domain and define the meaning of terms and relationships between them. On the other hand, PSMs define the way of how to achieve the goal of a task, and the data flow between its subtasks [9].

In this work, we are interested in ontologies since we are focused on the static aspect of the knowledge. An ontology has been formally defined in several ways. In [10], an ontology is defined as "an explicit specification of a conceptualisation", while in [3] an ontology is viewed as " a formal specification of a shared conceptualisation". In both definitions, the term conceptualisation refers to an abstract model of some phenomenon in the world by having identified the relevant concepts of that phenomenon [16]. In the following subsection, we introduce the ontological model we have used for this work.

2.2 Our ontological model

In this work, an ontology is seen as a specification of a domain knowledge conceptualisation([17]). Ontologies are represented here by means of *multiple hierarchical restricted domains* (MHRD). This representation has been used previously in a similar sense by other authors (see, for instance, [5]). In particular, we have used the notion of **Partial, Hierarchical, Multiple and Restricted Domain** (PHMRD) ([11]), that can be specified as a set of concepts which are defined through a set of attributes. A concept has two types of attributes: specific attributes (derived from its nature) and inherited attributes (derived from its relationships with other concepts of the domain). This ontological model allows for three types of relationships among whatever two concepts: taxonomic (allowing for multiple inheritance), mereological and temporal. Taxonomic relationships are assumed to hold all the irreflexive, the antisymmetric and the transitive properties, while mereological relationships are assumed to hold all of them except for the transitive one ([3]).

In addition to this, the ontology representation schema adopted here includes "structural" axioms, that is, axioms that result from the relations *concept has attribute*, *concept 1 is a class of concept 2, concept 1 is a part of concept 2,* and *concept 1 occurs after concept 2*. Moreover, this schema also embodies other axioms derived from some properties concerning interconceptual relationships in taxonomic, mereological and temporal organisations. We must clarify that the fact of defining ontologies without non-structural axioms does not mean that this sort of axioms cannot be defined by users as a (part of the) specification of a conceptualisation. What we do is to split up the classic definition of ontology (i.e., the one including structural and non-structural axioms) into two parts so that we term ontology to the whole specification of a conceptualisation excluding non-structural axioms.

In order to implement temporal relationships, the Fuzzy Temporal Constraint Network (FTCN) model ([4]) has been used. This model formalises the computational representation of general situations in which an arbitrary number of events are specified. A FTCN is a couple $<X,L>$, where $X= \{X_0, X_1, ..., X_n\}$ is a finite set of variables (concepts in our case) and $L= \{L_{ij} \mid i, j \leq n\}$ represents a finite set of fuzzy binary temporal constraints. The variable X_0 represents a precise origin, in our case, when the time is supposed to start (i.e. the first process of the temporal chain occurs). Therefore, each constraint L_{0i} defines the absolute occurrence of X_i.

Figure 1 shows a graphical representation of an ontology as defined here. Each concept is expressed in a rectangle. Inside every rectangle we can see the information of the corresponding concept, its name and its specific attributes. Relationships are represented by the arrows between the concepts that take part of the relationship. The three types of relationships appear in Figure 1, taxonomic (IS-A: Material is a class of Damage), mereological (PART-OF: Damage is part of a Natural Disaster) and temporal (AFTER: Damage occurs after Phenomenon). The inherited attributes of a concept are not explicitly expressed in Figure 1 and they can be obtained through the union set of the specific attributes of its taxonomic parents.

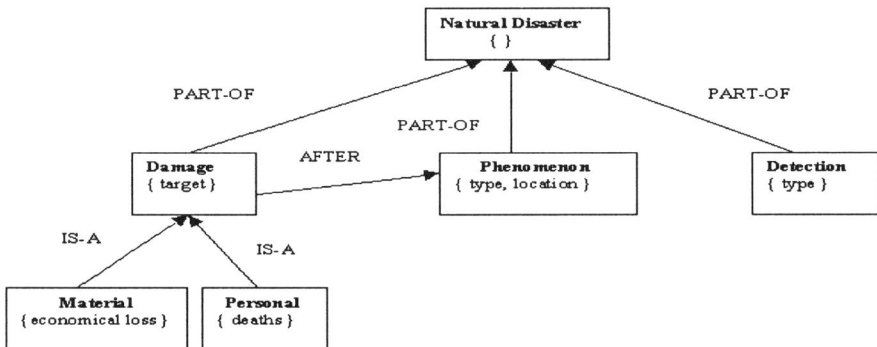

Figure 1: An example of ontology

3. A Framework for Integrating Knowledge

When we try to integrate knowledge that belongs to different people who work co-operatively, we must overcome some problems which are implied by this co-operation. Some of these are:

a) *Redundant Information.* Two different experts might attempt to describe the same part of the domain knowledge. Given this eventuality, it would be desirable that the system was capable of managing this possible situation, so that redundancies could be avoided.

b) *Use of synonym terms for a concept.* Apart from dealing with redundant information, different experts can employ different terminologies for the same concept. In other words, there can be a correspondence between different terms employed for a given concept ([15]). During the ontology construction process, the information concerning the use of synonym terms for a concept must be stored and managed, since a particular terminology should not be imposed to any expert during the Knowledge Acquisition process. However, an ontology would strive towards 'consensual knowledge', that is, a fixed terminology. Synonyms are possible but, ideally, everybody should agree on the terminology.

c) *Inconsistent knowledge.* This aspect has a double nature. An ontology can be internally or externally inconsistent. We say that an ontology is internally inconsistent when there is some part of it that is inconsistent with another part of itself. For instance, an ontology is internally inconsistent if there is at least one property concerning relationships between concepts that is not satisfied. An ontology can be externally inconsistent with respect to another ontology, that is, both descriptions of the same domain can be incompatible.

We assume that the system is supplied with ontologies free of internal inconsistencies in order to avoid the evaluation of them once they have been built. For it, and given that each ontology can be built in a particular way, users must introduce their own ontologies by using a specific format for the ontology file. The specification of this format can be summarised as follows. It is comprised of the concepts which are part of the ontology. Each concept is defined through its attributes, its name and its parent concepts, either mereological, taxonomic or temporal ones. The successfully parsing of the ontologies defined according to this model are granted to be consistent. From now, when we refer to inconsistency we refer to external inconsistency.

As we mentioned previously, the aim of this work was the design and implementation of a tool for building knowledge from the knowledge supplied by a set of users. In order to achieve this goal, the system must be able to solve some possible consistency conflicts between the candidate ontologies to be integrated until a specific instant. In particular, each time a user adds or modifies his/her private ontology, such knowledge will have to be incorporated into the integration-derived ontology, which is denoted by $O_{int}(t)$. It is also remarkable that more than one user might decide to send its knowledge contribution to $O_{int}(t)$ at the same time. This made it necessary that the system was able to distinguish amongst pieces of knowledge belonging to different users. In this sense, a user-oriented integration principle has been followed, which basically states that 'the knowledge in $O_{int}(t)$ at a specific instant will have to be consistent with that included in every private user ontology $(O_i(t))$ that takes part of the integration process for every previous instant'.

In order to obtain the integration of the knowledge specified in users' ontologies (i.e., only the ontologies that belong to experts), the following algorithms have been used:

3.1 Ontological_Integration

Let $O_i(t)$ be the i-th ontology that is intended to be incorporated into $O_{int}(t)$; n = number of ontologies to integrate (the previously existing ontologies as well as the new incoming ones). Let candidates(t) be the set of ontologies to be integrated.

For i=1 to n
   If (there is any ontology $O_j(t)$ belonging to candidates(t) such that $O_i(t)$ and $O_j(t)$
     belong to the same user) then (remove from candidates(t) the oldest ontology)
   End-if
End-for
subset= *Select_Ontologies*(candidates(t))
i=1
While $i \leq$ Card(subset) do
Ontological_Inclusion($O_i(t), O_{int}(t)$) {this algorithm is defined below}.
End-while
End

3.2 Ontological_Inclusion

Let $O_j(t)$ be the j-th ontology that is intended to be incorporated into $O_{int}(t)$; let *topic* be the topic which the final user requests information for; and let $O_i(t)$ be the ontology whose root is *topic-according to-group i* in $O_{int}(t)$.
Begin
Add $O_j(t)$ to $O_{int}(t)$ as a mereological child concept, so that its root is *topic-according to-user j*
End

3.3 Select_Ontologies

Let candidates(t) be the set of candidate ontologies to be integrated. Let compatible$_i$(t) be the set of ontologies $O_j(t)$ belonging to candidates(t) that are compatible with $O_i(t)$.

For i=1 to Card(candidates(t))
  For i=j to Card(candidates(t))
    compatible$_i$(t)=compatible$_i$(t) $\cup$ O_j(t) if compatible(O_i(t), O_j(t))
  End-For

End-For
Return the best subset according to the desired criterion (i.e., the subset with higher number of ontologies)
End

where
compatible(x,y) is true if and only if (not(inconsistent(x,y) or equivalent(x,y)));
equivalent(x,y) is true if and only if for each concept belonging to x, there is another from y such that :
a) both of them have the same attributes and parent/children concepts
b) they are not temporally inconsistent concepts.

inconsistent(x,y) is true if and only if there are at least 2 concepts, one belonging to $O_i(t)$ and the other to $O_{int}(t)$, such that one of the following conditions holds:
a) They both have the same name, the concepts do not have any attribute in common and their respective parent/children concepts (if there were any) have the same attributes.
b) They both have the same attributes, there is no other concept, being a parent of one of them with the same attributes as those of any parent of the other concept. The same property holds for the children.
c) They are *temporally inconsistent*

Where given two (different) concepts c(t) and c'(t), these are said to be *temporally inconsistent* if there is a concept c''(t) which belongs to the same ontology as c(t), whose name is the same as the name of c'(t) and there is a concept c'''(t) which belongs to the same ontology as c'(t), whose name is the same as that of c(t) such that one of the following conditions holds:

a) c(t) is a temporal parent concept of c''(t) and c'''(t) is a temporal parent concept of c'(t).
b) c(t) is a temporal child concept of c''(t) and c'''(t) is a temporal child concept of c'(t)

The next step is to transform the integrated ontology into a new ontology that can be accessed by users in a global way, adapting the terminology that appears in the integration-derived ontology to users' terminology. This process involves to adapt the terminology of each ontology which is part of the integrated ontology to the terminology used by the user who is going to see the result of the integration process. This is possible in case the user has his/her own ontology on that specific topic. Otherwise, the terminology used as a reference will depend on the user's preferences. The process of adapting different terminologies implies the need for a synonym detection and management mechanism. In our approach, synonym concepts are detected by taking into account attribute and structural equivalencies.

Despite the automatic adaptation of the terminology, when the name of a concept is changed in order to adapt the terminology for a certain user, the original term assigned to that concept is kept (as an alternative name) in order to provide the best possible information to the user, so (s)he will be able to decide which term is more appropriate for a concept. Once the terminology has been unified, the ontologies that are members of the integration-derived ontology must be merged in order to provide the user with a single ontology. We have followed the following algorithm for this task:

3.4 Ontological_Transformation

Let $O_{int}(t)$ be the integration-derived ontology; let $O_i(t)$ be the i-th mereological child of $O_{int}(t)$, i=1,...,n; and n= number of mereological children of $O_{int}(t)$.

For i=1 to n
    For each concept c(t) belonging to $O_i(t)$ do
      If there is any concept c'(t) belonging to $O_{int}(t)$ such that
        equivalent_concepts(c(t),c'(t)) or (c(t) and c'(t) have the same name)
      then merge_attributes_and_relationships(c(t), c'(t))
      else link c(t) to its parents in $O_{int}(t)$
    End-for
End-for
where *equivalent_concepts(x,y)* is true if and only if :
a)   both concepts have the same attributes and parent/children ;
b)   they are not temporally inconsistent.

We can see that new knowledge is generated from previously existing knowledge because, as it is supported in [12], it is easier to generate knowledge from different source ontologies (belonging to the users) than generating it from scratch (i.e., starting from having no information at all).

4. A WWW-based Tool for Integrating Knowledge

The WWW can be accessed at anytime and from anywhere. The WWW also facilitates the communication between people from different (geographical) locations and co-operative work can be performed asynchronously, that is, people who are co-operating do not need to work at the same time in order to perform co-operative tasks efficiently. The aim of the designed and implemented application was to provide a system and framework for ontology integration and personalisation that allowed each user to take advantage of the knowledge supplied by other users with the purpose of increasing his/her own knowledge.

4.1 Users

The starting point of the system is a set of users working in an intranet/internet and generating ontologies co-operatively, although this co-operation is totally transparent for each user because they do not know whether their knowledge is shared with other users. Nevertheless, a user is never allowed to modify another user's work (i.e., knowledge), but users receive the global benefits from all the users' (knowledge) contributions represented by ontologies. The system differentiates among three types of users, namely:

- *Normal user*: This is an information consultant, that is, a person who wants to actualise his/her information about a topic at a specific moment.
- *Expert user*: An expert represents a system worker, that is, a person who generates knowledge for the system in such a way that normal users are able to look it up. Every expert user may combine his/her own contribution with that of other experts about the same topic. An expert is also allowed to see other experts' work but (s)he cannot modify those ones' work.
- *Administrator*: This is the figure in charge of keeping the system working correctly. Another responsibility left to the administrator is the management of both the users and the topics that the system is to deal with. The dialogue between the administrator and the users is via email, which is a compulsory information the user has to facilitate through the registration form.

4.2 Knowledge integration

The integration of knowledge in our system is produced when a user, either normal or expert, applies for it. It may happen that at that specific moment there exist some expert users working on the same topic, so the knowledge stored in the server becoming obsolete. That is, it would not contain the latest versions of the users' works or new contributions. We had to make a decision with respect to this situation. There were two main ways of facing this problem: (1) integrating the knowledge stored in the server up to the moment of the request, or (2) actualising the knowledge the server contains. This last solution was adopted. It implies to know what users are working with a specific topic at each instant, that is, we need to keep track of the active users.

Each time that an integration request is made, the system checks a users register in order to know if the server's knowledge needs to be actualised. If this happens, that is, if there is at least one expert user who is working on the topic, the new information must be retrieved by the system. In order to achieve this capability, new elements have been added to our first architecture, so becoming more complex. Finally, the process of knowledge integration is briefly described as follows: Check whether the server is up to date. If it is not, actualise it. Finally, the knowledge integration is the last step.

When the user manages to connect to the URL connected to our system, (s)he will be requested to log into the system. Once this process has been completed, the possibilities for experts and normal users are different. An expert user can actualise the server, either because he/she wishes to do it or due to an automatic server update. Once the knowledge has been integrated, the following step is to personalise the information. At this stage, the user has the chance for redefining the terminology that is assigned to the concepts belonging to the derived ontology. Then, the user will have better information about the topic than the one he/she had before the request was made. Therefore, he/she will be able to decide the terminology more accurately. In addition to this, they are offered the possibility of changing the name that the integration process has given to a concept. The new name could be the one assigned by another user, who must have taken part in the integration process, or a different one that the user thinks to be more appropriate.

4.3 Knowledge visualisation

A significant goal of the tool was to give the best and largest possible information to the users in a comfortable way. This has been the main reason for providing the system with a very flexible visualisation, allowing the users to see what they want at each instant. To be more precise, the following visualisation options are facilitated by the system:

- Complete ontology: This option shows the hierarchy defined by the integrated and personalised ontology.
- Concept exploration: This option allows the user to visualise a specific concept in terms of attributes and relationships with other concepts belonging to the same ontology.
- Expanding taxonomies: This option visualises the existing taxonomies with respect to a specific set of attributes of a concept.

5. A Practical Example

In this section, we show an example of the integration framework we have described across this work. The application domain chosen for this example is the management of knowledge about natural disasters. We have focused our example to a small part of the domain. Let us suppose that, at a specific moment, there are two expert users working in the system on the topic "Natural Disasters". Moreover, let us suppose that they are the only users who have supplied the system with two ontologies (Figure 2 and Figure 3) on that topic.

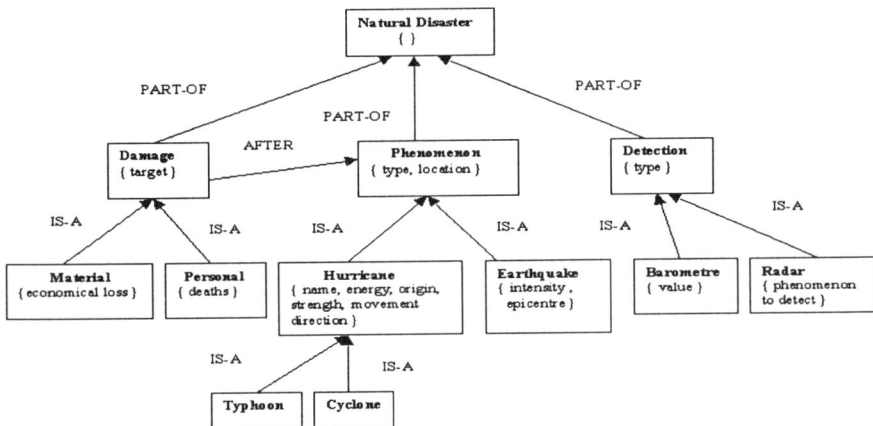

Figure 2: Ontology corresponding to expert 1

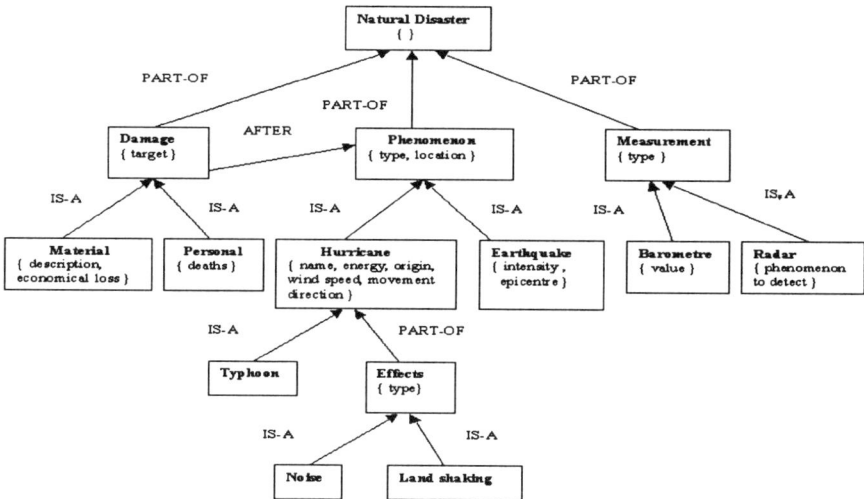

Figure 3: Ontology corresponding to expert 2

Let us suppose now that a normal user logs into the system via the WWW, and (s)he wants to check for the existing knowledge about the topic "Natural Disasters". Once the integration request has been submitted, the integration process is executed as it has been explained in Section 3 and Section 4. We stated in that section that the terminology is adapted to the user's terminology but this occurs only for expert users because we do not know the preferred terminology of a normal user. Therefore, a decision must be made in order to integrate the terminologies of the different ontologies. In this case, we have taken the decision to choose the ontology with the highest number of concepts as the reference ontology for the integration process. The final result of the process is shown in Figure 4, which shows the transformed ontology, and it corresponds to the application of the Ontological_Transformation to the integration-derived ontology. This integration-derived ontology has been achieved by applying the Ontological_Integration algorithm to the two source ontologies shown in Figure 2 and Figure 3, respectively.

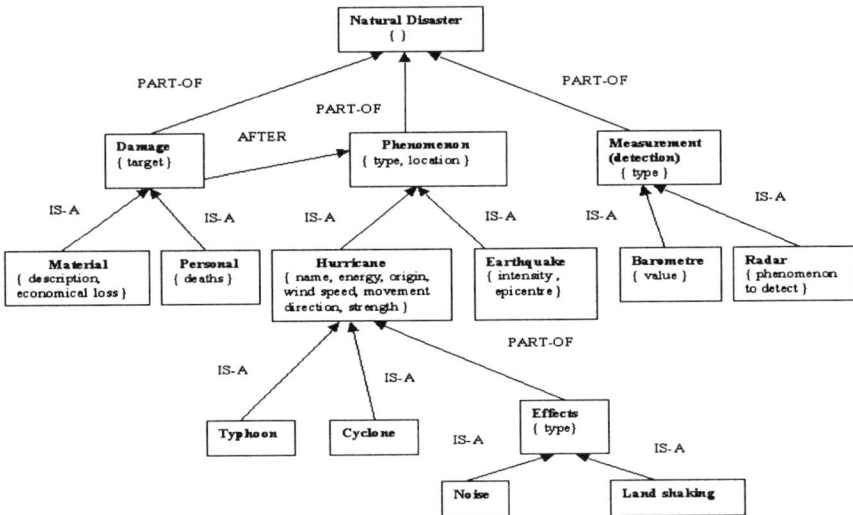

Figure 4: The transformed ontology

In Figure 4, we can see how the problems we mentioned earlier in Section 3 about co-operative work, redundant information and synonym concepts are overcome. We can see that in Figure 2 and Figure 3 there are some concepts which appear in both ontologies, such as "Damage" or "Phenomenon". Both experts are describing the same part of the domain and we must treat it in order to supply the user with non redundant information, so those redundant concepts must be merged into single ones. A particular case is the concept "Hurricane". It appears in both ontologies, but its sets of specific attributes are different in the different experts' ontologies. It is obvious that both descriptions refer to the same concept, so both sets of attributes must be unified into a single set with all the attributes that describe the notion of "Hurricane".

Concerning the detection of synonym terms, we can see how this process works in the example. The ontology of the expert 1 has a concept named "Detection", while the ontology of the second user has a concept named "Measurement". Both concepts are determined by the system to be synonym because their sets of attributes are equivalent (they are the same in this case), and they have the same parents and children concepts. Therefore, the system concludes that both concepts are equivalent and it keeps one of the names for the concept ("Measurement" in this particular case) while the other name ("Detection") is considered an alternative one (it is shown in Figure 4 between brackets). The user will be who decides the final name for the concept since (s)he has the possibility of changing it.

Finally, Figure 5 represents how the transformed ontology would be displayed at the corresponding web page.

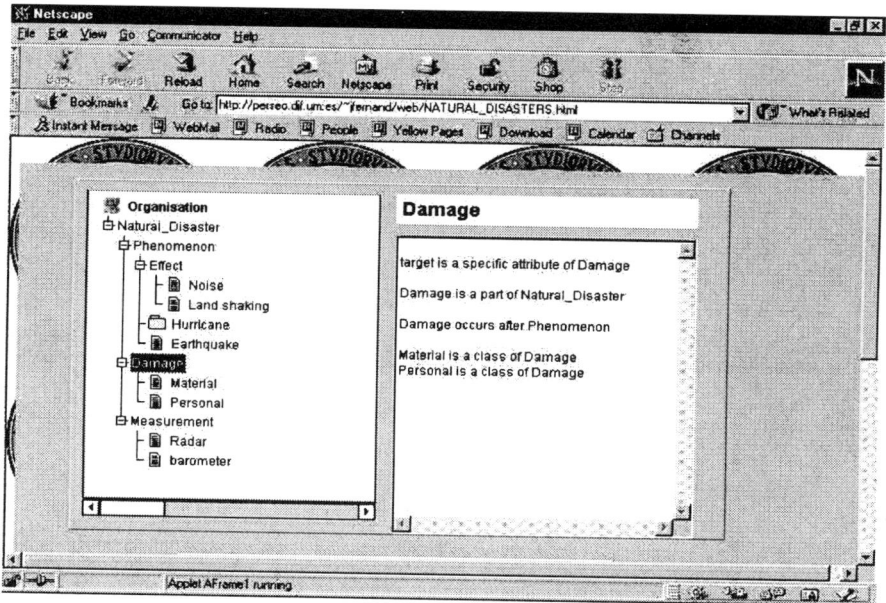

Figure 5: Visualisation of the transformed ontology

6. Conclusions

In this work, we have described a framework through which various (human) agents can build ontologies in a co-operative way. These agents may be at different geographical locations (i.e., WWW sites). The basic operation of the described process is the integration of a set of predefined ontologies. The inclusion of ontologies is related to the reuse of ontologies to create new ones, so reducing the cost of obtaining ontologies. In this work, the selection of the terminology for the integration-derived ontology depends on some parameters, particularly on two. The first one is the consistency of the ontologies that belong to the derived ontology with the last version corresponding to an agent ontology. The second parameter is the amount of knowledge which is contained in an ontology.

In [6], the author has presented a system for collaborative construction of consensual knowledge bases. Such a system is based on the peer-reviewed journals: before some piece of knowledge is introduced in a knowledge base, that piece must be submitted to and accepted by a given community. In order to achieve it, a protocol for submitting knowledge is defined. This consensus principle guarantees the consistency of the introduced knowledge and leads the collaborative dialog among the experts. An important concern underlying this approach is that the community must use the same terminology. In our approach, this restriction is overcome via a mechanism for synonym concepts management that allows each agent to operate with its particular vocabulary. In order to solve the problem of synonym concepts, we use an approach close to that used by [18]. However, it differs in the way that conflicts are detected. In our approach, the system is in charge of finding out which concepts are synonyms and which ones are not. This facility is not included in the other referred approach.

In [8], a different approach for integrating ontologies is presented. There, two concepts are introduced: merging and aligning. Merging is the process by which two ontologies are merged into a single ontology, and both source ontologies have to cover similar or overlapping domains. In our work, every ontology must represent the same domain in order to be eligible for being integrated, although the semantics of both processes is similar. Ontological alignment implies to link concepts from both source ontologies. Besides, the SMART algorithm ([8];[13]) merges concepts whose names match or are linguistically similar (for example, Military-Unit and Modern-Military Unit). However, internal and structural properties are not considered. In that system, the user always take the decision about what concepts are synonyms (and therefore merged) because SMART only intends to support the merging process, while our system performs automatic ontological integration. In the future, we want to increase the participation of users in the integration process by providing an interactive integration process as well as facilities for defining user's preferences.

Another feature of our approach is that through the co-operative dialog provided by the framework, the agents may check the current state of the co-operatively built ontology at a given instant in order to retrieve the maximum possible information to build their particular ontology. The management of synonym concepts adapts the information requested by the agent to his/her own terminology.

Nowadays, the framework only admits structural axioms, that is, axioms that are directly derived from the organisational structure of the ontology. Some examples are the axioms that are derived from taxonomic, mereological or temporal relationships, or those stating that a concept possesses a certain attribute. In the future, it is intended to extend the framework to admit other kinds of axioms, such as those connecting attributes from one or more concepts (independently of the application domain). Regarding attributes management, it is intended to use attributes with structures (attribute, value) instead of considering attributes uniquely. In that manner, additional types of axioms could be established.

Another future goal is to increase the number of possible relationships among concepts in ontologies (in addition to taxonomic, mereological and temporal ones). In this manner, we expect to improve the performance of the system (in terms of augmenting the kinds of potential situations reflected in ontologies) in the context of a Knowledge Management-oriented framework.

References

[1] Abecker, A. Aitken, S., Schmalhofer, F. et al (1998). KARATEKIT: Tools for the Knowledge-Creating Company. *Proceedings of the 11th Banff Knowledge Acquisition for Knowledge Based Systems Workshop. Banff, Canada, vol. 2, KM-1.1-KM-1.18.*

[2] Benjamins, V.R., Fensel, D. (1998). Community is Knowledge! In $(KA)^2$. *Proceedings of the 11th Banff Knowledge Acquisition for Knowledge Based Systems Workshop. Banff, Canada, vol. 2, KM-2.1-KM-2.18.*

[3] Borst, W.N. (1997). *Construction of Engineering Ontologies for Knowledge Sharing and Reuse.* PhD Thesis. University of Twente. Enschede, The Netherlands.

[4] Cardenas, M.A. (1998). *A constraint-based logic model for representing and managing temporal information* (in Spanish), PhD Thesis, University of Murcia, Spain.

[5] Eschenbach, C., and Heydrich, W. (1995). Classical mereology and restricted domains, *International Journal of Human-Computer Studies, 43: 723-740.*

[6] Euzenat, J. (1996). Corporate Memory through Cooperative Creation of Knowledge Based Systems and Hyper-Documents. *Proceedings of Knowledge Acquisition Workshop'96, p. 36-1 -36 -20, Banff, Canada.*

[7] Farquhar, A., Fikes, R. and Rice, J. (1997). The Ontolingua Server: a tool for collaborative ontology construction. *International Journal. Human-Computer Studies 46, 707-727.*

[8] Fridman-Noy, N., Musen, M.A. (1999). An Algorithm for Merging and Aligning Ontologies: Automation and Tool Support. *Proceedings of the 16th National Conference on Artificial Intelligence , Workshop on Ontology Management, Orlando, FL.*

[9] Gómez-Pérez, A., Benjamins, V.R. (1999). Overview of knowledge sharing and reuse components: ontologies and problem-solving methods. *In V.R. Benjamins, B.Chandrasekaran, A.Gómez-Pérez, N.Guarino and M.Uschold (Eds), Proceedings of the IJCAI-99 workshop on Ontologies and Problem-Solving Methods, Stockholm, Sweden.*

[10] Gruber, T.R. (1993). A translation approach to portable ontology specifications. *Knowledge Acquisition, 5:199-220, 1993*

[11] Martínez-Béjar, R., Benjamins, V.R. and Martín-Rubio, F. (1997). Designing Operators for Constructing Domain Knowledge Ontologies. *In E. Plaza and R.Benjamins (Eds.), Knowledge Acquisition, Modelling and Management, Lecture Notes in Artificial Intelligence 159-173, Springer-Verlag, Germany.*

[12] Musen, Mark A (1997). *Domain Ontologies in Software Engineering: Use of Protegé with the EON Architecture.* Report.

[13] Musen, M.A. , Fridman-Noy, N. (1999). SMART: Automated Support for Ontology Merging and Alignment. *Proceedings of the 12th Banff Workshop on Knowledge Acquisition, Modelling, and Management, Banff, Alberta, Canada.*

[14] Reimer, U. (1998). Knowledge Integration for Building Organizational Memories. *Proceedings of the 11th Banff Knowledge Acquisition for Knowledge Based Systems Workshop. Banff, Alberta, Canada, vol 2, KM-6.1-KM-6.20.*

[15] Shaw, M. L. G., Gaines, B. R. (1989). A Methodology for recognising conflict, correspondence, consensus and contrast in a Knowledge Acquisition System. *Knowledge Acquisition, 1(4):341-363.*

[16] Studer, R., Benjamins, V.R., Fensel, D. (1998). Knowledge Engineering, principles and methods. *Data and Knowledge Engineering, 25 (1-2):161-197, 1998.*

[17] Van Heijst, G., Schreiber, A. T., and Wielinga, B. J. (1997). Using explicit ontologies in KBS development, *International Journal of Human-Computer Studies, 45: 183-292.*

[18] Wiederhold, G. (1994). Interoperation, Mediation and Ontologies. *Proceedings. of 5th Generation Computer Systems'94 Workshop on Heterogeneous Cooperative Knowledge-Bases, Tokyo, Japan, p.33-48.*

Acknowledgements

This work has been possible thanks to the financial support of the Fundación Séneca, Centro de Coordinación para la Investigación through the program Séneca.

Redesign Knowledge Analysis, Representation and Reuse

Inés Arana[1], Hatem Ahriz[1] and Pat Fothergill[2]

[1]School of Computer and Mathematical Sciences, The Robert Gordon University, Aberdeen AB25 1HG, UK
[2]Department of Computing Science, Aberdeen University, Aberdeen AB24 3UE, UK

Abstract: DEKLARE is a framework which enables the elicitation, representation and reuse of redesign knowledge. In this paper we present MAKUR, an extension to DEKLARE which allows the capture and management of more redesign knowledge, thus contributing towards a better and faster redesign of families of products.

1. Introduction

Design is a very important stage in the manufacturing of a product. Redesign is a type of design where the aim is to produce a variant of a known product family in order to satisfy a slightly different specification. For example, a company may manufacture gas-taps for a variety of purposes. Their designs may all be very similar, but each of them will have some purpose-specific characteristics. If a new gas-tap design is needed, the designer may simply modify an existing design in order to satisfy the new requirements. Redesign is a fairly routine task to which designers devote much of their time.

DEKLARE (ESPRIT CIME project 6522) [1] is a methodology which supports engineering redesign. It consists of three tools:

- Design Analysis Methodology [2]: a "paper and pencil" tool which elicits a company's expertise in the redesign of a family of products. It captures not only the physical structure of the product and its functions but also the process by which a new design is created.
- Design Description Language [3]: a formal specification language that is used to represent the knowledge acquired through the Design Analysis Methodology, i.e. a way of structuring the knowledge so that it can be accessed both by the company and by specific software systems.
- Design Advisory System [4]: an interactive problem-solving tool, which provides particular design solutions for a product, given a set of requirements. It is a generic tool, and is customised by the use of a specific product model, which is described using the Design Description Language.

This paper reports on the MAKUR project, which extends DEKLARE in order to make use of more redesign knowledge.

2. Design Knowledge Acquisition

The Design Analysis Methodology was developed within DEKLARE in order to support the analysis of a product family from three viewpoints: physical, functional and process. Although alternative methodologies were investigated [5], these were not suitable for DEKLARE's purposes, because they either had little adaptation to design problems, were too costly or did not acquire all the (functional, physical or process) knowledge required. For example, CommonKADS [6] is a framework for the construction of knowledge based systems (and, therefore, support systems). However, it provides little adaptation for redesign and its application is expensive and time consuming.

DEKLARE's Design Analysis Methodology consists of a series of tasks which are carried out sequentially, as illustrated in Figure 1.

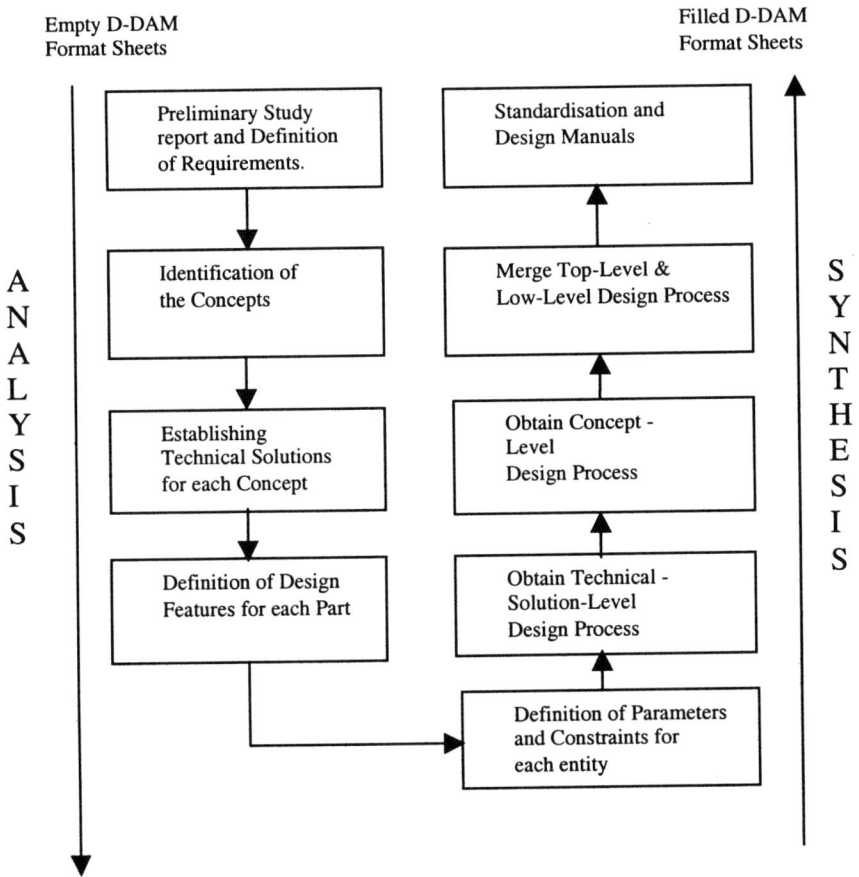

Figure 1: Major steps in applying DEKLARE's Design Analysis.

The result of applying this methodology is a set of format sheets which contain the company's "know how" for the redesign of a particular product family. The knowledge elicitation process relies on a variety of sources such as the designer, documentation regarding past designs and current regulations.

DEKLARE's Design Analysis first studies the design of a product family in a top-down fashion, in order to elicit background knowledge and a general understanding of the product. The application of this phase results in a set of design entities (e.g. features, concepts, parts, etc). Next, the relationships between these entities are discovered using a bottom-up approach. The result of analysing a design is a set of format sheets that encode a product's design knowledge. The following benefits have been reported by industrialists:

- It encourages standardisation and normalisation of design alternatives - i.e. functionally invariant solutions are eliminated. This results in a better understanding of the design.

- The design process is documented and can, therefore, be accessed by the company.

- Stock may be reduced, since some design alternatives are no longer considered.

- Machine set up time may decrease due to a reduction in the number of types of parts which are needed.

DEKLARE's Design Analysis Methodology suffers from the following problems: the type-subtype knowledge that it elicits is poor and knowledge about default values and design drivers is not captured. In the MAKUR project, we have extended DEKLARE's Design Analysis Methodology in order to ensure that more knowledge is elicited and captured in the format sheets. The MAKUR Design Analysis Methodology (MADAM) is illustrated in Figure 2.

Although some type-subtype knowledge was acquired using DEKLARE's analysis methodology, this was very limited: a subtype was seen as a property of its supertype instead of as a subclass. For example, in the gas-tap testcase used by DEKLARE, the body (part of the gas-tap) may be of different types. One of these subtypes is g_17500. In DEKLARE, g_17500 is an attribute of the body. This has two disadvantages:

- Information about each subtype is not acquired explicitly.

- This methodology does not allow for the acquisition of type-subtype knowledge where the hierarchy has three or more levels (e.g. type/subtype/subsubtype).

We have modified DEKLARE's Analysis Methodology to encourage the capturing of type-subtype knowledge, since we have observed that this kind of knowledge is used by designers. This has meant changing both the elicitation process and the format sheets where the additional knowledge is to be written. Hence, there are two types of hierarchies:

- Type/subtype hierarchy: For example, "g_17500" is a type of "gas-tap body".

- Part/subpart hierarchy: This is a decomposition of the article into assemblies, parts and design features.

The use of such typing enables the inheritance of parameters and constraints from types, thus avoiding redundancy, facilitating reuse and ensuring consistency.

Moreover, it eases the acquisition of knowledge for future applications where typing is important.

Designers often use default parameter values wherever possible, since this encourages standardisation and usually these values are best (according to some design criterion such as ease of manufacture). The elicitation of default values has, therefore, been incorporated into MADAM.

In the design of some products, designers may use a driver (i.e. a property which aids in the selection of the best option e.g., cost, ease of installation, manufacturing time, etc.) in order to select an alternative. Moreover, in some applications, designers have "prioritised" technical solutions, i.e. potentially suitable alternatives are investigated according to the level of desirability which the designer has assigned to each of them.

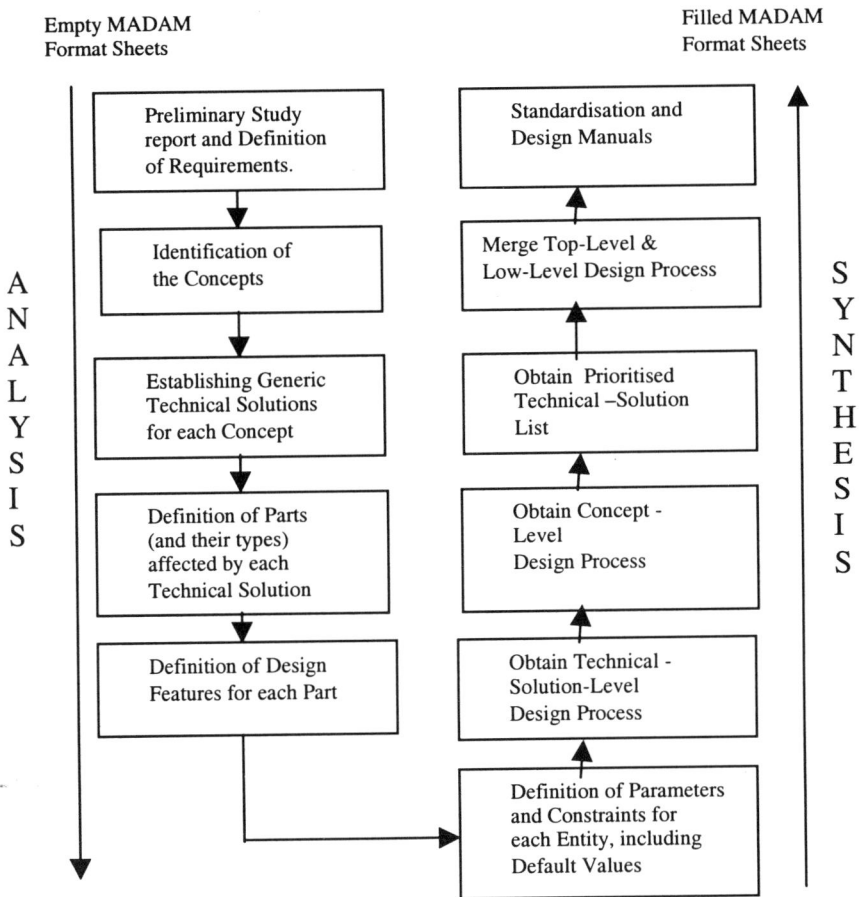

Empty MADAM
Format Sheets

Filled MADAM
Format Sheets

A
N
A
L
Y
S
I
S

| Preliminary Study report and Definition of Requirements. |
| Identification of the Concepts |
| Establishing Generic Technical Solutions for each Concept |
| Definition of Parts (and their types) affected by each Technical Solution |
| Definition of Design Features for each Part |

| Standardisation and Design Manuals |
| Merge Top-Level & Low-Level Design Process |
| Obtain Prioritised Technical –Solution List |
| Obtain Concept - Level Design Process |
| Obtain Technical - Solution-Level Design Process |
| Definition of Parameters and Constraints for each Entity, including Default Values |

S
Y
N
T
H
E
S
I
S

Figure 2: Major steps in applying the MADAM.

For example, in order to provide some cushioning effect for the gas-tap operation, it may be better (cheaper) to use an aluminium spindle and a spring rather than a brass spindle with a slot cut into it (which provides a spring effect). The first option (aluminium spindle and spring) is, therefore, investigated first and only if this option is not appropriate (i.e. some constraint is violated) the second alternative is tried. We have added the ability to capture design drivers (technical solution priorities) into MADAM. It should be noted that the knowledge about design drivers which it obtains is restricted, since it does not elicit all the knowledge required in order to calculate optimal solutions.

3. Representing Redesign Knowledge

The knowledge elicited through DEKLARE's Design Analysis Methodology needs to be expressed in electronic form before it can be used by a design advisory system. DEKLARE provides a Design Description Language for the representation of redesign knowledge. Although existing representation languages were investigated, these were unsuitable for DEKLARE's purposes. For example, STEP [7] was considered and rejected because its Process Specification Language (PSL) [8] is still being defined; OMT [9] considers a product from three (physical, functional and process) viewpoints, but it has different semantics than DEKLARE; the expressibility of functional and physical knowledge in DSPL [10] was insufficient for DEKLARE's purposes.

The DEKLARE Design Description Language enables the representation of three hierarchical models:

- Physical model: represents the product in terms of assemblies, the parts they are composed of and design features in each part.
- Functional model: describes the various functions that a product satisfies in terms of concepts and the technical solutions which implement a concept. The design features used in a technical solution relate this model with the physical one.
- Process model: describes the tasks which need to be carried out in order to obtain a design and the methods which can be used in order to perform a task. Tasks and methods are related to entities in the functional/physical model.

A model encoded in the Design Description Language is described in terms of a physical, a functional and a process model. All the model entities have parameters (represented by their type and their domain) which describe them. Constraints are used to express the relationships between parameters (and, therefore, between the entities to which they are associated). Hence, constraints are specified in order to integrate the three models.

The Design Description Language has been modified in order to reflect the changes introduced in MADAM. Hence, its data model has been extended, thus allowing for two types of hierarchies: abstraction (type-subtype) and detail (component-subcomponent). The MAkur Design Description language (MADD) is illustrated in Figure 3.

The changes introduced in MADD have led to the following classification of constraints:

- Level of desirability:
  - Soft constraints (constraint 2 in Table 1): describe relations which are desirable but not mandatory, i.e. they do not need to be satisfied for a solution to be valid. They may be used to indicate default values for parameters.
  - Hard constraints (constraints 1, 3 & 4 in Table 1): these must be satisfied for a solution to be valid.
- Scope:
  - Global constraints (constraints 2, & 4 in Table 1): generic constraints on a type apply to all its sub-types, since they are inherited.
  - Local constraints (constraints 1 & 3 in Table 1): only apply to a particular entity.

The set of format sheets obtained from the application of MADAM is encoded into MADD using the MAkur Graphic Editor (MAGE), a tool which not only facilitates the input of design knowledge but also checks the (physical, functional and process) models for consistency.

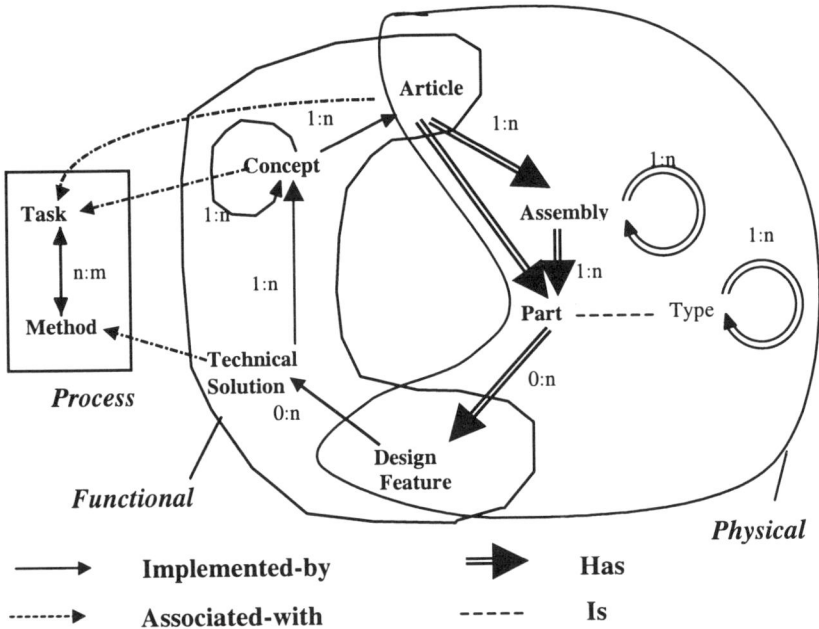

Figure 3: MAKUR Design Description Language.

4. Redesign Knowledge Reuse

Models written using MADD can be used by a computerised system to aid redesign. A generic design advisory system MADAS (MAkur Design Advisory System) is being developed in order to aid the design of families of products. A product-specific support system can be built by linking a particular product model (written in MADD) to MADAS.

The core of MADAS is a constraint engine, which ensures that the design is valid (i.e. its constraints are satisfied) at all times. As the designer makes decisions, their effects are propagated throughout the model, ensuring that the (partial) design is consistent. Hence, if the designer makes an unsatisfactory decision, MADAS detects the mistake and, therefore, prevents the designer from producing an incorrect design.

Table 1: An example of the use of constraints.

Part: SP	
Description: A type of spindle	
Type: Spindle	**Sub-Types:** SP1, SP2, SP3
Design Features: {...}	**Parameters**: {Diameter, Length, ...}
Constraints: 1. Diameter < 0.7 * GSH.Diameter 2. Length = 20 (*default*) (*inherited*) 3. SP **requires** MP 4. Weight <= 100 gr. (*inherited*) ...	**Sketch**:

MADAS can be integrated in the designer's current environment, and, therefore, be used together with other tools that the designer is familiar with, such as the CAD system, and the database.

5. Conclusions

Redesign is an area where advisory tools can help reduce the design time and produce better solutions. The quality of an advisory system is highly dependent on the knowledge it uses. Hence, in order to provide good redesign support systems, it is important to elicit and represent all the knowledge that designers use in this activity. The MAKUR project extends DEKLARE by obtaining and representing more redesign knowledge in order to provide (or improve on) the following benefits:

- Reduced variance (standardisation) in the various arrangements (technical solutions) which can provide the right functionality in a particular situation.
- Prioritisation of these solutions by considering drivers criteria (e.g., cost, ease of installation, etc.)
- No inconsistent solutions.

We believe that the MAKUR methodology can contribute towards better and faster redesign of product families and that it can be integrated in the designer's environment.

Acknowledgements

The MAKUR project is funded by the School of Computer and Mathematical Sciences, The Robert Gordon University, Aberdeen, UK.

We are grateful to Chris Mowatt, Angus Murray and David Whiteside (AMEC Process and Energy) for their useful input into this project.

References

[1] Forster, J., Fothergill, P., Lacunza, J.A., Arana, I., 1995, *DEKLARE: a methodological approach to re-design*, Proceedings of Conference on Integration in Manufacturing, Vienna, pp. 109-122.

[2] Forster, J., Fothergill, P., Lacunza, J.A., Arana, I., 1995, *DEKLARE: Knowledge Acquisition and Support System for Re-Design*, Proceedings of Expert Systems, The Fifteenth Annual Technical Conference of the British Computer Society Specialist Group on Expert Systems, Cambridge, England, pp. 23-40.

[3] Forster, J., Arana, I., Fothergill, P., 1996, *Re-Design Knowledge Representation with DEKLARE*, Proceedings KEML'96 – The Sixth Workshop on Knowledge Engineering: Methods and Languages, Laboratoire de Recherche Informatique, Intelligence Artificialle et Systèmes d'Inférences, Paris.

[4] Fothergill, P., Arana, I., Forster, J., 1996, *Constraint Management and Design Models in Supporting Re-Design*, Proceedings of the Sixth International Conference on Flexible Automation and Intelligent Manufacturing, Begel House Inc., pp. 363-372.

[5] DEKLARE Project Consortium, 1993, *Survey of Design Analysis Methodologies*, ESPRIT Project 6522, Report R1.1.

[6] Schreiber, G., Wielinga, B., de Hoog, R., Akkermans, H. et al, 1994, *CommonKADS: a Comprehensive Methodology for KBS Development*, IEEE Expert 9, 6, 28-37.

[7] Owen, J., 1993, *STEP: an Introduction*, Information Geometers Ltd.

[8] Knutilla, A., Schlenoff, C., Ray, S., et al, 1998, NISTIR 6160, National Institute of Standards and Technology, Gaithersburgh, MD.

[9] Rumbaugh, J., Blaha, M., Premerlani, W., et al., 1991, *Object Oriented Modelling and Design*, Prentice-Hall.

[10] Brown, D. C., Chandrasekaran, B., 1989, *Design Problem Solving: Knowledge Structures and Control Strategies*, Morgan Kaufmann.

Project Memory in Design

Nada Matta[1], Myriam Ribière[2], Olivier Corby[3], Myriam Lewkowicz[1] and Manuel Zacklad[1]

[1]   UTT/Tech-CICO, 12, rue Marie Curie BP. 2060, 10010 Troyes Cedex France, email: nada.matta, myriam.lewkowicz, manuel.zacklad} @univ-troyes.fr
[2]   SRI INternational, AI-Center - EJ213, 333 Ravenswood Avenue, Menlo Park, California 94025 USA, email: ribiere@ai.sri.com
[3]   INRIA-ACACIA, 2004 route des Lucioles, BP.93, 06902 Sophia-Antipolis cedex France, email: Olivier.Corby@sophia.inria.fr

Abstract: Learning from past projects allows designers to avoid past errors and solve problems. A number of methods define techniques to memorize lessons and experiences from projects. We present in this chapter an overview of these methods by emphasizing their main contributions and their critical points.

1   Introduction

Knowledge management, first considered as a scientist stake, becomes more and more an industrial stake. It is a complex problem that can be tackled from several viewpoints : socio-organizational, financial and economical, technical, human and legal [5]. It concerns theoretical and practical know-how of groups of people in an organization. A number of researchers studied knowledge management and defined techniques in order to build corporate memories Some researchers have studied knowledge management and consider organizational memory as an "explicit, disembodied, persistent representation of knowledge and information in an organization" [20]. They have defined techniques in order to build corporate memories. Some of those techniques provide methods in order to capitalize past experience. Let-us note for example, REX [10], MKSM [6], etc. These methods provide guidelines to interview experts, to extract, analyze and model knowledge from documents, etc.

Other studies focus on how to keep track of an activity and especially a project. In this type of studies, the challenge is how to capitalize knowledge without perturbing actors' activities and workspace. Main questions can then arise, as : how to directly extract knowledge directly from tools and documents ? How to keep track of the realization and the evolution of a project ? How to quickly model this knowledge and represent it in a way that can be easily accessible and usable by enterprise actors. Several methods are defined in order to help this type of traceability.

In this chapter, we study this second type of knowledge management. So, we focus on knowledge management of a design project in order to define, what we call, design project memory (PM).

First, we start by defining what is a project memory in design (Part 2), and then we present a number of knowledge capitalization approaches (Part 3). Finally, we discuss some guidelines that help to build a design project memory (Part 4).

2   What is a Project Memory ?

A project memory can be defined as "lessons and experiences from given projects" [15] or as "project definition, activities, history and results" [19]. So, we consider as crucial knowledge in project memory knowledge used and produced during the project realization. In this chapter, we analyze project memory in design. So our definition and study concerns only design projects.

Nowadays a design project is realized thanks to the contribution of several actors from different disciplines and belonging to one or several organizations. Once the project carried out, this project organization is dissolved, thus we called this sort of project organization "a virtual organization". Like any organization memory, a project memory must so consider :

- The project goal and context.
- The project organization : teams, participants, tasks, etc.
- References : rules, methods, directives, …
- The project realization : problem solving, solution, …

Considering these elements, we organize [12] a   project memory in two main parts:

a) Project characteristic memory

- Context : main objectives, environment, rules, instructions, etc.
- Organization : participants and their organization, task definition and distribution, planning, etc.
- Results : documents, prototypes, tests, etc.

b) Project design rationale memory

- Problem definition : type, description.
- Problem solving : participants, methods used and potential choices.
- Solution evaluation : rejected solutions and arguments, advantages and disadvantages.
- Decision : solution and arguments, advantages and disadvantages.

Project objectives, environment, rules and results are generally described in textual documents besides planning, processes. There is a number of tools as LEXTER [1], FX-Nomino [14], that allow to index textual documents. These tools are based on a terminological and linguistic analysis. They help to define specific lexicons.

Database management tools are generally used in order to represent formal documents such as planning, processes, etc.

Generally, in a project, this type of knowledge is memorized. But discussions, alternative choices, problem solving are fleeting knowledge in a project. The challenge is now to define methods and tools in order to represent the rationale of a project and to memorize it. Several methods are defined for this aim. We describe some of them in the following section.

3   Project Memory Definition Approaches

Several methods offer techniques to capitalize different parts of a project memory, like design rationale, project management and products.

3.1   Capitalizing design rationale

As we noted above, design rationale is generally a fleeting knowledge. Several methods have studied how to capitalize problem solving knowledge by emphasizing the problem treated, potential solving choices and arguments. We note for example IBIS, "Question, Options and criteria" (QOC), DRAMA and DRCS. Let us describe how design knowledge can be represented using these methods.

3.1.1   IBIS

IBIS [4] has been defined in order to help the management of complex problem solving. It proposes to structure problem solving on "Issue, Positions and Arguments" (0). This type of structure can be used to represent a project design rationale. The method has been qualified as process oriented approach [3], because of its narrative aspect. In fact, IBIS can be used as historical memory of design rationale. The gIBIS tool is defined as method support. It allows to represent trees of Issue/Positions/Arguments, in which objections and support arguments are illustrated by "+ "and "– "(0). Note also that unchosen positions are shown with the mark "?".

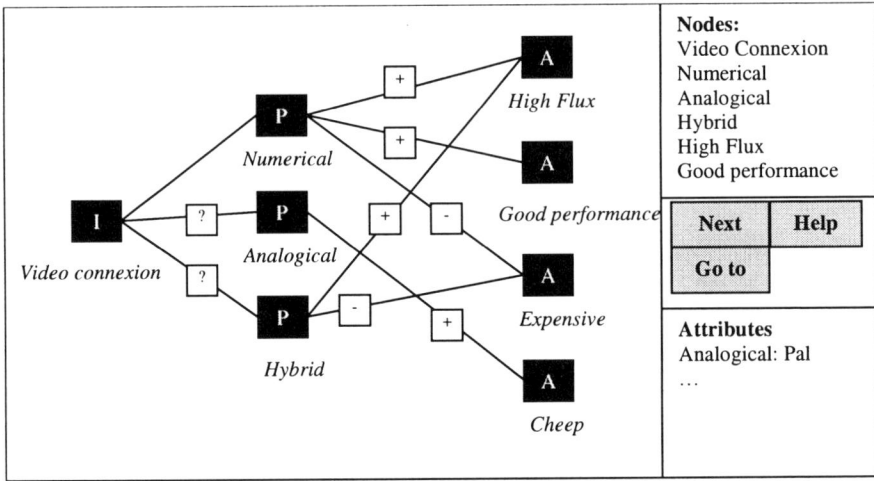

Figure 1: gIBIS editor

3.1.2 QOC

In QOC [9], design rationale is structured in Questions, Options and Criteria (0). This representation allows to characterize arguments by criteria and thus emphasizes influences in decision. So, QOC has been qualified as decision oriented approach [3]. Questions, options and criteria are organized as a decision tree. A description of arguments characterized by criteria can be also represented in the tree. The decomposition of an option into sub-options is shown as links between the decision tree and the decomposition one.

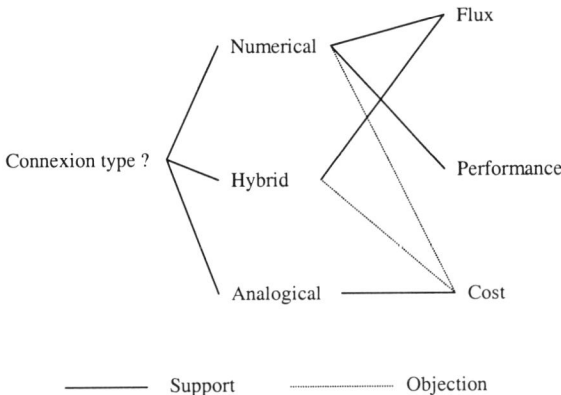

Figure 2: QOC representation: Question/Option and Criteria tree

In these two types of representation (IBIS or QOC), the identification of questions or issues are not obvious. Secretary has to distinguish these elements from discussions and meetings. But, the representation of options (positions) and their argumentation (arguments or criteria) allows to distinguish several choices given in order to solve a problem. They also emphasizes advantages and disadvantages (corresponding to the given problem) of the different solutions. Note also, that QOC can be in order to represent a chronological order and, hence, keep track of the design process.

3.1.3   Drama

In the same philosophy, we note the Drama System [2]. It allows to describe a problem solving as goals and options (0). A decision table summarizes criteria and choices (0) Hypertext links have been defined in the system. Data can also be extracted directly from databases and represented in different ways.

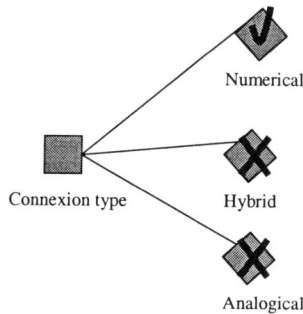

Figure 3: DRAMA's Goal/options representation

Criteria	Option 1	Option 2	Option 3
Performance	1	3	2
Flux	1	3	2
Cost	3	1	2
Installation	4	1	1
Score	0	-1	-2
Rate	3	1	2
Decision	Rejected	Accepted	Rejected

Figure 4: Drama's Decision table

3.1.4   DRCS

In DRCS [7], three models are defined in order to represent design rationale : "Intent, Version and Argumentation" models. The "Intent" model shows the question related to a given problem and the solving strategies. The "Version" model represents several options as different versions of a problem solution. Finally, the "Argumentation" model (0) emphasizes arguments that support or deny a "recommendation". These models are represented as a semantic network in which links show the roles of elements.

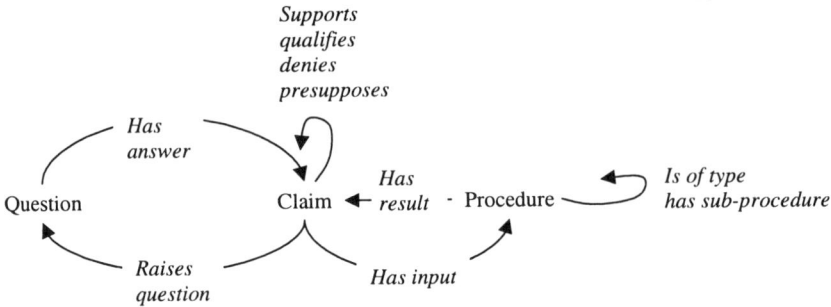

Figure 5: DRCS' Argumentation model

3.1.5   DIPA

The aim of the DIPA [8] model (from the French words "Données, Interprétations, Propositions, Accord", meaning "facts, interpretations, propositions, agreement") is to approach the cognitive dimension of reasoning. Design Rationale models are then enriched with problem solving method concepts from knowledge engineering. This link with problem solving methods seems to us a natural evolution in our researches of more realistic Design Rationale models suited to the complexity of real projects. The DIPA model has itself two declinations according to the situations that lead the actors to give priority to either analysis or synthesis processes.

In DIPA, each argument of an actor in a meeting is categorized by its role in the problem solving method. The model comes in two forms, according to the kind of process that it assists (analysis or synthesis) (Table 1). Actually, we could say that Design Rationale models have neglected the "information" phase of Simon's decision making process [18], and have only taken into account the solutions selection phase. Models from Artificial Intelligence do not have this weakness.

Table 1. Implementation of DIPA model for synthesis and analysis activities

DIPA	DIPA synthesis	DIPA analysis
Problem	Goal	Malfunction
Fact	Requirement	Symptom
Interpretation	Functionality	Cause
Abstract constraint	Constraint	Constraint
Proposition	Means	Corrective action
Concrete constraint	Constraint	Constraint
Agreement	Choice	Choice

In the DIPA model (**Error! Reference source not found.**), the reasoning progresses in three major steps:

- a problem description step plus collecting of data, considered as symptoms in analysis situations and as needs in synthesis situations;
- an abstraction step going from the collecting of problem data to their interpretation corresponding to a possible cause in analysis situations, and to a functionality in synthesis situations;
- an implementation step that going from an interpretation (cause or functionality) to the elaboration of a proposition that is a corrective action removing the symptom's cause (analysis) or the means suitable for the expressed functionality (synthesis).

We implemented the DIPA model to build the MEMO-Net GroupWare. This system consists of two modules, one for synthesis phases (named "design" in the interface), and the other for analysis phases (named "diagnosis" in the interface). Its goal is to allow a project team to solve problems met during design by alternating the two types of activity on a cooperative way. The exchange structure allows both to guide the solution process and to organize the arguments, particularly in argument capitalization aspects.

In the diagnosis module, members of the project team identify a dysfunction and evoke symptoms, causes or corrective actions. In design, once the goal is known, the actors evoke requirements, functionality and means.

Contributions are classified chronologically or according to DIPA model

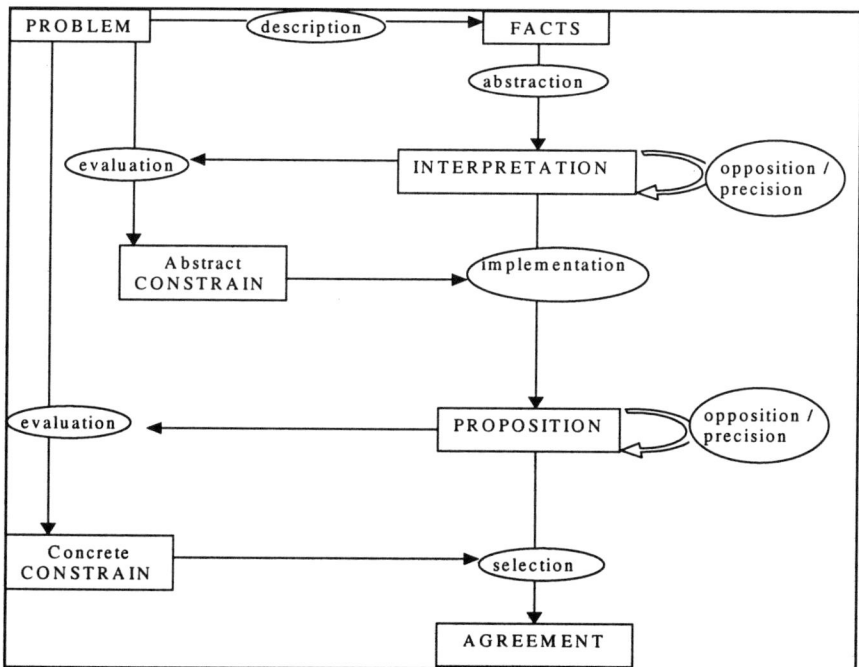

categories, or to the authors' names, their roles, or their department.

Figure 6: DIPA, a heuristic model of design reasoning for analysis and synthesis

The description of design rationale, as recommended in DRCS and DIPA is richer that the one defined in IBIS, QOC and DRAMA. But, actors in a design project have to learn another formalism of representation (similar to semantic network) which is not obvious, by comparison of the tree representation of QOC and IBIS.

3.2 Representing Project Management

A number of methods represent project management as task planing. For example, EMMA [11] recommends to present project management as a tree of goals and plans (0). Each goal is described by a name, the corresponding solution, resources and hypotheses of the solution. Potential plans (that allow to achieve the goal), collaborators (that provide and execute plans) and a description of modifications done (that illustrates the evolution of the goal) are associated to a goal. A plan is defined by its elaboration (a decomposition of sub-goals), the collaborators and the evolution of the plan. The plan chosen as a solution to reach the corresponding goal is called active plan.

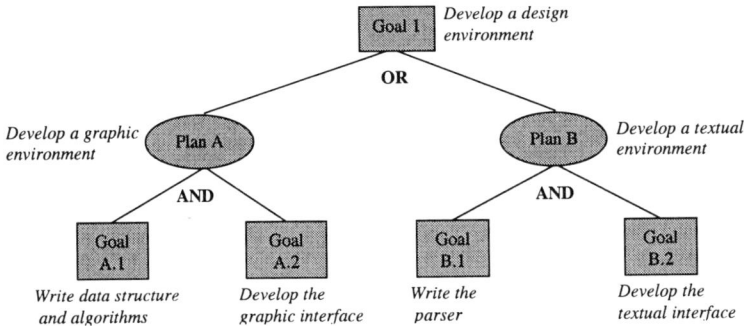

Figure 7: EMMA Representation: Goal/Plan tree

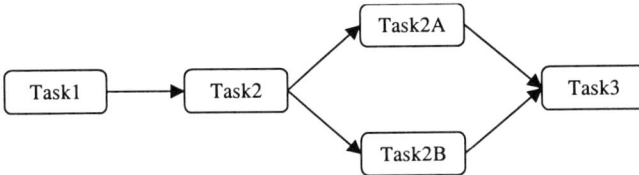

Figure 8: Planning of tasks represented with DRCS.

DRCS [7] represents the sequencing of tasks as shown in 0. The association of modules, actions and constraints to a task is presented as "synthetic model" (0).

has priority
has greater priority than
has subtask
has temporal relationship
is of type

Assertion ← *Has action* ——— Task ← *Has plan* ——— Module

Has attribute

Attribute — *Has value* → Constraint

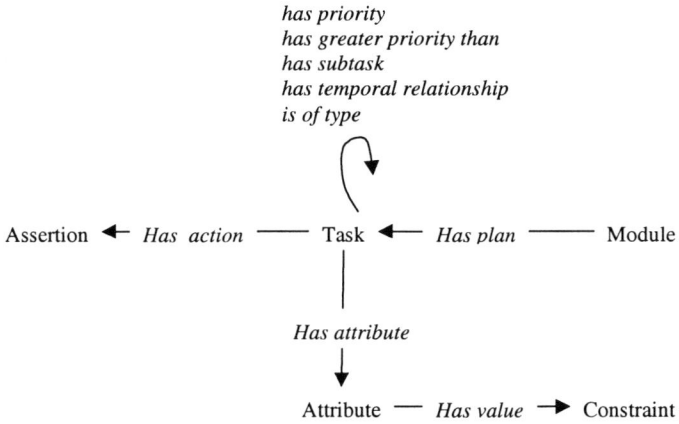

Figure 9: The synthetic model

After analyzing these types of approaches, we can note that a number of elements in the representation of a project, its management and its evolution must be emphasized: task planning, actors, links to solutions, resources, constraints and task. Representations, already used in design, as action tables, grant graphs, etc. can be used and be enriched by links to documents and other types of representations.

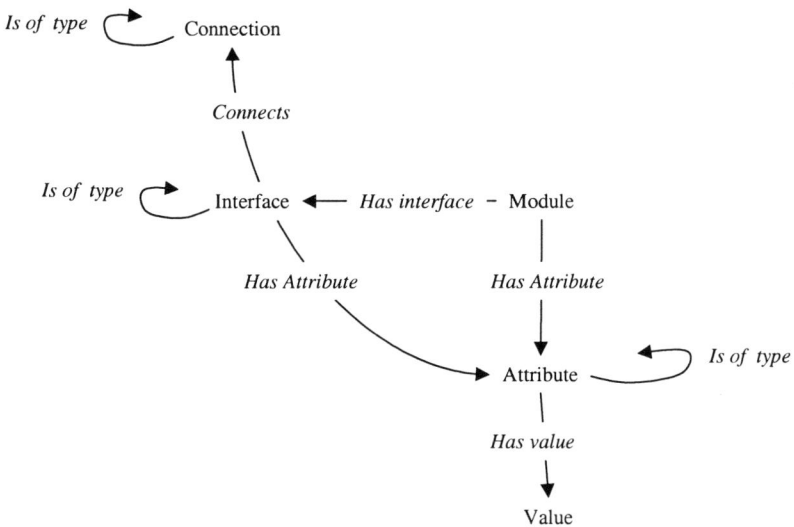

Is of type ⤴ Connection

Connects

Is of type ⤴ Interface ← *Has interface* – Module

Has Attribute   *Has Attribute*

Attribute ⤴ *Is of type*

Has value

Value

Figure 10: Representing a module with DRCS.

3.3   Representing the Artifact

DRCS [7] recommends to represent the artifact in modules. A number of links such as "Specializes, is-connected, decomposes, etc." allow to represent relations between modules. Each module is described by a number of attributes and an interface (0).

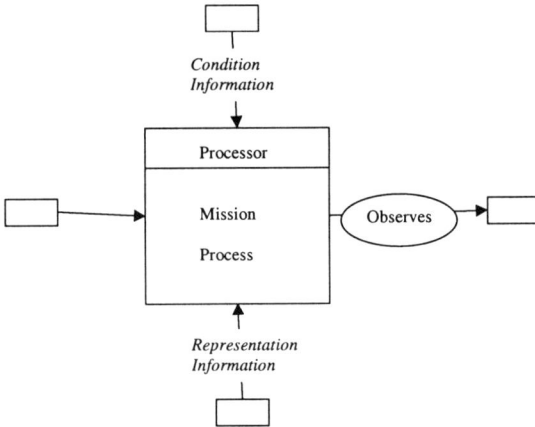

Figure 11: Elements used in SAGACE to represent a vision

Another approach, SAGACE [13], provides a structure in order to represent an artifact in three visions: "function", "organic" and "strategic". Three elements are defined in order to represent these three visions: "processor", "flow" and "observer" (0). These elements can be saved as a technical reference.

We propose also another kind of description using the *viewpoint notion*. This notion has been defined in [17] to represent knowledge from multiple experts. The description of an artifact is composed of the description of each component object and the relation between them. Those descriptions are providing by different participants from different fields. So they propose in [16] to index the description of each component object by a viewpoint. A viewpoint in design is defined as "a perspective of interest from which a participant can see the artifact". This viewpoint has two dimensions (0):
- A contextual dimension where the work context is described, in other words the focus of the participant during the object description,
- A personal dimension where the characteristics of the person describing an object are defined, like his name, his field, his skill level, his role in the organization.

Viewpoint description	
Focus	**Design view:** Mechanic **Task :** Shaft design **Step :** 4 **Objective**: stability improvement
View angle	**Participant:** Designer1 **Domain, level :** mechanic, expert

Figure 12: Viewpoint description: contextual and personal dimension

This definition of a viewpoint has two aims:

• Propose a partition of the artifact according to the object components and the possible viewpoints on these objects. Each viewpoint is a consistent and partial view of the artifact.

• Propose a user viewpoint access to the artifact. Each participant can then consult the artifact according to his/her interest and not get lost in the whole description of the artifact. For example he/she can access all the descriptions of a particular participant or all the descriptions of a certain step in the design process and in a certain design view, etc.

Tichkiewitch in [16] has proposed a multi-view model of an artifact. This model is based on the design life cycle. Each component is so, described under several views: skeletal view, structure view, topological view, geometric view, manufacturing view, etc. So the contextual part of the viewpoint definition contains one of the step in the design life cycle. In the example of the Figure 19, a design view, a task, a step and a design objective characterize the contextual part. The personal part contains the characteristics of the person responsible of the description (it could be also a group of person).

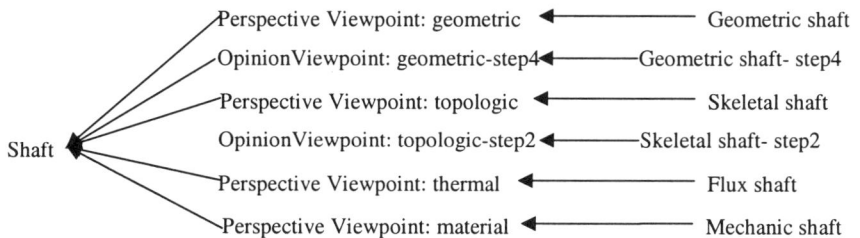

Figure 13: Viewpoint representation.

The model proposed in [17] define also two types of viewpoints:

• **Perspective viewpoints** that index the final states of descriptions, called consensual descriptions.

• **Opinion viewpoints** that index previous states of descriptions, called non-consensual descriptions.

This distinction between those two sorts of viewpoints allows keeping all the states of the artifact in the same model. Besides the design objective of a new description of a component like the "stability improvement" or "lower cost" could be indicate in the contextual part of the viewpoint (0).

Remark: The characteristics of the focus and the view angle in the viewpoint definition can be adapted to different design applications.

Table 2. A comparison of methods

Approach	Design rationale	Project results	Project management	Tools	Application types
IBIS	Tree: Issue/ Position/ Argument			gIBIS QuestMap	Design
QOC	Tree: Question/ Option/ criteria				Design
DRAMA	Tree: Goal/ Option + Table of criteria			DRAMA	Design and other type of applications
DIPA	Graph of Problem solving process: steps, opinions, arguments, roles and decisions			Memo-Net	Design, Help-Desk and other types of applications
DRCS	Graph: Entity/ Relation	Graph: Entity/ Relation	Graph: task Scheduling Graph: Entity/Relation	DRCS system	Concurrent engineering
EMMA			Tree: Goal/Plan	EMMA	Software engineering
SAGACE		Graph of Flux		Systémographe	All application types

3.4 Summary

First of all, we can distinguish which part of a project, methods help to represent and how. This comparison is based on the type of knowledge studied, the knowledge representation and application types. Table 2 presents this comparison.

4 Design project Memory Definition Guidelines

We have analyzed above, methods that help to capitalize different parts of a project especially design rationale. We have distinguished a number of elements to consider in a design project memory. We can extract from this analysis a number of guidelines to define the project characteristics and rationale.

4.1 Guidelines to Define Project Characteristics Memory

In design, there are a number of tools and techniques used to define design components and products. Our main objective is to respect designers' workspace and their representation world. So, our approach proposes to index project elements instead of reorganizing them, using new hybrid formalisms. Designer can easily understand knowledge so represented and has not to learn another representation type as for example semantic network, or classification tree, etc. Note that, index can be built by designres themselves and supervised by a knowledge engineer and the project manager. We recommend also the definition of a tool based on hypertext links, structuring techniques and web services in order to support the representation of elements in a design project memory.

First of all, documents (project specification, objectives, rules, ...) can be indexed by their goals and their nature. Note also that keywords and process graphs can be used to index this type of documents especially project specification.

As we saw in EMMA and DRCS, the project organization must emphasize tasks, their coordination and the collaboration of actors for their realization. So, a task (also called action in design) can be associated with collaborative actors (actor groups, role of each actors, its competence and its department or organization), task results   (design propositions), planning (start/end/stakes). A task can be decomposed in subtasks. Links to human resources, project management documents and databases provide detailed description of these elements.

More depth analysis must be done to study communication and relationships and to represent collaboration of actors in a project. We do not analyze in this chapter, this type of studies.

4.2 Guidelines to Define Project Rationale Memory

Design rationale is generally difficult to memorize. Meetings and discussions reports do not describe precisely problem solving and especially choices and their justifications. Methods like QOC, IBIS, DRAMA, DIPA, DRCS, have defined supports to capitalize these types of knowledge. As we noted before, these methods recommend to emphasize elements that describe problems, choices and arguments. So, a tree or a graph that leads to describe problem solving by emphasizing these elements is a good support to memorize project rationale. But, we think that the description of these elements directly when reporting a meeting or a discussion is not sufficient. Quality and project managers associated with the knowledge engineer must review these descriptions in order to enrich them and to characterize them.

Thus, we distinguish some elements to describe problems, options and arguments:
- Problem: type and description.
- Option: description and associated information (like cost, constraints, consequence, etc.).
- Argument: characteristic, description, advantages and disadvantages.
- Solution: definition, Implicated resources and consequence (as well as on the process and on the artifact).

Literal description and links can be used to describe these elements, in addition to an index using for example, problem types, argument characteristics, etc.

5   Conclusion

There are some approaches that deal with traceability of knowledge. Those approaches study how to extract knowledge from designers' workspace and to represent it in (so called) project memory. In this chapter, we discussed guidelines provided in this type of approaches by emphasizing their knowledge representation formalisms. We find different types of representation like decision making trees and tables, planning trees, semantic networks, problem solving process and flow graphs, etc.

Some representation techniques like semantic network and problem solving model are inherited from knowledge engineering. They are rich in terms of rational but they impose additive representation formalisms that are not obvious for a designer. Trees and tables are simpler to represent, but they describe only knowledge indices.

The main objective of our study is to respect designers' workspace and representation. Our challenge is then to define a formalism close to designers' representation. After studying some knowledge capitalization methods, we are going to analyze design models and techniques in order to define relations between models and knowledge representation. We plan also to develop a tool that supports the traceability of design projects. Having the same objectives, we aim at studying communication and relationships between designers within a collaboration with social science and linguistic researchers in our team.

6 References

[1] Bourigault D. and Lépine P. – Utilisation d'un logiciel d'extraction de terminologie (LEXTER) en acquisition des connaissances, *Acquistion et Ingénièrie des Connaissances, tendances actuelles*, Editions Cépaduès, 1996

[2] Brice A. – *Design Rationale Management (DRAMA)*, http://www.quantisci.co.uk/drama

[3] Buckingham Shum S. – Representing Hard-to-Formalise, Contextualised, Multidisciplinary, Organisational Knowledge. *Proceedings of AAI Spring Symposium on Artificial Intelligence in Knowledge Management*, P.9-16, 1997. http://ksi.cpsc.ucalgary.ca/AIKM97/AIKM97Proc.html

[4] Conklin J.E. and Yakemovic KC.B. – A Process-Oriented Approach to Design Rationale, *Human-Computer Interactions*, Vol.6, 1991.

[5] Dieng R., Corby O., Giboin A. et Ribière M. – *Methods and Tools for Corporate Knowledge Management*, in Proc. of KAW'98, Banff, Canada.

[6] Ermine J.L., Chaillot M., Bigeon P., Charenton B. et Malavielle D., MKSM a method for knowledge management, *Knowledge management : Organization, Competence and Methodology, Proceedings of ISMICK'96*, Schreimenmakers ed., Rotterdam, 1996.

[7] Klein M. – Capturing Design Rationale in Concurrent Engineering Teams, *IEEE, Computer Support for Concurrent Engineering*, January 1993.

[8] Lewkowicz M., Zacklad M., MEMO-net, un collecticiel utilisant la méthode de résolution de problème DIPA pour la capitalisation et la gestion des connaissances dans les projets de conception, IC'99, Palaiseau, 14-16 juin 1999, p.119-128.

[9] MacLean A., Young R.M., Bellotti V.M.E., Moran T.P., – Questions, Options, and Criteria: Elements of Design Space Analysis, *Human-Computer Interaction*, Vol.6, 1991.

[10] Malvache P. et Prieur P. – Mastering Corporate Experience with the REX Method, *Management of Industrial and Corporate Memory, Proceedings of ISMICK'93*, Compiègne 1993.

[11] McCullough D., Korelsky T. et White M. – Information Management for Release-based Software Evolution Using EMMA, *Software Engineering and Knowledge Engineering*, 1998.

[12] Matta N., Ribière M., Corby O., Définition d'un modèle de mémoire de projet, INRIA Report, N.3720, June, 1999.

[13] Penalva J.M. – SAGACE, la modélisation des systèmes dont la maîtrise est complexe, *ILCE*, EC2 (Ed), Montpellier, 1994.

[14] Poitou, J.P. – Documentation is Knowledge: An Anthropological Approach to Corporate Knowledge Management. *Proceedings of ISMICK'95*, Compiègne, France, 1995, p. 91-103.

[15] Pomian J. – *Mémoire d'entreprise, techniques et outils de la gestion du savoir*. Ed Sapientia.

[16] Ribière M. and Matta N. – Virtual Entreprise and Corporate Memory, *Building, Maintaining and Using Organizational Memories, ECAI-98 Workshop*, Brighton, August, 1998.

[17] Ribière M. – *Représentation et gestion de multiples points de vue dans le formalisme des graphes conceptuels*, Rapport de thèse en Sciences, spécialité informatique, Avril 1999.

[18] H.A., Simon, "Administrative behavior : a study of the decison-making processes in administrative organization," The MacMillan Cy, New York, 3rd edition, 1977.

[19] Tourtier P.A. – *Analyse préliminaire des métiers et de leurs interactions.* Rapport intermédiaire, projet GENIE, INRIA-Dassault-Aviation, 1995.

[20] Van Heijst G., Schreiber A. Wielinga B., Using Explicit Ontologies in KBS Development. International Journal of Human Computer Studies, Vol. 46, 1997.

Micro Knowledge Management: A Job Design Framework

Michel J. Leseure and Naomi J. Brookes

Department of Manufacturing Engineering, Loughborough University, Loughborough, LE11 3TU, UK

Abstract: This paper deals with the introduction of Micro Knowledge Management tools and techniques in operational units, and explores human hurdles to this process. A framework based on the context of knowledge interactions is proposed to describe the various types of job design practices that can constrain/facilitate the introduction of Micro Knowledge Management tools. Case studies used to illustrate this framework suggest that the functionality of Micro Knowledge Management systems is also highly contextual. The paper concludes with a typology of these systems.

1. Introduction

The tension between collaboration and competition has moved to the forefront of business research in recent years [1]. When applied to a shop floor, the paradoxical tension between collaboration and competition generates a number of questions. Should shop floor employees be competing with one another or should they work as a collaborating team? The incentives for competitive behaviour are well known: stimulate productivity and profitability for the company; promotions, pay raises and increase in personal status for the individuals. The incentives for collaborative behaviour are: sustainability of know-how for the company; well being through a friendly work atmosphere, access to valuable knowledge, employability and professional pride for the individuals.
Micro Knowledge Management focuses on the capture, validation and diffusion of shop floor knowledge through the use of modern technologies derived from a variety of disciplines: e.g. information technology, artificial intelligence, cognitive science. Although the economic benefits of more efficient exchanges of knowledge in the workplace are obvious, it is important to recognise that the application of these technologies may conflict heavily with the existing value systems and work habits of employees. Micro Knowledge Management is a technology-enabled change in work practices driven by an effectiveness motive. The introduction of such a change programme is delicate and can end up in failure. This paper is

primarily concerned with the following question: "What is the relationship between job design and Micro Knowledge Management?".

2. Managing Change

Despite the overwhelming body of literature on change management, criticisms about modern change management abound. Schaffer and Thompson report that 70% of corporate re-engineering programmes fail [2]. Beer and his colleagues state that "change programmes do not change anything" [3].

After a four-year study of organisational change in six large corporations, Beer, Eisenstat and Spector concluded that most change programs do not work because they are guided by a theory of change that is fundamentally flawed [3]. Beer, Eisenstat and Spector refer to the traditional approach as the "fallacy of programmatic change", where top managers push down a programme of change which was developed without any concertation of the employees who are eventually asked to implement this program.

The fallacy of programmatic change finds its roots in the popular and implicit belief that work redesign is always feasible [4]. When adopting a new technology, the financial evaluation of its benefits usually outweighs the potential difficulty (or impossibility) of introducing this technology into an existing work system. Consistent with Beer and his colleagues, a number of authors have also pointed out that companies are usually happy to modify work practices in order to support a new technology, but are usually reluctant to do the opposite (i.e., to modify a technology to fit within existing work practices) [5]. As an exception to this rule, Scherer and Weick [6] report a case study where a company realised the existence of a conflict between their ongoing job empowerment initiative and the introduction of a centralised production information system. This system gave precise task assignments to the employees and eliminated any local control they had on the production process. The company modified the information system so that it only provided useful information to the employees to facilitate their individual decision-making (e.g. daily production planning).

Consistent with the view that both technologies and work systems can be modified to obtain a viable, effective configuration, this paper does not prescribe how to change work practices to accommodate Micro Knowledge Management. Instead, its aim is to develop a framework based on an analysis of the contextual requirement regarding exchanges of knowledge and to propose, for each context, the best configuration between job design and Micro Knowledge Management techniques. Although the readers will be familiar with Micro Knowledge Management techniques thanks to the other chapters of this book, they may not be so familiar with the theory and practice of job design. The next Section is a brief review of this discipline.

3. A Review of Job Design Theory & Practice

3.1. Job simplification

The need for explicit, managed job design policy stems from the complexity of collective forms of work. It is therefore not surprising that the first job design theories were formulated during the industrial revolution, at a point in time where job shops changed from small craft shops to gigantic, massively staffed factories. The initial school of job design, *job simplification*, had for purpose to de-skill jobs by reducing them to their most elementary tasks. Job simplification is achieved by the application of two principles:

- Horizontal division of labour: The making of a complex product should be broken down into a series of simpler tasks, and each of these tasks will be assigned to a single worker. This principle was promoted by the economist Adam Smith [7] and the engineer Charles Babbage [8].
- Vertical division of labour: Not only should workers be assigned to unique, elementary tasks, but they should not have any discretion regarding how to perform this task. This decision will be made by supervisors/engineers who will study the most efficient way to carry out the task. This second principle was promoted by Taylor's *The Principles of Scientific Management* [9] and by Gilbreth with the formulation of the principles of time-and-motion studies [10].

Although initially met by sceptical comments, these principles quickly became popular, for instance with the success story of the production of the Model T by Henry Ford in 1914. The impacts of job simplification principles are still well rooted in many modern organisations and in business thinking. Parker and Wall state that job simplification is not only the reference point of job design today, but also the default option to any managers having to design jobs [4].

It is only in the 1950s that the premises of job simplification were questioned. Two different schools of thought emerged:

- In the United States, researchers became concerned with individual job designs, and its relationship to concepts such as job satisfaction and motivation. These efforts led to the *job enrichment* concept.
- In England, researchers focused instead on group work design, and introduced the concept of *self-managed teams*.

3.2. Job enrichment (or empowerment)

Herzberg and his colleagues can be credited for putting forward recognition and responsibility as two variables of job design [11]. Their ideas were developed further by Hackman and Oldham who proposed a model of five core job characteristics [12]:

- *Skill variety*: the degree to which the job requires different skills
- *Task identity*: the degree to which the job involves completing a whole, identifiable piece of work rather than simply a part.
- *Task significance*: the extent to which the job has an impact on other people, inside or outside the organisation.
- *Autonomy*: the extent to which the job allows jobholders to exercise choice and discretion in their work.
- *Feedback from the job*: the extent to which the job itself (as opposed to other people) provides jobholders with information on their performance.

In this model, the successful combination of the five characteristics leads to high internal work motivation, high "growth" satisfaction, high general job satisfaction and high work effectiveness.

From a practical standpoint, this model led to the formulation of the *job enrichment* school. Wall and Parker [4] compiled from a broad range of papers the following list of standard recommendations:

- Arrange work in a way that allows the individual employee to influence his or her own working situation, work methods, and pace. Devise methods to eliminate or minimise pacing.
- Where possible, combine interdependent tasks into a job.
- Aim to group tasks into a meaningful job that allows for an overview and understanding of the work processes as a whole. Employees should be able to perceive the end product or services as contributing to some part of the organisation's objectives.
- Provide a sufficient variety of tasks within the job, and include tasks that offer some degree of employee responsibility and make use of the skills and knowledge valued by the individual.
- Arrange work in a way that makes it possible for the individual employee to satisfy time claims from roles and obligations outside work (e.g., family commitments).
- Provide opportunities for an employee to achieve outcomes that he or she perceives as desirable (e.g., personal advancement in the form of increased salary, scope for development of expertise, improved status within a work group, and a more challenging job).
- Ensure that employees get feedback on their performance, ideally from the task as well as from the supervisor. Provide internal and external customer feedback directly to employees.
- Provide employees with the information they need to make decisions.

3.3. Self-managed teams

The practical concept of self-managed teams was derived from socio-technical systems theory. This body of theory is based on the premise that any organisational system is the superposition of two independent but related systems: a human system and a technical system. In order to reach effectiveness, it is important that these two systems are jointly optimised [13]. This broad principle has led to the practical school of *self-managed teams*. Parker and Wall [4] give the following list of specifications to describe this school of job design:

- Group interdependent tasks to make a meaningful set and to involve a balance between less popular and desirable tasks.
- Provide clear performance criteria for the team as a whole.
- Provide clear feedback on group performance.
- As far as possible, leave methods of working to employee discretion.
- Allow employees to control variances at the source, but ensure that they have the necessary knowledge, skills, and information to intervene.
- Allow the group to control equipment, materials, and other resources, making them responsible for their prudent use.
- Increase the skill level of employees to allow flexible responses to uncertainties.
- Ensure that selection, training, payment systems and so forth are congruent with the work design.
- Regularly review and evaluate the work design.

3.4. Action theory

Action theory is an alternative school of job design which was originally developed by German researchers [4; 14]. Work is defined as being action oriented, i.e. as being a collection of elementary activities. There are two important features of actions:

- An action encompasses a number of activities: defining a goal for the action, translating it into plans, executing these plans, and finally, receiving feedback from the action.
- Any action is subject to an individual cognitive regulation. Some actions become highly routinised (e.g. an assembly line workers' task) whilst some can never follow a rigid, identical structure (i.e., a commercial engineer's sales tactic).

The practical recommendations made from action theory are to a great extent similar to those of job enrichment, with two exceptions.

First, the golden rule of action theory is that the active involvement of an employee is only possible if this employee is dealing with the entire action, and not part of it.

Second, action theory is probably less prescriptive than any other school in terms of practical recommendations. The tasks and context of executions of these tasks are the determinant of the final job design to implement. For this reason, action theory has been coined as a *dynamic* form of job design.

4. Job Design & Knowledge Management

A review of the job design literature and of the knowledge management literature shows that there are few guidelines for managers in terms of designing work environments that are conducive to the implementation of new knowledge-based work practices.

The research questions stressed by job design researchers are:

- How should job design theories be adapted to accommodate modern, knowledge-rich work environments?
- What organisational factors can be used to promote and manage a culture based on collaboration and interaction?

At the time of writing this paper, there have been no definitive answers to these research questions [4]. There is, however, one consistent research finding about knowledge workers: the use of team seems beneficial and is highly recommended [15].

This collaborative, team-based aspect of knowledge management is confirmed by the few exploratory case studies published in the knowledge management literature, such as the knowledge network of Buckman Laboratories [16] or the team-based approach to product development of GKN Westlands Aerospace [17].

Therefore, the evidence of the supremacy of teams is still anecdotal. This lack of empirical basis to the development of a generic model is also a problem in job design research [4]. Similarly, the concept of teams uncovers a great variety of forms and is not context-free [18]. This paper aims to close this gap between observation and practice with a contextual model.

5. Knowledge Contexts in Operational Units

The goal of this Section is to define what are the different contexts of Micro Knowledge Management. Cummins and Blumberg are amongst the few job design researchers who have identified key organisational contingencies [19]. They found that three factors influence job design practices:

- *Technical interdependence*: the degree of required co-operation to make a product.
- *Technical uncertainty*: the amount of information processing and decision making required to execute a task.
- *Environmental uncertainty*: The extent to which a market is stable.

As indicated in the introduction, the real novelty in the knowledge management process is the emphasis on collaborative behaviour in the workplace. However, the extent to which interaction between employees is needed will vary greatly from one operational unit to another.

In this paper, technical interdependence and environmental uncertainty are merged to create a first contextual variable: the required frequency of knowledge interactions. Technical interdependency creates the background for knowledge interaction. Environmental uncertainty adds a time (frequency) dimension and controls how often these interactions need to take place.

Technical uncertainty is the second dimension of the context of micro-knowledge management. A low uncertainty describes a system that is primarily based on the reuse of routinised knowledge. A high uncertainty occurs in systems where problems and products vary too broadly to be handle simply through reuse of knowledge. In high technical uncertainty environments, new knowledge needs to be created.

Figure 1 illustrates the contextual framework of this paper by combining these two dimensions. The two dimensional area in Figure 1 is a continuous spectrum where all types of contexts can be observed. The four quadrants defined in this figure are therefore not "ideal types" or "archetypes". Instead, they are "representative profiles". The purpose of this paper is to position Micro Knowledge Management tools onto this framework, whilst taking into account the different social systems that will necessarily exist at different points of the knowledge spectrum.

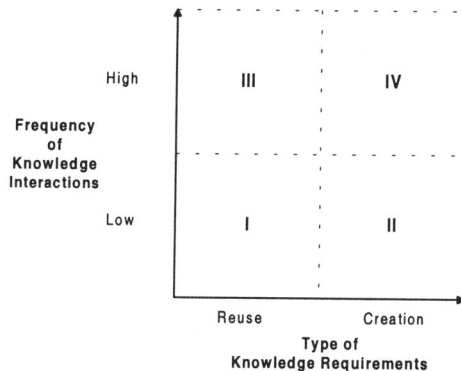

Figure 1: Micro-Knowledge Management Contextual Framework

6. Job Design Framework

6.1. Theoretical framework

Figure 2 indicates how job design theories map onto the knowledge framework. The arrows indicate which job design practices are likely to be appropriate when one moves in a specific direction.

In quadrant I, workers usually do not have to create knowledge - their jobs are usually simple, codified, and based on routinised application of basic skills. Improvements and modifications normally come from engineers and project managers, with little or no consultation of the line workers. In such environments, such as the moving assembly line of Henry Ford, job simplification is a very appropriate and efficient form of job design.

In quadrant II, output and effectiveness is dependent on the ability of operational employees to create new knowledge. This requires the ability to detect a problem and to use advanced skills to formulate and implement a solution. Whereas the use of routinised knowledge in quadrant I prompted the introduction of job simplification, the lack of routines, both in the problems encountered and the methods to solve them, precludes this approach in quadrant II.

At the extreme corner of quadrant II is the individual craftsman: a highly skilled individual, not dependent on any other parties, relying on specialist knowledge used autonomously and not on networking. In this case, the individual job perspective provided by the job enrichment school is ideal to build a conducive, motivating work environment.

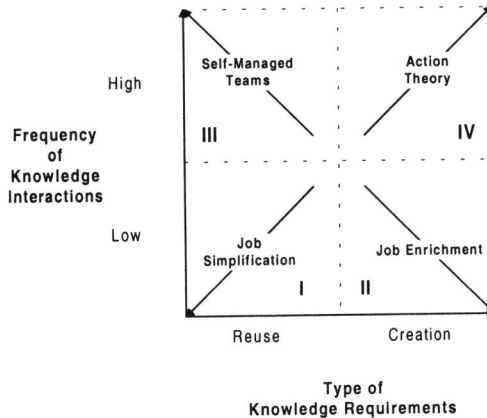

Figure 2: Job Design Framework

In quadrant III, the productivity of the operational units is mainly dependent on routinised work procedures and reuse of knowledge, but a higher level of interaction is needed that in quadrant I. Examples of rationales for more interactions in quadrant III are:

- union pressures regarding physical strain, and the introduction of concepts such as job rotation and cellular manufacturing,
- the decision to decrease supervisory costs by giving more discretion to the team,
- a socio-economic environment based on group work.

In this quadrant, the principles of self-managed teams find a perfect application field.

Newman reports that lean production (a concept that relies extensively on teamwork) improves productivity at the cost of innovative capability [20]. This confirms the opposite, diverging directions of job enrichment and self-managed teams, and the fact that the task of seeking the benefits of both types of job design system is not trivial.

In the fourth quadrant, the requirements of knowledge creation and collective work are merged together. The problem in this quadrant is to find how to smoothly combine the independence of the specialist and the requirement of frequent, extensive interaction. With its emphasis on the preservation of elementary action, action theory offers a mean to preserve the benefits provided by job enrichment in quadrant IV. With its open, dynamic view on job design, action theory also allows the planning and implementation of adequate networking and teamwork, as needed. In other words, action theory provide a theoretical background to support the implementation of flexible, open, dynamic, boundary-less forms of work systems in quadrant IV.

The positioning of the self-managed teams school of job design may sound critical, and at odds with the general idea in knowledge management that "the group is good". The framework in Figure 2 is not a criticism of the concept of team work in innovative work environment, but a criticism of the literature on team work that tend to treat teams as a unitary concept. Researchers in job designs have already stressed that teams can take many forms. For instance, a basic typology of work groups proposed by Bryant, Farhy and Griffiths [21] can be mapped onto the knowledge framework of Figure 2 (the variables used by these authors are skill complexity and task interdependence; they easily superpose with the variables of Figure 1).

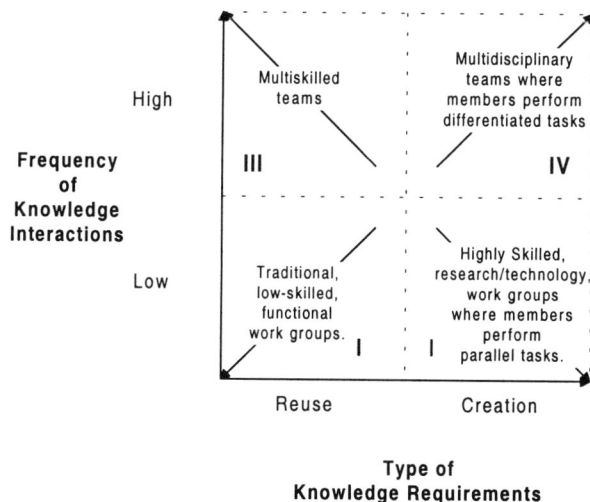

Figure 3: A Typology of Teams (adapted from [21]).

6.2. Case studies

The job design framework was populated with 14 industrial case studies. The case studies are either borrowed from the literature (in which case, they are named by a letter, e.g. case A) or were visited and interviewed by the authors (named by a number, e.g. case 1). Table 1 briefly describes the cases. A qualitative scale was used to position each case on the knowledge grid. The results are presented in Figure 4.

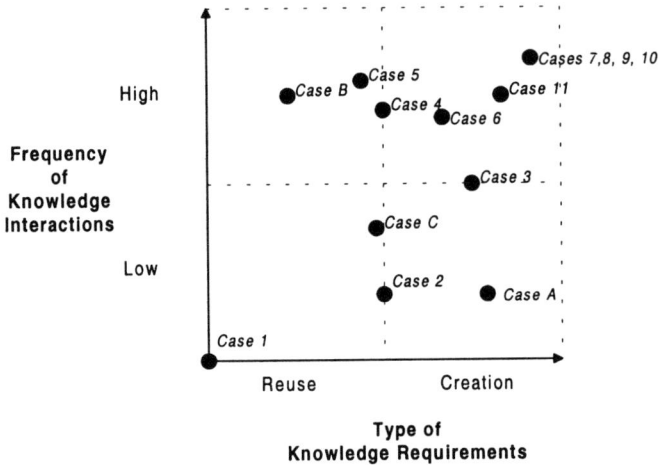

Figure 4: Knowledge Context of the Case Studies

The actual job design practices were compared to those suggested by the model (e.g., is case 11 job design system compatible with action theory? or is case 3 job design system a compromise between job enrichment and action theory?). The case studies where job design systems matched the model predictions are indicated in the "Fit" column in Table 1. Only one case (case 10) did not fit the model, but this misfit can be explained in the light of special circumstances (reorganisation and staff turnover following a bankruptcy).

7. IT Implications

The job design framework promotes an inner understanding of the mechanisms of workplace human interactions and indicates which work designs are important in which context. During the analysis of the case studies, it became clear that this human systems approach revealed important specifications for the implementation of IT systems. Some IT functionality were desirable and would compensate for weaknesses in the human systems. Other IT functionality, used in the wrong context, could be detrimental. The insights provided by the case studies follow.

Table 1: Case Studies and their Contributions

Case	Fit	Description	Contribution
A	Yes	Automotive craft shop as described by [22].	Confirms that knowledge creation environments based on individual skills are viable.
B	Yes	Assembly team in a Toyota assembly plant as described by [23].	Confirms the fact that self-managed team reuse routinised knowledge and rarely innovate.
C	Yes	Assembly teams at Volvo Udevella factory as described by [22].	Consistent with case A & B.
1	Yes	Packaging unit of a small mustard manufacturer.	Confirms the existence of workshops relying solely on knowledge reuse.
2	Yes	Small naval dockyard specialising in sailboat repair and maintenance.	Consistent with the observations of cases A and 1.
3	Yes	Average size design unit of a medium to high technology defence company.	Confirms (1) that interaction is possible without teams and (2) that action theory does not conflict with interactions.
4	Yes	Small engineering team responsible for the acquisition of capital goods.	Confirms that some engineering tasks can be based on reuse of knowledge, and that self-managed teams can perform this function.
5	Yes	Medium size design unit at an aerospace component manufacturer.	Consistent with the conclusions of case 4.
6	Yes	Medium size network of industrial companies distributing a novel robotics system.	Confirms that the self-managed team approach fails to function effectively in knowledge creation environments.
7	Yes	Small size unit of skilled fitters and project managers in a custom machine manufacturer.	Confirms that action theory forms of work design are effective in quadrant IV.
8	Yes	"Time-based competition" pilot project in case 7.	Indicates that project teams with an action focus are efficient but can fail to reuse organisational memory.
9	Yes	Same as case 7, but after significant growth and an evolution toward job simplification.	Confirms the inability of job simplification principles to manage the creation of knowledge.
10	No	Same as case 9, after extensive revision of job design toward action theory.	Benefits of action theory were not observed.
11	Yes	Unit of fitters in a competing company of case 10.	Consistent with case 7 but removes the conflict of case 10.

- Knowledge Representation: This is the first design decision that precedes the development of any type of Micro Knowledge Management Systems. The quality of this representation (i.e. its ability to capture and diffuse operational knowledge) has a critical impact and the usefulness of the IT system. For instance, in the case of the robotics system distribution network (case 6), the network failed to find means of representing knowledge relative to the technical risk of integration (e.g. compatibility of output with traditional peripheral feeding and vision systems).
- Knowledge Capture: The process of capturing individual *or* group knowledge.
- Knowledge Validation: Whereas the process of capture can be relatively unconstrained, a phase of knowledge validation is necessary to eliminate noise or "sabotage" input [24] in the system. This process is contextual, as different systems on the knowledge grid naturally validate knowledge differently.
- Knowledge Storage: The process of archiving knowledge for future access. The complexity of this functionality is a direct function of the complexity of the representation scheme.
- Diffusion of Knowledge (support work environment): The process of distributing knowledge and allowing access. This process of diffusion can aim to support a work system by facilitating interactions in environments not conducive to socialisation (e.g., quadrant II) or by delivering operational objectives in a usable format (e.g. quadrant III).
- Off-line Knowledge Analysis: The advantage of the availability of a large knowledge base is the possibility of analysing it in more detail. At the time of writing this paper, this is an unexplored potential of Micro Knowledge Management, if one excludes knowledge discovery tools. In quadrant IV, off-line knowledge analysis tools could be used to detect operational problems and help resolve conflicts.
- Protect Work Environment: This is another unexplored area of Micro Knowledge Management. The knowledge-rich environment of quadrant II unveils problems of knowledge ownership that are not addressed today. Both human and technical systems can threaten the integrity of actions in quadrant IV, and an advanced IT system could help to limit this threat. IT systems (e.g. demand forecasting, aggregate planning and CIM) are used in quadrant II to protect and insulate self-managed teams from a knowledge creation situation that they would not be able to cope with.
- Automation of Routinised Knowledge: This is the key application of Micro-Knowledge Management in Quadrant I.
- Support Knowledge Reuse: This is a main benefit of Micro-Knowledge Management in quadrant IV.
- Support Self-development and Learning: Some work environments rely extensively on the ability of workers to keep current and to learn new methods and technologies (i.e. quadrant II & IV). Intranets can provide opportunities for learning, but other systems such on-demand learning seem to have a real

potential. The notion of "virtual apprenticeship systems" could be a direction to try to improve the self-development capabilities of individuals.

Whereas some IT functionality are relatively context-free (i.e. capture and diffusion) others are very context-dependent. Figure 5 proposes a typology of the different types of Micro Knowledge Management Systems that will prevail in different areas of the knowledge grid.

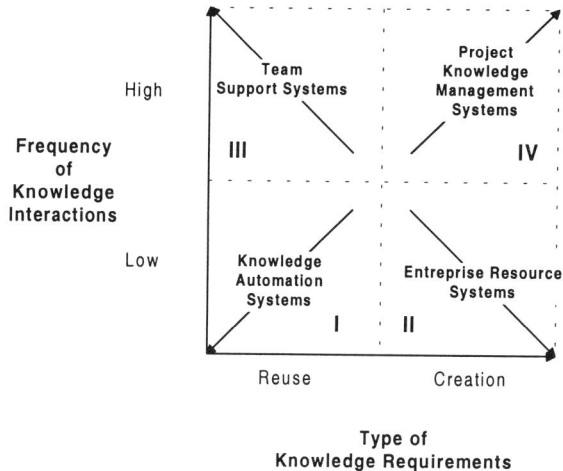

Figure 5. A Typology of Micro Knowledge Management Applications

8. Conclusion

From the lessons learnt from the case studies, four types of Micro Knowledge Management Systems (Figure 5) are defined to address the diversity of work environments where Micro Knowledge Management can be applied (Figure 2).

1. **Knowledge Automation Systems**: This is the most standard application of Micro Knowledge Management, and is equivalent to knowledge engineering. Job simplification is about de-skilling jobs and knowledge automation is about removing decision making and knowledge creation from jobs. Job simplification provides an environment for replacing manual labour with automated machines, and knowledge automation replace human, routinised reuse of knowledge by computer generated decision. This type of knowledge management follows the logic behind Taylorism and Fordism.

2. **Enterprise Resource Systems**: These systems aim to support work environments that rely extensively on individual performance for innovation and knowledge creation. These systems should aim to (1) support access to

information and self-learning materials and (2) to compensate for the individuality of tasks by providing means of informal networking and knowledge exchange. These systems should not jeopardise the motivation and productivity of experts.

3. **Self-Managed Teams Support Systems**: These systems have two main objectives. The first objective is to provide an environment where a self-managed team can operate. This is done primarily through advanced forecasting and production planning systems. The second objective is to give autonomy regarding the management of effectiveness within the group. Examples of such systems are the one used in lean production environments.

4. **Project Management Support Systems**: The complexity of the jobs in quadrant IV is such that most companies will adopt a project organisation to handle them. This raises specific issues regarding the feasibility of creating, maintaining and using some form of "project memory". Too often, project teams fail to reuse past knowledge and tend to "reinvent the wheel". Micro Knowledge Management Systems can provide solutions to (1) capture, store and diffuse this project memory, (2) to help to eliminate unnecessary innovation by enabling reuse, and (3) to enhance knowledge creation by providing tools and frameworks to integrate and validate the knowledge of various experts.

References

[1] von Krogh, G. & J. Roos, eds., 1996, *Managing Knowledge*, Sage Publications: London
[2] Schaffer, R. and H. Thompson, 1992, "Successful change programs begin with results," *Harvard Business Review*, January-February, pp. 80-89.
[3] Beer, M., R. Eisenstat and B. Spector, 1990, "Why change programs don't produce change", *Harvard Business Review*, November-December, pp. 158-166.
[4] Parker S. and T. Wall, 1998, *Job and Work Design*, Sage Publications London.
[5] Goodman, P., R. Devadas, and T. Griffith-Hughson, 1988, "Groups and productivity: analyzing the effectiveness of self-managing teams", in J. Campbell, R. Campbell & Associates, eds, *Changes in Organization: New Perspectives on Theory, Research, and Practice*, Josey Bass, San Fransisco; and Kelly, J., 1978, "A re-appraisal of socio-technical systems theory", *Human Relations*, 31, pp. 1069-1099.
[6] Scherer, E. and S. Weick, 1996, "Autonomy and control in decentralized systems", in R. Koubek & W. Karwowski, eds., *Manufacturing Agility and Hybrid Automation*, Louisville, KY: IEA.
[7] Smith, A., 1776, *The Wealth of Nations*.

[8] Babbage, C., 1835, *On the Economy of Machinery and Manufacturers*, Charles Knight, London.

[9] Taylor, F., 1911, *The Principles of Scientific Management*.

[10] Gilbreth, F., 1911, *Brick Laying System*, Clark, New York.

[11] Herzberg, F., B. Mausner and B. Snyderman, 1959, *The Motivation to Work*, John Wiley, New York; and Herzberg, F., 1966, *Work and the Nature of Man*, World, Clevevand, OH.

[12] Hackman, J. and G. Oldham, 1975, "Development of the job diagnostic survey, *Journal of Applied Psychology*, 60, pp. 159-170.

[13] Cherns, A., 1987, "The principles of socio-technical systems revisited", *Human Relations*, 40, pp. 153-162.

[14] Frese, M. and D. Zapf, 1994, "Action of the core of work psychology: a german approach", in H. Triandis, M. Dunette and J. Hough, eds., *Handbook of Industrial and Organisational Psychology*, (Vol 4, 2nd ed., pp. 271-340), Consulting Psychologists, Palo Alto, CA.

[15] Morhman, S., S. Cohen and A. Morhman Jr, 1995, *Designing Team Based Organizations: New Forms of Knowledge and Work*, Jossey-Bass, San Fransisco; and Clegg, C., P. Coleman, P. Hornby, R. Maclaren, J. Robson, N. Carey and G. Symon, 1996, "Tools to incorporate some psychological and organizational issues during the development of computer-based systems, *Ergonomics*, 39, pp. 482-511.

[16] Pan, S. and H. Scarbrough, 1999, "Knowledge management in practice: An exploratory case study", *Technology Analysis & Strategic Management*, 11(3), pp. 359-374.

[17] Ratcliffe, P., 1997, "Knowledge transfer at the design-manufacture interface", *Advances in Manufacturing Technology XI*, *Proceedings of the National Conference on Manufacturing Research*, Glasgow Caledonian University, September 9-11, pp. 645-652.

[18] Dunphy, D. and B. Bryant, 1996, "Teams: panaceas or prescriptions for improved performance", *Human Relations*, 45(5), pp. 677-699.

[19] Cummins, T. and M. Blumberg, 1987, "Advanced manufacturing technology and work design", in T. Wall, C. Clegg and N. Kemp, eds., *The Human Side of Advanced Manufacturing Technology*, pp. 37-60, John Wiley, New York.

[20] Newman, V., 1999, "The limitations of knowledge management", Unpublished presentation at the BPRC Seminar on Knowledge Management in Manufacturing, November 29, BPRC, University of Warwick, UK.

[21] Quoted in [18].

[22] Womack, J., D. Jones and D. Roos, 1990, *The Machine that changed the world*, Rawson Associates: New York.

[23] Monden, Y., 1983, *Toyota Production System, Practical Approach to Production Management*, Industrial Engineering and Management Press: Atlanta, Georgia.

[24] Carter, C., 2000, "The strange death of the professional engineer: a sojourn into the management of knowledge", Unpublished Working Paper, University of Leicester Management Centre.

Building the KDD Roadmap:
A Methodology for Knowledge Discovery

J.C.W. Debuse, B. de la Iglesia, C.M. Howard and V.J. Rayward-Smith,
School of Information Systems, University of East Anglia, Norwich, NR4 7TJ, UK
{jcwd,bli,cmh,vjrs}@sys.uea.ac.uk

Abstract: Knowledge discovery in databases (KDD) is an iterative multi-stage process for extracting useful, non-trivial information from large databases. Each stage of the process presents numerous choices to the user that can significantly change the outcome of the project. This methodology, presented in the form of a roadmap, emphasises the importance of the early stages of the KDD process and shows how careful planning can lead to a successful and well-managed project. The content is the result of expertise acquired through research and a wide range of practical experiences; the work is of value to KDD experts and novices alike. Each stage, from specification to exploitation, is described in detail with suggested approaches, resources and questions that should be considered. The final section describes how the methodology has been successfully used in the design of a commercial KDD toolkit.

1. Introduction

In recent years world-wide interest in Knowledge Discovery in Database (KDD) and Data Mining has soared. Organisations store vast amounts of information about their products, customers and many other areas of their business; these collections of data are termed data warehouses. The idea that these huge databases can be mined for interesting patterns has appealed to a wide range of organisations. Typical KDD projects may investigate customer behaviour, plan direct marketing, detect fraudulent activity or identify machine faults [1, 2].

Organisations who aim to partake in KDD projects will require a considerable amount of expert knowledge about both the data and the KDD methods so that high quality, valid and interesting results can be obtained. In the years since KDD was introduced, the number of approaches, algorithms and software packages has grown rapidly, not to mention the size of the databases that are being collected. The expansion of the subject has made it increasingly difficult for KDD engineers to keep track of the techniques available to solve a particular task. This methodology aims to collate this wealth of expert input and present it in such a way that it can easily be applied when designing new projects or reviewing existing projects. The stages of the KDD process are expressed here in the form of a roadmap with a feedback loop. The iterative nature of the problem allows results from one stage to be improved by returning to previous stages and making minor adjustments to some of the parameters and decisions.

The target audience of this methodology is broad, making it suitable for all levels of KDD users. Users can fall into a number of categories but two important extremes should be considered. Firstly, there is the data owner who maintains and works with the data and has extensive knowledge of its usefulness and limitations. The data owner is likely to seek results of the desired quality, present results in a suitable format and be able to interpret the result in the context of the problem domain. At the opposite end of the spectrum is the KDD engineer; an expert in the field of knowledge discovery with experience in selecting and applying a range of algorithms. KDD engineers may also be familiar with the various database systems and formats, but would not have in-depth knowledge of every organisation's data. In an ideal world, these skills would be available from a single resource but in reality for most projects a compromise will be found. The most common solution for single-resource projects would be a data owner with a collection of well-chosen software packages. For larger projects, a team would have to be formed where each member brings to the group a unique set of skills. Personnel at management level should also be considered where there may be one or more people controlling the higher-level attributes of the project such as objectives, budget, time and quality.

2. The KDD Process

Outline descriptions of different KDD processes can be found in [3, 4, 5] with an early methodology described in [6]. An alternative methodology that is much more focussed on business process modelling can be found in [7]. The methodology presented here is experienced-based and is geared towards the decisions that must be taken and the options available at each stage to accomplish a given task.

The KDD process can be divided into eight major stages which are described in detail in the following sections.

- Problem specification.
- Resourcing.
- Data cleansing.
- Pre-processing.
- Data mining.
- Evaluation of results.
- Interpretation of results.
- Exploitation of results.

We present the KDD process in the form of a roadmap which is illustrated in Figure 1. The process has one major route which is to get from the starting data to discovered knowledge about the data. Like most long and difficult journeys, there will be a number of stops along the way. A stop represents one of the eight KDD stages named above, each of which is made up of a number of smaller tasks. Each stage, and task, is optional although in most circumstances at least one task from each stage will be necessary. When applying the roadmap, it is unlikely the project will run directly from start to end; stages will usually have to be repeated using a different set of decisions and parameters. For this reason, the process is deemed iterative which is denoted by the inner feedback route on the figure. The iterations performed during the KDD process are vital to the success of the KDD project. In addition to reviewing and repeating a major stage, some or all of the smaller tasks within the stage can be repeated.

When beginning each stage of the process a number of prerequisites must be met. The prerequisites ensure that all of the required information and tasks have been completed in the previous stages before moving onto the next. The tasks accomplished during a stage are listed when completing that stage to form the output list. This output list should meet the prerequisites of the following stage.

An important analogy can be drawn between the KDD process and well-established software engineering processes [8]. Both processes share this iterative nature and emphasise the importance of the early stages of the processes. It is well known in software engineering that at least sixty percent, if not more, of a project should be allocated to the analysis and design of the software; if this is carried out correctly, it should reduce the work involved in the programming and testing stages. As we shall describe in the coming sections, the early stages of the KDD process share the same importance. If the project is carefully analysed, specified and managed, the decisions to be made at later stages will be clearer and it will be easier to identify the critical stages of the project.

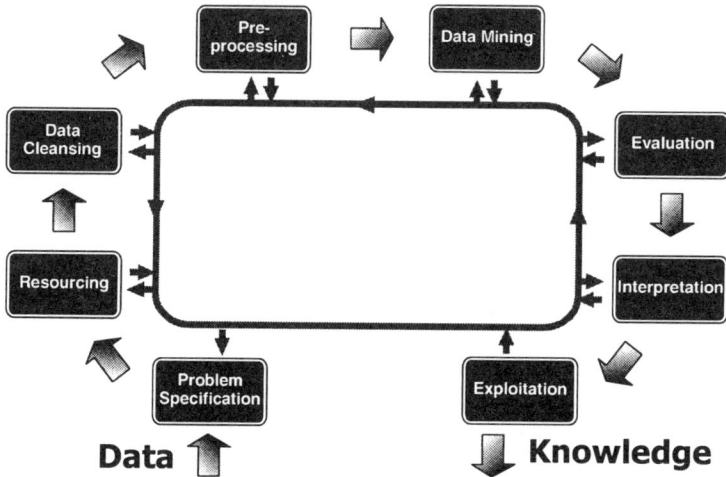

Figure 1: Outline stages of the KDD process.

2.1 Problem specification

The main purpose of the problem specification stage is to take all of the loosely defined ideas that have been involved with the project proposal and construct a tightly defined specification for the problem. The processes performed within this stage include determination of project tasks, availability of data and the software and hardware requirements. At the end of this stage, there should be a sufficiently detailed problem specification for a senior project manager to determine the feasibility of the KDD problem.

2.1.1 Inputs to problem specification phase

The input to this phase consists of a number of loosely defined ideas and proposals that will have prompted the initiation of a KDD project. The ideas should at least cover the application area, the problem to be solved or investigated, the currently available resources including equipment, personnel and data. All of these items and more will be established in broad terms at this point.

2.1.2 Quality management documentation

Many organisations incorporate quality management systems to provide some form of guarantee to their customers that any service or product they provide is of the highest standard and has been carefully monitored at each stage of the project lifecycle. A KDD project is often a major investment for a company and the quality of this project should be supervised as with any process. For each stage of the KDD process, a controlled document should be updated with any decisions, results, assumptions and parameters that have been used. In addition to providing management personnel and customers with a detailed structure of the how the project has been carried out, it also becomes a valuable asset to the KDD engineer. The iterative nature of most projects will mean backtracking and repeating previous stages with different parameters and data; the precise information that will have been recorded at each stage will allow the user to reverse any decisions if necessary and find a new route through the process. Each item in the documentation should be logged with the resources used, versions of the data and software and time stamped for project management and time allocation issues.

2.1.3 Preliminary database examination and the data dictionary

A data dictionary should be constructed for each data source used in the project as part of the quality management documentation. The information will be used and updated during most stages of the KDD process. Within the problem specification stage, a dictionary is created for all data sources currently available to the project; in cases where data is not yet available, the information should still be documented but must be validated after the resourcing stage. The content of the databases will help determine which high and low level tasks are feasible, sensible specifications for hardware and software, and which data mining algorithms can be used. The following characteristics are determined.

- The number of records.
- The number of fields.
- The proportion of the database which is missing.
- The proportion of the fields and records which contain missing values.
- The format of the database; i.e. flat files, DBMS, paper.
- The accessibility of the database(s) including access speeds.
- Noise level determination such as the number of contradictions.

2.1.4 Database familiarisation

Familiarisation is an extension of the previous section that looks at the internal data values rather than the structure of the database itself. As before, if the databases are not yet available, this stage must be completed after resourcing. Access to domain experts can significantly help the understanding of the data during this familiarisation process. The following items are typical of those examined during this stage. All information is added to the data dictionaries for the project.

Field type determination. There is a variety different of terms used to describe the different types of field; we will use the following numerical and categorical terms. There are two types of numeric fields; numeric discrete for describing integers (such as number of dependants), and numeric continuous for describing reals (such as temperature). Categorical data also falls into two categories; categorical ordinal for which values have an implied ordering (small, medium and large), and categorical nominal for when there is no ordering (such as male, female). For some fields, the type will be obvious; however, data in other fields may be represented using an encoding which could change the apparent type, e.g. colours could be represented by numbers but have no ordering.

Field semantics. Knowledge of the meaning of a database field may influence the KDD process considerably. Knowing, for example, that two fields are really measurements of the same value could immediately help reduce the number of features. Familiarisation with field value semantics is also important in identifying outliers, erroneous values and missing value representations. Any ambiguities in field names or values should be addressed at this stage.

Reliability. Certain fields, or even values within a field, can have varying levels of associated reliability. Understanding these levels can help pre-processing and interpretation. Patterns that rely on unreliable fields should be considered less useful than those involving strongly reliable data.

Visualisation. Plotting data using histograms and x-y plots can significantly, and rapidly, increase the overall understanding of the database. Identifying patterns based on combinations of fields or spotting similarities between fields can easily be performed using visualisation techniques. As before, if the data is not yet available, this is deferred until after the resourcing stage.

2.1.5 High level task (HLT) determination

The goal or goals of the KDD project must be determined which will be prediction and/or description [4]. We refer to these goals as high level tasks; low level tasks will be discussed in Section 2.1.6. The aim of descriptive tasks is to present the resulting knowledge in an understandable form whereas the main aim of predictive tasks is to predict unknown values; it is often the case that the results from description tasks are also used for prediction.

2.1.6 Low level task (LLT) determination

Given the nature of the problem, the choice of low levels tasks is first limited to those which are feasible; classification problems will require a class to be assigned to each record and time series analysis must include temporal information. One or more low level tasks can then be selected from the set of feasible tasks which will depend on the application area and the high level goals.

The following are examples of LLTs that could be performed during the data mining stage of the process.

Classification. Classification uses the attributes of a newly presented record to assign one of a predefined set of class labels [9]; classification tasks can be either prediction or description. Prediction tasks commonly involve "black-box" approaches which do not easily disclose the attributes that contributed towards a classification decision. A total classification will classify all classes within the database whereas a partial classification, sometimes referred to as nugget discovery, will concentrate on just the classes of interest.

Clustering. Records in the database are grouped together into clusters according to some measure of similarity based on the data within each record [9]. Clustering techniques can be used for description and prediction.

Regression. Used primarily for prediction, regression techniques construct a model that maps every record in the database onto a real value [10]. In some cases it may be possible to decompose the model and express the attributes in a form useful for description.

Dependency modelling. A model is produced which describes significant dependencies between attributes [11]. The model is normally used for description tasks because of the attribute information but can be used for prediction if required.

Time series analysis. Records in the database will have some form of temporal information available which allows the records to be ordered in some way. Patterns describing changes over certain time periods, or repeated patterns are sought [12]. These patterns can be used for both description and prediction.

Visualisation. The database, or a subset thereof, is presented graphically in a way that patterns and hence knowledge can be easily identified. Visualisation methods are commonly used in addition to other methods to identify the key features in complex patterns and are best suited to description tasks [13, 14].

Although the problem and associated data may be clearly suited to one of the above approaches, it is possible to convert some low level tasks into others. Consider a database where a set of small events is collected in the period leading up to a major event. Essentially this is temporal data and could be handled using a time series approach. However, if the fields describing the smaller events are used as the input fields and the major event becomes the output field then a classification approach could be applied.

Once the set of low level tasks has been decided upon, it is important to define an interest measure to compare the results of these tasks. There is no single interest measure that exists for all problems. A measure should be chosen that best suit the characteristics of the discovered knowledge. Interest measures can be based on properties such as the accuracy or confidence of the results for classification, or the spread of values in clustering [15].

2.1.7 Hardware and software requirements

Before starting the practical aspect of the project, it is vital to ensure the equipment available is suitable for performing the required tasks. Limitations of hardware and software could dramatically affect the project plan. If the equipment does not meet the required specifications, one of two choices has to be made. The project could be redesigned to require lower specifications by applying more intense pre-processing techniques such as sampling or feature selection, the net result being a smaller, simplified problem. Alternatively, new equipment could be purchased to reduce the search time. However, it would not be feasible to increase resources for techniques such as complex algorithms whose execution time is exponential in relation to the problem size; sometimes another approach has to be found.

Example areas to consider when examining the hardware and software requirements for the project are as follows.

- Standard data access and processing software such as database management systems (DBMSs), online analytical processing (OLAP) tools and spreadsheets. This software can be used to extract the working databases from the data warehouse and data marts and to cleanse and pre-process.

- Specialist KDD software to support pre-processing tasks including feature selection and discretisation. Software will also be required to support the data mining operations selected for the high and low level tasks.

- Hardware issues such as memory and storage requirements will also be an issue. Many data mining algorithms allocate vast amounts of memory and scaling up to larger databases is sometime impractical. The iterative nature of the process often involves storing multiple and intermediate copies of the working database, the size of which should be estimated to determine the required amount of disk space.

- A number of system performance matters to consider include the data access speed both locally and remotely, the speed of the processor for executing the data mining algorithms and whether the system is a shared resource. Updating hardware goes some way to improving the performance of the system, but careful choice and configuration of the software such as the operating and database systems can help significantly.

2.1.8 Feasibility determination

Determining the feasibility of a KDD project is a key stage that decides whether the project can continue with the current set of characteristics. Areas that should be examined to determine the feasibility include, but are not limited to, the following.

Data availability and quality. KDD cannot take place without data, but data availability and data quality are important factors. If data mining is carried out with the intention of using patterns to make predictions about events and those predictions are only of value if made by a certain time, the test data must be available from its source when needed. Data quality is an important feature that can be affected by missing values and reliability issues. If an excessively large proportion of the database is missing then data mining may become infeasible; however, if only a small portion of data is missing then several approaches can be applied as described in Section 2.3.3.

Equipment and performance. As discussed in Section 2.1.7 the hardware and software components can render a project infeasible. Pre-processing techniques such as feature selection, sampling, discretisation and clustering can reduce the dimensionality of the data so that effectively it can be dealt with as 'macro' rather than 'micro' level data.

Personnel. Although not discussed here in depth, it is important that the right people with the right data mining skills and sufficient domain knowledge are available when required throughout the project.

Size of database regions of interest. In some projects, nugget discovery is used to search for patterns describing only a small region of the database; possibly just one class. If the number of records in the class of interest is particularly small in relation to the entire database, the project may be infeasible.

High and low level task feasibility. The high and low levels tasks described in earlier sections can be dependent on the available data and equipment. For example, it is more difficult to carry out descriptive data mining if you only have access to neural network software which is more suited to predictive tasks. Similarly, if the available data does not have an associated element of time then time series analysis cannot be carried out.

Cost. An estimated cost for continuing with the proposed project constitutes the final component of the feasibility measure. Calculating factors such as the risk analysis or return on investment are critical to the success of the KDD project.

Once all of the building blocks of the problem specification have been put in place and the feasibility accepted, the project can continue.

2.1.9 Summary of the problem specification stage

The result of following through the details in the problem specification stage will be a tightly defined project specification, which contains the following components.

- A list of resource requirements, including cost, time, personnel, hardware and software. The list should be presented to management level personnel for approval.
- The high and low level tasks expected to be undertaken during the project.
- A data dictionary.
- A feasibility measure of the project.
- A detailed quality management document detailing all of the steps taken and decision made during this first stage.

2.2 Resourcing

The resourcing stage builds on the information collected during the problem specification stage. More precisely, it uses decisions made to gather the resources, including the data mining algorithms, required for the project. One of the most critical resources to obtain, and often the most time consuming, is any specified data. In some projects data may not be available when the project was proposed and the problem specification written down; in extreme cases the data may not even exist in a form which is useful to the project. For example, some of the data required for the project could still exist only on paper and may need to be converted to electronic format before continuing.

The central reserve of data within any large organisation would ideally be a data warehouse. These warehouses are vast stores of data that have been collected about all aspects of the organisation; usually over a number of years. Data for a KDD project may be need to be retrieved from one or more data warehouses, or from smaller 'data marts'. A data mart is a subset of the data warehouse which consists of related data from a particular area of the organisation for a selected group of users. Data may potentially be easier to source from a working data mart than a massive data warehouse which will generally contain far more information than is required or manageable by the KDD software.

Once data has been made available from all of these sources, it should be prepared and converted into a consistent format suitable for the KDD software. The format should be directly accessible by most of the KDD software and should have consistent field name structure, delimiters, type information and identifiers for missing data. Other data transformation procedures that should be considered for preparing a consistent data source include banding of data and determination of micro/macro level. Controlling banding of data refers to the examination of similar fields from different data sources. Consider a field representing age as young, middle-aged and elderly; different data sources may have different bounds on these categories which needs to be resolved to a consistent representation. Another point is the granularity of the data - does the database consist of micro-level or macro-level records? Micro-level records are the raw individual records whereas macro-level records represent a cluster, or grouping of similar records usually with some count of the number of records represented. Again, it is important for this granularity to be consistent over all data sources.

When all of the data is available and has been transformed into a useable format, we are said to have constructed the operational database for our KDD project. For databases that were not available in the problem specification stage, which should be those gathered during this stage, it is important to carry out the preliminary examination and familiarisation tasks described in Sections 2.1.3 and 2.1.4 respectively. The remaining resource issues deal with finding, and possibly employing, the appropriate personnel; obtaining the necessary equipment; and deciding upon a time plan for the project goals.

2.3 Data cleansing

The aim of the data cleansing stage is to prepare and clean the data for subsequent stages of the process. The cleansing stage involves such operations as searching for and removal of errors, sampling, dealing with missing and unreliable values, and possibly balancing. Although these tasks can be categorised as pre-processing, they differ from those in Section 2.4 in two key ways. Firstly, an element of learning can take place in the pre-processing stage that does not happen in cleansing. For example, the results from feature selection can expose important information about patterns in the database. Secondly, the cleansing stage is usually only performed once for a given database, whereas the pre-processing techniques can be applied a number of times.

2.3.1 Outlier handling

Outliers are generally classified as either errors or groups of interest [16]. If most of the outliers turn out to be of interest, they probably relate somehow to the overall goal of the KDD project. Within the data cleansing stage, only those outliers classified as errors are dealt with.

Domain knowledge is vital in the treatment of erroneous outliers. Knowing the domain and range of the fields within the database will help identify the problems in addition to understanding the semantics of the values. Domain knowledge can also help rectify the errors by suggesting feasible corrective actions.

2.3.2 Random sampling

Sampling in the context of the cleansing stage refers to the process of constructing one or more subsets of records. Sampling can take place either as a data reduction technique because the number of records exceeds the available resources, or to split the database into training and testing sets for validation purposes. In both cases, sufficient records should be available. The size of the chosen subsets will depend upon the size of the database and the resources available. For very large databases, it may be sufficient to select only 10% as the training database whereas for smaller problems a 50-50 split may prove better. Common sampling techniques include 'random' sampling where a number of records are selected at random from anywhere in the database; 'first n' sampling which simply selects the first n records from the database; and '1-in-n' sampling where every nth record is selected.

For very small databases alternative evaluation techniques such as bootstrapping or n-fold cross-validation [17] should be used. Cross-validation divides the database into n random subsets and the data mining algorithm run n times. On each run, one of the partitions is used for testing, the other n-1 will be used for training; the overall interest of the patterns will be calculated by taking the average of the results from all n runs.

2.3.3 Handling missing and unreliable data

The approach to handling missing and unreliable data should be determined and performed during the data cleansing stage. In the resourcing stage we stated that a decision should be made to represent all missing values using a common notation. However, when we come to apply the pre-processing and data mining algorithms there must be some way to account for these occurrences either by further transformation and treating them as normal values, or by using a set of algorithms that are missing value aware. Unreliable data can also be present in databases but tends to be considered less than missing values, usually since it is much harder to obtain unreliability figures and process the results.

When dealing with missing values it is important to understand the context in which the values are deemed missing. When considering the results of a questionnaire, if a question that can only take a true/false answer has no answer then the value is truly missing. Sometimes, the fact that the answer is missing could be interpreted as an additional value within the domain of the question and represented by an appropriate identifier within the field.

If the data mining algorithms to be used later in the process are capable of dealing with missing values, the cleansing stage has only to record this decision and ensure the representation selected during resourcing is suitable. If the missing data is not to be handled primarily during the data mining stage, there are two main approaches to 'repairing' the database at this point. A more detailed discussion of the following techniques can be found in [18].

Removal of data structures. This method is the most extreme measure that can be used to eliminate the problem of missing values. Any record and/or field that has a certain proportion of missing values is simply removed from the database. In the situation where missing values can be handled to some degree in later stages, it is possible to only remove records/fields that have a large amount of missing values; say over 50% missing. Although simple, this technique has the drawback of removing much valid information from the database and in domains where there are values missing from nearly all fields, this could potentially leave very little data to work with.

Estimation of missing values. Missing values are estimated using a variety of available techniques; these vary from the simple (using the arithmetic mean or a moving average) to the more complex (such as training a decision tree or neural network to predict the missing values using the available data [19]). While this approach does not remove information as above, the values being inserted could be vastly different from the true values (if available) and patterns discovered in later stages would be based on these.

If information is added to, or removed from, fields using one of the above methods, the changes should be recorded in some form and used in conjunction with the results of the data mining stage. Where values have been estimated, these fields should be treated as less reliable than those with complete original values.

When fields are known to be unreliable, they should be flagged and taken into consideration when evaluating the results of data mining. The major difficultly working with real-world data is assigning a reliability score to each field. For example, it may be feasible to estimate the reliability of readings taken from a monitoring system on a modern factory machine, but it would be more difficult to apply the same process to historical data where the reliability was unknown. A further complication is storing the reliability measures alongside the data at the record, field and even element levels; however, advancements in database systems technology, such as object-oriented structure and complex data types, are making this less difficult.

2.3.4 Database balancing

The database or databases to be used may need to be 'balanced' at this stage. Balancing is the process of reducing the ratio between the number of records belonging to a minority class and those belonging to the majority class. As this description suggests, the approach is extremely useful for partial classification tasks (nugget discovery) where the class of interest is considered rare. Generally, balancing algorithms work in one of two ways.

Data deletion. Records that do not belong to the class of interest are discarded at random until the proportion of records in the chosen class is sufficiently large. This approach has the disadvantage of throwing away large quantities of data.

Data duplication. Records that belong to the class of interest are duplicated at random until the proportion of records in the chosen class is sufficiently large.

The disadvantage of this approach is that weaker patterns may be distorted to appear more important than they really are and the level of noise in the database could increase. Increasing the number of records will also introduce a performance issue for the data mining algorithms.

One method that may prove beneficial when balancing is to experiment with databases made up of different ratios of classes. For example, begin training with a well-balanced database, then progressively use a less balanced version until the database performs sufficiently well.

2.4 Pre-processing

The pre-processing stage of the KDD process is usually the first place where learning takes place and potentially useful patterns can be found. Although the entire process is iterative, the pre-processing is one of the main stages that is applied a number of times; second only to the data mining stage. If one of the outcomes of the problem specification stage required the size of the data to be reduced, the pre-processing stage is the most likely place for this to happen. Algorithms used in the previous stages will not have been as complex and resource intensive as those in the coming stages. It is not necessary to make use of all pre-processing techniques but in many cases, they can provide valuable information about the data that can improve the performance of the mining algorithms. The following operations may be performed at this stage.

2.4.1 Feature construction

Feature construction techniques, such as those described in [20, 21], can be used to build new fields using one or more constructive operators. Examples of such operators include aggregation to summarise a set of fields, the difference between two fields, or breaking down a complex encoded field into its simpler constituent parts. Domain knowledge can also be used to build new fields or modify existing ones. Ideas for feature construction often originate from the data visualisation stage previously discussed in Section 2.1.4 where patterns and relationships from graphs can be represented by new fields. Constructing fields does introduce a few drawbacks; the complex nature of the new fields could make the resulting patterns more difficult to understand, and adding additional fields to a database will obviously add performance overheads.

2.4.2 Feature selection

Feature selection, or feature subset selection (FSS), is the process of selecting a highly predictive subset of fields from the entire database and discarding those fields that appear weak in comparison. If training and testing sets are being used, feature selection is applied only to the training database. Details of feature selection algorithms can be found in [22, 23, 24]. Because feature selection is used to select the most powerful features with respect to the output field, the benefits are two-fold. Firstly, the fields which relate strongest to the target concept are identified and can be monitored during data mining. Secondly, by reducing the number of fields in the database (which may have been a requirement from the problem specification) the performance of the data mining algorithms will be improved in terms of speed and possibly accuracy and simplicity.

2.4.3 Discretisation

A variety of discretisation techniques are described in [25]. Some data mining algorithms may only work with discrete data; others may benefit in performance from the simplified format. The advantages of using discretisation are similar to those of feature selection, that is an increase in speed and simplicity. However, both techniques can suffer if a poor quality algorithm is used since the data mining stage is designed to work only with the data provided as a result of pre-processing. A large number of discretisation algorithms exist which can be grouped into a number of categories.

Local or global. Local discretisation is applied to a localised region of the database whereas global methods are applied to the entire database.

Supervised and unsupervised. Supervised methods will use the class value (where applicable) to help group together sets of values within the fields; unsupervised methods will only use the values in the current field

Static or dynamic. Dynamic methods discretise all features simultaneously, whilst static approaches discretise each field individually in turn.

2.5 Data mining

The core stage of all KDD projects is the data mining stage where the patterns and knowledge are extracted from the database. There is a wide range of data mining algorithms available; some are quite similar differing by only minor changes in the algorithms whilst others work in very different ways. Each algorithm will have its own advantages and disadvantages in terms of the accuracy, efficiency and applicability suit different formats of database. Ideally, if enough software packages are available, multiple algorithms can be used to perform the same task. Results from one algorithm could even influence the way others are used, producing better overall results.

Typically, each data mining algorithm has a number of parameters which should be considered before execution. Most software packages make available sensible defaults for these parameters but given the extreme variation between databases it is important to understand how the parameters affect the search. Parameters generally fall into one of two categories.

Algorithm parameters. These parameters control the execution of the algorithm and the shape of the search space affecting its overall performance. Parameters such as the topology of neural networks or the neighbourhood operators in simulated annealing are examples of algorithm specific parameters.

Problem parameters. These parameters allow the user to manage the options related to a particular instance of a problem and to have some control over the format of the extracted knowledge. For example, the number of clusters required in a clustering algorithm or the maximum number of fields that may appear in a rule for a classification task.

Data mining is performed on the training portion of the operational database using the chosen algorithms and parameters. When the algorithms have finished running, any knowledge discovered is examined to determine whether training has been successful. At this stage only preliminary evaluation techniques are used based on the results of the training set; full evaluation of the results on testing sets is covered in Section 2.6. If the evaluation results on the training data are not satisfactory using the chosen interest measure, the algorithms are re-run with a different set of parameters. The first category of parameters that should be revised is the algorithm specific ones. These changes will help improve the algorithm's chances of finding solutions that satisfy the desired solution attributes; increasing the length of time for which the algorithm is allowed to run is also a valid change. If, after making these changes, the results are still unsatisfactory then we revert to modifying the problem parameters. Changing these parameters will alter the type of solution the algorithm is trying to find which may help if the original set of problem parameters were looking pattern types that do not exist in the database. Possible changes to the problem parameters may include lowering the desired accuracy or allowing the algorithm to use more complex rules. In circumstances where changes to both sets of parameters do not improve the results, then more drastic action may be required such as choosing a different data mining algorithm or returning to previous stages and reformatting the data.

2.6 Evaluation of results

There are a range of approaches which can be used for evaluating the results of the data mining stage. The decision as to which method to use will depend on the data mining goal and the interest measure chosen to compare results; these will have been decided upon in Section 2.1.6. If the KDD problem includes a testing database, it can be used within this phase to evaluate the performance of the discovered knowledge with previously unseen examples. If no test database is available, this phase is effectively merged with the data mining stage since the training database is used for both production and testing of patterns (see Section 2.3.2). The following areas are considered when evaluating discovered knowledge.

Performance on testing database. If separate training and testing databases were constructed in Section 2.3.2, the testing database is used here to evaluate the performance of the discovered knowledge using a suitable interest measure (in most cases the same as that used on the training database). Where sufficiently accurate results are obtained on the test data, we can proceed to the next stage otherwise there are various reasons why the patterns may not have performed well. If the algorithm was trained for too long, or searched for highly accurate patterns, it may have been overfitted on the training data. Overfitting is where the discovered patterns are not general enough to include examples that differ only slightly from those found in the training data. Returning to the data mining stage and adjusting the algorithm, and possibly the problem parameters can help rectify this situation.

Simplicity. The structure and complexity of the solutions will contribute towards the overall evaluation of the discovered patterns. If the project's high level task was descriptive then obtaining clear and understandable patterns are vital, although the format will depend on to whom the results are presented. A KDD engineer will normally understand the raw output of the algorithms, domain experts will understand a slightly tidied-up format of the results but higher level management would require clear, summarised patterns that can easily be communicated.

Application area suitability. The suitability of the discovered knowledge to the application area is important in determining the success of the KDD project. After performing data mining on the database, the resulting patterns may be of an unsuitable form or of insufficient quality. Applications areas which use the extracted knowledge to evaluate critical decisions will require very accurate rules that can be justified; if the results do not meet the required criteria the project must be revised accordingly. In the worst case this may involve returning to the problem specification stage and rethinking the overall goals of the project; in less drastic situations it may involve returning to pre-processing or data mining.

Visualisation. Visualising the results of data mining and comparing patterns with the original database can yield potentially useful characteristics of the data. The types of problems that can be understood further by using this approach include determining areas of the solution space where rules are performing poorly and identifying how rules can be simplified to become more general and easier to comprehend.

2.7 Interpretation of results

Interpreting the results is an advanced stage of evaluation that is performed by domain experts. The majority of the evaluation measures in Section 2.6 can be carried out by KDD engineers using software tools and their knowledge of the project. The crucial test for discovered knowledge is to satisfy the domain experts who can justify the results using their much deeper knowledge of the problem domain. Patterns that differ greatly from the knowledge of the domain expert should be carefully analysed to explain such an anomaly. The most likely cause of such differences would be an error during part of the KDD process; commonly during the cleansing and pre-processing stages. These phases are when the database undergoes a number of major transformations that are difficult and time consuming to validate by eye. However, in rare cases the differences may represent important knowledge regarding application areas that have not been completely understood by the domain experts.

The key outcome from the interpretation stage is a set of patterns representing knowledge that has been validated and justified by domain experts. Frequently, domain experts are able to justify patterns predicting certain outcomes when presented to them, but would not necessarily have known to look at that particular pattern due to the large number from which to choose. The discovered knowledge can then be summarised and illustrated in a suitable form that can be presented to higher level management, with the aim of exploiting this information.

2.8 Exploitation of results

The final stage of the KDD process involves exploiting the knowledge discovered from performing the previous stages. So far, the results from data mining will have undergone thorough evaluation and been validated in considerable detail by domain experts; the knowledge is now believed to be of good quality, valid and suitable for the application area. The overall aim is to maximise the return on investment in the project, which apart from any additional benefits, helps to justify the undertaking of this, and future, KDD projects. Exploitation integrates results into the working environment and by putting them to good use will yield benefits for the organisation. Any changes implemented within the organisation may require a major transformation of the day-to-day working practices and the process of integration should be carefully planned. Implementing the changes may involve a cost to the organisation that should have been taken into consideration when initially determining the feasibility of the project. However, it is all too common for KDD projects to be carried out and the results not exploited to their full potential.

The extracted knowledge can be integrated and exploited within the organisation in many ways. Ideally, the process should involve maximum benefit with minimum risk. It may prove beneficial to utilise gradual integration by performing trials with the new procedures either in practice or using simulation software. For a purely descriptive high level task, the knowledge can be used to change the working procedure of a particular team, or initiate an entire new project. When the results are to be used for predictive purposes, whether using black box techniques or a set of descriptive patterns, this needs to be embedded into current systems. If some form of expert system already exists and is being used to make predictions, the additional knowledge can be added directly to improve performance. In other cases, new software will need to be written to embed the knowledge; some data mining software can export code so the patterns can easily be used within proprietary software.

When the knowledge has been suitably exploited and the KDD project is nearing the end of the process, it is important to look back and review several key issues. Probably the most important area to examine involves a final cost benefit analysis for the project which should be delivered to management level personnel who originally approved the work. Decisions, software and project personnel should be appraised to examine the efficiency of the process and compared to the initial problem specification stated in the quality management documents. This documentation can also be useful if KDD is planned to be an ongoing investment for the organisation. Future projects can refer back to the detailed documentation, which can provide a useful starting point and source of reference. Once experience is gained in the handling of KDD projects there is the potential to automate the process and re-run the procedure on a regular basis. Any changes found between runs of the process may provide crucial information to the company and allow them to keep up to date with the environment in which they operate.

3. Using the Roadmap in the Design a of KDD Toolkit

The previous sections have shown how following the proposed methodology for knowledge discovery can simplify the task of designing and carrying out a KDD project. At many stages during the process, the practitioner relies on software to be available to perform a given task. As part of a continuing interest in the field of KDD, the authors have worked with the Lanner Group to design and implement the DataLamp Predictor toolkit [26].

The methodology played a key role in the design of the toolkit. The software uses a visual stream-based interface to represent routes through the KDD roadmap. Each node in a stream belongs to one of the eight stages of KDD and represents an activity that can be performed on the database; the links between nodes signify the data flow between the stages in the process. The methodology helps define the order in which tasks can be performed and the critical parameters for each stage. When a new node is linked to an existing stream, the system must determine the validity of the connection. Validation rules from the different stages of the methodology are built into the software and ensure that any prerequisites for a new node are met by the outputs of the previous stage before completing the link.

Users can be guided through the KDD process using wizard-style dialogs that examine the database at each stage and propose tasks that should be performed according to the characteristics of the data. For example, at the cleansing stage if the database contains a sufficient number of records, sampling would be suggested, if not cross-validation may be suggested at the data mining stage.

The quality management document described in Section 2.1.2 can be partially automated using the software. When a database is first loaded into the toolkit, the data is scanned to construct the data dictionary; the user can examine this information to validate their understanding of the data and to ensure the correct data is being used. The documentation is updated each time the user runs a stream of KDD tasks. The system logs the decisions and parameters used at each stage, along with run times and any intermediate and final results. These reports, detailing runs of the program with different parameters, can be examined to identify the critical decisions that affect performance. Reports are created in standard HTML format to enable users to readily share information on their organisation's Intranet.

References

[1] C. Westphal and T. Blaxton, *Data Mining Solutions: Methods and Tools for Solving Real-World Problems*. John Wiley, 1998.

[2] C. X. Ling and C. Li, "Data Mining for Direct Marketing: Problems and Solutions", *Proc. of Fourth Int. Conf. on Knowledge Discovery and Data Mining (KDD-98)*, AAAI Press, 1998.

[3] J. C. W. Debuse, *Exploitation of Modern Heuristic Techniques within a Commercial Data Mining Environment*, Ph.D. Thesis, University of East Anglia, 1997.

[4] U. M. Fayyad, G. Piatetsky-Shapiro and P. Smyth, "Knowledge Discovery and Data Mining: Towards a Unifying Framework", *Proc. of Second Int. Conf. on Knowledge Discovery and Data Mining (KDD-96)*, AAAI Press, 1996.

[5] U. M. Fayyad, G. Piatetsky-Shapiro, P. Smyth *et al.*, "From Data Mining to Knowledge Discovery: An Overview", in *Advances in Knowledge Discovery in Databases*, Chapter 1, AAAI Press / The MIT Press, 1995.

[6] R. J. Brachman and T. Anand, "The process of knowledge discovery in databases", in *Advances in Knowledge Discovery in Databases*, U. M. Fayyad, G. Piatetsky-Shapiro, P. Smyth *et al.* Eds., Pages 37-58, AAAI Press / The MIT Press, 1995.

[7] P. Chapman, J. Clinton, J.H. Hejlesen *et al.*, The current CRISP-DM process model for data mining, http://www.crisp-dm.org, 1998.

[8] R. S. Pressman, *Software Engineering: A Practitioners Approach*, McGraw-Hill, 1992.

[9] M. J. A. Berry and G. Linoff, *Data Mining Techniques for Marketing, Sales and Customer Support*, John Wiley, 1997.

[10] A. Berson and S. J. Smith, *Data Warehousing, Data Mining and OLAP*, McGraw-Hill, 1997.

[11] R. Agrawal, H. Mannila, R. Srikant et al., "Fast Discovery of Association Rules", in *Advances in Knowledge Discovery in Databases*, U. M. Fayyad, G. Piatetsky-Shapiro, P. Smyth *et al.* Eds., Pages 307-328, AAAI Press / The MIT Press, 1995.

[12] G. Das, K-I Lin, H. Mannila *et al.*, "Rule Discovery from Time Series", *Proc. of Fourth Int. Conf. on Knowledge Discovery and Data Mining (KDD-98)*, AAAI Press, 1998.

[13] C. Ware, *Information Visualisation: Perception for Design*, Morgan Kaufmann, 2000.

[14] C. Brunk, J. Kelly and R. Kohavi, *MineSet: An Integrated System for Data Mining*, *Proc. of Fourth Int. Conf. on Knowledge Discovery and Data Mining (KDD-97)*, AAAI Press, 1997.

[15] R. J. Bayardo and R. Agrawal, "Mining the Most Interesting Rules", *Proc. of Fifth ACM SIGKDD Int. Conf. on Knowledge Discovery and Data Mining (SIGKDD-99)*, AAAI Press, 1999.

[16] J. C. W. Debuse and V. J. Rayward-Smith, "Knowledge Discovery issues within the financial services sector: the benefits of a rule based approach", *Proc. of the UNICOM Data Mining / Data Warehousing Seminar*, UNICOM, 1998.

[17] S. M. Weiss and C. A. Kulikowski, *Computer Systems that Learn*, Morgan Kaufmann, 1991.

[18] C. M. Howard and V. J. Rayward-Smith, "Knowledge Discovery from Low Quality Meteorological Databases", *in Knowledge Discovery and Data Mining*, M. A. Bramer Ed., IEE, 1999.

[19] A. Gupta and M. S. Lam, "Estimating Missing Values using Neural Networks", *Journal of the Operational Research Society*, Vol. 47, 1996.

[20] A. Ittner and M. Schlosser, "Discovery of relevant new features by generating non-linear decision trees", *Proc. of Second Int. Conf. on Knowledge Discovery and Data Mining (KDD-96)*, AAAI Press, 1996.

[21] C. J. Matheus and L. A. Rendell, "Constructive induction on decision trees", *Proc. of the Eleventh Int. Joint Conf. on Artificial Intelligence*, Morgan Kaufmann, 1989.

[22] P. A. Devijver and J. Kittler, *Pattern Recognition: A Statistical Approach*, Prentice-Hall Int., London, 1982.

[23] G. H. John, R. Kohavi and K. Pfleger, "Irrelevant features and the subset selection problem", *in Machine Learning: Proc. of the Eleventh Int. Conf.*, W. W. Cohen and H. Hirsh Eds. Morgan Kaufmann, 1994.

[24] M. Pei, E. D. Goodman, W. F. Punch *et al.*, "Genetic algorithms for classification and feature extraction", *Proc. of the Classification Soc. Conf.*, 1995.

[25] J. Dougherty, R. Kohavi and M. Sahami, "Supervised and unsupervised discretization of continuous features", *Proc. of the Twelfth Int. Conf. on Machine Learning*, A. Prieditis and S. Russell Eds., San Fancisco CA, 1995.

[26] Lanner Group Ltd, "DataLamp Predictor User Guide", www.lanner.com, 1999.

Chapter 3

Theory to Practice

Supporting Virtual Communities of Practice
J. Davies

Managing Micro- and Macro-level Design Process Knowledge across Emergent
Internet Information System Families
A. Tiwana

The Development of Case-Based Reasoning for Design - Techniques and Issues
S. Potter, S. Culley, M. Darlington and P. Chawdhry

An Internet-Based Approach to the Capture and Reuse of Knowledge in Design
P. Rogers N.H.M. Caldwell and A.P. Huxor

Ontology-Driven Knowledge Management: Philosophical, Modelling and
Organizational Issues
S.B. Shum, J. Domingue and E. Motta

Supporting Virtual Communities of Practice

John Davies

BT Advanced Communications Technology Centre
Adastral Park
Ipswich
IP5 3RE
UK

Abstract: The importance of knowledge sharing and reuse in knowledge management in order to share best practice and prevent duplication of effort has led to much interest in the concept of communities of practice. Trends towards flexible working and globalisation have led to interest in the use of technology to support geographically dispersed communities. In this paper, we describe a system which facilitates and encourages the sharing of information between communities of practice within (or perhaps across) organisations and which encourages people to interact where there are mutual concerns or interests. We discuss the extent to which knowledge flow through an organisation is supported and describe a number of factors that we have found relevant to the success or otherwise of virtual communities of practice.

1. Introduction

The importance of knowledge sharing and reuse in knowledge management in order to share best practice and prevent duplication of effort has led much interest in supporting communities of practice. Recent ethnographic studies of workplace practices indicate that the ways people actually work often differ fundamentally from the ways organisations describe that work in manuals, organisational charts and job descriptions. The term 'community of practice' [1] describes the informal groups where much knowledge sharing and learning takes place and has been increasingly applied in the knowledge management context. Essentially a community of practice is a group of people who are 'peers in the execution of real work' [2]. They are typically not a formal team but an informal network, each sharing in part a common agenda and shared interests. In one example it was found that a lot of knowledge sharing among photocopier engineers took place through informal exchanges, often around the coffee point. As well as such local, geographically-based communities, trends towards flexible working and globalisation has led to interest in supporting geographically dispersed communities using Internet technology. The challenge for organisations is to support such communities and make them effective.

Knowledge management tools must give users the ability to organise information into a controllable asset. Building an intranet-based store of information is not sufficient for knowledge management; the relationships within the stored information are vital. These relationships cover such diverse issues as relative importance, context, sequence, significance, causality and association. The potential for knowledge management tools is vast; not only can they make better use of the raw information already available, but they can sift, abstract and help to share new information, and present it to users in new and compelling ways

In this paper, we describe the Jasper system which facilitates and encourages the sharing of information between communities of practice within (or perhaps across) organisations and which encourages people – who may not previously have known of each other's existence in a large organisation – to make contact where there are mutual concerns or interests.

We go on to discuss some of the factors we have found to be important in the creation and sustenance of viable virtual communities of practice, based on our trials of Jasper with different user communities.

2. Jasper

Jasper is WWW-based knowledge sharing environment comprised of a system of intelligent software agents that hold details of the interests of their users in the form of user profiles. Jasper has the capability to summarise and extract key words from WWW pages and other sources of information and it then shares this information with users in a community of practice whose profiles indicate similar interests.

Jasper agents are used to store, retrieve, summarise and inform other agents about information considered in some sense valuable by a Jasper user. This information may be from a number of sources: it can be a note typed by the user him/herself; it can be an intra/Internet page; or it can be copied from another application on the user's computer.

Each Jasper user has a personal agent which holds a user profile based on a set of key phrases which models that user's information needs and interests. As we will see below, Jasper agents modify a user's profile based on their usage of the system, seeking to refine the profile to better model the user's interests.

When a user finds information of sufficient interest to be shared with their community of practice, a 'share' request is sent to Jasper via a menu option on his or her WWW browser. Jasper then invites the user to supply an annotation to be stored with the information. Typically, this might be the reason the information was shared or a comment on the information and can be very useful for other users in deciding which information retrieved from the Jasper store to access. The user can also specify at this point one of a predefined set of interest groups to which to post the information being stored.

In the case of WWW-based information the URL of the WWW page is then added to the Jasper store. Similarly, when the user wishes to store some information from a source other than WWW, he or she can enter the information in a text box on their WWW browser and can again supply a relevant annotation. The information thus entered could be from a document in another format or might be a note or snippet of knowledge that the user wishes to enter directly themselves. This information is converted to a WWW HTML page on the user's Jasper server and stored as before.

When information is shared in this way, the Jasper agent performs four tasks:

i. an abridgement of the information is created, to be held on the user's local Jasper server. This summary is created using the ProSum text summarisation tool. The summariser extracts key theme sentences from the document. It is based on the frequency of words and phrases within a document, using a technique based on lexical cohesion analysis [3]. Access to this locally held summary enables a user to quickly assess the content of a page from a local store before deciding whether to retrieve (a larger amount of) remote information.

ii. the content of the page is analysed and matched against every user's profile in the community of practice. If the profile and document match strongly enough, Jasper emails the user, informing him or her of the page that has been shared, by whom and any annotation added by the sharer.

iii. the information is also matched against the sharer's own profile. If the profile does not match the information being shared, the agent will suggest phrases that the user may elect to add to their profile. These phrases are those reflecting the information's key themes and are automatically extracted using the ProSum system. Thus Jasper agents have the capability to adaptively learn their user's interests by observing the user's behaviour.

iv. for each document, an entry in the Jasper store is made, holding keywords, an abridgement of the document, document title, user annotation, universal resource locator (URL), the sharer's name and date of storage.

In this way, a shared and enhanced information resource is built up in the Jasper store. Given that users must make a conscious decision to store information, the quality of the information in the Jasper store is high - it is effectively pre-filtered by Jasper users. Furthermore, each user leverages the assessment of the information made by all the other users.

2.1 Sharing & retrieving explicit knowledge in Jasper

We have seen in the section above how Jasper allows a user to store information of interest using an enhanced, shared bookmark concept. This facility goes well beyond the bookmarks familiar from WWW browsers such as Netscape Communicator, in that in addition to the reference to the remote WWW document, a summary of the document, an annotation, date of storage and the user who stored the information are recorded in a shared store. Furthermore, Jasper can be used to store and organise information from many sources and in many formats (rather than only WWW-based information).

In this section, we discuss the various ways in which Jasper facilitates access to and the automatic sharing of the information thus stored.

Email notification

As described above, when information is stored by a Jasper agent, the agent checks the profiles of other agents' users in its particular community (the set of users who contribute to that particular Jasper community). If the information matches a user's profile sufficiently strongly, an email message is automatically generated by the agent and sent to the user concerned, informing the user of the discovery of the information. Thus in cases where a user's profile indicates that they would have a strong interest in the information stored, they are immediately and proactively informed about the appearance of the information.

Keyword retrieval – Accessing information and people

From their Jasper home page, a user can supply a query in the form of a set of key words and phrases in the way familiar from WWW search engines (see Figure 1). The Jasper agent then retrieves the most closely matching pages held in the Jasper store, using a vector space matching and scoring algorithm [4].

In addition to these pages from the Jasper store, the agent can also retrieves a set of pages from an organisation's intranet and from the WWW. The agent then dynamically constructs an HTML page with a ranked list of links to the pages retrieved and their abridgements, along with the scores of each retrieved page. In the case of pages from the Jasper store, any annotation made by the original user is also shown. Note that the user can ask their agent to search for other users by selecting the appropriate check box (see Figure 1). We will have more to say about this capability to identify other *users* as well as *information* in Section 2.3 when we look at accessing tacit knowledge via other users using Jasper.

What's new

A user can ask his Jasper agent "What's new?" The agent then interrogates the Jasper store and retrieves the most recently stored information. It determines which of these pages best match the user's profile. A WWW page is then presented to the user showing a list of links to the most recently shared information, along with annotations where provided, date of storage, the sharer and an indication of how well the information matches the user's profile (the thermometer-style icon in Figure 1).

In addition, a series of buttons are provided so that the user can:

i.   add their own comment or annotation to information stored by another user;
ii. indicate interest or disinterest in a particular piece of information – this feedback will be used to modify the user's profile;
iii. examine a locally held summary of the information before deciding to download all the information;
iv. ask their Jasper agent to identify other users with an interest in the information under consideration.

This What's New information is in fact displayed on the user's Jasper home page, so that whenever they access the system, they are shown the latest information. A typical Jasper home page is shown in Figure 1.

2.2 Adaptive agents

We have already mentioned that Jasper agents adapt to better understand their user's interests over time. There are two types of event which trigger the profile adaptation process.

As discussed above, when a user is sharing some information, if the sharer's profile does not match the information being stored Jasper will automatically extract the main themes from the information using ProSum. The user's agent then suggests to the user new phrases that they may wish to add to their profile. The user can accept or decline these suggestions.

Similarly, when information stored by another member of the community is retrieved by a user using one of the methods described in section 2.1, a feedback mechanism is provided whereby the user can indicate interest or disinterest in the information by clicking on a button (indicated by ☺ or ☹ as shown in Figure 1). Again, the agent will suggest to the user phrases that should be added to or removed from the profile.

2.3 Finding people & tacit knowledge in Jasper

In Section 2.1, we focused on the *technical* aspects of Jasper and on the sharing and storing of explicit knowledge. Explicit knowledge we take to be that knowledge which has been codified in some way. This codification can take place in many different media (paper, WWW page, audio, video, and so on). In the context of Jasper, by explicit knowledge, we mean the information shared in Jasper, along with the meta-information associated with it such as the sharer, the annotations attached to it, and so forth. We now turn to the *social* aspects of the system and tacit knowledge.

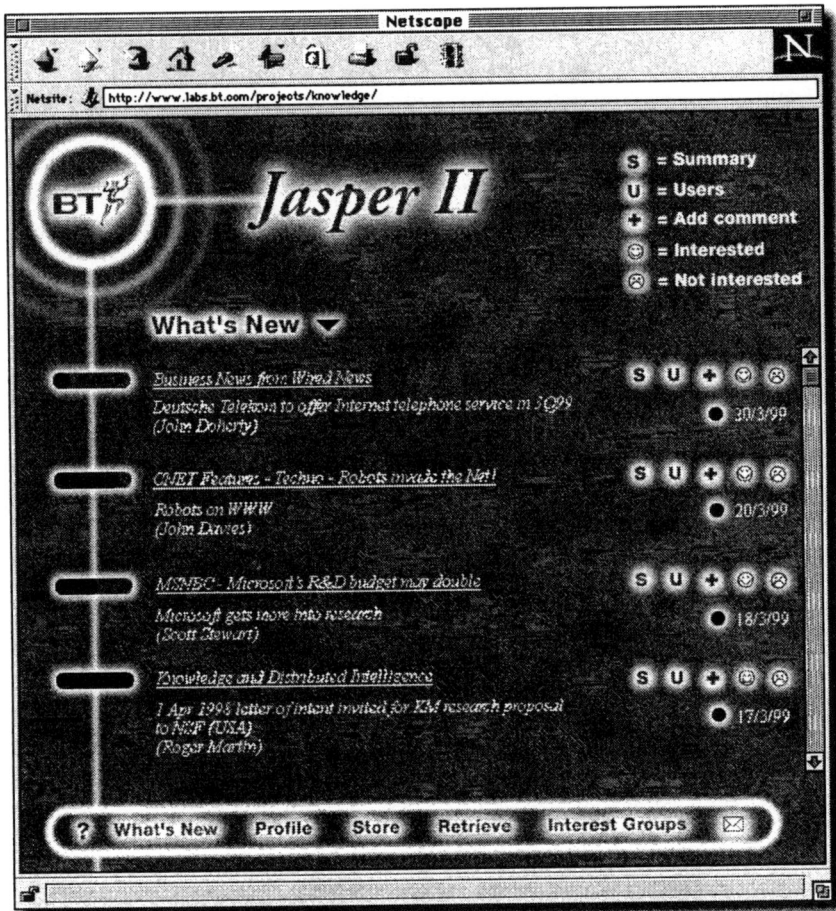

Figure 1: A typical Jasper home page

A large amount of the knowledge within an organisation may of course not be codified: it may be personal, context-specific and difficult to write down, and may be better transmitted through a master-apprentice "learning by watching and copying" arrangement. Such knowledge is referred to as *tacit* knowledge [5]. When tacit knowledge is difficult to make explicit (codify), we need to find new ways of transmitting the knowledge through the organisation. Failure to do so can lead to loss of expertise when people leave, failure to benefit from the experience of others, needless duplication of a learning process, and so on.

One way in which a system such as Jasper can encourage the sharing of tacit knowledge is by using its knowledge of the users within a community of interest to put people who would benefit from sharing their (tacit) knowledge in touch with one another automatically.

One important way we gain new insights into problems is through 'weak ties', or informal contacts with other people [6, 7]. Everyone is connected to other people in social networks, made up of stronger or weaker ties. Stronger ties occur between close friends or parts of an organisation where contact is maintained constantly. Weak ties are those contacts typified by a 'friend of a friend' contact, where a relationship is far more casual. Studies have shown that valuable knowledge is gathered through these weak ties, even over an anonymous medium such as electronic mail and that weak ties are crucial to the flow of knowledge through large organisations. People and projects connected to others through weak ties are more likely to succeed than those not [8, 9].

Though Jasper does not explicitly support weak ties, initial trials of Jasper have shown a number of features that support a community:

- People contributing information are more likely to make informal contact with others using Jasper.
- Jasper can identify those people who could be sources of information.
- The store of URLs, with associated annotations and other meta-information, becomes a long-term memory, for the community.

User profiles can be used by the Jasper system to enable people to find other users with similar interests. The user can request Jasper via their WWW client to show them a list of people with similar interests to themselves. Jasper then compares their profile with that of every user in the store and returns to the WWW client for viewing by the user a list of names of users whose interests closely match their own. Each name is represented as a hypertext link which when clicked initiates an email message to the named user. Profiles in Jasper are a set of phrases and the vector space model can be used to measure the similarity between two users. A threshold can then be used to determine which users are of sufficient similarity to be deemed to 'match'.

This notion is extended to allow a user to view a set of users who are interested in a given document. When Jasper presents a document to the user via their WWW client using the "What's new?" facility (see above), there is also a hyperlink presented which when clicked will initiate a process in the Jasper system to match users against the document in question, again using the vector cosine model. Jasper determines which members of the community match the relevant document above a predetermined threshold figure and presents back to the user via their WWW client a list of user names. As before, these names are presented as hypertext links, allowing the user to initiate an email message to any or all of the users who match the document.

In addition, as we have already discussed in section 2.1, a user can carry out a keyword search on other users and thus identify users with an interest in a particular subject.

In this way, Jasper, while not claiming to actually capture tacit knowledge, provides an environment which actively encourages the sharing of tacit knowledge, perhaps by people who previously would not otherwise have been aware of each other's existence.

2.4 Organisational knowledge flow

In the organisational literature, it is common to distinguish between different types of knowledge along two axes: tacit and explicit knowledge, as discussed in the previous subsection; and group and individual knowledge.

Clearly, knowledge management is concerned with the sharing of knowledge and hence the transfer of individual knowledge to group knowledge. This much is uncontentious but the means by which this may be achieved is less so. Nonaka [10] is typical of much of the literature when he says "While tacit knowledge held by individuals may lie at the heart of the knowledge creating process, realising the practical benefits of that knowledge centres on it externalisation...", where by externalisation is meant the conversion of tacit knowledge to explicit. Therefore Nonaka is keen to attempt facilitation of this conversion process.

However, as we have discussed above, we view tacit knowledge as essentially of a different type from explicit knowledge. It follows from this that tacit knowledge cannot necessarily be 'externalised'. Space does not permit a full discussion of the epistemological issues here and the reader is referred to [11] for further details.

Given that the transfer of tacit to explicit knowledge is at best problematic, we have concentrated in Jasper on the transfer of individual to group knowledge, as shown in Figure 2. Explicit knowledge is transferred via the sharing of information (along with additional meta-information) in Jasper. Tacit knowledge transfer is not directly supported in Jasper but is facilitated by allowing and encouraging users to contact others with shared interests or concerns.

3. There's No Algorithm for Community[*]

We have experience of using Jasper with a number of communities. As the title of this section indicates, there is no blueprint to guarantee a successful virtual community. In our experience, there are however a number of relevant factors, some of which are inter-related. We discuss these factors below:

---

[*] This succinct (and accurate) observation originates, I believe, from Peter Kollock of University of California at Los Angeles.

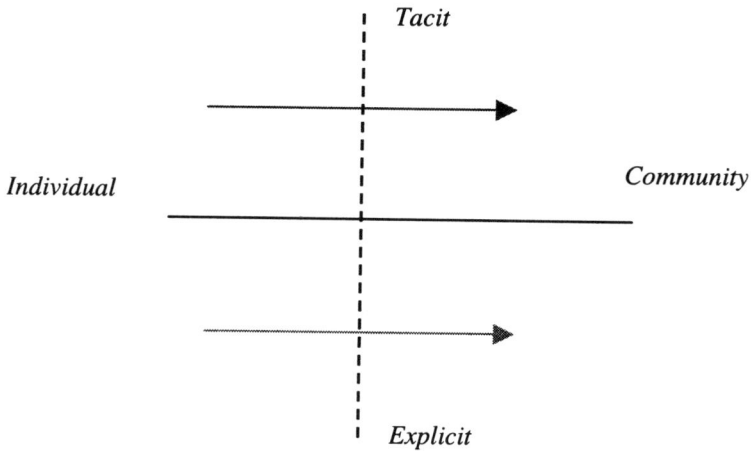

Figure 2: Knowledge flows in Jasper.

Reward collaborative behaviours. This can be done in a number of ways. Collaboration in appropriate communities can be written into job descriptions and/or personal objectives and thus become part of an organisation's appraisal processes. Less formal approaches could include the award of monthly prizes for the most effective contributions to a community, perhaps selected by a senior manager. Care needs to be exercised if using purely numeric measures (an example of this would be to measure the number of information items a user stores in Jasper). There can be a tendency for users to exploit these kinds of metrics by flooding a system with items of questionable quality.

Identify a senior management champion and gain local management support. We have found it almost universally true that senior management buy-in and active promotion is very helpful. In one trial, users were asked if increased usage of Jasper by their immediate manager would affect their usage. The results of this inquiry are shown in Table 1 below.

Table 1: Usage of Jasper as affected by manager's usage.

	$\%^{age}$
Probably Use it less	0
Probably use it as now	21
Probably use it somewhat more	32
Probably use it a lot more	17
Not applicable as my manager is an active user	30

A degree of care must be taken, however, to ensure that involvement from senior figures does not jeopardise the informal nature of the communities. Studies at

Xerox found that informality was key to creating trust and a greater tendency to information exchange in many situations [2].

Consider job types and cultures. In practice, it is usually the case that many different cultures exist within the overarching culture of a large organisation. One needs to understand the culture (and of course information needs) of candidate job groups in order assess the areas where communities of practice are likely to be most beneficial. In one organisation, we trialled a collaborative working solution with a community of engineers and a community of salespeople: The engineers' community flourished, while that of the salespeople withered on the vine. When we looked a little more closely at the nature of the communities, we uncovered the attributes summarised in Table 2. Of course, this summary would not necessarily apply universally: it does not even apply to all the members of the groups we were working with but nevertheless explains the success of one group and relative failure of the other.

Table 2: Job groups and culture.

Engineers	Salespeople
Job spanned multiple customers	Rarely concerned with more than one customer
Motivation tied to best quality solutions	Motivated by unique solutions and 'being first'
Schooled in TQM and reuse of best practice	Competitive by personality (and selection)
	Compensation schemes reinforce competitive attitudes

Technology drivers. It has become something of a commonplace in KM circles that "technology is only 20% of the solution, the rest is culture". While the sentiment that we need a wider perspective than just technology is indisputably correct, it reveals the assumption of a dichotomy between technology and organisational culture which does not exist. Rather, technology-based tools are among the many artefacts entwined with culture, whose use both affects and is affected by the prevailing cultural environment. A holistic view is required and technology often plays a larger part in cultural factors than is sometimes acknowledged. We should never ignore the possibility of the introduction of a new technological capability driving new behaviours. An example of this would be a geographically dispersed team exhibiting a low level of collaborative behaviour: the introduction of the right tools can increase the level of collaboration significantly. A good discussion of the inter-relationship between technology and culture can be found at [12].

Identify facilitators. In a number of communities, we have seen great benefit in assigning 'early adopters' a facilitation role. In Jasper, for example, this would typically involve the facilitator sharing items of the right content (and at the right technical level) on a regular basis, particularly during the early stages of the tools use.

Offer training and motivation. Clearly, training of users is a key issue and should have encompass both the technical aspects of using the tools in questions and also some explanation and motivation to the individual for the support of communities of practice at the personal and organisational level.

Build trust. Trust between member of a community of practice is of course essential if people are to participate fully (indeed, it is highly doubtful whether a group of individuals comprise a genuine community without it). In some trials, members of communities will be well known to one another. Where this is not the case, it may be appropriate to bring people (physically) together to begin to build mutual trust and to set expectations for the community. It should always be made clear that no kind of censure will result from 'incorrect' participation in the community. Trust is of course a highly complex issue in its own right and the reader is referred to [13] for further discussion.

Perceived Barriers to Participation

In one community of around 70 users, we asked users a series of questions around the issue of what they considered to be the main barriers to active participation in the community.

Frequent Jasper users responses indicated that the main barrier to sharing more information was the need to be selective ('concern about relevance'). Most respondents indicated they were not concerned about adding too many items or overloading others and they were not confused about which items to add.

Around 30% agreed with the statement that they felt more comfortable when they saw what others were putting on the system. One user indicated they were initially motivated to share but had become frustrated with the quality of material being shared by others and so became less willing to share, perhaps making the case for a system moderator. When asked explicitly about the use of a moderator, 33% replied that they would support this idea. In other communities, however, the concept of a moderator has met with resistance. This would seem to indicate that the desirability (or otherwise) of moderation is dependent on the type of community.

Further work

There is scope for much further research in the area of the usability and acceptability of tools for supporting virtual communities. We are currently considering further trials with different types of community and the use of ethnographic and other methods to monitor and analyse these trials.

4. Conclusion

We have described Jasper, a WWW-based tool for supporting virtual communities. The ubiquity of Internet technology and awareness of the importance of communities of practice for sharing knowledge and (social) learning on the one hand, and globalisation and more flexible working practices on the other, have driven the increasing interest in tools to support virtual communities. However, as we have seen, non-technical issues are often key in determining the success or otherwise of particular communities and we discussed some of the key factors we have observed in this process.

Acknowledgements

Thanks to Tom Boyle and Dan Moorhead for valuable discussions on the nature of communities of practice. Some of the factors discussed for successful virtual communities originate with Tom's work. Steve Lakin championed and analysed a large Jasper HR community.

References

[1] Wenger, E., 1998, *Communities of Practice*, Cambridge University Press, Cambridge, UK.

[2] Seely-Brown, J. & Duguid, P., 1991, *Organisational Learning and Communities of Practice*, Organisation Science, Vol. 2, No. 1, also available at http://www.parc.xerox.com/ops/members/brown/papers/orglearning.html,

[3] Davies, N.J. & Weeks, R., 1998, *ProSum – profile-based text summarisation*, First Automatic Text Summarisation Conference (SUMMAC-1), May 1998, Virginia, USA.

[4] Salton, G., 1989, *Automatic Text Processing*. Reading, Mass., USA: Addison-Wesley.

[5] Polyani, M., 1966, *The Tacit Dimension*, London: Routledge & Paul.

[6] Granovetter, M., 1974, *The Strength of Weak Ties*, American Journal of Sociology, 78, pp 1360-1380.

[7] Granovetter, M., 1982, *The Strength of Weak Ties: A Network Theory Revisited*, in 'Social Structure and Network Analysis', Marsden, P. and Nan, L. (Editors), Sage Publications, California.

[8] Constant, D., Sproull, L. and Kiesler, S., 1996, *The Kindness of Strangers: The Usefulness of Electronic Weak Ties for Technical Advice*, Organisation Science, Vol 7, Issue 2, P119-135, 1996.

[9] Hansen, M.T., 1997, *The Search-Transfer Problem: The Role of Weak Ties in Sharing Knowledge Across Organisation Subunits*, Working Paper, Harvard Business School.

[10] Nonaka, I., 1994, *A Dynamic Theory of Organisational Knowledge Creation*, Organisation Science, 5, 1.

[11] Cook, S. & Brown, J.S., 1998, *Bridging Epistemologies*, forthcoming.

[12] http://www.deepwoods.com

[13] Maxwell, C., 2000, *The Future of Work – Understanding the Role of Technology*, BT Technology Journal, 18, 1, Kluwer, Netherlands.

Managing Micro- and Macro-level Design Process Knowledge across Emergent Internet Information System Families

Amrit Tiwana[1]
[1]Robinson College of Business, Georgia State University, Atlanta, atiwana@gsu.edu

Abstract: The Internet poses new challenges for information systems development, as it also does in many other practices that worked well in the preceding era that fail as adherents refuse to abandon what was once tried, tested, and now obliviously believed. This chapter integrates various diverse theoretical perspectives from manufacturing, innovation management and diffusion, and social systems research to propose integration of knowledge management as the primary basis for systems development practice. Influences of homophilous networks are analyzed to call for the rejection of the "big-bang" notion of systems delivery. A new approach to viewing an information system as a family instead of an artifact is proposed; challenges to traditional ISD practices in the Internet-centric environment are identified and system level decomposition followed by process knowledge management at both a micro and macro systemic level is explicated.

1. Introduction

Emergence of the Internet has changed many industries, their practices, norms, and assumptions in fundamental ways. Among the many affected practices is information systems development (ISD). While Internet ISD is not fundamentally different from traditional ISD, several intervening influences send followers of conventional methods down the road to failure or anti-competitive systems. This chapter argues for the need for integrating the notion of process knowledge management in ISD practice, and further elucidates how this may be done using the two theoretical lenses of emergence theory and new product platforms. Knowledge-based systems development is called for in favor of the rejection of ISD practices that are reminiscent of the mainframe and desktop-computing era. After examining the unique needs of Internet information systems, it is further argued that Internet-centric ISD must focus on developing a family of products instead of single instances—simultaneously leveraging knowledge created through the process of preceding *versions*.

1.1 The Internet as a fruitfly for ISD

Fine notes that geneticists observe genetic evolution and genetherapy effects on humans using *Drosophila* (fruitflies) as subjects [7]. The reason: While this species has a complex genetic makeup akin to humans, they go from egghood to parenthood to death in a span of two weeks. The time frame over which one evolutionary cycle could be analyzed in humans, geneticists can observe several thousand in fruitflies. Fine further suggests that industries that resemble such rapid evolutionary behavior can be thought of a *fruitfly industries*—industries with exceedingly high rates of change (called "clockspeeds") wherein certain phenomenon are more readily observable. For complex information systems development, the Internet arguably is *the* fruitfly.

The Internet has created a fast-paced, even if an arguably inefficient, test bed for ideas that would otherwise take much longer to test and validate. Stepping aside from the limelight of the so-called "Internet economy" that these firms have created, there are several theoretically intriguing phenomena that can be observed. While many budding entrepreneurs burn their fingers on this test bed, it is the ideal place for researchers to test their epistemological and theoretical beliefs regarding ISD in new light. Arguing both for and against a rapid pace of development over which IS organizations have little control, Mullins and Sutherland [24] warn that such rapid technological change can be both a blessing and a curse.

2. Internet Information Systems: Fundamentally Different or *Deja Vu*?

The Web has changed the fundamental ways in which software products and services are distributed, beginning with new distribution channels for business-consumer transactions and mechanisms for mass delivery of information (such as that through news sites).

It has already been speculated that the emergence of the Internet as a platform-environment for software will force software developers to rethink the architecture of their products ([20] p.201). This paper further argues that as architecture of such software products evolve, the methodologies and practices associated with the development of such systems will also necessarily be vastly different from those used before.

The first question that arises is whether Internet information systems represent systems development of a fundamentally new kind, or whether they are simply a fast paced instantiation of "conventional" systems development environments. This chapter argues that the latter is the case; however, there are several characteristics of Internet software products that distinguish them from conventional software products. These distinctions do not stem from any fundamental differences, but from the inherently higher "clockspeeds"—or rate of change—in these systems.

Several factors distinguish traditional ISD and Internet ISD: varying influence of homophilous social networks, higher levels of uncertainty and unpredictability without the luxury of time to minimize them through extensive validation, the need for rapid *in situ* innovation assimilation and integration (often after design freezes and occasionally right before final release), failure of the "big-bang" delivery model, indistinctness of boundaries between IS use and ISD, and finally the oft-observed coasting mentality prevalent in large scale information systems development. I argue that many fatalities in Internet information systems development come from the mindless reapplication of "older" ISD practices and methodologies that are reminiscent of the mainframe and desktop systems era. Such "superstitious application of old knowledge to new problems [31]" is partly the reason why an embarrassingly high majority of Internet software development projects are either off schedule, exceed budgets, or perform poorly after being implemented. These differentiators of Internet information systems and the characteristic concept of systems emergence in the context of Internet-centric development are examined in the following sections.

2.1 Homophilous influence networks, innovation diffusion, and network effects

Innovation diffusion literature provides notable insights into emblematic differences between traditional IS and Internet IS. Networked social systems created by the use of an information system through the Internet activate homophilous influence networks through interpersonal communication between members of the social system [2, 23, 32].

Whether this is a network of purchasers who can discuss issues (such as reverse auctioneer Accompany.com's buyer chat networks), networks of existing clientele (such as Dell Corporation's customer discussion area), or networks of both customers and non-customers/future customers (such as Amazon.com's reader review system), these are arguably easier to form when no spatial or temporal barriers akin to those in brick and mortar stores exist.

While preceding research has applied this concept to explain the cumulative diffusion patterns that give an innovation its characteristic S-curve shape, it is used to emphasize the faster rate of rejection, acceptance, or criticism that an Internet IS can encounter when exposed to such a homophilous network.

Application service providers—firms that rent information systems to clients on a usage basis while simultaneously refining them—represent a diametric opposite of "delivered and installed" information systems. The emergence of ASP-like firms represent what have been referred to as "foreign invaders" in technology innovation literature [34]. Such market players are distinguished by their knowledge and capabilities that make it possible to alter the established technological path. Common patterns that indicate the emergence of such "invaders" involve [34]:

1. Major modifications in the way the technology's components and subsystems are integrated into a system
2. Integration of new subsystems
3. Movement of well-developed capabilities from one technology sector to another.

Knowledge gained over the course of systems development and maintenance must address these concerns and create sufficiently relevant varieties of knowledge to help the IS organization self-create these patterns; in other words, engage in "creative destruction [4, 9]."

2.2 Uncertainty and unpredictability

Internet ISD is vulnerable to high levels of uncertainty and unpredictability. Frequent reports in technology trade press serve to highlight how Internet systems that were supposed to simplify or enhance work in organizations were unpredictably counterproductive. Uncertainty arises from two incongruent challenges: (1) the firm's customers' inability to articulate their needs and (2) managerial inability to translate incremental and discontinuous technological innovations into product features that benefit the consumer [24]. Unfortunately, while developing an Internet information system such as an electronic commerce site, it is often difficult to accurately guess *who* the users might be. Even if a sample population is selected and surveyed, it may already be too late in the development process when it is determined that that sample population was ill-representative of the actual user population.

In the context of Internet information systems, there are six characteristics that exacerbate uncertainty that surrounds design, as described in **Table 1**. Possible "fixes" based in process knowledge management are suggested alongside. These are further developed later in this chapter.

Table 1: Contributors to uncertainty and the initial role of KM

Contributor to uncertainty	*Role of knowledge management*
Need to cope with uncertainty over fundamental design assumptions	Explication of design assumptions through KM tools that encourage assumption surfacing [30]
Need to match customer needs and solutions made possible by new technology	Incorporation of external knowledge gained through alliance with or observation of competing systems [12]
Uncertainty over new market paths chosen Uncertainty regarding technological choices made and given up	Analysis of paths and consequences in (1) past projects, (2) competing projects, and (3) results driven incrementalism [41]
Uncertainty regarding the level of resources committed to uncertain market segments	System segmentation and versioning through extensive modularization [36]
Uncertainty regarding the timing of such resource commitments	Incremental decompositional development driven by knowledge of results from preceding increments [40]; modularization [27]

2.3 Rapid assimilation of innovation in design

Innovation in the Internet industry has occurred at a rate faster than most firms have been able to assimilate. Technological discontinuities—innovations in technology that are not incremental, hence are harder to incorporate into a stable, integrally architected system—can quickly make both information systems and technological capabilities archaic, and even irrelevant [11, 21]. This also holds true for underlying technology and technology standards that form the basis of new products [37]. Indeed, an ISD team must move quickly enough to incorporate these innovations in its family of products after first deciding whether that should be done in the first place. Failing this, the ISD organization's products might "run the risk of becoming obsolete" overnight [20]. Indeed, "greater the uncertainty, greater is the need for learning [1]." Recent research on innovation management has specifically highlighted this concern in the context of IS organizations.

> "Information systems organizations operate in turbulent times, faced with the need for creating new knowledge and competencies in view of rapidly changing technologies [1]."

The inability of organizations to cope with uncertainties has often been attributed to non-existent though much needed social and organizational mechanisms for supporting collective knowledge management [1], and not just stable systems drag (as argued by [42]). Much of this knowledge that could otherwise be used as an effective weapon against uncertainty is often never used as it either cannot be applied at the needed moment or has lost its context in an overly optimistic attempt to "capture" it. Correspondingly shortened product life cycles [21] in many markets take away the luxury of time only to further exacerbate this problem. Technological innovations—both incremental and discontinuous—can therefore be expeditiously assimilated in an Internet information system through better application and management of process knowledge.

2.4   The "big bang" fallacy and information systems emergence

Emergence, in the context of organizations, refers to their "state of constantly seeking stability, while never achieving it [42]." Truex et al. have suggested that emergence theory challenges some of the basic premises on which we have long practiced information systems design.

> "...every feature of social organizations—culture, meaning, social relationships, decision processes, etc.—are continually emergent, following no predefined pattern. While temporal regularities might exist, these are only recognizable by hindsight, as they are always in transition but never arriving [42]."

If we are to accept emergence theory, then conventional systems development based on the "waterfall model" (developed and popularized in [45]) and its many variants must *necessarily* fall apart. The waterfall model and the systems development life cycle (SDLC) ISD method, as I have argued elsewhere [41] follow the notion of the "big bang" deliverable—there is an extended period without contact between the organizational users and the ISD group following requirements gathering; then, several months later, a final version of the system is delivered.

Initial Investigation

Requirements Generation

Window of Isolation

Design

Development

Test and Deploy

Post-Deployment Review

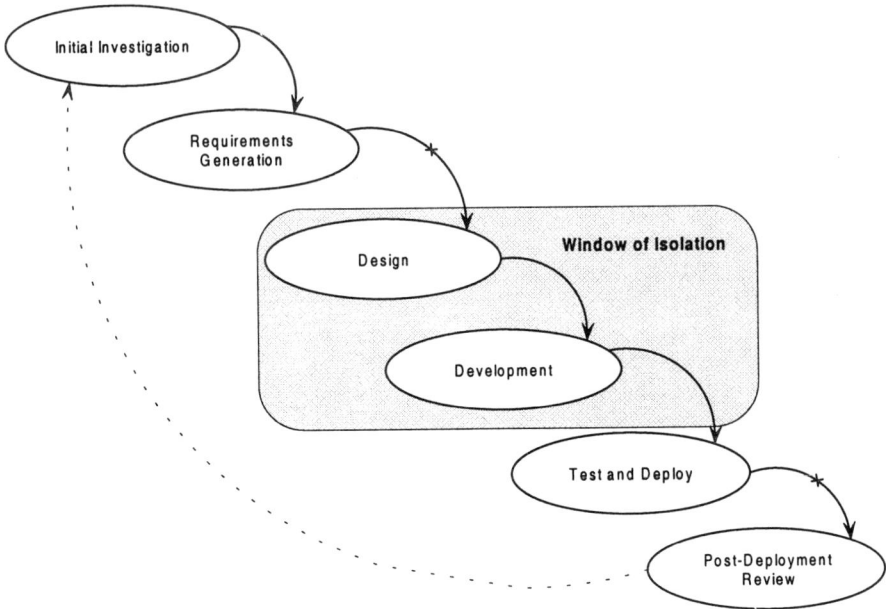

Figure 1: The waterfall ISD model creates a jeopardous *window of isolation*

As shown in **Figure 1**, there is an extended "window of isolation" (ranging from several months to years) between the requirements generation and deployment stages. During this time frame, characteristics of the origination and its business environment, as well as its needs might evolve as much as to make the original set of requirements obsolete. Methods such as prototyping, end-user computing, and pilot implementation have cumulatively challenged Yourdon's notion of "big-bang" delivery wherein an information system is delivered by the ISD group after a long period of non-interaction separating requirements analysis from the eventual delivery of the system. If emergence theory were taken into account, and the observed rapidity of organizational and business environment were to be accounted for, there would be increasing likelihood that the original goals of an ISD initiative and those at the time of delivery will be incongruent. In such cases, Emergence theory suggests that the information system will inhibit rather than facilitate organizational change through its oft-rejected goals of low maintenance and long life span [42].

This inflexibility created by ISD methodologies has been referred to as "stable systems drag" [42] and methodological rigidity elsewhere [40]—wherein *the organization* must adapt to both its emergent environment as well as its outmoded and possibly incongruent information systems. Not surprisingly, such methodological rigidity is also the key point of failure in systems intended to support knowledge management [40]. Information systems that do not impose such encumbrance on the adopting organization must—out of necessity—be optimized for diametrically opposite goals than have be pursued in stable systems design—high maintenance and expectations of short life spans.

In Internet environments, opportunities for interactions—direct and systemic— between users and the ISD group are more frequent than in "delivered" information systems. Berger and Luckmann's seminal work [3] on the social construction of reality further suggests that the perceived and believed needs of users are socially constructed through complex and sustained interactions between the people who develop an information system and those that use it. As such, there is a "constant negotiation of fact, opinion, and meaning [42]" that allows organizational members to arrive at their decisions in the design process. In Internet-centric environments, such formation, negotiation, and reformation of meaning and needs tends to be more pronounced because every instance of the system's use can be viewed as an opportunity for learning and relearning about the very purpose that the IS is *supposed* to serve.

Organizations are self-referential and use their own identity as the primary point of reference when they reconstruct themselves [18]. Since such socially constructed reality forms the basis for the next version of the organization, its social and technological systems, and its emergent behavior, it would be likely that technologies and stimulants such as knowledge management that help "view" such evolution of its outcome artifacts (such as products or services) will facilitate incrementalism in emergent system needs and in its design.

Table 2: Traditional and emergent perspectives applied to knowledge-based ISD activity (based on [3, 29, 30, 42, 44])

Characteristics	Traditional ISD assumptions	Emergent ISD assumptions	Impact of knowledge-based "family" orientation
Shared construction of reality [42]; shared understanding [29, 30]	Economic advantages of prolonged initial analysis and requirements gathering [30, 42]	Persistent analysis [42]	Shared understanding created through evolution of incremental versions within an IS family; requirements emergence is supported by knowledge of past incremental versions
Self-referentialism [3, 42]	Abstract requirements; Complete and well-articulated specifications [42]	Incomplete and "usefully ambiguous" specifications	Approaches, successes, war-stories, and failures from past approaches are traceable in past ISD

2.5 The amalgamation of IS usage and ISD

The distinction between IS and ISD blurs in the scenario where systems evolve with rapidity as is often the case on the Internet—whether due to internal forces (e.g., changing organizational needs) or external forces (e.g., compliance with new standards). Systems in this case have an explicitly incorporated adaptability orientation; the underlying architecture must then be conducive to continuous redevelopment. Application service providers (ASPs) that "sell" computer applications to subscribers on a per-use basis while continuously developing them at the same time, represent one such case. In the case of ASP services, it can be argued that ASP service *adopters* are shifting the burden of such continuous redevelopment on the shoulders of the ASP. In traditional systems, this burden (or its expense) often fell on the shoulders of the client organization, following the system's delivery.

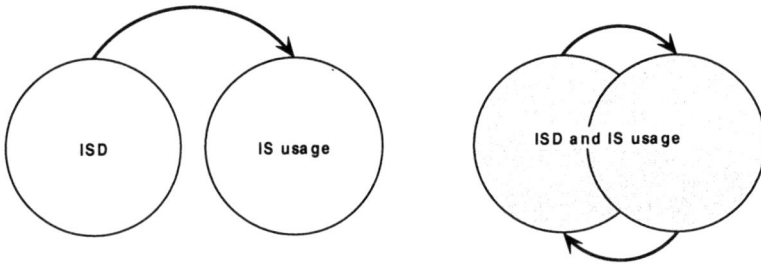

Figure 2: Blurring of the boundaries between ISD and IS usage

Constant monitoring of the process of self-organization is necessary for detecting disruptions in the "normal" process of incremental innovations [34]. There is an extensive history of cases in which companies and even entire industries have been unable to adapt to highly disruptive technological changes. In the context of Internet IS, such changes/innovations might be discontinuous (such as the invention of the graphical Web browser itself) or incremental (such as changing technology, bandwidth, and normatively accepted standards). In either case, the information system must change in synchrony with the organization for either change to be a successful adaptation to the environment. These disruptions or discontinuities demand greater managerial participation [34] and greater levels of maintenance than were traditionally considered necessary.

2.6   Inflexibility, unlearning, and the "coasting" mentality

Problems in trying to upkeep an information system do not stop with the inability to collect a stable set of requirements or with need to learn. The more problematic aspect is the need to *unlearn* what might have already been learned as soon as it is no longer applicable, becomes invalid, or simply outmoded. Typically, problem solving is grounded in an experience-based body of heuristics that guide both actions and places that are scanned for solutions, and follow established ways of learning [16]. To describe how these established ways no longer represent capabilities, Leonard refers to these as being core *rigidities*. Such rigidities can often become obliviously embedded beliefs that are so transparently applied that unlearning them can indeed be a challenge [41].

Ill-considered approaches and outmoded practices can destroy hard won capabilities of the firm and lead to ineffective solutions while impeding learning [11, 21]; the same holds true for an ISD group as for any other. Once an information system reaches an initially acceptable level of success, management often tends to allocate maintenance-level resources to its upkeep—a disaster prone step that Meyer and Utterback [21] refer to as *the coast mentality*. This observation in manufacturing literature is very similar to the mentality of trying to minimize maintenance costs in information systems that Truex et. al [42] warn about. Indeed, Drucker forewarns that one cannot manage change, one can only be ahead of it [6]. The process of learning, unlearning, and redevelopment must accompany the use of an information system; the boundaries between ISD and IS use therefore begin to blur beyond recognition.

3. Process Knowledge in Information Systems Development

The overarching strategic goal of networks is to continuously achieve a fit among core capabilities, complementary assets and learning opportunities. Viewed as social networks, there lies embedded enormous potential for two-way learning between IS developers and users as they interact over electronic networks such as the Internet. Each instance of the information system's use can be considered an opportunity for learning. The precursor for achieving such fit is to establish effective linkages among sources of knowledge about underlying technologies and among firms that are coevolving with them [35], and strengthening relationships that facilitate the movement and use of tacit knowledge [34]. Not all users would, arguably, be good sources of new knowledge that could facilitate refinement of a system.

Meyer and DeTore [19] further suggest that integration of markets, products, and embodied technologies is arguably the most difficult yet important challenge facing firms seeking continued growth. Increasingly information systems design, development, implementation, and deployment require sharing of significant amounts of expertise among IT professionals and end users. There are, however, inherent problems associated with such knowledge sharing because both stakeholder groups need to possess a certain level of shared understanding of their environment before fruitful interaction is possible. Consequently, IT professionals rely on surrogate methods for building this shared understanding and use mechanisms such as tools, architectural blueprints, models, etc. The following sections address this issue by examining various levels at which knowledge can be assimilated and applied, suggesting a set of approaches for decomposing systems, and creating linkages between such decomposable "components" and rich process knowledge associated with them.

3.1 Information systems development and process emergence: opposing forces?

The IS organization must develop a basis for a "family" that is robust and flexible and that provide the foundation for rapid and cost effective development of specific information system instances or versions. Knowledge, in that case, can be shared across generation across such families. Various approaches for process knowledge management such as post hoc and ad hoc rationale management [25], conceptual process modeling [14], and method rationale in method engineering [33] have been proposed in recent information systems research. We suggest that such process knowledge should not only encompass design processes but also those of implementation and integration.

3.2 From integrality to modularity

Of interest is the shift from integral architectures in information systems to modular architectures. **Figure 3** illustrates the essential difference between modular and integral system designs. While integral architectures are reminiscent of mainframe systems, custom developed in-house software systems, and tightly integrated software systems that are perhaps still the best choice in certain application domain, modular systems do not exhibit such tight coupling between their constituent parts.

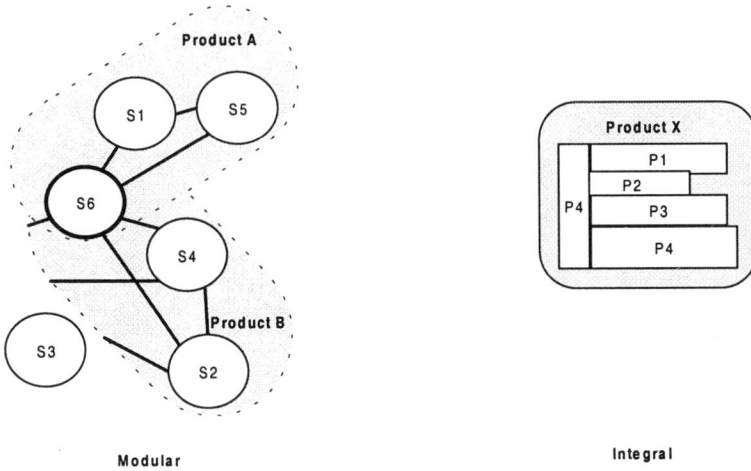

Modular   Integral

Figure 3: Modular versus integral systems design affords differing opportunities for design process knowledge management

As **Table 3** illustrates, the characteristics of integral and modular system architectures do not necessarily overlap. Integral architectures have been successfully deployed in the design of products and systems that enjoy design stability and long life spans. Modular products, in contrast, are better suited for products and systems that are inherently unstable or have life spans so short that they project an image of instability. Internet products would fall in the latter category.

Table 3: Integrality versus modularity (based on an extension of [7, 43])

Characteristic of system components	Integral	Modular
Function of each	Multiple	Singular
Spatial relationships	Tight	Loose
Interchangability	Low	High
Individual upgradeability	Low	High
Interfaces	Nonstandardized	Standardized
Failure points	Nonlocalized	Localized
Scope for knowledge management	Narrow	Broad
System-level decomposability	Low	High
Number of subsystems	Low	High

It is now widely recognized that proprietary software systems, transformed by client/server computing and network-centric technologies, have given way to open systems [20]. Further, Internet software products often need to be built within a short time frame, and are complex systems (especially in the electronic commerce applications domain), developers have frequently been using back-end functionality from other commercially available subsystems to finally integrate these components into the final information system.

Modularity in systems allows for easier integration of breakthrough technologies [20] and shifting standards at a rate hitherto impossible in more tightly integrated systems. In Internet information systems, flexibility is key. A flexible system is characterized by the ability to incorporate new functionality and even new modular components without any significant integration cost [20]. Meyer and Lehnerd [20] note:

"Achieving modularity while minimizing the number of interfaces between subsystems is the essence of elegance in software design."

Modularity in design is therefore an essential but insufficient guarantor of flexibility of a system. If modularity is designed into the system right from its inception, the task of managing process knowledge can then be accomplished at a component and sub-component (module) level rather than at an unmanageably lower level. It can therefore be suggested that modularity in design enables easier, if not necessarily better management of process knowledge.

3.3 Analyzing modular complexity

Software systems are increasingly becoming too complex to feasibly develop on a piecemeal basis, and this complexity is further heightened by the fact that subsystems used in many of these come from different developers. Though everything might work independently, problems arise when these subcomponents are integrated and interact. Technology integration literature suggests that high levels of knowledge are needed to successfully accomplish technology integration [10]. Further, this knowledge itself evolves and might rapidly become obsolete or inapplicable over the course of integration. Technology integration alone stands to immensely gain from deliberate attempts at managing certain types of design process knowledge over the course of evolution of an Internet information system family. **Table 4** describes complexity as viewed from the perspective of linkages between subsystems and modules (and associated knowledge) within a modular information system.

Table 4: System interfaces in modular Internet software products

Interface type	Functionality	Associated knowledge
Internal	Link key subsystems within the core software product	Interaction Integration Choice
Cross-system	Link the final system with other systems	Interaction effects
Expansion	Link the family and externally acquired modules to expand functionality	Limitations Optimizing solutions Satisficing solutions

Integration complexity inherent in Internet information systems raises the value of associated knowledge to such high a level that Meyer and Lehnerd [20] hint that software firms might benefit more from selling this knowledge than they might from developing the product itself. This case can be observed in the case of enterprise resource planning (ERP) software vendors who often charge more for their implementation services than for the product itself; the premium paid for their service can be equated with the fair market value of their knowledge. Meyer and Lehnerd even go on to speculate that the potential for selling this knowledge is more than many would conceive.

> "The software firm of the future will have the potential of seeing its program modules used in hundreds of different applications, many of which it will not control [20]."

Indeed, their prediction is already manifested by the emergence of Application Programming Interfaces (APIs) and the infamous *Microsoft versus United States* antitrust case that partially hovered on the issue of the firm's overly strong hold on its products inner details, specifically APIs. Salability and market value of complex knowledge has also been suggested in extant knowledge management literature (e.g., [5, 26]).

Correspondingly, opportunities for managing design and process knowledge are exceedingly abundant in modular architectures than they have ever been in integral ones.

3.4 Lessons from product platforms and families

New product development (NPD) research has partially focused on developing families of products as opposed to the development of a single product (e.g. [13]). NPD literature refers to this idea as product platforms—an example of this in IS being ERP systems that can be customized to specific instantiations within a given implementation. In manufacturing literature, the term product platforms refers to a core set of components that can be used as a basis for future derivative products, primarily through extensions within the base platform. Modularity is a key characteristic of product families that allows for the creation of a stable, long-lived platform [38]. Such platforms provide a common set of design rules and implemented subsystems and subsystem interfaces that form a common structure from which a stream of derivative products/services can be efficiently developed [20, 38]. Product platforms consist of individual subsystems and subsystem interfaces which themselves become the key focus of both investments and innovation.

While this has been explicitly recognized in manufacturing literature, this recognition is somewhat subtle in ISD. Effective platform renewal and product family management "generates successive generations of market driven products that are the continuing basis for the firm's specific commercial offerings [44]." We can then attempt to piece together insights provided by product platform literature, innovation management, and emergence theory to better understand how process knowledge can be managed at two different levels of abstraction in the context of Internet information system families.

4. Integrating Knowledge Management and IS Evolution

When a system's design and development begins, the focus of the ISD team must be on understanding the needs of lead users. From this understanding, initial specifications for the first version of the software product must be arrived at. Even in this initial stage of design, various alternatives are considered for each design decision, only one of which is finally adopted. Once this decision is made, without an explicit mechanism to record them, knowledge about other alternatives that were considered is irretrievably lost [30]. Further, knowledge about why each given alternative was chosen comes from a variety of stakeholders and participants. The rationale underlying these decisions is often lost once they are reached [29]. This observation has led to extensive research in the area of design traceability in information systems (for a review, see [28]).

4.1   From information system entities to families

It is proposed that Internet information systems must discontinue the philosophical association of systems as independent physical artifacts that can be built and delivered at a given point in time. Instead, an Internet information system must be built as a family[1] of successive products that will continually replace the preceding one. While some of these product "versions" might be breakthrough, most will be incremental refinements of earlier versions. As previous research on product platforms suggests, the vital focus is then on developing and sharing key components (or subsystems) within the family of products [21]. Series of generations or iterations within the family create new process knowledge that can be used as a facilitator for making design decisions in consecutive versions. The shared knowledge-based commonality leads to efficiency and effectiveness in the design and development of newer versions of the original information system.

Further, as each incremental version of the product (whether or not released to public) progresses toward the final version (or subsequent versions), opportunities abound for both creating new knowledge about the process and recording the context of related design processes. This process of emergence in the case of an Internet software system is illustrated in Figure 4. Version numbers are chosen arbitrarily solely for expository purposes. The project begins at time T_0 wherein the initial analysis and design leads to the first version, V_1, of the system artifact at time T_1. Emergent learning and feedback at each stage/version finally leads to Version 1.00 corresponding to V_4 at time T_4.

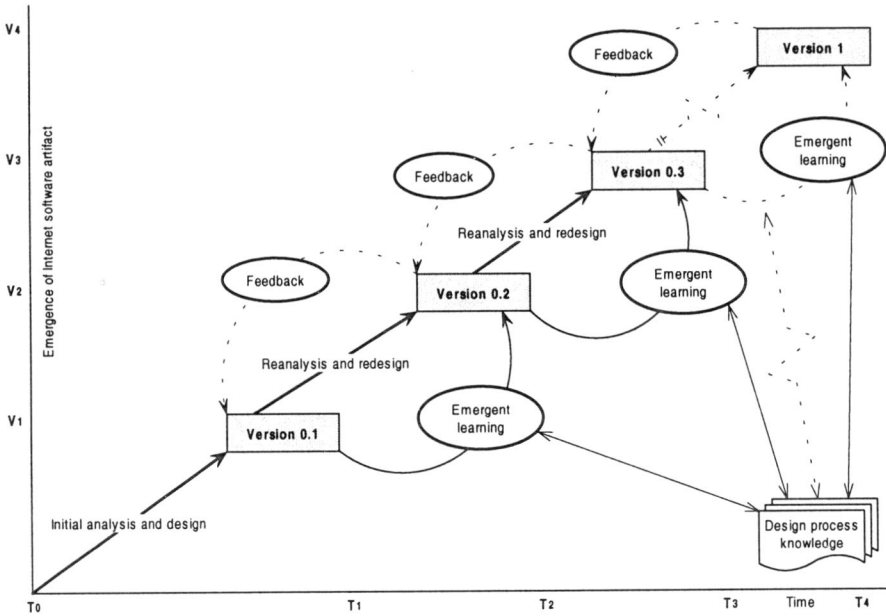

Figure 4: Evolution of an Internet software artifact and creation of design process knowledge

4.2 Micro- versus macro-level knowledge management

Early planning in ISD (such as that preceding version 0.1 in **Figure 4**) must be coupled with high levels of modularity inherent in the design. Modularity, Meyer and Utterback [21] suggest, allows the firm to upgrade subsystem-level components with newer and better variations from other suppliers as and when needed. Using a product family approach, the firm's competencies, a cohesive set of skills and techniques, and knowledge are used to design, develop, distribute, support, and deliver continually incremented versions of the information system [39]. Such knowledge management can either be at a micro-level or a macro-level (see **Figure 5**).

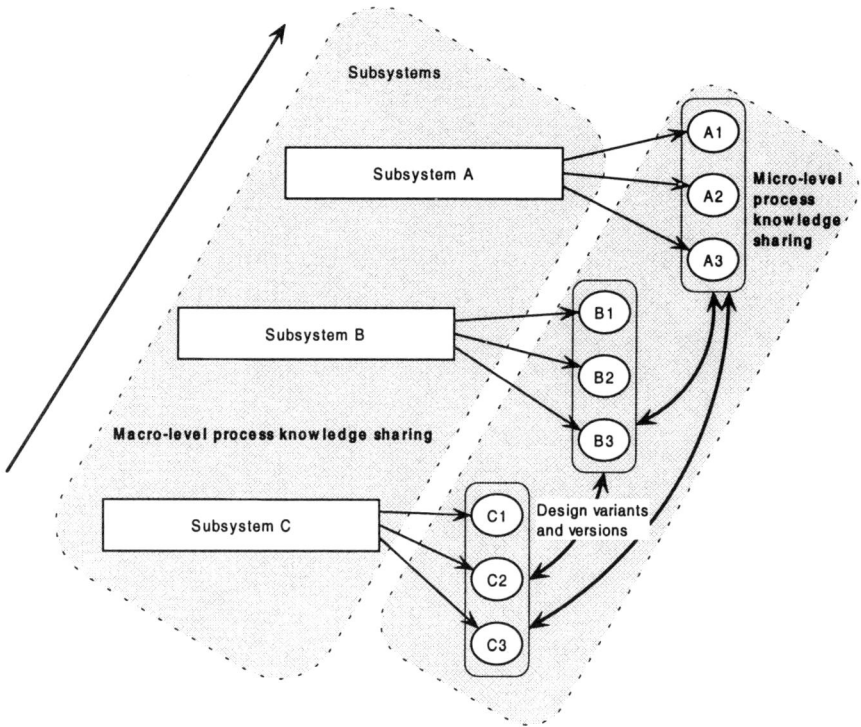

Figure 5: Micro- and macro-level knowledge management in ISD

When knowledge is shared at a macro level, the focus is on managing knowledge about how subsystems or modules interact when assembled to build a given system. Accordingly, knowledge sharing is limited to the subsystem level, with a focus on platform renewal rather than subsystem renewal. However, if this knowledge is managed at a micro level, the focus shifts towards interactions between components and subsystems when integrated in various possible combinations. Accordingly, instantiations of modules in various systems (e.g., A1, A2, and A3 for subsystem A) and their many versions are the focal object. Such knowledge sharing requires more effort in the part of system designers, and allows knowledge gained and lessons learned (successes, failures, and limitations) from each version to be applied to subsequent versions of both the original subsystem and to other similar subsystems. Such management of process knowledge then focuses on the evolution and emergence of the properties of subsystems and components rather than the platform (in this case, the sum of subsystems A through C).

Markets are dynamic—they evolve or are newly born, and demand new generations of platforms and component technologies that provide new features, higher levels of performance, or better value [19]. The question that arises is whether the firm must primarily focus on managing knowledge about its Internet IS platform or its components, or both. The two choices are not necessarily exclusive. Arguably, both foci must be present even though one must be the dominant mode. The choice of the balance point is highly context dependent and will largely determined by the relative importance of maintaining the platform versus maintaining the set of subsystems in a given organization. For example, a software consulting company developing Internet commerce applications for various customers might choose to focus on a macro-level strategy whereas a software developer selling secure credit card transaction components to electronic commerce site developers might choose the micro-level strategy.

With a macro-level strategy, building blocks of product platforms can be integrated with new subcomponents to rapidly address new market opportunities while reducing the time cycle for new product introductions [8, 17, 20, 22]. A micro-level strategy, on the other hand, will enable a given set of subsystem-level components to evolve for integration with various types of information systems. An example of such evolution is the added ability of a credit card processing component to work with new types of Internet mark-up languages (e.g., XML).

By integrating knowledge over the course of an information system family's evolution, newer and better aligned versions can be developed more efficiently and flexibly, while retaining critical linkages with ensuing market needs that each version must address.

5. Contributions and Implications

The contribution of this work lies in the proposed approach for implementing knowledge management in the context of Internet-centric ISD. This chapter has demonstrated that Internet information systems are not fundamentally different from traditional IS. They, however, exhibit certain characteristics of systems at a much more amplified level. The significance for managing knowledge both at a micro and macro level has been established through an Emergence Theory-based analysis. An analysis of varying levels of modularity and their impact on the ability and feasibility for managing process knowledge has been presented. Through a comparison with incremental innovation in product platforms, an approach for managing the evolution of an Internet software artifact/system as a family of incremental products is proposed along with a mechanism for asynchronously and atemporally sharing process knowledge.

This has two key implications. First, Internet information system development cannot be carried out the same way as traditional large-scale projects have been executed. Second, managing process knowledge is a key long-term success factor in Internet-centric ISD and this is possible through an explicitly and intentionally followed product family development approach that focuses on strategically linking KM and the organization's objectives.

5.1 Limitations and directions for future research

While product families offer a unique opportunity for managing process knowledge in information system families, modularity in information systems design still lacks sufficient empirical evidence.

Internet information systems are as characteristic of software artifacts as they are of IS services. Viewing platforms in such services businesses as monolithic entities can be self-limiting. Monolithic service platforms will induce organizational rigidity over time and make the firm vulnerable to competing services from more flexible entrants [15]. Approaches for determining when and how process knowledge becomes outdated, and methods for identifying such knowledge and unlearning it are little understood and call for further research.

Managing process knowledge comes at some cost that integratively includes time and resources expended in creating, explicating, and maintaining it. Further research is needed to establish the cost-benefit tradeoffs involved in such engagement. Further, qualitative and empirical investigation of the effects of knowledge management in Internet ISD would serve to further validate this approach.

References

[1] R. Agarwal, G. Krudys, and M. Tanniru, "Infusing Learning into the Information Systems Organization," *European Journal of Information Systems*, vol. 6, no. 1, 1997, pp. 25-40.,

[2] F. Bass, "New Product Growth for Consumer Durables," *Management Science*, vol. 15, no. 1, 1969, pp. 215-227.

[3] P. Berger and Luckmann, *The Social Construction of Reality: A Treatise in the Sociology of Knowledge*, New york: Anchor Doubleday Books, 1966.

[4] M. Boisot and D. Griffiths, "Possession is Nine Tenths of the Law: Managing a Firm's Knowledge Base in a Regime of Weak Appropriability," *International Journal of Technology Management*, vol. 17, no. 6, 1999, pp. 662.,

[5] T. Davenport and L. Prusak, *Working Knowledge*, Boston: Harvard Business School Press, 1998.

[6] P. Drucker, *Management Challenges for the 21st Century*, New York: Harper Business, 1999.

[7] C. Fine, *Clockspeed: Winning Industry Control in the Age of Temporary Advantage*, Reading, MA: Persues Books, 1998.

[8] M. Fisher, K. Ramdas, and K. Ulrich, "Component Sharing in the Management of Product Variety: A Study of Automotive Braking Systems," *Management Science*, vol. 45, no. 3, 1999, pp. 297.

[9] R. Glazer, "Measuring the Knower: Towards a Theory of Knowledge Equity," *California Management Review*, vol. 40, no. 3, 1998, pp. 175-194.

[10] M. Iansiti, *Technology Integration: Making Critical Choices in a Dynamic World*, Boston, MA: Harvard Business School Press, 1998.

[11] M. Iansiti and A. MacCormack, "Developing Products on Internet Time," *Harvard Business Review*, no. September-October, 1997, pp. 108-117.

[12] A. Inkpen, "Creating Knowledge Through Collaboration," *California Management Review*, vol. 39, no. 1, 1996, pp. 123-140.

[13] B. Kogut and N. Kulatilaka, "Options Thinking and Platforms Investments: Investing in Opportunities," *California Management Review*, no. Winter, 1994, pp. 52-71.

[14] C. Landauer, "Process Modeling in Conceptual Categories," in *Proceedings of Hawaii International Conference on System Sciences*, Maui, 2000, pp. CD-ROM.

[15] D. Leonard, *Wellsprings of Knowledge*, Boston: Harvard Business School Press, 1998.

[16] L. Leonard, M. Reddy, and J. Aram, "Linking Technology and Institutions: The Innovation Community Framework," *Research Policy*, vol. 25, no. 1, 1996, pp. 99-100.

[17] E.Y. Li, "Software Testing in a System Development Process: A Life Cycle Perspective," *Journal of Systems Management*, vol. 41, no. 8, 1990, pp. 23-31.

[18] N. Luhmann, "The Autopoiesis of Social Systems," in Felix Geyer and Johannes van der Zouwen, ed., *Sociocybernetic Paradoxes: Observation, Control and Evolution of Self-steering Systems*, London: Sage Publications, 1986, pp. 172-193.

[19] M. Meyer and A. DeTore, "Product Development for Services," *Academy of Management Executive*, vol. 13, no. 3, 1999, pp. 64-76.

[20] M. Meyer and A. Lehnerd, *The Power of Product Platforms: Building Value and Cost Leadership*, New York: The Free Press, 1997.

[21] M. Meyer and J. Utterback, "The Product Family and the Dynamics of Core Capability," *Sloan*, no. Spring, 1993, pp. 29-47.,

[22] M. Meyer and J. Utterback, "Product Development Cycle Time and Commercial Success," *IEEE Transactions on Engineering Management*, vol. 42, no. 4, 1995, pp. 297-304.

[23] D. Midgley, "A Simple Mathematical Theory of Innovative Behavior," *4*, vol. 1, no. 31-41, 1976.

[24] J. Mullins and D. Sutherland, "New Product Development in Rapidly Changing Markets: An Exploratory Study," *Journal of Product Innovation Management*, vol. 15, no. 3, 1998, pp. 224-236.

[25] L. Nguyen and P. Swatman, "Complementary Use of ad hoc and post hoc Design Rationale for Creating and Organizing Process Knowledge," in *Proceedings of Hawaii International Conference on System Sciences*, Maui, 2000, CD-ROM.

[26] I. Nonaka and H. Takeuchi, *The Knowledge-Creating Company: How Japanese Companies Create the Dynamics of Innovation*, New York: Oxford University Press, 1995.

[27] H. Post, "Modularity in Product Design, Development, and Organization: A Case Study of Baan Company," in R. Sanchez and A. Heene, ed., *Strategic Learning and Knowledge Management*, Chichester: John Wiley & Sons, 1997, pp. 189-208.

[28] B. Ramesh, "Factors Influencing Requirements Traceability Practice," *CACM*, vol. 41, no. 12, 1998, pp. 37-44.,

[29] B. Ramesh and V. Dhar, "Supporting Systems Development using Knowledge Captured during Requirements Engineering," *IEEE Transactions on Software Engineering*, 1992, pp.

[30] B. Ramesh and A. Tiwana, "Supporting Collaborative Process Knowledge Management in New Product Development Teams," *Decision Support Systems*, vol. 27, no. 1-2, 1999, pp. 213-235.

[31] D. Robey and M. Boudreau, Accounting for Contradictory Organizational Consequences of Information Technology: Theoretical Directions and Methodological Implications, *Information Systems Research*, Vol. 10, No. 2, 1999, pp. 167-185.

[32] E. Rogers, *Diffusion of Innovations*, Fourth ed., New York: Free Press, 1995.

[33] M. Rossi, J.-P. Tolvanen, B. Ramesh, K. Lyytinen, and J. Kaipala, "Method Rationale in Method Engineering," in *Proceedings of Hawaii International Conference on System Sciences*, Maui, 2000, CD-ROM.

[34] R. Rycroft and D. Kash, "Managing Complex Networks--Key to 21st Century Innovation Success," *Research Technology Management*, no. May-June, 1999, pp. 13-18.

[35] R. Saviotti, "Technology Mapping and Evaluation of Technical Change," *International Journal of Technology Management*, vol. 10, no. 4/5/6, 1995, pp. 423.

[36] C. Shapiro and H. Varian, "Versioning: The Smart way to Sell Information," *Harvard Business Review*, no. November-December, 1998, pp. 16-114.,

[37] C. Shapiro and H. Varian, *Information Rules: A Strategic Guide to the Network Economy*, Boston, MA: Harvard Business School Press, 1999.

[38] M. Tatikonda, "An Empirical Study of Platform and Derivative Product Development Projects," *Journal of Product Innovation Management*, vol. 16, no. 1, 1999, pp. 3-26.

[39] D. Teece, "Profiting From Technological Innovation: Implications for Integration, Collaboration, Licensing, and Public Policy," *Research Policy*, vol. 15, no. 6, 1986, pp. 285-306.

[40] A. Tiwana, "Custom KM: Implementing the Right Knowledge Management Strategy for Your Organization," *Cutter IT Journal (formerly American Programmer)*, vol. 12, no. 11, 1999, pp. 6-14.

[41] A. Tiwana, *The Knowledge Management Toolkit: Practical Techniques for Building a Knowledge Management System*, Upper Saddle River, NJ: Prentice Hall, 2000.

[42] D.P. Truex, R. Baskerville, and H.K. Klein, "Growing Systems in an Emergent Organization," *CACM*, vol. 42, no. 8, 1999, pp. 117-123.

[43] K. Ulrich, "The Role of Product Architecture in the Manufacturing Firm," *Research Policy*, vol. 24, 1995, pp. 419-440.

[44] S. Wheelright and K. Clark, *Revolutionizing New Product Development*, New York: Free Press, 1992.

[45] E. Yourdon, *Modern Structured Analysis*, Yourdon Press computing series, Englewood Cliffs: Yourdon Press, 1989.

---

[1] The term "product family" is used in a slightly different context from that found in product platform and manufacturing literature. In manufacturing literature, the family refers to a series of parallel spin-off products derived from the same set of core technologies. However, in this case, it primarily refers to only the series of consecutive iterations without simultaneous parallel versions.

The Development of Case-Based Reasoning for Design – Techniques and Issues

Stephen Potter, Stephen Culley, Mansur Darlington and Pravir Chawdhry

Engineering Design Centre, Department of Mechanical Engineering, University of Bath, Bath, United Kingdom, BA2 7AY. Email: S.E.Potter@bath.ac.uk

Abstract: Case-Based Reasoning (CBR) offers a problem-solving strategy for situations in which complete knowledge of the task is not available and expertise is lacking – which is a common state of affairs for engineering design tasks. Moreover, the perceived similarity of this strategy to the manner in which humans might address certain design tasks is appealing, and so there has been a substantial amount of research into CBR systems for design.

This chapter introduces CBR, and describes its characteristics and the issues surrounding its use. In order to give a practical illustration of these concepts, a CBR tool for the design of fluid power circuits that has been developed by the authors is also described.

1. Introduction

In general, the engineering design task is an extremely specialised one, and, as a consequence, it is the preserve of engineers with expert skills. However, by its very nature, expertise is time-consuming and difficult to accrue, and so is often in short supply within an organisation. This difficulty could be addressed, to some extent, by the automation of some, or all, aspects of the design task. This automated 'designer' could then be duplicated and distributed at will to the locations where there is the greatest need.

If human designers are to be emulated in this fashion, the automatic designer requires design knowledge. A typical 'expert systems' approach to design automation relies on the generalised knowledge of a human expert being encoded explicitly in the form of, for example, design rules. Unfortunately, even when a willing human expert is available, rarely will this sort of heuristic knowledge be easy to acquire directly through knowledge engineering.

However, an alternative source of knowledge often *is* available – examples of designers' work. Case-Based Reasoning (CBR) [1, 2] is an approach to automated problem-solving which exploits the knowledge contained within previous problem-solving episodes. As a consequence, it has been recognised as a useful strategy for automating design, and, indeed, a number of such applications have been developed.

This chapter is divided into two parts. The first part describes CBR in general terms, and then discusses the use of this approach for addressing design problems, as well as dealing with the issues that surround its use in design contexts. In the second part of the chapter, these ideas are clarified through the description of one particular CBR system that has been developed.

2. Case-Based Reasoning and Design

The basis of CBR lies in the assumption that similar problems will, in general, be solved in similar ways – if this is so, then a problem may be (almost) solved by the re-use of a known solution to a similar problem. Accordingly, a CBR system relies on the use of explicit experiences – or *cases* - of particular problem-solving episodes in the past to suggest solutions to new problems.

2.1 The Case-Based Reasoning approach

A typical CBR system might operate in the following manner. The system maintains what can be thought of as a memory of cases, each case consisting of, say, the description of some problem and its corresponding solution. When posed a new problem, this memory is searched (using a *similarity metric*) to find that case that describes a problem which has most in common with the new one. The solution to this retrieved case is proposed as the solution to the new problem. In this way, a solution is suggested by analogy to a previous problem. Next, this solution is evaluated. If it is not a satisfactory solution to the new problem, then it must be modified in some way to make it more appropriate: this might be done using generalised knowledge of the domain and of the task.

This process is illustrated in Figure 1, and represents the 'conventional' approach to CBR. Alternative strategies have been proposed – one will be seen later in this chapter – but all have in common the idea of exploiting examples of previous problem-solving episodes.

In CBR, then, a significant portion of the knowledge of the system resides in the form of the *explicit* problem-solving cases. This contrasts with an 'expert system' approach to the same task in which all the procedural knowledge of the task will be in the form of generalised heuristics.

2.2 The Case-Based Reasoning approach to design

The CBR paradigm, then, is based upon the use of previous, successful problem-solving episodes to solve a new problem. This is analogous to the re-use of previous design experiences to help to develop new design solutions, a problem-solving strategy that seems to present quite a natural, *human* approach to design. As a result a great deal of research has been devoted to the development of CBR for design.

The practical appeal of this approach for design tasks arises due to the fact that for most complex domains no comprehensive and detailed body of explicit design knowledge will be available. Even under such conditions, CBR can produce complete, useful design solutions. Furthermore, since design cases are the source of knowledge, the system can 'learn' during its lifetime through the incorporation of new examples into the case-base. The cumulative effect of this should be an improvement in the performance of the system.

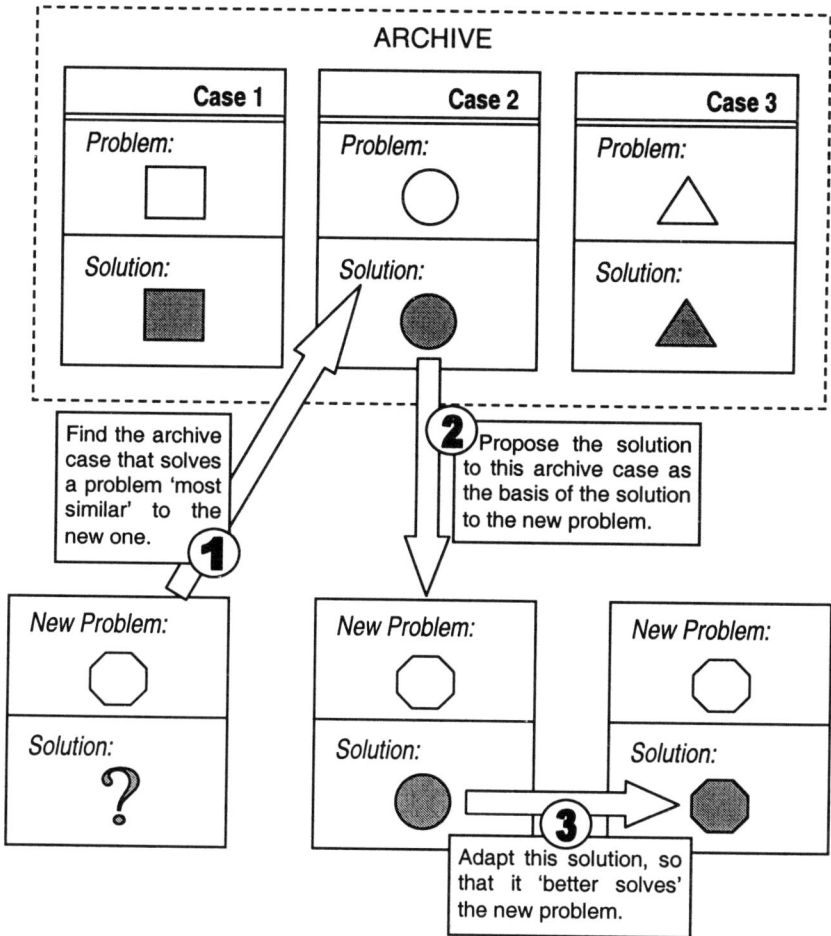

Figure 1: The basic Case-Based Reasoning model of problem-solving.

CBR tools have been developed for tasks as diverse as the design of mechanical devices, architectural design and software design. Pu [3] and Maher and Gómez de Silva Garza [4] provide overviews of this research in greater detail than space permits here, and Maher, Balachandran and Zhang [5] go into the themes and implementation issues in greater depth. The following discussion draws on these works and summarises some of the points their authors raise.

2.2 Case-Based Reasoning for design – issues

The implementation of a CBR system for a design task raises a number of interrelated issues, which must be addressed in some form if a system is to be implemented. This section discusses these issues.

2.2.1 Automation or support?

In building a CBR design tool, is the intention one of *automating* the design task, or of *supporting* a human designer in the performance of the task? This can have implications for precise form of reasoning adopted, the sort of knowledge that is stored and manipulated, and the way in which it is presented to the user of the system. (In the following discussion, the focus is placed primarily on the automation of design, but in certain circumstances, it might be more appropriate – and useful - to develop a support system.)

2.2.2 What is the CBR design process?

It is necessary to devise some plausible CBR mechanism that operates upon the sort of knowledge that is available so as to provide useful design automation or support. In other words, how are design solutions constructed or design support effected? Pu [3] suggests that design solutions can be developed in two general ways: either by the *adaptation* of an old solution (as in the model of CBR described above), or by the *composition* of elements from several old solutions to produce a new solution. A system implementing this second form of reasoning will be described later.

2.2.3 What constitutes a case?

Within design domains, rarely will there be a clearly defined notion of a 'case' - the nature and content of the information provided and subsequently generated will differ from one design episode to the next. However, a CBR approach demands that a decision be made about the sort of information that must be present in all cases. This will be a factor of the information that is available, and will influence, and be influenced by, the nature of the CBR mechanism that might be applied to the task.

Beyond design problems and solutions, it might also be useful to include and reason with information about the process whereby the solution was derived; recall of an applicable design process in a particular context might be more useful than recall of a specific solution. In other domains, it might be helpful to store parts of design episodes in cases, if it is recognised that this is the level at which they are reused. Furthermore, the retrieval of information from *unsuccessful* design episodes might provide useful lessons for the users of design support systems.

2.2.4 How are cases represented?

Once the information that constitutes a case has been recognised, some manner of representing this in a computer system must be devised. This representation must allow the manipulation of the case as demanded in the adopted CBR mechanism.

Cases are usually indexed and accessed based upon the particular design problem they address. Hence there is a need to describe this problem in a fashion which will allow relevant cases to be recalled at relevant times – the representation must capture the essentials of the problem and express the relevant features that make it distinct from other problems in the case-base. A typical approach is to define a set of descriptive attributes, relevant to the design task in hand and with which all appropriate design problems can be described. Associated with each attribute is the range of acceptable values that it can assume; a particular design problem will be described in terms of a suitable value for each attribute.

With a large number of design cases available, there may be no need for the adaptation of solutions, and so it might be appropriate simply to store the solutions as graphical images, and display these when recalled. However, in situations in which solutions must be adapted, or their elements selected and recombined, it becomes necessary to represent them at a level appropriate to this manipulation.

2.2.5 How is the similarity of two cases judged?

The recall and use of previous design episodes is based on judgements of the similarity of old design problems to the new problem. However, an appropriate similarity metric can be difficult to formulate and express. The features of the problem description that are relevant to determining similarity must be identified (and irrelevant features disregarded), and weighted in such a manner as to arrive at some comparable (and hence, numerical) figure expressing the degree of similarity. This quantitative measure does not seem to be particularly 'natural' knowledge, and a workable metric can often only be attained through a process of trial and error, gradually modifying the weightings until satisfactory performance is achieved.

2.2.6 How are case solutions adapted?

If a strategy of solution adaptation has been adopted, whenever a case solution is retrieved on the basis of an inexact match, then some adaptation mechanism must be invoked to make the solution acceptable. This involves identifying the failures of the solution in the current context and then rectifying these. Obviously, the system needs to 'know' how to do this, and failure diagnosis and repair is not a straightforward task. The system would seem to require a relatively complex domain model and heuristics to perform this, the sort of knowledge that is often hard to come by in design situations. The fewer the number of cases available, relative to the number of potential problems, the smaller is the chance of finding a good match, and so, the greater the burden on the adaptation knowledge – it has to do more of the design work.

2.2.7 How are new solutions composed?

An alternative approach is to construct a new solution from elements of old solutions. Here, knowledge is required to identify and isolate useful elements of solutions, and then suggest how these might be reconstituted to form a complete solution. Again, knowledge of this sort might not be readily available.

2.2.8 What is the source of cases?

In most design domains, it would seem optimistic in the extreme to expect a ready-made, sufficiently large corpus of cases, each described in an appropriate manner, to be available. If CBR is to be applied, though, just such a case base must be developed. 'Raw' examples from different sources might be described in quite different ways and to quite different standards of completeness. Hence, the development of a consistent case-base may involve a certain amount of manual interpretation of existing design information - which may result in errors being introduced into the system.

Furthermore, cases from different sources might be of varying quality, with different organisations producing design solutions to different degrees of rigour and 'goodness' according to their needs. This may have an effect on the overall quality of the system using these designs.

2.2.9 Is the system to 'learn'? If so, how?

One of the advantages of CBR approaches is that the performance of systems can be incrementally improved through the incorporation of additional cases into the case-base. This may be done by including cases from external sources, which would first have to be described in the appropriate manner. Problems arise here if a new case is not properly describable using the existing case representation. If this happens, to incorporate the case, the representation must be modified, which would presumably mean re-describing all existing cases; it may prove easier to simply discard the new case.

Alternatively, new solutions generated by the system itself could, along with the current design problem, be used to generate new cases. Presumably, however, there would need to be some external check of the solution to ensure that it does indeed solve the stated problem.

3. A Case-Based Reasoning Design Tool

To provide a more concrete illustration of some of the above ideas, and how these issues might be addressed in practice, this section describes a CBR system that the authors have developed for the automatic *conceptual design* of *fluid power circuits*. Conceptual design is an early stage of the design process, during which an initial solution or solutions to the design problem is proposed, in which "the means of performing each major function has been fixed, as have the spatial and structural relationships of the principal components" [6]. This solution(s) forms the basis for subsequent development and elaboration into a complete design. The conceptual design stage is often considered to be among the most difficult stages of the design process.

This particular task is typical of many to which CBR has been applied in that examples of design episodes are available where the heuristic knowledge of how to produce designs is lacking. The CBR strategy adopted here is one of *composing* a solution from elements of a number of previous solutions (Figure 2 – compare the select and adapt model shown in Figure 1). This approach has been adopted because it was felt that the knowledge required to implement it was more readily available than that required to implement an alternative strategy.

Before describing this process in detail, the nature of this design task itself will be discussed, followed by a description of the information which is considered to constitute a case and the case-base that has been developed.

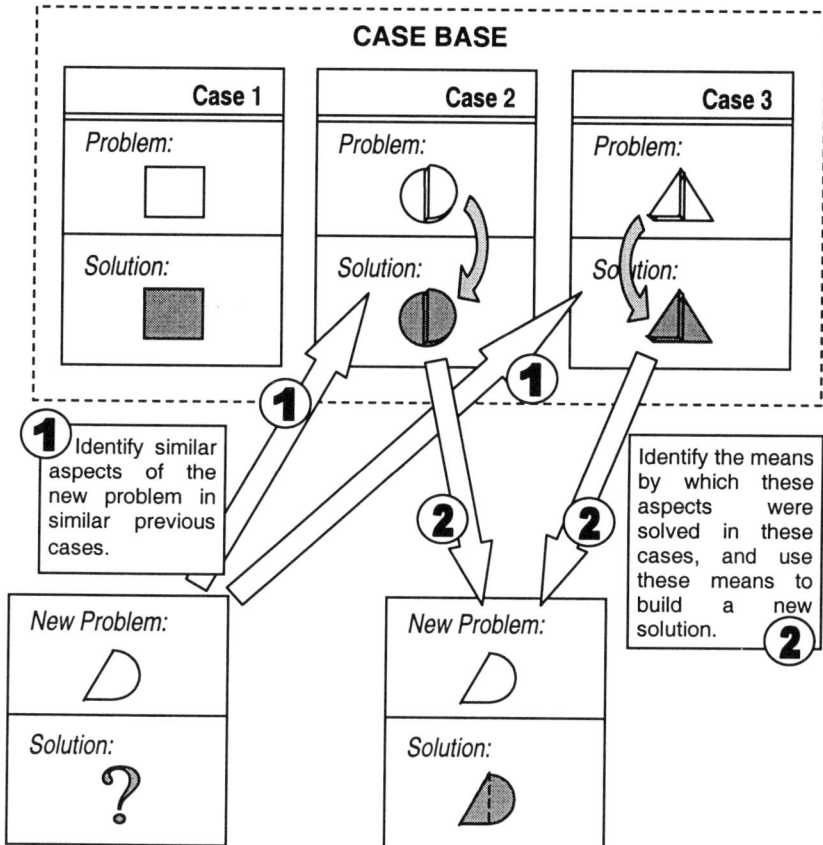

Figure 2: A CBR approach to the composition of design solutions.

3.1 The conceptual design of fluid power circuits

The task in question is the conceptual design of fluid power circuits. These circuits convert rotational mechanical energy into fluid energy (via a pump), which is then transmitted by connecting pipes to some remote point of actuation, at which it is transformed back into mechanical energy. In addition to pumps and actuators, and the connecting pipe-work, there are a number of other standard domain components which serve to control the pressure or flow rate of the fluid, or to direct its flow. Figure 3 shows a schematic of simple fluid power circuit.

At the conceptual design stage, then, the task is to translate a customer-supplied design *specification*, which describes the desired functionality in abstract terms, into a configured circuit that is able to supply this functionality. With this in mind, and considering the sort of information that was available, it was decided that a case should consist of a design specification and the corresponding circuit design solution. The next task, then, is to introduce some computer-tractable way of representing this case knowledge.

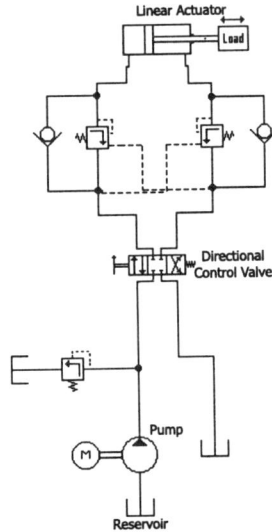

Figure 3: An example fluid power circuit.

3.2 Case representation - design specifications

The design specifications, in particular, can be difficult to express consistently. For this task, there is a great deal of information, presented at different levels of description, that might usefully be considered to form part of a specification. However, some formal, uniform method of describing specifications must be developed. Here, this representation consists of a set of 14 attributes; to describe a particular problem, an appropriate choice of one of a corresponding set of values must be made for each attribute. Amongst this set, it was recognised that certain attributes seemed to be describing some aspect of the performance or characteristic of the system to be designed as a whole (these are termed *characteristic attributes*). In contrast to these, the remaining attributes refer more directly to the functions that the system must fulfil (*functional attributes*). This distinction in the nature of the attributes is exploited later in the CBR mechanism that has been devised.

An example of a characteristic attribute is *maximum load*, that is the maximum load that the circuit must move – this can take the value *low* ($<1\times10^4$N), *medium* (1×10^5-1×10^6N) or *high* ($>1\times10^6$N). An example of a functional attribute is *hold load stationary*, that is, the requirement for some ability to hold the load stationary at any point in its motion. This can assume a value of either *yes* or *no*. Table 1 shows the attributes, their accepted values, and their types.

Table 1: Specification attributes, their values and their types.

attribute name	permitted values	type
maximum load	high, medium, low	characteristic
maximum speed	high, medium, low	characteristic
plane of motion	horizontal, non-horizontal	characteristic
continuously variable speed	yes, no	functional
hold load stationary	yes, no	functional
smooth accelerations	yes, no	functional
hold load on failure	yes, no	functional
load-independent speed	yes, no	functional
control extend speed	yes, no	functional
control retract speed	yes, no	functional
solution requires motor	yes, no	functional
control inertia	yes, no	functional
energy efficiency paramount	yes, no	characteristic
control accuracy	high, low	characteristic

3.3 Case representation - design solutions

As mentioned above, it was decided to adopt a composition approach to the construction of design solutions. For this to be done, it would be necessary to describe solutions in terms of the useful elements that they contain.

In the domain under consideration, the majority of design solutions share a basic framework – the solution *template* (Figure 4(a)). The template embodies the functionality integral to all solutions (it includes a power source in the form of a basic pump, an actuator, and the facility to direct flow from the pump to the actuator). Moreover, the template has a number of labelled *slots*, into which additional components can be placed in order to provide the precise functionality demanded by the specification. Hence, by reference to these slots, the template also provides a convenient context for describing complete solutions: a solution consists of the template along with the set of additional components, their corresponding slot positions and orientations within the slot. Any slots without components are considered to be filled by a direct pipe connection.

Although, in theory, any component in this domain could be placed in any slot, in practice only certain components used in particular ways are found to provide *useful* functionality. The next step in describing solutions, then, is to identify this set of useful components and their positions and the orientations in which they are used. The combination of component, slot and orientation defines a *solution element*: through inspection of the available design cases, 17 such elements have been defined. Figure 4(b) and Figure 4(c) show two of the identified solution elements. Since each element can usefully appear only once in a solution, and given the constant template, a particular solution can be described by labelling each element as either *present* or *absent*, as appropriate.

So, this combination of solution template and elements provides a means for describing existing cases and for constructing new solutions.

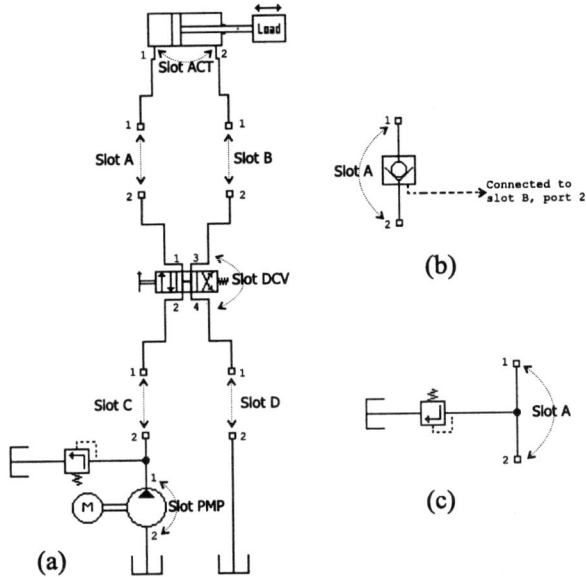

Figure 4: (a) The solution template; (b) solution element *POCV_A* - a *pilot-operated check valve* in slot A, and; (c) solution element *PRV_A* - a *pressure relief valve* in slot A.

3.4 The case-base

A number of successful examples of the design of fluid power circuits have been collected into a case-base. Each case consists of a design specification and the corresponding design solution, both described according to the devised representations. The cases have been collected from several sources: textbooks, training material and industrial design. Many of the source design episodes were found to be poorly documented or otherwise lacking, so a certain amount of interpretation has been necessary to fully describe each case.

In total, the case-base contains 30 cases, providing weak coverage of the range of potential design problems in this domain. Nevertheless, this is felt to represent a realistic ratio between available examples and potential problems in engineering design contexts.

3.5 A CBR mechanism for fluid power circuit design

Since a total of only 30 cases are available, it would be unlikely that the solution retrieved to address any new specification will represent an satisfactory solution as it stands. However, adaptation knowledge is also lacking in this domain, so the approach adopted here is to construct solutions using elements of the existing solutions in the case-base. As will be seen, each of these elements appears to have solved some aspect of the design problem that is also a feature of the current problem, and, moreover, has done so in a similar context. In order to build solutions in this manner, additional knowledge is required to enable useful elements to be identified – here, this additional knowledge takes the form of *explanation rules*. As will be seen, these rules contrast with the synthetic rules that might be used in an expert system approach to design.

3.5.1 Explanation rules

Analytical knowledge of the domain is required in order to decompose the case solutions into useful elements. To represent this, the authors have introduced the concept of explanation rules. Each of these rules has the form:

$$\{x_1, x_2, \ldots, x_n\} \Rightarrow f$$

where x_i is a member of the set of solution elements, and f is one of the *functional* attributes of the specification. The rule may be read as, 'this set of solution elements can be used to provide functional attribute f '. These rules allow the construction of explanations of the manner in which a particular solution achieves its functionality (expressed in terms of the functional attributes). In the case-base of design examples, each design solution is paired with the specification from which it was generated. The rules allow each element present in the solution to be explained as satisfying (or contributing to the satisfaction of) one or more of the demanded functional attributes.

A set of explanation rules for the restricted domain in question was collected through interrogation, in the context of the case-base, of a human familiar with the domain. This was done by asking for an 'explanation', in terms of one of the specified functional attributes, for each of the solution elements in each of the example solutions in the case-base. In total, there are about 30 such rules, with each functional attribute referred to by at least one. As an example, the rules for one particular functional attribute, *hold load stationary*, are as follows:

        {POCV_A} ⇒ hold load stationary

        {CBV1_A&B} ⇒ hold load stationary

        {DECV_A&B} ⇒ hold load stationary

Given a new design problem, in which it is required that *hold load stationary* be satisfied, these rules indicate that there are three different ways in which this might be done – by using element *POCV_A*, or element *CBV1_A&B*, or else element *DECV_A&B*. There is no indication, however, of which element provides the *best* way of holding the load stationary when faced with a particular design problem. Hence, these are *not* design rules; what is lacking is some indication of the *context* in which the elements will achieve the function: these are *analytical* rather than *synthetic* rules. The basis for making the choice of which solution element to use lies in the memory of example design cases, in which the successful use of elements in particular contexts can be seen.

3.5.2 A Case-Based Reasoning algorithm

This set of explanation rules and the case-base of designs are used to solve new configuration design problems in the following manner. Initially, a new solution will consist solely of the template. A new design specification is presented to the system. For each *functional* attribute that is demanded (i.e., has the value *yes*), a search is made of the case-base to identify those examples in which this attribute is asked for (and, hence, satisfied by the solution). If the attribute is satisfied in more than one example, then the *best* case is that which has a specification that is the most similar *in its entirety* to that of the new problem.

Once this best case is found, its design solution is examined, and, using the explanation rules, the *largest* set of solution elements in the circuit that provides the functional attribute under consideration is identified. If it does not already exist in the new solution, each of the elements in this set is then added to the solution. If, however, no example of the satisfaction of a particular functional requirement exists in the case-base, then an explanation rule associated with the requirement is selected by default to suggest the element(s) to use. The process continues in this way for each required functional attribute. Thus, the process is one of finding solution elements that are recalled as having been used successfully to provide some required functionality in a context which is similar to that of the current problem. This algorithm is shown in Figure 5.

All that remains to be described is the manner in which the *similarity* of two specifications is judged. For this purpose, a simple 'Hamming distance'-type measure is applied to determine the 'distance' of the specification of each case example from the new specification. The number of values of corresponding attributes (of both types) that differ between the two specifications is the distance between them. That example that is 'closest', in other words, which has a specification having the fewest differences from the new specification, is considered to be the best matching case. This is an unsophisticated, domain-independent measure, and takes advantage of the somewhat simplistic attribute-value manner in which the specifications are represented here. Better solutions would probably follow from the use of a more sophisticated matching algorithm. However, this would require additional task knowledge of quite a complex nature.

3.6 Results

The approach described above has been implemented in a system using the case-base of 30 examples. As would be expected, given the use of the explanation rules in the algorithm, this system provides solutions in which all the desired functionality is embodied in some identifiable form. To this extent, all the solutions that are generated *appear* to be plausible. To illustrate the operation of the system, its response to a test specification in which three functional attributes - *continuously variable speed*, *smooth accelerations* and *control inertia* – are demanded will now be described.

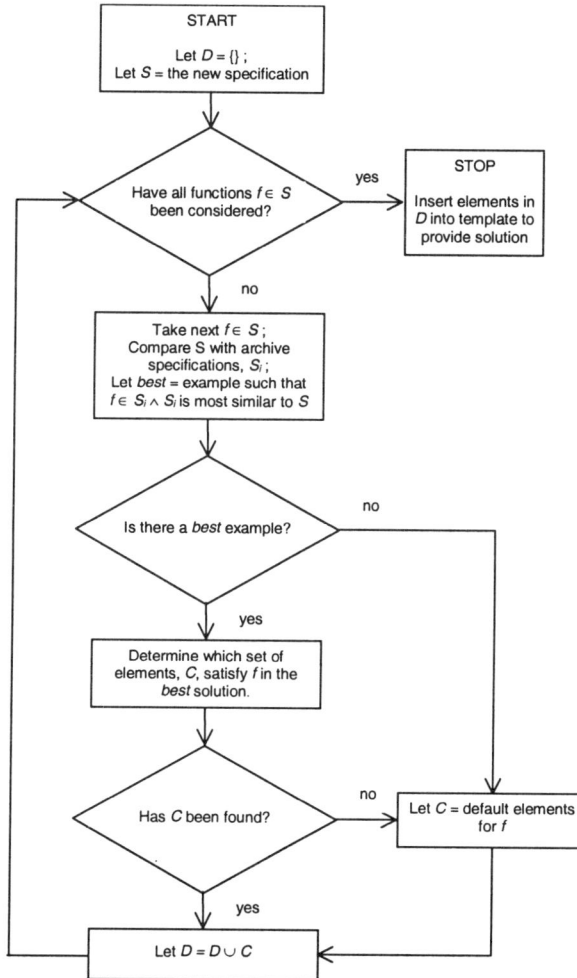

Figure 5: A Case-Based Reasoning algorithm.

Considering each of these functional attributes in turn:

- *continuously variable speed* (i.e., the solution must provide the facility for specifying any speed during operation) – case number 11 is the closest match (a 'distance' of 3 separating the specifications) satisfying this attribute. The solution to this case is examined. The solution contains the element *PROP_DCV* (a *proportional directional control valve* element), and, from the explanation rule:

   {PROP_DCV} $\Rightarrow$ continuously variable speed

   it is known that this element provides the functionality, and so it is added to the template.

- *smooth accelerations* (i.e., accelerations with the system are to be jolt-free as far as possible) – with case 13 providing the match this time (a 'distance' of 4), the presence of the element *CBV_A&B* (*counterbalance valves*) in the solution, and the known rule:

$$\{CBV1\_A\&B\} \Rightarrow \text{smooth accelerations}$$

dictate that this element is added to the solution.

- *control inertia* (i.e., control the sometimes unwelcome effects of inertia in the circuit) – in this instance the solution to case 6 (a 'distance' of 3 away) is examined, and the rule:

$$\{CBV1\_A\&B, CVC\_A\&B, PRV2\_A\&B\} \Rightarrow \text{control inertia}$$

is used to add the elements *CVC_A&B* (*check valves*) and *PRV2_A&B* (*pressure relief valves*) to the solution (*CBV1_A&B* already having been added to provide *smooth accelerations*). For this particular function, the explanation rules also state:

$$\{CVC\_A\&B, PRV2\_A\&B\} \Rightarrow \text{control inertia}$$

$$\{CVC\_A\&B\} \Rightarrow \text{control inertia}$$

$$\{PRV2\_A\&B\} \Rightarrow \text{control inertia}$$

In other words, subsets of the elements from the first rule can be used to satisfy the function. The strategy is to search the entire solution to find the *largest* set of elements that provide *control inertia* in order to ensure that it is provided in the new solution.

In this manner, then, a circuit solution is constructed, in which some means of producing all the desired functionality has been provided. By inspection, this seems to present a plausible solution to the stated design problem.

3.7 A Case-Based Reasoning design tool - summary

The CBR approach described here is an attempt to harness the design knowledge that resides in the form of a set of design examples. A new design problem describes the functionality required of a solution. The examples are searched to determine which solution elements have been used to provide this functionality in similar contexts in the past. These elements are then used to construct a circuit solution to the new problem. This approach is an attempt to exploit a relatively small set of examples by recombining parts of them to solve new problems. It relies on system functionality being embodied locally in sub-sets of solution elements (and the relationships between functions and solution elements being known). If this is not the case (as in a domain in which, for example, the functionality of a design is a global feature of the system as a whole, rather than of localised elements of it), then this approach cannot be applied.

Since the crucial element in all design tools is the knowledge embodied within the system, it is useful to summarise the different types of knowledge that are required to implement this form of CBR:

- The requirements representation, the 'typing' of each parameter.
- The solution representation, including the template.
- The explanation rules.
- The case-base of design examples.
- The CBR algorithm itself.
- The matching algorithm (effectively domain- and task-independent here).

Some of this knowledge is quite detailed, and not always easy to acquire and express. However, it provides an environment in which the knowledge in the previous examples can be exploited and used to develop new design solutions.

4. Conclusions

This chapter has introduced and described the technique of *Case-Based Reasoning*, which has been applied to a number of design processes. It has often been recognised that previous design episodes constitute an extremely valuable source of design knowledge and CBR offers an approach whereby this source can be exploited and the knowledge transferred to generate solutions to new problems. A number of issues surround such an approach, however, and these must be addressed in some form if a working system is to be implemented.

To better illustrate some of the ideas and decisions that must be made, the latter half of this paper is devoted to the description of a CBR tool developed by the authors for the conceptual design of fluid power circuits. This tool adopts a strategy of selecting elements from previous solutions and combining them into a new solution, making use of the limited number of available cases to guide this process.

Acknowledgements

The work described in this paper forms part of the research undertaken in the Engineering Design Centre at the University of Bath, and is funded by the UK Engineering and Physical Science Research Council grant GR/L26858.

References

[1] Riesbeck, C. K. and Schank, R. C., 1989, *Inside Case-Based Reasoning*, Lawrence Erlbaum, New Jersey.
[2] Kolodner, J. L, 1993, *Case-Based Reasoning*, Morgan Kaufmann, San Mateo, CA.
[3] Pu, P., 1993, Introduction: issues in case-based design systems, *Artificial Intelligence for Engineering Design, Analysis and Manufacture*, 7(2), 79-85.
[4] Maher, M. L. and Gómez de Silva Garza, A., 1997, Case-based reasoning in design, *IEEE Expert*, March-April, 34-41.
[5] Maher, M. L., Balachandran, M. B. and Zhang, D. M., 1995, *Case-Based Reasoning in Design*, Lawrence Erlbaum, New Jersey.
[6] French, M. J., 1985, *Conceptual Design for Engineers*, The Design Council, London.

An Internet-based Approach to the Capture and Reuse of Knowledge in Design

Paul A. Rodgers[1], Nicholas H.M. Caldwell[2] and Avon P. Huxor[3]

1. Department of Design, Napier University, 10 Colinton Road, Edinburgh EH10 5DT, p.rodgers@napier.ac.uk
2. Engineering Design Centre, University of Cambridge, Trumpington Street, Cambridge CB2 1PZ, nhmc1@eng.cam.ac.uk
3. Centre for Electronic Arts, Middlesex University, Cat Hill, Barnet, Herts. EN4 8HT, a.huxor@mdx.ac.uk

Abstract: Designers in modern design settings are confronted with several pressures. They are faced with increasing demands to reduce development costs, shorten time to market periods and at the same time improve the total quality of the product which inevitably leads to an increase in the complexity of design projects. In an effort to meet these pressures, and others, designers now rely upon vast quantities of knowledge to support their design tasks. This knowledge must be in a format that the designer can locate, access, use and re-use easily. To this end, this paper describes a practical approach that facilitates methods for capturing, storing, applying and re-using knowledge in design activities. The approach taken is based on an "AI as writing" metaphor which has been implemented in an Internet-based "knowledge server" architecture. The implementation, WebCADET, provides effective mechanisms for capturing, using and re-using knowledge in modern conceptual design tasks.

1. Introduction

Modern design and development activities require large amounts of knowledge, information and data that is well beyond the capacity of any individual designer [1, 2]. Thus, it is now common to find in many design and development projects large teams of individuals with a wide range of skills and backgrounds, including designers, manufacturers, financial analysts, business managers and so on. The effectiveness and efficiency of the design process is influenced by the procedures used by those involved, their personal knowledge and experience, their ability to assimilate the wider experiences of other more senior personnel and their own inventive and intuitive powers. Maximising designers' effectiveness is particularly important in today's climate of ever increasing pressures. For example, fierce competition from overseas has increased demands for shorter product development and lead times, lower costs, and higher levels of product quality [3].

In attempting to design and develop high quality products effectively and efficiently, designers have to utilise a wide variety of sources of design knowledge [4] such as those illustrated in Figure 1 and apply them in many different design life cycle tasks.

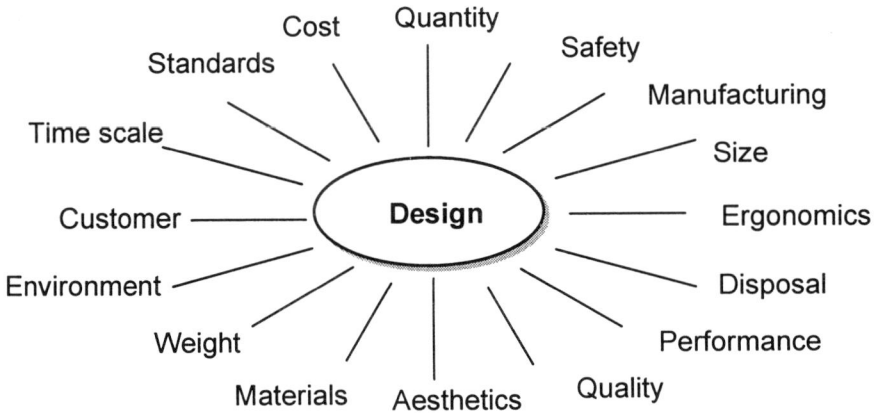

Figure 1: Knowledge and Information Life Cycle Elements in Design

The active capture and reuse of knowledge is now emerging in many large companies [5, 6] and it is an area of increasing research significance. The management of knowledge in design organisations, specifically, has been a recent focus for much research [7, 8, 9]. An important objective of much of this research is to provide effective and efficient mechanisms for capturing, using and reusing design knowledge [10, 11] that will go some way towards supporting the development of high quality products.

2. Knowledge Types in Design

Knowledge is a word that is used in everyday conversation without much difficulty, but is a very difficult concept to define and classify. This is reflected in the fact that a whole branch of philosophy (epistemology) is dedicated to questions on the subject of knowledge. Design requires the assimilation and application of many types of knowledge.

Design organisations acquire and utilise at least two major categories of knowledge. There is knowledge related to the product itself, and knowledge concerned with how the product will be manufactured [12]. These categories may be further divided into the different types described in the following four sections.

2.1 Explicit and tacit knowledge

Knowledge can be split along an explicit and tacit dimension. Explicit knowledge is knowledge that has been captured and codified into manuals, procedures or rules whereas tacit knowledge cannot be easily articulated and thus exists only in people's minds [13]. It can be very difficult to make tacit knowledge explicit. However, this is precisely what is required if you wish to understand, explain, test or teach it.

2.2 Declarative and procedural knowledge

Procedural knowledge is concerned with how to do some action, such as the actions and decisions that drive the design and manufacture of products. This type of knowledge comes predominately from practical experience and formal knowledge. Procedural knowledge, in the form of design guidelines, is known to provide effective support for designers [14].

Declarative, or factual, knowledge is knowledge about some entity [15]. Declarative knowledge comes predominately from the applied sciences and can be transformed into rules and methods for designing. The limitation of these rules is that they usually only apply to one part of the design problem, and do not indicate how the result can be integrated into the overall design solution.

2.3 Heuristic and algorithmic knowledge

In the context of design, a heuristic can be described as a "rule of thumb" which may lead to an acceptable result with good probability. A heuristic procedure has a chance of finding something suitable, but there is no guarantee that it will be found always and by everyone [16]. An algorithm, on the other hand, is a set of steps which will lead to a solution if one exists [17]. It is widely acknowledged that designers do not take an algorithmic approach in their design activities, but adopt more of a heuristic approach [18].

2.4 Deep and shallow knowledge

Knowledge can be either deep or shallow. Shallow knowledge expresses the know-how of an expert without explaining the fundamental issues, whereas deep knowledge expresses theoretical fundamentals of a domain [17].

2.5 Summary of knowledge types

Decision making in design involves all of the different knowledge types listed above. For example, shallow knowledge is generally used early in the design process followed by increasingly deep knowledge levels [19]. Thus, any knowledge support system, if it is to be truly effective throughout the design process, must be flexible to facilitate all these types of design knowledge. For instance, during the early stages of the design process activities such as identification of need(s), problem analysis, market research, concept generation and evaluation requires shallow, broad design knowledge. Typically, this will include an initial analysis of the competition, an investigation into which national and/or international standards and specifications the product must meet, literature and patent reviews and customer feedback from the market [4].

Whereas during the more detailed stages of the design process, after many critical decisions have been made, the knowledge needed becomes more specific (deep) and may be better described as algorithmic in nature. Generally, this includes the use of design knowledge which has been represented in design methods and tools such as parametric CAD analysis models, optimisation programs and domain specific rules and procedures [20].

3. WebCADET Knowledge

The WebCADET system is a re-implementation of prior research [21]. Amongst the findings of this research was the utility of heuristics to measure the success or failure of a design concept in meeting specific desirable attributes. WebCADET retains this emphasis on heuristics and design guidelines as the knowledge types of interest as they have been shown to be a particularly effective approach in design support [22].

The basic unit of knowledge within WebCADET is termed a "rule-text". Each rule-text focuses upon capturing guidance concerning the acceptable design parameters (dimensions, materials, texture, colour, shape, etc.) that a proposed concept design must possess in order to meet a desirable design criterion element, be it an aesthetic, ergonomic, safety or other issue. Moreover, each rule-text represents design knowledge specific to a particular context and so many rule-texts may co-exist for a given criterion and product combination. An example rule-text relating to shaver design is shown below in Figure 2.

The rule-text in Figure 2 is uniquely identified by the **rule_id** slot as *shaver_comfortable_to_hold_1*. This indicates that it contains design knowledge relevant to shavers in general and specifically that the knowledge relates to the desirable attribute that the shaver should be comfortable to hold when in use. The **name** and **precondition** slots aid faster knowledge retrieval during a WebCADET session.

The **conditions** slot is the most important part of the rule-text as it contains the design knowledge expressed as a set of if-then-else conditional expressions. The values in the *then* and *else* components represent pass and fail scores defined using a pair-wise comparison method [23] for rating the importance of each design parameter. Which value is used during execution depends on whether the condition in the *if* component is met. The mnemonics *eq*, *gt* and *lt* abbreviate the conventional equality and relational operators *equal to*, *greater than* and *lesser than* respectively. The *iaiof* operator abbreviates the *is an instance of* relation and links the condition to a set of facts, such as the shaver texture which must be a comfortable handle texture so that the user will be able to hold the shaver effectively. The *has_aspect* operator supports similar links to additional design knowledge.

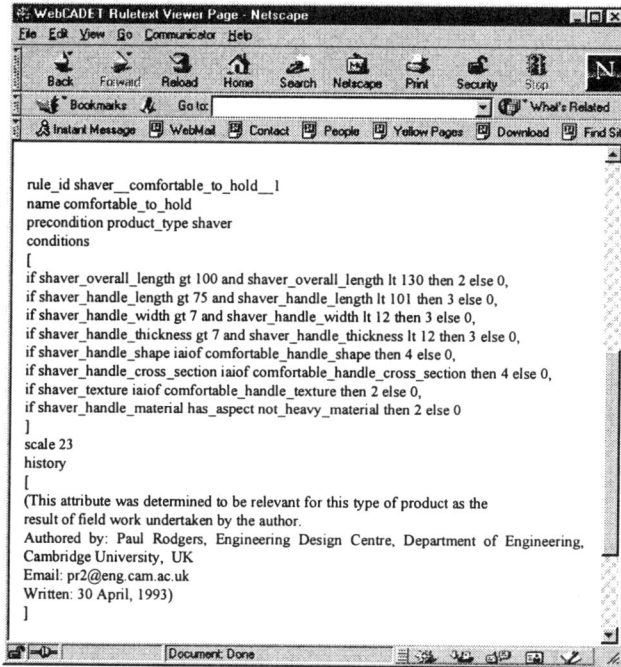

Figure 2: Rule-Text Example

The **scale** slot is used to normalise the accumulated scores in execution to a value between zero and one. The **history** slot contains the original rationale for the creation of the rule-text and, as an original rule-text is modified by other rule-text authors, will accumulate their reasons for the modifications. This slot illustrates the role of provenance in the "AI as writing" approach (described in the next section), and also serves to place the rule-text in a specific context to explicitly associate the knowledge with its human originator. Each pairing of **rule_id** and entry in the **conditions** slot provides an index to a justification in the explanations database (including references to design manuals, design literature, and increasingly on-line resources).

4. "AI as Writing" Approach

The traditional view of artificial intelligence (AI) seeks to create autonomous expert systems which might replace a colleague who is unavailable or has left the organisation [24]. The approach taken here, however, is based on an alternative view, in which symbolic AI is seen as an extension to writing, taking "textuality" to a new level, but nonetheless better understandable as a "technology of the word"

[25]. This approach also recognises the importance of transparency and modifiability of the knowledge-base content. Hill [26] suggests that our representations are commitments to particular ways of thinking about a domain, and thus users should be able to identify the provenance of the knowledge and be able to challenge it. In other words, the provenance of the knowledge embodied in the system must be made accessible to the reader (user), who can contest and modify its contents in the light of their own experience and needs.

To this end, the "AI as writing" approach assumes that design problems are best solved by a combination of human designer(s) and AI design support systems which will empower designers to develop successful solutions by augmenting their intellectual capacity and productivity. This view, which has recently become more widespread [27, 28], considers the machine as a mediator between people in a similar fashion to writing. Viewing AI systems as a form of writing sees their role as not one to simulate human thought, but to act as a medium for thought. That is, just as written texts act as a medium for spoken words, so the knowledge medium acts as a medium for what can be thought. The AI as writing approach resembles certain elements of Brown and Duguid's work [29] in that they also view text (documents) as a powerful resource for constructing and supporting strong communities of individuals. Viewing AI tools in this way, as a form of writing, leads us to explore a number of conventional features in writing and publishing including:

- *transparency* - it is possible to follow the argument (rule-text) as written;
- *quotability* - it is possible to re-use complete and partial rule-texts to create new rule-texts;
- *authorship* - authors are responsible for their submitted rule-texts in terms of their applicability and accuracy and may gain recognition for their contribution;
- *communities* – specialist groups of rule-text authors (ergonomists, architects, software engineers) will be encouraged to form robust "virtual" communities.

Figure 3 shows an original rule-text on the left concerning whether a phone will be operable with one hand or not. This has been initially authored by Paul Rodgers, 30 April 1993. This rule-text has been read by other users who have considered the initial contents to be inappropriate for their specific region or end-user group. Modified versions of the initial rule-text have been created by different users (authors), such as that of Ming Xi Tang, 18 October 1995, Avon Huxor, 26 September 1996 and Nicholas Caldwell, 12 May 1998. The changes may, for example, have been made due to the different markets in which each user is working. For example, markets such as for children or the elderly who have specific requirements or because of cultural or regional differences [30]. Users (authors) are able to submit their rule-texts for publication via the WebCADET server. Thus, a market in authored rule-texts will develop as users (readers) seek and apply rule-texts that are appropriate to their particular design scenarios. Some rule-texts will be used more than others due to the reputation of their author(s), amongst other reasons, whilst others may fail as they are deemed inappropriate. By adopting this approach, participants in the process of knowledge externalisation [31] can be viewed in terms of their roles in a publishing model, that is, as editors, authors and readers.

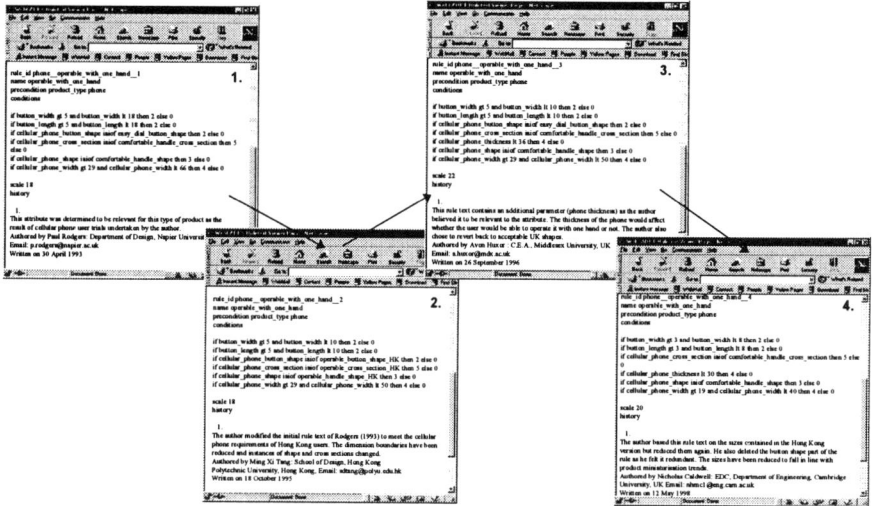

Figure 3: "AI as Writing" Model

The remaining sections of this paper illustrate how these features have been applied in the development of WebCADET, a web-based knowledge server, built to aid the capture and use/reuse of knowledge for supporting and guiding designers in their activities.

5. WebCADET Architecture

In architectural terms, WebCADET possesses the traditional components of a knowledge-based system, namely a knowledge base, an inference engine, and a user interface. The tool has two main operational features, knowledge guidance (use/reuse) and knowledge capture. WebCADET differs in that it adopts the "knowledge server" paradigm pioneered by Eriksson [32]. In a knowledge server implementation, the knowledge base and the inference engine reside on a conventional web server and the user interface is exported on demand to remote users and displayed on standard web browsers (Figure 4).

This approach has a number of advantages. The system can be accessed by any individual who has access to a standard web browser and a network connection regardless of their geographical location. The emerging omnipresence of the Internet in modern design working styles means that most designers will be accustomed to web page usability and this reduces the effort required to master the user interface of a knowledge server system. For system builders and system maintainers, the architecture eliminates distribution overheads in updating the system or its knowledge base because such tasks are only required on the centralised servers.

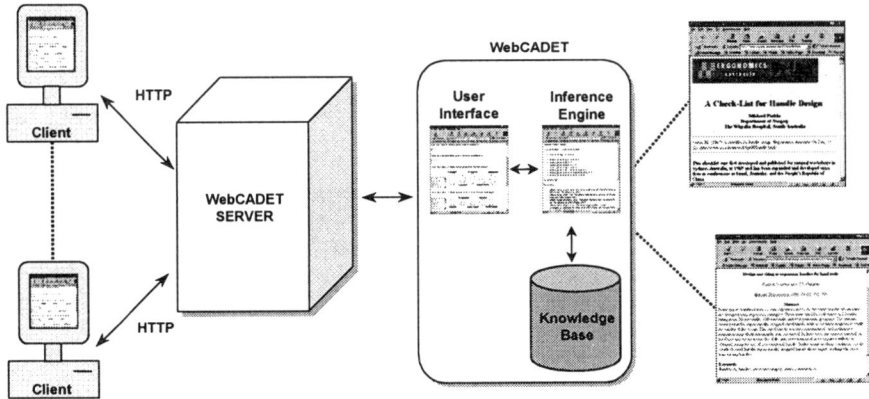

Figure 4: "Knowledge Server" Architecture

6. Knowledge Use (Guidance)

The interactive design guidance mode has been developed to help designers evaluate their concept ideas in the early stages of the design process. The PDS elements of Pugh [4] have been used as a basis for decomposing WebCADET's guidance knowledge into manageable-sized chunks, rather than have the designer deal with the entire problem at once [33]. To access the relevant task at hand, the user selects the PDS element relating to their guidance query (e.g. aesthetics, ergonomics, environment, etc.) followed by the attribute they wish to check and finally the rule-text version. For any PDS element there is likely to be a range of relevant attributes and rule-text versions to select from (Figure 5).

Upon selecting the rule-text the designer wishes to use, the attribute and the product combination (i.e. operable with one hand and phone) is used to determine the set of relevant parameters from the knowledge base. Each parameter has a predetermined design property and the appropriate HTML encoding specific to the parameter and solution space is generated.

It is acknowledged that this approach has its limitations in terms of supporting truly creative design as some aspects of domain knowledge and problem understanding are known beforehand. However, the emphasis of this work is aimed at supporting the type of design which is classed as either variant or routine [34]. This type of design activity is thought to make up at least 75% of total design work actually undertaken [35]. The utility of WebCADET is that it acts as both a prompt and reminder of relevant heuristics, some of which may be unknown or simply forgotten even by experienced designers and it will also assist in the training of novice designers within a company.

Next, the user inputs a description of their concept design idea by completing each slot in the form to check the ergonomic suitability of it. Specifically, the designer is checking if the concept phone idea will be operable with one hand. Each condition is processed using the parameter data to evaluate how well the proposed concept meets the specific criteria for the design attribute (Figure 6).

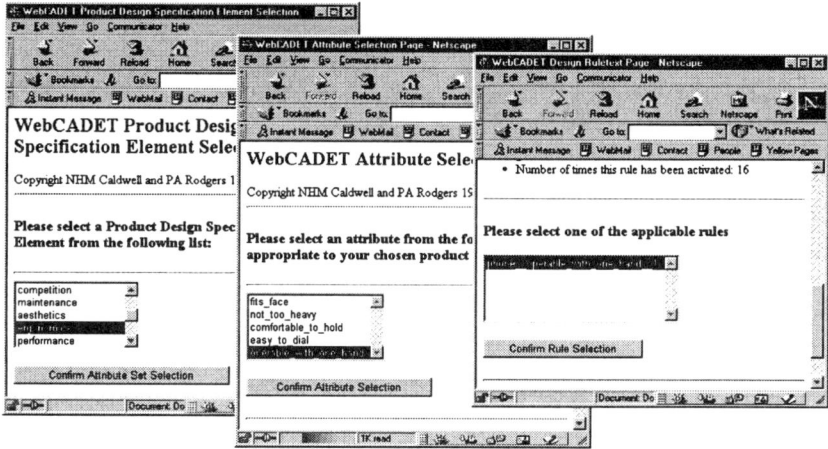

Figure 5: PDS Element/ Attribute/ Rule-Text Selection Example

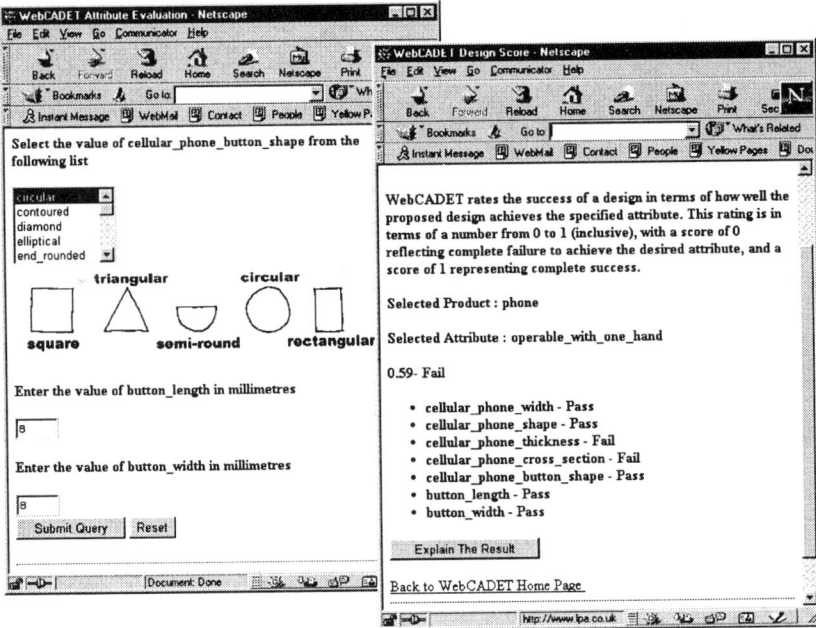

Figure 6: Concept Evaluation Example

The user can request an explanation from the WebCADET server as to why specific parameters are deemed unsuitable (e.g. in this case, why the phone thickness and cross-section failed). As illustrated in Figure 7, the justification sources can be either links to internal or external knowledge (e.g. on-line documents, heuristics, people, etc.). This has certain similarities with the approach taken by Leake et al [36] in the reasoning facility of their stamping advisor CBR system.

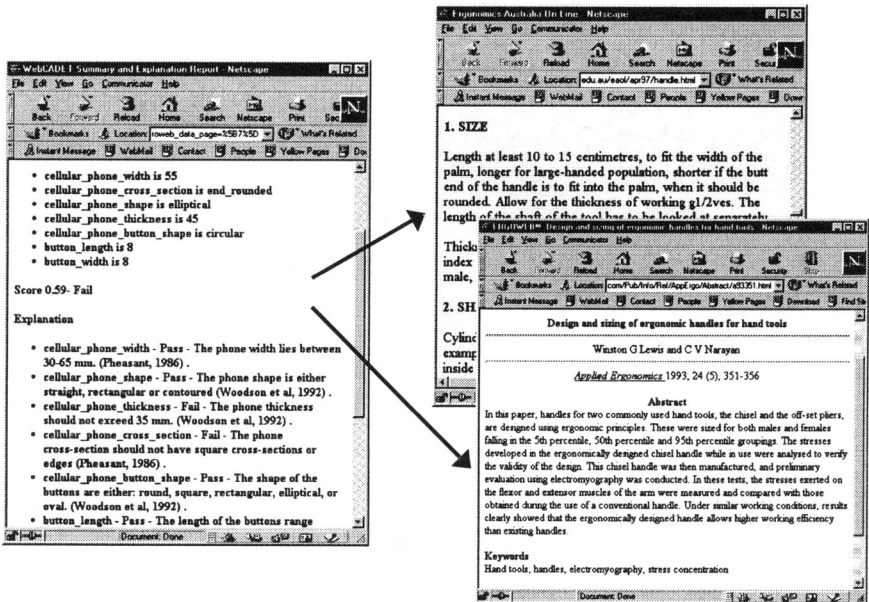

Figure 7: WebCADET Explanations and Rationale (On-line links)

The explanation facility serves as a form of rationale for supporting design decision making tasks. This is a particularly important facility when designers are faced with making a decision between alternative concept solutions [37, 38]. This facility addresses some of the key knowledge management issues faced by design companies in that the tool should alleviate work duplication, error repetition and also help promote the sharing of knowledge amongst designers by locating either internally stored or externally stored best practice and knowledge.

7. Knowledge Capture

In the underlying theory of the "AI as writing" paradigm, the knowledge base of WebCADET is viewed as an evolving rather than a static resource. The corpus of product rule-texts accumulates through the sharing of knowledge by and between members of the design community. The transparent nature of this approach empowers individuals (authors) within and among organisations by returning responsibility and capability for creating and sharing knowledge to them. This approach, however, requires an effective and efficient knowledge capture facility that will allow designers to author and edit the contents of the knowledge base in a number of ways. Presently, designers (authors) can create and edit four main parts of the knowledge base including the two shown in Figures 8 and 9.

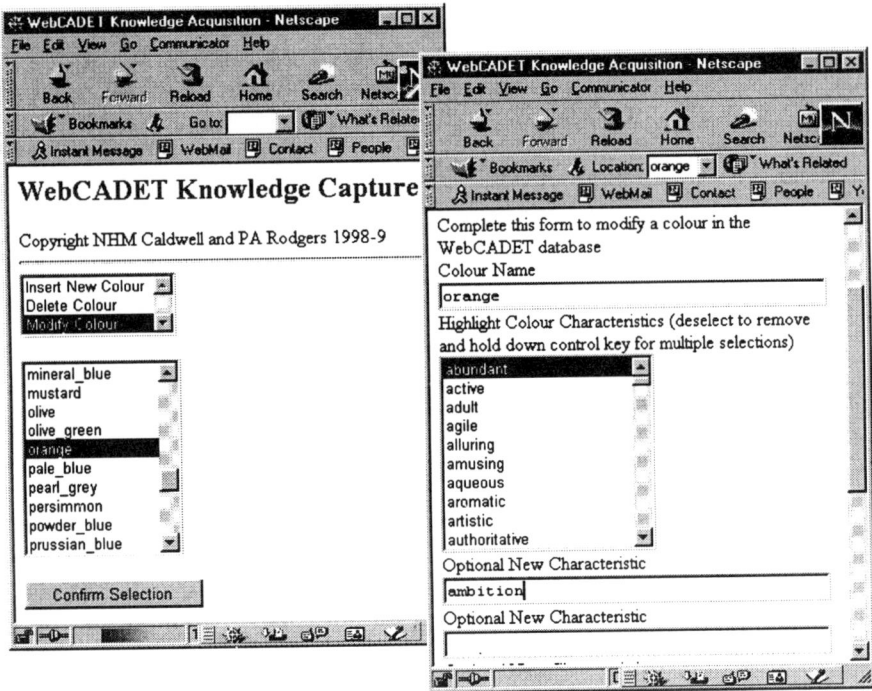

Figure 8: WebCADET Colour Database Modification Example

Colour database modification allows the user to add and remove colours, and to select and modify associations such as conservative, healthy, stylish and so on for chosen colours. It uses a standard colour scheme proposed by Kobayashi [39]. Figure 8 shows the colour orange being modified to include the associative meaning "ambition" as suggested by www.avcweb.com.

It is in the rule-text construction mode, however, that the AI as writing approach is best illustrated. In this mode, the user can create a complete new rule-text or modify an existing rule-text. In each case, WebCADET will generate a new **rule_id** based on the product, attribute and number of existing rule-texts for that combination and pre-load a clause capture form with any previously supplied parameter names, data-set names, property names and citations as shown in Figure 9. For example, the large window of figure 9 shows the user modifying the clause *cellular_phone_width* of the rule-text *comfortable_to_hold*. The user justifies the rule-text clause change as depicted in the smaller window on the right hand side of Figure 9. In this instance, the acceptable phone width has been increased to serve users who may have dexterity problems whilst holding and using products. This feature of the capture mode records the user's rationale for altering the original rule-text.

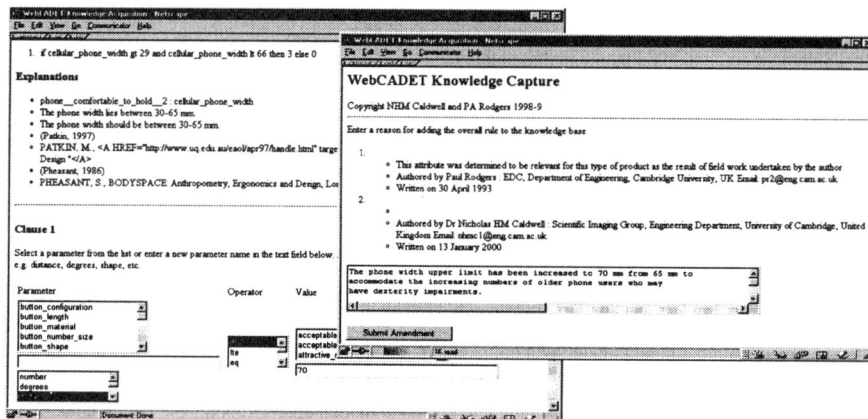

Figure 9: WebCADET Rule-Text Capture/ Modification Example

8. Conclusions and Future Work

This paper has described an Internet-based knowledge capture and guidance tool, WebCADET, which has been developed to provide designers with relevant knowledge during the early stages of design, guide them in their concept evaluation tasks and also support the capture and creation of new design knowledge.

To date, several WebCADET knowledge servers have been deployed. The primary server is located at the University of Cambridge and is mainly used to test new WebCADET functionality. A second WebCADET server is currently undergoing further development, in light of testing in undergraduate design education trials, and will be deployed at the Design Department of Napier University. This server will act as a repository for support in the domain of 3-D product examples such as phones, corkscrews and toothbrushes. A third WebCADET server is presently being tested at the University of Middlesex to determine whether "virtual environment" heuristics can be stored in WebCADET's formalism. An initial rule-text set is currently being generated for this server. A fourth knowledge server has been deployed within the School of Design, Hong Kong Polytechnic University and this is being used in a collaborative project with the Design Department of Napier University and the Engineering Design Centre of Cambridge University as a test-bed for rule-text evaluation and comparison in a distributed design project [40].

Finally, the fifth WebCADET server is presently being developed for deployment within the corporate Intranet of a UK-based material handling equipment manufacturer. Recently, an initial user evaluation of WebCADET has been undertaken within this organisation and the results are being used to enhance the system and its overall effectiveness [41]. Ongoing work with this company will include further testing and development of the tool.

References

[1] MacLeod, A., McGregor, D.R. and Hutton, G.H., 1994, "Accessing of information for engineering design", *Design Studies*, 15(3), pp. 260-269.

[2] Court, A.W., Culley, S.J. and McMAHON, C.A., 1997, "The influence of information technology in new product development: Observations of an empirical study of the access of engineering design information", *International Journal of Information Management*, 17(5), pp. 359-375.

[3] Ertas, A. and Jones, J.C., 1993, *The Engineering Design Process*. John Wiley & Sons Inc., New York.

[4] Pugh, S., 1991, *Total Design*. Addison-Wesley, Wokingham, UK.

[5] Lincoln, J.R., Ahmadjian, C.L. and Mason, E., 1998, "Organizational learning and purchase-supply relationships in Japan: Hitachi, Matsushita, and Toyota compared", *California Management Review*, 40(3), pp. 241-264.

[6] Brown, J.S. and Duguid, P., 1998, "Organizing knowledge", *California Management Review*, 40(3), pp. 90-111.

[7] Court, A.W., 1995, *The modelling and classification of information for engineering designers*. PhD Thesis, University of Bath, UK.

[8] Boston, O.P., Culley, S.J. and McMahon, C.A., 1996, "Designers and suppliers - modelling the flow of information", *Proceedings of ILCE '96*, Paris, France.

[9] Marsh, J.R., 1997, *The capture and structure of design experience*. PhD Thesis, Cambridge University Engineering Department, UK.

[10] Hamilton, J., Clarkson, P.J. and Burgess, S., 1997, "The modelling of design knowledge for computer supported aerospace design", *Proceedings, 11th International Conference on Engineering Design (ICED '97)*, Tampere, Finland, 1, pp. 235-240.

[11] Frank, D., 1999, "The importance of knowledge management for BMW", *Proceedings, 12th International Conference on Engineering Design (ICED'99)*, Munich, Germany, 1, pp. 33-40.

[12] Nonaka, I. and Takeuchi, H., 1991, *The Knowledge Creating Company: How Companies Create the Dynamics of Innovation*, Oxford University Press, New York.

[13] Polanyi, M, 1974, *Personal Knowledge Towards a Post-Critical Philosophy*. University of Chicago Press, Chicago.

[14] Roozenburg, N.F.M. and Eekels, J., 1995, *Product Design: Fundamentals and Methods*. John Wiley and Sons, Chichester, UK.

[15] Johnson, P., 1992, *Human Computer Interaction*. McGraw-Hill, London.

[16] Garnham, A. and Oakhill, J., 1994, *Thinking and Reasoning*. Blackwell, Oxford.

[17] Boy, G., 1991, *Intelligent Assistant Systems*. Academic Press, London.

[18] Stauffer, L.A. and Ullman, D.G., 1988, "A comparison of the results of the empirical studies into the mechanical design process", *Design Studies*, 9(2), pp. 107-114.

[19] Sanvido, V.E., Kumara, S. and Ham, I., 1989, "A top-down approach to integrating the building process", *Engineering with Computers*, (5), pp. 89-103.

[20] Wood, W.H., Yang, M.C., Cutkosky, M.R., Agogino, A.M., 1998, "Design information retrieval: Improving access to the informal side of design", *CD-ROM Proceedings of ASME Design Engineering Technical Conference (DETC'98)*, Atlanta, Georgia, USA.

[21] Rodgers, P.A., Patterson, A.C. and Wilson, D.R., 1993, "A computer-aided evaluation system for assessing design concepts", *Manufacturing Intelligence*, (15), pp. 16-19.

[22] Coyne, R.D., Rosenman, M.A., Radford, A.D., Balachandran, M. and Gero, J.S., 1990, *Knowledge-Based Design Systems*. Addison-Wesley Publishing Co., Reading, Massachusetts.

[23] Cross, N., 1994, *Engineering Design Methods: Strategies for Product Design*. John Wiley and Sons, Chichester, England.

[24] Allen, L., 1986, "The cost of an expert", *Computerworld*, July 21, pp. 59-68.

[25] Ong, W. J., 1986, "Writing is a technology that restructures thought", *In* G. Baumann (Ed.), *The Written Word: Literacy in Transition*. Clarendon Press, Oxford, pp. 23-50.

[26] Hill, W.C., 1989, "The mind at AI: Horseless carriage to clock", *AI Magazine*, (9), pp. 29-41.

[27] Fischer, G. and Nakakoji, K., 1992, "Beyond the macho approach of artificial intelligence: Empower human designers - do not replace them", *Knowledge-Based Systems*, 5(1), pp. 15-30.

[28] Wood, W.H. and Agogino, A. M., 1996, "Case-based conceptual design information server for concurrent engineering", *Computer-Aided Design*, 28(5), pp. 361-369.

[29] Brown, J.S. and Duguid, P., 1996, The social life of documents, *First Monday*, 1(1).

[30] Hoft, K. and Ito, Y., 1999, "A method for culture - and mindset-harmonised design", *Proceedings, 12th International Conference on Engineering Design (ICED'99)*, Munich, Germany, 1, pp. 313-316.

[31] Sinding-Larsen, H., 1988, "Information technology and the externalisation of knowledge", *In* H. Sinding-Larsen (Ed.), *Artificial Intelligence and Language: Old Questions in a New Key*. Tano, Oslo, pp. 77-89.

[32] Eriksson, H., 1996, Expert systems as knowledge servers, *IEEE Expert*, 11(3), pp. 14-19.

[33] Liu, J. and Brown, D.C., 1994, "Generating design decomposition knowledge for parametric design problems", *In* J.S. Gero and F. Sudweeks (Eds.), *Artificial Intelligence in Design '94*. Kluwer Academic Publishers, Dordrecht, pp. 661-678.

[34] Dym, C.L., 1994, *Engineering Design: A Synthesis of Views*. Cambridge University Press, Cambridge, UK.

[35] Pahl, G. and Beitz, W., 1996, *Engineering Design: A Systematic Approach*. Springer-Verlag, London.

[36] Leake, D.B., Birnbaum, L., Hammond, K., Marlow, C. and Yang, H., 1999, "Integrating information resources: A case study of engineering design support", *Proceedings of the 3rd International Conference on Case-Based Reasoning (ICCBR-99)*, Springer-Verlag, Berlin (in press).

[37] Brown, D.C. and Grecu, D.L., 1997, "An 'AI in design' view of design", *Worcester Polytechnic Institute Computer Science Dept. Technical Report*, WPI-CS-TR-97-12

[38] Lee, J., 1997, "Design rationale systems: Understanding the issues", *IEEE Expert*, 12(3), pp. 78-85.

[39] Kobayashi, S., 1990, *Colour Image Scale*. Kodansha International, London.

[40] Rodgers, P.A., Blessing, L.T.M., Caldwell, N.H.M., Huxor, A.P., 1999, "Exploring the effectiveness of network-based tools in distributed conceptual design activity", *In* N.P. Juster (Ed.), *The Continuum of Design Education*. Professional Engineering Publishing Ltd., London, pp. 151-161.

[41] Rodgers, P.A., Caldwell, N.H.M., Clarkson, P.J., Huxor, A.P., 2000, "Managing knowledge in dispersed design companies", *In* J.S. Gero (Ed.), *Artificial Intelligence in Design '2000*. Kluwer Academic Publishers, Dordrecht, (in press).

Ontology-driven Knowledge Management: Philosophical, Modelling and Organizational Issues

Simon Buckingham Shum, John Domingue and Enrico Motta

Knowledge Media Institute, The Open University, MK7 6AA, Milton Keynes, UK
{s.buckingham.shum, j.b.domingue, e.motta}@open.ac.uk

Abstract. Our approach to knowledge management, called *Ontology-Driven Document Enrichment*, aims to support knowledge creation and sharing by *enriching* documents through formal models. These reflect a particular *viewpoint* over a set of resources, which is shared by a *community of practice*. In the first part of this chapter we describe the basic tenets of the approach and we briefly illustrate the technologies which we have developed to support it and the application domains in which it has been applied. In the second part of this chapter we evaluate our approach in terms of a number of issues, which have been raised in the research community. These pose important research questions for designers of knowledge management technologies that make use of formal knowledge models. These questions address issues of *epistemology* (what happens to the knowledge as it is formalized?), *socio-technology* (how to integrate knowledge technologies into work practice?), and *usability* (do end-users understand, *and use*, the formal scheme?). In the chapter we show how these issues are tackled in our approach and we emphasize the multi-faceted nature of knowledge management solutions.

1. Introduction

Over the past three years we have developed an approach to knowledge management (KM), which relies on *knowledge modelling* technology [1] to support document-centred knowledge creation and sharing. The basic tenet of our approach, which is called *Ontology-Driven Document Enrichment,* is that knowledge creation and sharing in an organizational setting can be facilitated by *enriching* documents through formal models. These models reflect a particular *viewpoint* over a set of resources, which is shared by a particular *community of practice* [2, 3, 4]. The viewpoint in question is formalized by means of an *ontology* [5], which is used to drive the collaborative construction of the formal model. In simple terms, an ontology can be seen as providing a vocabulary for describing a range of models. For instance, in our *PlanetOnto* project [6, 7], we have developed an ontology which focuses on the main concepts and relations required to characterize life in an academic department. The ontology is centred around six *key classes*: *events, people, organizations, stories, projects* and *technologies*. In our view, these entities provide the right 'ontological dimensions', which enable us to characterize life in an academic department from the viewpoint of a visitor, who is interested in finding out about the academic contributions made by the department. This viewpoint (and the resulting ontology) is of course very different from that of a financial administrator, or from that of a personnel officer. The PlanetOnto ontology is used to drive the acquisition of knowledge from lab members and to produce conceptual structures which can be used to annotate lab resources semantically. As a result, we can provide semantic retrieval services and we can specify agents which make use of the codified knowledge to deliver personalised news services or to request missing information from lab members.

Our knowledge management approach is currently being applied to a variety of academic and industrial domains. These include enabling intelligent delivery of courseware, managing best practice in the aerospace industry, delivering reusable components for engineering design, managing medical guidelines [8], structuring academic debate [9] and supporting electronic publishing processes [6, 7]. The results have so far been very encouraging and have led to the deployment of knowledge management solutions in several organizations.

In the first part of this chapter we will illustrate our approach to ontology-driven document enrichment, in particular focusing on the key activities and the supporting technologies. Having presented our approach we will then reflect on our work by analysing it in terms of a number of issues which have been raised in the Human-Computer Interaction community. These pose important research questions for designers of knowledge management technologies that make use of formal knowledge models. Specifically, we will consider questions of *epistemology* (what happens to the knowledge as it is formalized?), *socio-technology* (how to ensure user acceptance for the technology?), and *usability* (do end-users understand, *and use*, the formal scheme?). We will show how these issues are tackled in our approach and we will emphasize the multi-faceted nature of knowledge management solutions.

2. Ontology-driven Document Enrichment

2.1 Enriching vs. translating

We say that our approach is *ontology-driven* to emphasize that the construction of the knowledge model is carried out in a top-down fashion, by populating a given *ontology* [5], rather than in a bottom-up fashion, by annotating a particular document.

An ontology-driven approach to model construction affords several advantages. Instantiating an ontology is usually simpler and speedier than developing a model from scratch. In addition, because an ontology makes explicit the conceptualization underlying a particular model, it becomes easier to maintain, reuse and interoperate the model with other components. Finally, reasoning modules can be associated with an ontology and these are then applicable to all models built by instantiating the ontology in question. For instance, in the case of medical guidelines [8] one can envisage building ontology-specific guideline verification tools, which can then be used to verify individual guidelines developed by instantiating the same generic guideline ontology.

Because our model construction process is ontology-driven, we prefer to use the term "enrichment", rather than "conversion" or "annotation", to refer to the process of associating a formal model to a document (or set of documents). In general, a representation, whether formal, graphical or textual, can be enriched in several different ways - e.g., i) by providing information about the context in which it was created, ii) by linking it to related artefacts of the same nature, or iii) by linking it to related artefacts of a different nature. Although in our document-centred knowledge management work we provide multiple forms of document enrichment, such as associating *discussion spaces* to documents [10], in this paper we will primarily concentrate on the association of formal knowledge models to documents.

Thus, an important facet of an ontology-centred approach to document enrichment is that the formalised knowledge is not meant to be a translation of what is informally specified in the associated document. The knowledge model typically plays a different role from the associated text. For instance, in the medical guideline scenario the knowledge model helps to verify that all the kinds of knowledge expected to be found in a document describing a medical guideline are indeed there. Another example can be found in our work on enhancing bibliographic repositories through the collaborative construction of a knowledge model [9]. The role of the knowledge model is not to simply mirror the contents of the bibliographic repository. Its role is to capture domain experts' *interpretative meta-knowledge* about these constructs, by for instance, adding assertions such as "theory X contradicts theory Y", or "software W modifies approach Z". Such information may not be in the document at all; if it is, it remains unavailable for machine interpretation without a knowledge level description.

In the next section we discuss the various activities prescribed by our approach.

2.2   Methodological steps

2.2.1   *Identify use scenario*

At this stage the services to be delivered by the knowledge management system are defined. In particular, issues of feasibility and cost are investigated. Addressing the latter involves answering questions such as: "What is the added value provided by the knowledge model, considering the non-trivial costs associated with the development and instantiation of an ontology?", "Is there the need for a 'full-blown' knowledge model and for going beyond the facilities provided by off-the-shelf search engines?", "What additional reasoning services will be provided, beyond deductive knowledge retrieval?". Addressing feasibility issues requires assessing (among other things) whether or not it is feasible to expect the target user community to perform document enrichment or whether specialized human editors will be needed. This latter solution introduces a significant bottleneck in the process and moreover assumes that to introduce a central editor in the model development is actually feasible. This is definitely not the case in some of our application domains. For instance, in the scholarly discourse scenario our aim is to allow members of academic communities to express their own views about academic contributions, whether their own or whether proposed by other scholars. In this scenario it is not feasible to envisage a central editor mediating between an author and a shared knowledge base. In addition, we also believe that aiming to enable non-experts to take part in the development of knowledge models is a good research objective per se, not only on the grounds of cost-effectiveness and sustainability, but also because of the perceived benefits deriving from being engaged in the modelling activity [11]. Of course, successfully enabling non-experts to engage in knowledge modelling is far from a trivial proposition. There are subtle trade-offs involved here, which we address in Section 2.2.4, when discussing the ontology instantiation task.

2.2.2 Characterize viewpoint for ontology

The previous step was concerned with assessing the feasibility of the approach and defining the functionalities of the envisaged system. This step focuses on the ontology: having decided on an ontology-based approach, it is important to characterize the particular viewpoint that the envisaged ontology will impose on the documents. For instance, in the electronic publishing domain we focus on the events characterizing academic life, while in the scholarly discourse domain the focus is on academic debate. Clearly the distinction between this step and the steps immediately before and after can in some cases be fuzzy. For instance, the specification of the viewpoint in the scholarly discourse domain is tightly integrated with the characterization of the use scenario. However, it is useful to explicitly distinguish a viewpoint specification task for two reasons. The first one is that it is feasible to envisage scenarios in which different viewpoints can be taken as the starting point for the ontology development process. Examples abound in the literature, especially with respect with highly generic ontologies - compare for example the *Cyc* ontology [12] with the work by Sowa [13]. Another reason is methodological: it is useful to separate the issues concerning the functionalities, scope and purpose of the envisaged system from the specification of the functionalities, scope and purpose of the ontology. For instance, in the electronic publishing scenario the knowledge management system is concerned with providing semantic retrieval capabilities and services supporting story identification and personalised news feeds. Within this scenario the chosen viewpoint focuses on modelling academic events and the key 'actors' in these events: technologies, people, organizations and projects.

2.2.3 *Develop the ontology.*

Having defined a particular viewpoint over a domain (in this case, a set of documents), ontology development is largely a technical enterprise - i.e., the issues here concern modelling and formalization, rather than scope and purpose.

In our scenarios we have followed a task-independent, middle-out approach [14], and we use the selected viewpoint to help us to identify the *key concepts* in the class of models we want to construct - e.g., technologies, events, people and organizations in the electronic publishing scenario.

Another important issue concerns who constructs the ontology? As pointed out earlier, we want to support scenarios in which knowledge models are constructed collaboratively. But what about the ontology itself? Is this constructed collaboratively? Our answer is negative. In all the projects we have carried out so far, we have centralised the ontology development. The main reason for this choice is that a careful design of the ontology is crucial to ensure the success of any particular document enrichment initiative. The ontology specifies the selected viewpoint, circumscribes the range of phenomena we want to deal with and defines the terminology used to acquire domain knowledge. In our experience small errors/inconsistencies in any of these aspects can make the difference between success or failure. Moreover, ontology design requires specialist skills which are normally not possessed by the members of our target user communities.

Although the ontology is developed centrally we should emphasise that the development lifecycle is iterative and participatory [15]. Versions of the ontology under construction are given to users for comments and feedback. In fact, within a number of our applications we have applied our knowledge management methodology reflexively to achieve this, embedding an embryonic ontology within our discussion environment [10]. Also, issues raised by our users during the knowledge base population phase (see next section) will often result in changes to the ontology.

Our development lifecycle means that the ontologies we produce need to be easy to understand and to instantiate. These user-centred considerations have priority over any other design criterion and inform our two main modelling guidelines.

- *Minimal ontological definitions.* That is, to try and provide only the minimal set of attributes needed to define a particular class. This approach has the advantage that, when populating the ontology, users will not be faced with lots of irrelevant attributes.

- *User-centred definitions.* This guideline requires that the terminology used by the ontology needs to be easy to understand for a user who is not necessarily a knowledge engineer. There are two aspects here: heavily technical modelling concepts – e.g., sophisticated modelling solutions for representing time – ought to be avoided. Moreover, the terminology should be as *context-specific* as possible. For instance, while we can talk about "agents performing events", when describing events in general, we should use the class-specific terminology, "awarding body assigns awards", when talking about an award-giving type of event. This latter guideline implies that the underlying modelling language should support slot renaming along *isa hierarchies* – i.e., inherited slots should get subclass-specific names.

2.2.4 *Perform ontology-driven model construction*

We are acutely aware that many schemes for registering shared resources and providing structured descriptions founder on the crucial 'capture bottleneck' - the envisaged beneficiaries of the system simply do not have the motivation or time to invest in sharing resources to reach a critical mass of useful material. Sobering lessons on this theme have been drawn for group and organizational memory systems [16], and indeed, for any system that requires users to formalise information [17]. Why should we succeed where others have failed? Our working hypothesis is that our domains have unique characteristics lacking in domains in which collaborative development has failed.

- *Co-operation rather than competition.* We are selecting domains where co-operation, meant here as "willingness to share knowledge", is either a basic work premise or is enforced by external constraints. For instance, academic analysis and publishing require scholars to read, refer to and praise/criticise each other's work. In sum, the dynamics of academic publishing requires co-operation. A similar situation occurs in the medical guidelines scenario. Institutions and individual scientists may compete with each other, but the outcome of consensus conferences (by definition) are shared knowledge resources. In other scenarios, for instance when constructing an organizational memory, other forces (e.g., directives from higher management) may force co-operation, even when competition would be the norm.
- *Benefits outweigh costs.* Motivation is a crucial aspect. Motivation essentially boils down to a cost-benefit analysis. For instance, in the scholarly discourse scenario, we assume that the basic motivation of an academic is to disseminate his/her work. Hence, having completed a new document, the author will want to maximise its 'digital presence' on the net by carefully encoding its contributions and connections to the existing literature.
- *Compatibility with organizational work practices.* This requires the seamless integration of our document enrichment model with existing work practices. For instance, in the case of electronic publishing, the ontology is used to enrich news items, which are submitted either through email or through a web-based form. Hence, at least for those users who submit through the latter mechanism, instantiating the ontology becomes an additional form-filling activity, carried out using the same medium (i.e., the web browser) and at the same time. Analogously, in the case of scholarly discourse, at least in some academic communities, authors are used to submitting papers to digital repositories and providing metadata. Filling an ontology-derived form should then be perceived as a small, additional step.

2.3 Customise query interface for semantic knowledge retrieval

At this stage the appropriate query interface is designed, in accordance to the use scenario and the expected functionalities. To support this step we have developed a flexible form-based interface, called *Lois*, which can be customised for each specific application domain. The aim of Lois is to allow an user to pose queries to a knowledge model at a level which abstracts from the underlying modelling language. This goal has been accomplished by developing an interface which allows users to select 'key concepts' in the ontology and then construct a query by navigating the structure of the ontology (i.e., by following relations between concepts). This navigation leads to the creation of a query as a *list of rows*, which are linked by logical connectives. For instance, Figure 1, which is taken from our electronic publishing domain, shows a query which asks for a member of the KMi academic staff, who is involved in the development of software visualization technology. The first row of the query was created by a) selecting the "Member of KMi" button, b) selecting "concept-sub-type" in the "Attributes" window and c) selecting "kmi-academic-member" in the "Attribute Types" window. The second row was created by selecting "More about kmi-member1" in the pop-up menu underneath "Organisation", then selecting relation "has-research-area" in the "Attributes" window, then "kmi-research-area" in the "Attribute Types" window, and then "res-area-software-visualization" in the "Values" window. The items in each window pane dynamically show the choices relevant to the currently selected item in the pane immediately on the left. For example, the window "Attributes" in Figure 1 shows all the attributes appropriate to the currently selected item under the "Concepts" window, which is "More about kmi-member1". Analogously, once we select item "has-research-area" in the "Attributes" window, then window "Attribute Types" is automatically updated, to show the types of research areas known to the underlying knowledge base.

The Lois interface is created automatically once the key classes for a knowledge model have been specified. In the example, these are: "Story Event Type", "KMi Technology", "KMi Project", "KMi Member" and "Organization". These act as the entry points to the space of possible Lois queries. In other words, Lois can cover any query, as long as this can be specified as a path through the knowledge model, starting from one of the key classes. The usability of the Lois interface also depends on the terminology used by the underlying ontology. The knowledge engineer is therefore required to employ class and slot names which can be understood by the Lois user. If this requirement is satisfied, then the user only needs to learn to construct queries through the accumulation of rows.

2.4 Develop additional reasoning services on top of the knowledge model

Once a knowledge model has been produced, then it becomes possible to provide additional intelligent functionalities and ensure that the benefits outweigh the costs.

These reasoning services tend to be application specific. For instance, in the scholarly publishing scenario, we are planning to develop specialised agents, whose goal is to identify emerging scholarly perspectives, using heuristic knowledge and machine learning techniques. For instance, an agent could discover a 'European perspective' on a particular issue, if a structural pattern in the knowledge model - e.g., use of formal methods - also matched the geographic location of the relevant researchers. In the electronic publishing domain we have designed two agents, which reason about the contents of the knowledge model to identify new, potentially 'hot' stories and to provide personalised news feeds [6, 7].

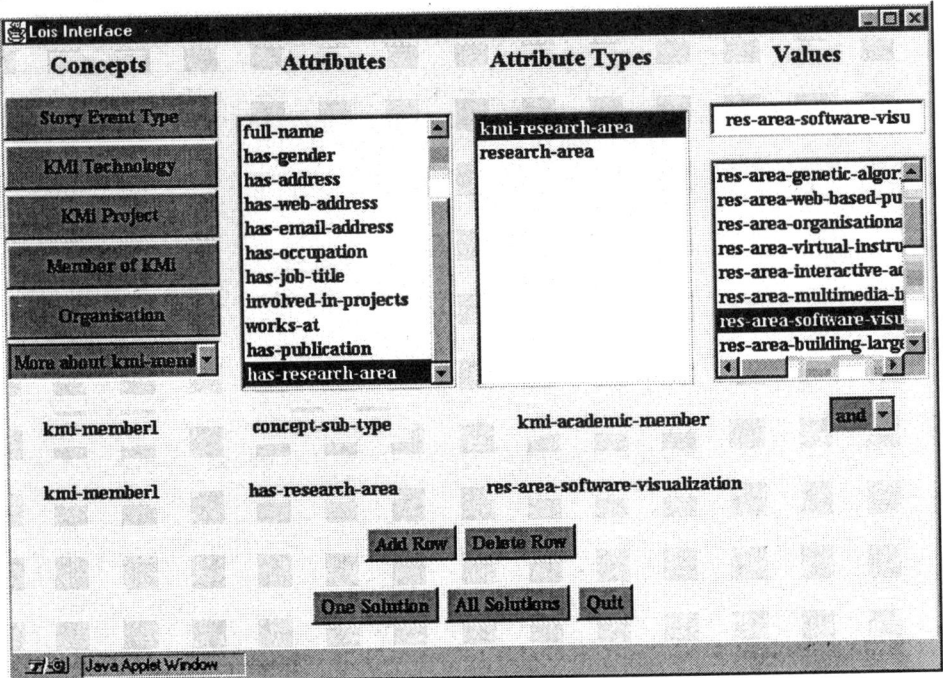

Figure 1: Finding a KMi researcher who works on software visualization.

3. Technologies for Ontology-driven Document Enrichment

We have developed a number of modelling support technologies, to enable ontology-driven document enrichment. These include:

- *OCML.* An operational knowledge modelling language, which provides the underlying representation for our ontologies and knowledge models [1].
- *WebOnto.* A tool providing web-based visualization, browsing and editing support for developing and maintaining ontologies and knowledge models specified in OCML [18].

- *Lois*. A flexible form-based interface for knowledge retrieval, which has been described in Section 2.3.
- *Knote*. A form-based interface for populating an ontology – see [19] for details.

The technologies listed above focus on modelling support. However, our end-user solutions normally integrate various forms of document-enrichment, to support alternative knowledge creation and sharing scenarios. For example our repository for enriched medical guidelines [8] integrates formal knowledge models with *D3E technology* [10], which supports a tight integration of discourse and debate facilities within documents.

4. Epistemological and Usability Issues

As pointed out earlier, one of the basic tenets of our approach to ontology-driven knowledge management is that the users themselves (the members of a community of practice) should be able to engage in the construction of the ontology-centred, formally represented organizational memory. This is important for *psychological* reasons (being part of the development process means that users 'own' the resulting resource), *pedagogical* reasons (we have already mentioned the benefits derived from being engaged in modelling) and *practical* reasons (centralised editing introduces a bottleneck in the whole process). Of course, collaborative model building by users who are not themselves expert in knowledge modelling is hardly a risk-free enterprise. In the previous sections we have already pointed out that several organizational and modelling variables need to be taken care of, to ensure the feasibility of the approach. Nevertheless, we have assumed the basic feasibility of a modelling approach to knowledge management. In other words, the implicit assumption here is that a modelling approach to knowledge management is feasible, once the relevant organizational and modelling issues are addressed. A number of researchers have however raised fundamental questions concerning the role of formalized systems in knowledge management. They have highlighted the perils of reductionism, the loss of context associated with codification and the danger of imposing a common terminology (e.g., a common ontology) across heterogeneous communities of practice. In what follows we will illustrate these issues, discuss the implications they have for formalized knowledge management and suggest some guidelines, which can be used to minimize the risks associated with these enterprises.

4.1   Codification

The process of 'objectifying' knowledge introduces mutations in what is represented. McLuhan's famous quotation, 'the medium is the message' reminds us that content and form are inextricably linked, but overstates the case a little. We see the medium *shaping* the message, as follows:

- A codification continuum exists, from tacit, individual, pre-understanding, to personal externalisation of concepts, to verbalisable and written communication, to increasingly fine-grained and abstracted symbolic representations. As we move from tacit, individual, pre-understanding to shared, formal, computer-based representations, we express our thoughts in an increasingly structured way, providing the computer with greater access to the content of the information. The intellectual effort required to transform knowledge representations from one state to another can lead to new insights, since the particular representation used forces us to make certain information explicit that was previously implicit. Typically, 'information chunks' have to be broken down into smaller units of particular classes, given names, classified and structured. Having to reason about these can clarify our thinking.

- However, as we move from tacit pre-understanding to fine-grained symbolic representations, we strip away details of the context(s) in which that knowledge was displayed and/or has meaning. It is usually difficult, and often impossible, to reverse the direction and recover tacit pre-understanding from symbolic representations. Formalization should be then understood as an *interpretative act,* that is, in a situation, from a perspective, for a purpose. The transformation brings about ontological changes that distort knowledge and alienate it from the person possessing it in particular ways, effecting a gradual shift in the definition of knowledge and expertise from an *ability* to a symbolically encoded *fact*.

Because each transformation is an act of interpretation, there is no such thing as objective knowledge representation, or indeed objective classification or codification of any sort (in software or any other medium). This leads to the view that information and communication systems cannot be thought of as neutral; in their formal structures and operations they embody the goals and perspectives of their developers.

4.2 Interpretation and communities of practice

If codification proceeds through successive reinterpretations, the interpretative process is placed centre-stage in our account of the interplay between ontology-based technologies and KM. Representations require an agent (human or machine) capable of interpreting them. Once information has been interpreted, the agent has knowledge for action (if only to decide that the information is irrelevant). Since people have unique backgrounds, bringing different assumptions to any reading that they do, meaning and significance depend on the interpreter. Thus, an expert may be able to glance at a spreadsheet and immediately spot a statistical trend that a junior member of staff has missed; a seasoned manager may glean implicit messages from reading a memo whereas a newcomer might miss these.

Boland and Tenkasi [20] refer to "interpretative strategies" as the way of reading and writing, listening and speaking that one's background and perspective moulds. Interpretative strategies are also moulded by the *communities of practice* to which one aligns oneself. Members share an identity and 'speak the same language' in the sense of approaching situations with significantly common agendas and skills. When one group attempts to interpret another without this common ground, we often witness failures in communication and misinterpretations. Robinson and Bannon [21] develop this theme in more detail in their concept of 'ontological drift', the changes in meaning that a representation takes on as it is passed between different design communities.

Before considering the implications for ontological KM technologies, we turn to one further source which reflects on the problems caused for end-users when they are forced to modify their work practices to align with formal representations in technology.

4.3 "Formality considered harmful"

There is no question that modelling a domain using a formal scheme/notation can clarify vague thinking through the intellectual discipline of reasoning within a constrained scheme. However, formalisms are a double edged sword. Shipman and Marshall [17] document a number of persistent problems that they have observed with a wide variety of interactive systems that have required their end-users to work with formal representations. These include groupware (e.g. where users struggled to classify messages), hypertext (e.g. overheads of structuring ideas as semantic networks of nodes and links), and knowledge-based systems (e.g. failure to properly use classification schemes). From this eclectic set of technologies and use contexts, they distil four generic problems for end-users associated specifically with the use of formalized representations:

- **Cognitive overhead**: the additional effort required to translate ideas into the template or notation provided by the system disrupted the work in hand; the formalism imposes new tasks such as chunking, labelling, and linking (for an example see [22]);
- **Tacit knowledge**: even if a formalism is based on an analysis of the structure of end-users' natural activity, making them think and work in those terms explicitly can actually disrupt work—the tacit has been made explicit, and so does not play the same role that it did originally (for an example see [23]).
- **Premature commitment to structure**: this problem relates to systems that enforce the declaration of explicit structure between elements before the user is confident of it, or even able to articulate it.
- **Situational structure**: formal structures can eliminate ambiguities that are extremely important for end-users. They may be important in order to facilitate ongoing collaboration between different communities of practice who have equally legitimate, but conflicting perspectives. Legal documents are a classic example of formalisms with a carefully crafted ambiguity.

Shipman and Marshall propose several strategies for tackling these problems, some of which we draw upon in our own work.

5. Implications for Ontological Technologies

In the previous section we have highlighted a number of issues which affect the use of formal representation in knowledge management. In sum, any ontology-centred approach to collaborative model building needs to address the following issues:

- Knowledge transformations are not neutral; in particular, codification strips out context and "formality can be harmful".
- The meaning of an interpretation is circumscribed to a particular community of practice.
- The introduction of technology can interfere with work-practices.

We address these issues in the following three sub-sections.

5.1   Managing codification

The introductory analysis of the codification process has one clear message: formalization changes what it seeks to represent, in particular it strips out context, it can remove important nuances of meaning. One solution proposed by Marshall and Shipman is *incremental formalization*. This may take the form of automatic extraction and classification of concepts to assist the user in updating the resource with new material, or simply the provision of a user interface that supports the user in declaring more explicit structure (a detailed examination of tool requirements to support this process for graphical notations is provided by [24]).

Our approach to this problem relies on a tight integration of different representations of knowledge. For instance, a colleague of us here at the Knowledge Media Institute, Dr Marek Hatala, has developed an environment called CEDAR, which makes it possible to perform incremental 'ontological annotation', by associating sections of web documents to formal, ontology-derived structures. Moreover, it is important to emphasize that a crucial aspect of our knowledge management work is that we assume that knowledge creation and sharing, although mediated and facilitated by formal structures, is fundamentally *document-centred*. In other words, codification does not imply loss of context in the *whole knowledge management system*. In addition, as pointed out in Section 3, end-users solutions often include various types of support for knowledge creation and sharing, for instance both through formal models, and through structured discussion spaces. In a nutshell, we recognize that formal models do not play an exclusive role in KM technologies. Multiple forms of representation and document enrichment are required to address knowledge management problems in the workplace.

5.2   Tailoring ontologies for communities of practice

Much work on ontologies tackle large, generic universes of discourse, which are applicable to innumerable domains [12, 13]. Not surprisingly, practical applications of these ontologies have reported mixed results. Cohen et al. [25]

found that these very generic ontologies provide less support and are less useful than domain-specific ones. The latter scored a constant 60% rate of reuse in the study, in contrast with the poor 22% rate of reuse scored by the generic ontologies. Cohen et al. point out that these generic ontologies are more important to help knowledge engineers to organize knowledge bases and more specific ontologies along sound ontological lines, than in directly supporting the development of end-user models. Another way to look at the problem is in terms of the target community an ontology is trying to address. Generic ontologies try to cater for alternative communities of practice and therefore are difficult to reuse. In contrast with this approach, because all our ontology development efforts are grounded around specific knowledge management or system development scenarios, our end-user ontologies tend to be highly customised for the target community of practice. This policy has two consequences for the ontology development process: i) members of the target community are normally heavily involved in the definition of the use scenario and the viewpoint specified by the end-user ontology; ii) the end-user ontology normally builds on several layers of ever more generic ontologies. This second point is related to the need to maximise both reusability and usability. To have reusability we need to have very generic ontologies; to have usability, ontologies need to be fine-tuned for user communities. The result is a careful, fine-grained ontology refinement process, in which ever more specific ontologies build on more generic ones. For example, let us consider the ontology we developed for supporting the management of pressure ulcers. In this scenario we used our knowledge management technology to help healthcare practitioners to look after patients at risk from pressure ulcer, according to the criteria specified by a medical guideline. Here, a specific ontology for pressure ulcers was developed, which in turns builds on a generic ontology for medical guidelines, which in turn builds on a generic medical ontology. At a higher level of abstraction, generic ontologies about events, technologies, organization, time-based concepts and others were also used. In total, ten ontologies were developed or reused and the resulting 'depth' of the ontology inclusion hierarchy spans eight levels of granularity, from a 'base ontology', which underlines any ontology development activity, level 1, to the pressure ulcer ontology, level 8.

5.3 Coping with 'technological interferences'

There are two aspects to technological interferences: in some cases they are caused by inappropriate technology - e.g., hostile user interfaces; in other cases they happen because the technologies in question are not integrated with current work-practices. In Sections 2.2.3 and 2.2.4 we presented our approach to tackling these issues. Our KM technologies are user-centred and we use modelling guidelines which attempt to maximise the usability of the end-user ontology - see Section 2.2.3. Integration with current work-practices is a fundamental part of our development process. The interfaces to the knowledge management tools are augmentations of *tools of work* enabling staff to interact with the corporate knowledge base via their day-to-day tools. This is possible because our technologies our web-based. As a result interoperability overheads are reduced, as long as existing work-practices are centred on company intranets.

6. Conclusions

There is more to knowledge management than any particular technology. Successful deployments of knowledge management systems in the workplace rely on a careful analysis of the context: organizational, epistemological, psychological and representational issues need to be addressed. In this paper we have presented our approach to knowledge management, called *Ontology-Driven Document Enrichment*, and we have analysed it in terms of various philosophical, modelling and organizational issues, drawn both from our own practical experience and from the knowledge management literature. We have now used this approach in a dozen different cases with a variety of industrial and governmental users. The results are very encouraging and appear to validate the solutions to the various knowledge management issues presented in this chapter.

References

[1] Motta E. (1999). *Reusable Components for Knowledge Models.* IOS Press, Amsterdam.

[2] Brown, J. S. & Duguid, P. (1991). Organizational Learning and Communities of Practice: Toward a Unified View of Working, Learning and Innovation. *Organization Science*, 2, pp. 40-57.

[3] Brown, J. S. & Duguid, P. (1998). Organizing Knowledge. *California Management Review*, 40, 3, pp. 90-111.

[4] Wenger, E. (1998). *Communities of Practice: Learning, Meaning, and Identity.* Cambridge: Cambridge University Press.

[5] Gruber, T. R. (1993). A Translation Approach to Portable Ontology Specifications. *Knowledge Acquisition*, 5(2).

[6] Domingue, J. & Motta, E. (1999). A Knowledge-Based News Server Supporting Ontology-Driven Story Enrichment and Knowledge Retrieval. In D. Fensel and R. Studer (editors), *Proceedings of the 11th European Workshop on Knowledge Acquisition, Modelling, and Management (EKAW '99)*, LNAI 1621, Springer-Verlag.

[7] Domingue, J. & Motta, E. (2000). Planet-Onto: From News Publishing to Integrated Knowledge Management Support. *IEEE Intelligent Systems*, May/June 2000.

[8] Motta, E., Domingue, J., Hatala, M. et al (1999). Ontology-Driven Management of Medical Guidelines. *PatMan Project Deliverable D6*, Knowledge Media Institute, The Open University, UK.

[9] Buckingham Shum, S., Motta, E., & Domingue, J. (1999). Representing Scholarly Claims in Internet Digital Libraries: A Knowledge Modelling Approach. In S. Abiteboul

and A.-M. Vercoustre (Ed.), *Proc. of ECDL '99: Third European Conference on Research and Advanced Technology for Digital Libraries.* Paris, France, September 22-24, 1999: Springer-Verlag (Lecture Notes in Computer Science). Available at: http://kmi.open.ac.uk/projects/scholonto/.

[10] Sumner, T., & Buckingham Shum, S. (1998). From Documents to Discourse: Shifting Conceptions of Scholarly Publishing. *Proc. CHI 98: Human Factors in Computing Systems*, (Los Angeles, CA), pp. 95-102. ACM Press: NY. Available at: http://kmi.open.ac.uk/techreports/papers/kmi-tr-50.pdf

[11] Wideman, H.H. & Owston, R.D. (1993). Knowledge Base Construction as a Pedagogical Activity. *Educational Computing Research,* 9(2), pp. 165-196.

[12] Lenat, D.B. & Guha, R.V. (1990). *Building Large Knowledge-Based Systems: Representation and Inference in the Cyc Project.* Addison-Wesley.

[13] Sowa J. F. (1995). Top-Level Ontological Categories. *International Journal of Human-Computer Studies*, 43(5/6), pp. 669-685.

[14] Uschold M. & Gruninger M. (1996). Ontologies: Principles, Methods and Applications. *Knowledge Engineering Review*, 11(2), pp. 93-136.

[15] Greenbaum, J. & Kyung, M. (1991). *Design at Work: Cooperative Design of Computer Systems.* Hillsdale, NJ, Lawrence Erlbaum Associates.

[16] Selvin, A. (1999). Supporting Collaborative Analysis and Design with Hypertext Functionality. *Journal of Digital Information*, 1, (4).

[17] Shipman, F. M. & Marshall, C. C. (1999). Formality Considered Harmful: Experiences, Emerging Themes, and Directions on the Use of Formal Representations in Interactive Systems. *Computer Supported Cooperative Work*, 8, 4, pp. 333-352.

[18] Domingue, J. (1998). Tadzebao and WebOnto: Discussing, Browsing, and Editing Ontologies on the Web. In Gaines & Musen (Eds.), *Proceedings of the 11th Knowledge Acquisition for Knowledge-Based Systems Workshop*, April 18th-23th, Banff, Canada. Available at http://kmi.open.ac.uk/people/domingue/banff98-paper/domingue.html.

[19] Motta E., Buckingham-Shum, S. & Domingue, J. (2000). Ontology-Driven Document Enrichment: Principles, Tools and Applications. *To appear in the International Journal of Human-Computer Studies.*

[20] Boland Jr, R. J. & Tenkasi, R. V. (1995). Perspective Making and Perspective Taking in Communities of Knowing. *Organization Science*, 6, 4, pp. 350-372.

[21] Robinson, M. & Bannon, L. (1991). Questioning Representations. In L. Bannon, M. Robinson, & K. Schmidt, *Proc. of ECSCW '91: 2nd European Conference on Computer-Supported Collaborative Work* (pp. 219-233). Amsterdam Sept 25-27: Kluwer

[22] Buckingham Shum, S., MacLean, A., Bellotti, V. et al (1997). Graphical Argumentation and Design Cognition. *Human-Computer Interaction*, 12, 3, 267-300.

[23] Fischer, G., Lemke, A. C., McCall, R., et al (1991). Making Argumentation Serve Design. *Human-Computer Interaction*, 6, 3&4, pp. 393-419.

[24] Buckingham Shum, S. & Hammond, N. (1994). Argumentation-Based Design Rationale: What Use at What Cost? *International Journal of Human-Computer Studies*, 40, 4, pp. 603-652.

[25] Cohen, P., Chaudhri, V., Pease, A. & Schrag, R. (1999). Does prior knowledge facilitate the development of knowledge-based systems? In *Proceedings of the Sixteenth National Conference on Artificial Intelligence, AAAI 99*, Orlando, FL, USA, 221-226.

Chapter 4

Tools

Micro-Modelling of Business Processes for Just-in-Time Knowledge Delivery
U. Reimer, B. Novotny and M. Staudt

Just-in-Time Knowledge Support
E. Dehli and G.J. Coll

STEP PC: A Generic Tool for Design Knowledge Capture and Reuse
N. Prat

WebGrid: Knowledge Elicitation and Modelling on the World Wide Web
M. G. Shaw and B. R. Gaines

Mapping for Reuse in Knowledge-Based Systems
J.Y. Park and M.A. Musen

Mining Very Large Databases to support Knowledge Exploration
N. Mackin

Micro-Modeling of Business Processes for Just-in-Time Knowledge Delivery

Ulrich Reimer, Bernd Novotny and Martin Staudt

Swiss Life, Information Systems Research Group
Postfach, CH–8022 Zürich, Switzerland
⟨first name⟩.⟨last name⟩@swisslife.ch

Abstract: The paper deals with the local support of business processes as seen from the perspective of an individual office worker. We call business activities viewed on such a more microscopic level "office tasks". Office tasks related to processing of contracts in the insurance business are complex and highly dependent on legislation and company-specific regulations. On the other hand, the increasing competition strongly calls for higher efficiency as well as higher quality standards. The EULE system developed at Swiss Life meets these demands by providing computer-based guidance and interactive support for office tasks dealing with private life insurances. Realized as a knowledge-based system EULE consists of a shell and a knowledge representation language to construct the models of the actual office tasks. It is therefore independent of the actual application domain and can be employed for other kinds of office tasks as well.

1. Introduction

Some time ago, Swiss Life, as many other companies, reorganized its customer support. Office workers are no longer specialists dealing with certain kinds of office tasks only, but meantime must be generalists who deal with all kinds of tasks. The number of different kinds of office tasks (like changing the beneficiary, increasing the risk sum, surrender of a contract) is rather high (about 60), and many of them only occur very sporadically. An office task has an attached process description which specifies how it should be performed. These processes are typically quite complex because there are many laws and company regulations that will influence the task if it is to be performed properly. As a consequence, the work of this new generation of office workers is quite demanding and urgently calls for an appropriate support in order to meet Swiss Life's high quality standards. The following observations can be made with respect to the characteristics of the office work:

- The experience of the office workers ranges from novices who have just joined the company to experts with many years of training.

- About 80% of the office work consists of standard cases with medium complexity. The remaining part consists of very complicated tasks that require a lot of experience.

- There are about 60 kinds of office tasks. Most of them seldom occur, so an office worker does not have the opportunity to acquire certain routine.

- It is the idea that new employees are trained when they first join the company. However, in order to have enough participants, the training courses actually take place only two or three times a year so that most of the training is nevertheless on the job and not before.

- The way a certain office task is properly executed changes from time to time due to new products, new company regulations, or new legal restrictions.

The situation characterized above shows that there is a potential for increasing the quality of office work and decreasing the average time needed per office task by providing the office workers with an appropriate support system for executing its process steps. Such a system must show the following functionality:

Support of Office Work

The system supports all kinds of office workers, from novices to experts. The novice gets his training while working with the system and can properly execute the processes attached to office tasks from the very beginning due to the active support and guidance of the system. The expert profits less from the system's guidance but from getting relieved from routine tasks, like writing letters and sending memos. He can request support from the system whenever the need arises (e.g. when dealing with an unusual case).

During the specification phase for such a system (which took many cycles), it became evident to us that additional requirements will have to be met by the system to become a success:

Just-in-time knowledge delivery

A user is always given exactly that kind of information he needs in a specific situation so that he never needs to ask for it (also called *just-in-time knowledge delivery* [1]).

Adaptation

Users with more experience need less guidance than novices. The system offers exactly so much guidance as needed by the user.

Maintainability

New legislation, and new company regulations (e.g., due to new products) make it necessary to adapt the office task descriptions regularly. As office task descriptions must be up-to-date to ensure proper office work it is of paramount importance that necessary changes can be done without much effort. Ideally, this is the case when the updates are made by the insurance experts themselves without involving people from the computing department (except maybe for special tasks like defining links to database fields).

Just-in-time knowledge delivery unifies business process support with the central knowledge management issue of supplying people with the knowledge they need to do their work properly. The user adaptation functionality originates from the requirement to support novices as well as experts. Maintainability ensures that keeping the system alive will be feasible w.r.t. time and money.

Current workflow management systems which also aim at automating processes, fail in these concerns due to the limitations of their modeling languages. These languages, e.g., do not allow to represent detailed legal aspects in a declarative way, nor do they support any associated inferences. However, both would be needed to support an office worker in making the right decisions. Instead, workflow modeling languages concentrate on the procedural flow between the different participants of a process. This is a *macro level* view of business processes while the requirements mentioned above require a support on a *micro level*.

Therefore, a system with a functionality that meets the above requirements was developed by the Information Systems Research Group of Swiss Life. It is called EULE and is situated in the triangle of Artificial Intelligence (AI), Knowledge Management (KM), and Business Process Modeling (BPM). It is an AI system because it is realized as a knowledge-based system. It contributes to the KM efforts of a company because its knowledge base encodes knowledge which is crucial for the company, and thus preserves it and helps to make it available where needed. EULE has an impact on BPM because it provides (formal) models of processes for performing office tasks in a much more detailed view than it is usually the case with the models resulting from a BPM approach. The level of detail of the EULE models is indeed needed for the active decision support the system must provide.

Attempts to bridge BP models and whole workflows are actually hard to find in practice. In so far our approach is also a way of operationalizing (very detailed) BP models so that they can be executed. The above aspects are discussed in Section 3.

2. EULE: A Knowledge-Based, Cooperative System for Supporting Office Work

In order to build a tool like EULE which is able to support the direct execution of arbitrary office tasks, a model-based approach is most appropriate: For each office task we formulate a model of all involved data objects, the process steps and the relevant regulation contexts. This model is then compiled to process descriptions which run on top of the shell-like EULE system components. In the following, we summarize the main features of EULE both from the user and the modeler perspectives and add some remarks concerning the implementation.

2.1 The user's view of EULE

EULE visualizes the process to be executed for an office task as a graph (cf. Fig.1). Its nodes contain a sequence of activities the user must perform, while its links are associated with conditions that must be fulfilled for the activities in the subsequent nodes to become relevant. The conditions result from federal law and company regulations. An office worker starts working with EULE by selecting an office task and entering task-specific data as requested by the system (EULE takes most of the data needed from various data bases and only asks the user for data which is not available elsewhere). For each activity encountered the user can request an explanation, why

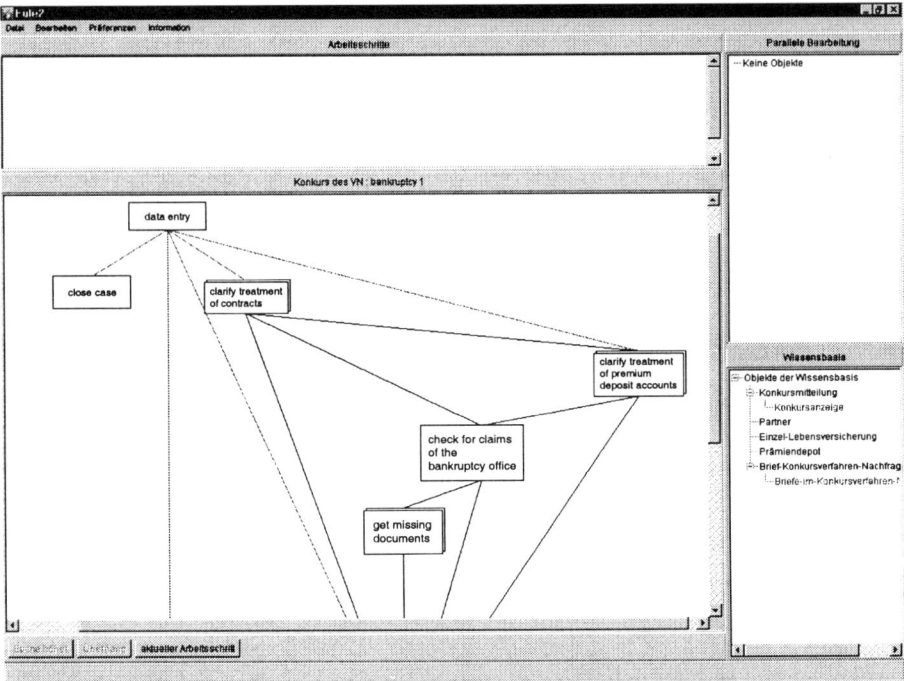

Figure 1: The top-most level of the office task 'policy-holder-goes-bankrupt'

it is necessary and why it is to be performed in the requested way. The explanations also give references to the underlying legislation and company regulations.

A node may contain a graph again, instead of a sequence of activities only. This nesting to an arbitrary depth allows to give even complex office tasks a comprehensible structure. While being guided through the graph (and its sub-graphs) the user sees only those activities which are actually relevant for the current office task instance. However, the user is free to browse through the whole graph and all the activities in its nodes, including the explanations attached to them, in order to get an overall view of the office task. This is especially helpful when the user is not familiar with that office task and wants to learn more about it.

Usually, during the execution of an office task process several letters (e.g. to policy holders, to beneficiaries, or to federal agencies) need to be written. As the EULE system already knows all the task-specific data it is able to automatically generate the required letters. In this way, the time needed for an office task is considerably reduced. Thus, EULE not only increases the correctness and quality of office work but also helps in accelerating it and is therefore helpful for experts, too.

To be able to provide a less experienced user with more detailed information, EULE can ask so-called *novice mode questions* which allow the user not only to enter 'yes' or 'no' but also 'I do not know'. Depending on his experience a user may already know the right answer (e.g., if a contract is legally valid or not) and can directly enter 'yes' or 'no'. If the user is not sure he can answer with 'I do not know' and (only then) gets a couple of additional questions which help him to find out which case holds. Of course, novice mode questions can be nested. An office task is completed when a terminal node in its top-most graph is reached.

Office Task Example: "Policy holder goes bankrupt"

To illustrate how strongly an office task may be influenced by law and company regulations let us discuss the case of a policy holder going bankrupt. A fraction of the top-most level of the corresponding office task graph is illustrated in Figure 1. The Bankruptcy Office publicly announces all cases of bankruptcy. Everybody who holds properties of the bankrupt is obliged by law to report them to the Bankruptcy Office. The properties will be sold by auction to pay off at least a part of the bankrupt's debts. The property an insurance company may hold of a bankrupt is an insurance where the bankrupt is policy holder. Whenever an office clerk is informed of an opened bankrupt proceeding, he must check if such an insurance exists (cf. node "data entry" in Fig.1). When the bankrupt is not policy holder of an insurance the office clerk destroys the notification of bankrupt and closes the case (cf. node "close case"). Otherwise, the office clerk must report the insurance to the Bankruptcy Office, unless an exception is defined by law. This is done inside the node "clarify treatment of contracts" which contains a whole subgraph. One exception states that a provident insurance is exempt from seizure. Thus, provident insurances cannot be sold by auction and therefore need not be reported to the Bankruptcy Office. Instead, the office clerk informs the bankrupt that he remains policy holder. An insurance where the spouse, the children or both are the only beneficiaries is also exempt from seizure. By law such an insurance is transferred to the beneficiaries as soon as the policy holder goes bankrupt. The office clerk informs the beneficiaries that they are the new policy holders (this and the following activities correspond to nodes not shown in Fig.1). The beneficiaries have the right to accept or reject the transfer of the insurance. They accept the transfer by sending a certificate issued by the Debt Collection Office to the insurance company. The reception of the certificate is a necessary precondition for Swiss Life to issue a new policy and to send copies to the new policy holders. By sending a note to the insurance company the beneficiaries can reject the transfer of the insurance. When an office clerk receives a note of rejection he reports the insurance to the Bankruptcy Office because it is up to the Bankruptcy Office to decide what to do with the insurance.

2.2 Creating and maintaining office task process descriptions

As has been stated in Section 1 the guidelines of how to properly execute an office task change whenever new products are introduced, or legislation or company regulations change. The latter case happens quite often. Therefore, it is important that the knowledge represented in EULE can easily be modified. Maintenance should be done directly by the insurance experts as far as possible – without a major involvement of the computing department. According to our experience this can only be achieved when the knowledge is represented in a high-level language that is especially tailored to the application domain and the view insurance experts have on their domain.

The high-level language we have developed for EULE is called HLL. Its syntax does not need to take computational efficiency issues into account (with respect to reasoning and integrity checking) because a representation in it is compiled down into efficiently executable code (but, of course, HLL must be computationally tractable). Thus, HLL can be tailored towards comprehensibility without any compromise. In the following, we give an overview of HLL by discussing a few examples of how knowledge is represented in it.

Activities:
EULE distinguishes between three kinds of activities:

- A *working step* is an atomic activity which occurs as an element inside a node of an office task graph. It consists of a textual description to explain the user what to do, and either requires an acknowledgement of completion of the activity, or contains a request to enter certain attribute values into a concept instance in the knowledge base of EULE (e.g. the name of a beneficiary, or the premium amount). For the latter purpose the user has an especially designed object editor available.

- A *simple activity* is a sequence of working steps. Each working step can be associated with a condition when it is to be performed. A simple activity corresponds to a node in an office task graph that does not again contain a graph.

- A *composite activity* is a set of partially ordered simple and/or composite activities, i.e. it forms an acyclic graph. A composite activity either corresponds to a whole office task or to a node of an office task graph which again contains a graph.

An example of a simple activity as it is defined in HLL is given in Figure 2.

Laws and regulations:
An important requirement for HLL is to support the representation of law and regulations that influence how an office task is properly executed. A law defines a right or an obligation and specifies when it holds. HLL therefore knows of two predefined concepts 'right' and 'obligation'. A certain (sub)section of a law is then represented by introducing an appropriate specialization of 'right' or 'obligation' and by describing the conditions under which a certain instance of it exists. This instance is automatically generated when its existence condition holds. A generated right or obligation instance may then cause the precondition of some activity to become fulfilled. An example is depicted in Figure 4 which represents the law given in Figure 3. Figure 4 also shows examples of concept definitions.

Preconditions for activities:
Any kind of activity can have a precondition that must be fulfilled for the activity to be permitted. Preconditions are derived from company regulations and often refer to the existence of a right or obligation which is the cause to perform the activity. An example of a precondition is given in Figure 5.

Activity
  Name:
      obtain-bankruptcy-announcement
  Description:
      "Enter required data about the bankruptcy announcement"
  Arguments:
      K: bankruptcy-announcement
  Type:
      simple
  Working Steps:
      (ObjectEditor, "Enter Data about Bankruptcy Announcement",
          (**Action**: Edit K
              (**Text**: "Where does the information originate?"
                Fields: information-source ⟨force⟩
              );
              (**Text**: "Specify the status of the proceedings"
                Fields: status-proceedings ⟨force⟩
              );
          EndAction)
      );
End

Figure 2: Example of a simple activity

VVG 81

Right of pre-emption of the spouse and the descendants

   ⋮

  *2 The beneficiaries are obliged to notify the insurer of the transfer of the insurance
  by presenting a certificate issued by the Debt Collection Office or the Official
  Receiver. If there are several beneficiaries, they must appoint a representative to
  accept the notifications incumbent upon the insurer.*

Figure 3: Fragment of a law

Each HLL construct has a reference to the knowledge source from where the knowledge being represented originates (see Figs. 4, 5). The reference can be used to give a user an explanation why a certain activity must be performed, and is used to show a knowledge engineer all constructs belonging to a specific source.

Of course, the linear syntax of HLL as discussed before is awkward to use, and can certainly not be used by insurance experts (as said before, it is our ultimate goal that insurance experts do (most of) the modeling themselves). Therefore, we plan to hide away the actual syntax by building a graphical modeling environment. Different types of constructs correspond to different graphical icons while the construct arguments correspond to slots in the icons. They can be filled with one of several predefined values or with instances of other construct icons (e.g. concept or activity icons which occur inside an activity icon).

Concept
    Name:
        obligation
    Type:
        ontological
    Properties:
        is-obliged (object, [1,*], natural-or-juristic-person);
        holds-against (object, [1,*], natural-or-juristic-person);
        is-fulfilled (boolean)
End

Concept
    Name:
        obligation-to-show-certificate
    Type:
        derived
    SuperConcepts:
        obligation
    Properties:
        is-obliged (object, [1,*], natural-person);
        holds-against (object, [1,*], natural-person);
        concerned-insurance (object, [1,1], life-insurance)
End

Law: VVG, Section: 81, Subsection: 2
InstanceDerivation
    For Derived Concept:
        obligation-to-show-certificate
    As:
        foreach i: life-insurance **and**
        foreach b: publication-of-bankruptcy
        if
            i.policy-holder = b.bankrupt **and** b.status = 'open' **and**
            i.beneficiary-clause = 'spouse-descendant' **and** exists-value(i.beneficiaries)
        then derive the Instance with:
            is-obliged = i.beneficiaries;
            holds-against = i.policy-holder;
            concerned-insurance = i
        end
End

Figure 4: Representing laws

The modeling tool should further provide several views on the knowledge base simultaneously, e.g. via a terminology browser, via a tool for editing and displaying office task graphs, and via a window which shows the various pieces of knowledge not by construct type but according to the knowledge source from which they originate. Thus, one can for example see all the concepts, preconditions and derivation rules that belong to one specific company regulation.

Regulation: GEP, No 29, 08/03/1993
Precondition
    Name:
        condition-to-clarify-treatment-of-vontracts
    For Activities:
        clarify-treatment-of-contracts
    Condition:
        exists b: bankruptcy-announcement **and**
        exists i: life-insurance
        such that
            b.bankrupt = i.policy-holder
        end
End

Figure 5: Activity precondition

Another, very important support a modeling environment should give is to allow a user to enter a whole pattern of HLL constructs in one step and then to fill in the place holders to tailor the pattern to the actual needs. According to our modeling experience there exist typical modeling patterns which occur frequently. They correspond to certain activity patterns in an office task. For example, a typical pattern in an office task is the postponed entering of data by the office worker. When some data usually to be entered at a certain point in the office task together with other data is not available at that time it might be possible to enter it later on. EULE must keep track of this by generating a specific request object in its knowledge base. At the latest possible time in the office task where the data is needed EULE checks for the existence of corresponding request objects, and if existent, prompts for input. The HLL fragment needed to model such a behaviour can be defined as a pattern "postponed data entry" which can then be included in a task model just by a mouse click. Of course, further details must then be added to make the pattern into a complete piece of HLL code.

The existence of modeling patterns is also known in other areas of software and model construction, as e.g. in the area of object-oriented software design where they are called design patterns [2].

2.3 Implementational aspects

From an implementation point of view the EULE system consists of four main components: the HLL compiler, the EULE knowledge base management system (KBMS), the EULE execution controller and the EULE GUI.

The EULE compiler transforms HLL descriptions into acyclic, nested graphs and generates the office-task-specific data structures. Its output consists of code fragments in Java and in a Prolog-style language which are interpreted by the controller and the KBMS, respectively. The controller governs the graph traversal at runtime, initiates GUI actions, e.g. editing of objects, and invokes the KBMS for data retrieval, deduction tasks and validity checks. Conditions attached to the links in a graph act as integrity constraints formulated as Datalog¬ clauses [3] and are tested on demand for being satisfied or violated. Knowledge about concepts and instances is formulated in the terminological and assertional components of a description-logic-like subsystem of the EULE KBMS [4]. Knowledge about law and regulations is encoded in Datalog¬ clauses where we distinguish integrity constraints that must not be violated, and deduction rules which derive new attribute values or whole concept instances when their condition part is fulfilled. With this combination EULE competes with the few implemented hybrid inference systems coupling description logic and deductive database technology [5]. However, opposed to earlier versions, the current EULE KBMS implementation relies on a deductive engine only, and limited terminological reasoning is exploited by the compiler and modeling support environment. For more details on EULE see [6, 7, 8].

3. The Impact of EULE on Knowledge Management and Business Process Modeling

The EULE knowledge base captures knowledge which is important to Swiss Life. Via EULE this knowledge is made available to office workers in such a way that they always get exactly that knowledge which is relevant in the current situation. Besides supporting office work, EULE's knowledge is useful for other people and for other purposes as well, e.g., for tutoring new employees, for inquiring about the effects of certain company regulations on office tasks, or for finding out how often a certain regulation took effect in an office task in the past. As a consequence, a system like EULE must not be considered isolated from other issues of knowledge organization and business process modeling. In the following, Section 3.1 discusses the integration of EULE into an organizational memory system, while Section 3.2 deals with the integration of EULE within a framework for business process modeling.

3.1 EULE as part of an organizational memory system

At the core of all knowledge management activities in an organization is the organizational or corporate memory (not necessarily (fully) computerized [9]). It is the central repository of all the knowledge relevant for an organization. Building up such organizational memories (OM) and making them available to people and application systems with quite converging needs is a big challenge which can only be met by an integration of approaches from various fields of computer science as well as business and administration science (cf. [10] and [11]).

There are two major roles an organizational memory can in principle play. In one role it has a more *passive* function and acts as a container of knowledge relevant for the organization (including meta-knowledge like knowledge about knowledge sources). It can be queried by a user who has a specific information need.

The second role an OM can adopt is as an *active* system that disseminates knowledge to people wherever they need it for their work. This second functionality is not just mere luxury but of considerable importance as users often do not know that an OM may contain knowledge currently helpful to them. Furthermore, querying an OM whenever the user thinks it might be possible that the OM contains relevant knowledge is not practical because the user does not always think of querying the OM when it might actually be helpful and because it would be too time consuming (as it interrupts the user's primary work and takes time for searching and browsing the OM).

For the OM to be able to actively provide the user with the appropriate knowledge it needs to know what the user is currently doing. EULE is an example of a system that solves this kind of problem. Such systems have been called *electronic performance support systems* in the literature (cf. [1]). A system like EULE plays an important role in utilizing the knowledge in an OM and should be a part of the organizational memory system. This involves the integration of the knowledge formalized in EULE with other formal knowledge in the OM. For example, the terminology in the EULE knowledge base should be consistent with all the terminologies (or ontologies) developed in various other contexts, e.g. as part of the business meta data for the data warehouse, or for a search facilitator in the intranet information system. This alone is a major unification and integration effort. Another example are the EULE office task representations which need to be integrated with other kinds of formal process representations, as for example workflow descriptions.

Furthermore, it is necessary to integrate the formalized knowledge in EULE with informal one (i.e., textual or semi-structured, like SGML, HTML). For example, the knowledge represented in EULE stems from written documents, like company regulations, memos, minutes of meetings. Often those documents contain more knowledge than has been formalized in EULE. Thus, a user of EULE may want to see the source of a company regulation that has an influence on the office task he is currently dealing with. The written document gives him additional background information which helps him to better understand the rationales for the regulation. For that reason, the knowledge represented in EULE has links to the sources which a user can follow to become more competent.

The Information Systems Research Group of Swiss Life is currently concerned with integration aspects of the above mentioned kind, aiming at building an OM from already existing but isolated knowledge sources in the company (see also [12]).

3.2 Embedding EULE in a business process modeling framework

As a support system for office tasks, EULE contains quite detailed task representations. Thus, EULE is a major component of the business process modeling framework in the company and must be well settled within it. Process descriptions typically evolve in a top-down fashion. They start with a high-level structure of business processes that span various organizational units, and then get broken down into more and more local views which at the same time become more detailed, until at the most detailed level EULE office task representations are obtained. The high-level descriptions are completely informal and are given in a textual and/or graphical notation. More detailed levels show more structure, typically accompanied by the utilization of a process modeling tool which offers predefined modeling patterns, or introduce whole reference architectures (see e.g. [13]). Even more detailed levels finally require formalized models. Consequently, EULE representations typically evolve from already existing, informal or semi-formal process descriptions by adding additional detail. New insights gained when building an EULE model may lead to a feedback and a revision of models on a higher level. A proper integration of all process modeling activities is therefore needed to allow the evolution and interaction over all levels. The idea is to come up with a layered modeling approach which allows to adopt the view with that granularity which is currently needed, while ensuring that changes made on one layer will automatically be propagated to those other layers where they are relevant, too. We intend to employ a meta schema as an important means to enable the transfer of models from one layer to the next.

Another possible integration requirement is EULE's interplay with a workflow management system (WFMS). The functionality of EULE does not much overlap with the functionality of a WFMS but is in fact rather complementary: EULE supports a single user, while a WFMS is specialized in coordinating tasks that are to be performed by a group of people. It knows which subtasks must be performed by which (kind of) people and manages the flow of control and information between them. Additionally, a workflow description is procedural while the office task descriptions in EULE are declarative (since EULE is knowledge-based). As already pointed out before, EULE would not be maintainable within acceptable time and cost limits if office task descriptions were given in a procedural language.

Finally, a working step in a workflow is atomic but typically corresponds to a task graph in EULE. Therefore, the main idea for coupling a WFMS with EULE is to associate subgraphs in an EULE office task with working steps of a workflow. A user will then start a new office task by selecting a workflow or by picking up a task in the in-box of his WFMS. When reaching a working step that is associated with an EULE subgraph the user can request EULE assistance. When entering the EULE subgraph the user is prompted for the information missing at that specific point. When the EULE subgraph is terminated EULE reports back to the WFMS that the current working step is finished. The WFMS selects the next working step according to the workflow description. If that working step has an EULE subgraph, too, the user may enter it again, or is transferred into it automatically when he indicates that he wishes to stay in the EULE support mode.

The integration of EULE with a WFMS is difficult because knowledge in the knowledge base of EULE is relevant for the workflow system and vice versa. A coupling of both systems requires a (partly) sharing of of their knowledge bases. A more detailed discussion of the coupling can be found in [14] and [15].

4. Experience and Related Work

There exist only few other systems with a functionality roughly comparable to EULE. Like EULE, they aim at supporting business processes by offering the user (access to) that information which he or she needs to successfully proceed with the current office task (see [16, 17]). Unlike EULE, those systems are not oriented towards detailed and comprehensively described office tasks but go for weakly structured workflows instead, as they are typical for tasks which comprise a great degree of unforeseeable variations.

A slightly different approach is presented in [18]. A user who wishes to perform a certain office task and does not exactly know how to do it correctly is presented cases of former task instances which are similar to the task at hand. For this, the user must first give the system a characterization of the current task and then gets task instances presented by the system from which the one which appears most appropriate is selected. Subsequently, that instance is manually adapted to the current needs.

While the processes in the approaches described in [16, 17, 18] are to a large extent underspecified, the processes supported by EULE are strongly constrained by Swiss Federal Law and internal company regulations.

The experience we gathered with EULE during a field study was extremely positive so that the system was finally deployed. The field study was conducted during August 1998 with the participation of seven office workers that had quite different experience and background. They used EULE to execute 311 office task instances. The field study was concluded with an interview of all participants and their team heads. Those five office workers who had considerably less experience than the remaining two said that they could extend their expertise during their work with EULE, and that they needed much less support from colleagues or their team head while using EULE. Six of the seven office workers said that EULE pointed out to them aspects to consider which they would otherwise have missed. Thus, EULE indeed helped to avoid incorrectly executed tasks.

The team heads of the participating office workers noticed a considerable relief from support they usually need to give their team members whenever they encounter a situation they do not know how to deal with. Furthermore, the necessity for the team heads to check the work done is reduced, additionally relieving them from unproductive tasks. Generally speaking, the overall quality of office work has considerably increased (less complaints, faster completion).

A rough cost-benefit analysis shows that we save at least 6 person years per year due to a relief of team heads, a faster completion of office tasks, and a considerably shortened training phase for new employees. The additional benefit of higher customer satisfaction (faster reaction to requests, less complaints) cannot be quantified but should not be underestimated. With costs for maintenance of the EULE shell and of the represented office task models of about 2 person years per year and an initial cost of 12 person years (for the software and 20 office task models (yet to be built)) we reach a break-even point after 3 years.

5. Conclusions

Office tasks related to processing of contracts in the insurance business are complex and highly dependent on legislation and company-specific regulations. On the other hand, the increasing competition strongly demands higher efficiency as well as higher quality standards. The EULE system developed at Swiss Life is designed to meet these demands by providing computer-based guidance and interactive support for office tasks. The system relies on a knowledge representation language which covers data and process aspects as well as the relevant legal aspects and company regulations. As a knowledge-based system EULE consists of a shell and a modeling language to construct the actual models of the office tasks to be supported. We sketched the embedding of EULE into a general organizational modeling framework, in particular w.r.t. workflow management and business process modeling in general, as well as organizational memories.

A field study showed that EULE indeed fulfills the high expectations we have in it. The system was set productive in mid-1999. The major future activities concerning EULE is the development of a tool that supports the modeling of knowledge needed for EULE. The ultimate goal for this tool is to enable insurance experts to do the modeling themselves, without the help of people from the computing department. Furthermore, we continue our work on integrating EULE into all the modeling activities going on at Swiss Life, aiming at an organizational memory as a central repository containing all the knowledge relevant for the company.

Acknowledgements. The following people have been involved to make the EULE system a success: Andreas Margelisch (with whom everything started), Priska Felber, Jörg Junger, Jörg-Uwe Kietz, Peter Lippuner, Claude Marksteiner, Georg Spirito, Thomas Vetterli, and Ulrich Weber.

References

[1] K. Cole, O. Fischer, and P. Saltzman. Just-in-time knowledge delivery. *Communications of the ACM*, 40(7):49–53, 1997.

[2] E. Gamma, R. Helm, R. Johnson, and J. Vlissides. *Design Patterns. Elements of Reusable Object-Oriented Software*. Addison-Wesley, 1995.

[3] S. Ceri, G. Gottlob, and L. Tanca. *Logic Programming and Databases*. Springer, 1990.

[4] W.A. Woods and J.G. Schmolze. The KL-ONE family. In F.W. Lehmann, editor, *Semantic Networks in Artificial Intelligence*, pages 133–178. Pergamon Press, 1992. (special issue of *Computers & Mathematics with Applications* 23(2-9)).

[5] A. Levy and M.-C. Rousset. Combining horn rules and description logics in CARIN. *Artificial Intelligence*, 104(1-2):165–209, 1998.

[6] U. Reimer, A. Margelisch, B Novotny, and T. Vetterli. EULE2: A Knowledge-Based System for Supporting Office Work. *ACM SIGGROUP Bulletin*, 19(1):56–61, 1998.

[7] U. Reimer, A. Margelisch, and B. Novotny. Making knowledge-based systems more managable: A hybrid integration approach to knowledge about actions and their legality. In Pareschi. R. and B. Fronhöfer, editors, *Dynamic Worlds:*

From the Frame Problem to Knowledge Management, pages 247–282. Kluwer, 1999.

[8] J.-U. Kietz and M. Staudt. *Integrating Access to Heterogeneous Data Sources into an Object-Oriented Business Process Modeling Language*. Swiss Life, Information Systems Research Group, 1999.

[9] E.W. Stein. Organizational memory: Review of concepts and recommendations for management. *International Journal of Information Management*, 15(2):17–32, 1995.

[10] A. Abecker, A. Bernardi, K. Hinkelmann, O. Kühn, and M. Sintek. Toward a technology for organizational memories. *IEEE Intelligent Systems*, 13(3):40–48, 1998.

[11] G. van Heijst, R. van der Spek, and E. Kruizinga. Organizing corporate memories. In *Proceedings of the 10th Banff Knowledge Acquisition for Knowledge-Based Systems Workshop*, Banff, Alberta, Canada, 1996. http://ksi.cpsc.ucalgary.ca/KAW/KAW96/KAW96Proc.html.

[12] U. Reimer. Knowledge integration for building organizational memories. In *Proceedings of the 11th Banff Knowledge Acquisition for Knowledge-Based Systems Workshop*, Banff, Alberta, Canada, 1998. http://ksi.cpsc.ucalgary.ca/KAW/KAW98/KAW98Proc.html.

[13] A.-W. Scheer. *ARIS - Business Process Modelling*. Springer, 1999.

[14] A. Margelisch, U. Reimer, M. Staudt, and T. Vetterli. Cooperative support for office work in the insurance business. In *Proceedings of the Int. Conf. on Cooperative Information Systems, Edinburgh, Scotland*, 1999.

[15] R. van Kaathoven, M. Jeusfeld, M. Staudt, and U. Reimer. Organizational memory supported workflow management. In A.W. Scheer, editor, *Electronic Business Engineering*, pages 543–563. Physica, March 1999.

[16] A. Abecker, A. Bernardi, and M. Sintek. Enterprise information infrastructures for active, context-sensitive knowledge delivery. In *ECIS'99 Proceedings of the 7th European Conference on Information Systems*, 1999.

[17] S. Staab and H.-P. Schnurr. Knowledge and Business Processes: Approaching an Integration. In *OM99 - Proceedings of the International Workshop on Knowledge Management and Organizational Memory (at Int. Joint Conf. on Artificial Intelligence, 1999)*, 1999.

[18] C. Wargitsch, T. Wewers, and F. Theisinger. Organizational-Memory-Based Approach for an Evolutionary Workflow Management System - Concepts and Implementation. In J.R. Nunamaker, editor, *Proceedings of the 31st Annual Hawaii International Conference on System Sciences*, volume 1, pages 174–183, 1998.

Just-in-Time Knowledge Support

Einar Dehli[1] and Gunnar J. Coll[2]

[1]Computas AS, Vollsveien 9, 1327 Lysaker, Norway, ed@computas.com
[2]Computas AS, Vollsveien 9, 1327 Lysaker, Norway, gjc@computas.com

Abstract: Just-in-Time Knowledge Support (JIT KS) is a new approach for organizations and employees to manage excessive information and ever-faster change. Most technology efforts in Knowledge Management to date have been restricted to making information *available*, thus enabling employees to seek knowledge in areas they are interested in. JIT KS, on the other hand, addresses the much harder problem of ensuring that relevant knowledge is spread throughout the organization and *applied* uniformly and consistently in mundane, day-to-day work, including situations where employees are not aware of or motivated to look up the knowledge. Successful JIT KS depends on having knowledge activation fully integrated in the organization's business processes and tools.

1. Why JIT KS?

This paper describes the evolution of what we call the Just-in-Time Knowledge Support (JIT KS) approach to Knowledge Management. JIT KS is the result of research dating back to the mid-eighties. It is currently manifest in a family of commercial tools and methodologies, known as Computas FrameSolutions™. Several mission-critical FrameSolutions applications have been successfully deployed in recent years, substantiating the usefulness of the JIT KS approach.

1.1 From *knowing that* to *acting on*

The notion that knowledge is a kind of justified, true belief is a traditional touch-stone in epistemological discussion, but remains problematic for a variety of reasons ([1], p 85). Our focus in this paper is on the problem of how to relate knowledge to behaviour, particularly in terms of performance quality in organizations as well as individuals. One thing is having knowledge, another is putting it to use. From school to professional life, what counts is the ability to mobilize knowledge in brief, often unpredictable time windows. Relevant knowledge that fails to be activated as and when a given task occurs, is of little value to its possessor. The philosophical questions pertaining to the relationship between knowing and acting are beyond the scope of this paper, and instead we choose to adopt a behavioural, constructivist perspective on knowledge. In this perspective, the problem is how to build concepts and mechanisms that demonstrably enhance knowledgeable behaviour in organizations, thereby building our understanding of the processes related to knowledge, identified by Nonaka and Takeuchi as socialization, externalization, combination and internalization [2].

When we say that an organization possesses knowledge, we want to imply more than the fact that it employs knowledgeable individuals. The knowledge organization proper is one that acts competently, timely and adaptively in its business context. At best, the gestalt principle applies: The organization produces synergies that make its knowledge more than the sum of its members' knowledge. At worst, the organization fails to perform competently in spite of abundant individual knowledge nestling in its cubicles. The critical issue is how well the organization can make appropriate knowledge available where and when a given task is executed.

Management as a general discipline is concerned with making organizations perform. Managing knowledge involves active tuning of the organization's ability to use its intellectual capital [3] to produce efficient and adaptive behaviour. The perspective is pragmatic, focussing on knowledge as a strategic resource.

1.2 Managing knowledge in a changing business environment

Accelerating change and mobility in several dimensions creates new challenges for the knowledge organization that wants to deliver consistently competent behaviour over time. Emerging career patterns make ambitious individuals more prone to seek change, and less loyal when a prospective new employer can offer an interesting opportunity. Partly, this is due to the necessity for competent people to seek change to maintain their own competence development momentum. Mobility is increasingly perceived as mandatory, if a person wants to stay sharp and learn new things, and it is often the best people that leave first. What suffers is the quality of jobs carried out by successive novices.

Also, new drivers, particularly from technology, force organizations to renew their business rules on a new scale. To adapt to new events, it may be necessary to redesign the organization continually, which means that people must become competent at new things faster and more frequently, with less availability of established "old hand" support. The tasks of individuals become more diverse and less permanent over time.

Domain knowledge as such is subject to accelerating revision, although its rate of change varies from the timeless to the fleeting. Particularly in technical fields, the critical combination is mastery of underlying principles, coupled with the ability to stay consistently up-to-date with the most rapidly changing knowledge strata. In addition to an increasing rate of change, the width of the research frontier also increases rapidly, as does the complexity of inter-disciplinary relationships, as old dividing lines fade. In administrative domains, rules and regulations steadily grow, and the prevalence of international legislation and conventions adds new complexity. The result is accelerating multi-dimensional growth of today's knowledge landscape, as a driver for change in the knowledge industry.

Software, paradoxically, is both one of the significant drivers for change and simultaneously perhaps the most powerful potential source of new solutions for coping with change. The use of software tools in this role has so far been a mixed blessing. It has introduced new learning challenges, as well as sometimes momentous maintenance challenges. Software as a carrier of business knowledge has often enough turned into yet another change problem, adding to the burden rather than alleviating it.

1.3 Dimensions of change

We maintain that the less-than-expected helpfulness of software in coping with change stems from a general neglect in the software industry to see that the drivers for change operate in several rather different dimensions, and that software systems must be designed to accommodate changes along each of these dimensions without cross-interference and undue dependencies. We also hold that different groups of people should be responsible for changes along each dimension, and that these must be enabled to operate independently of each other. The dimensions we propose are *system*, *contents* and *use*.

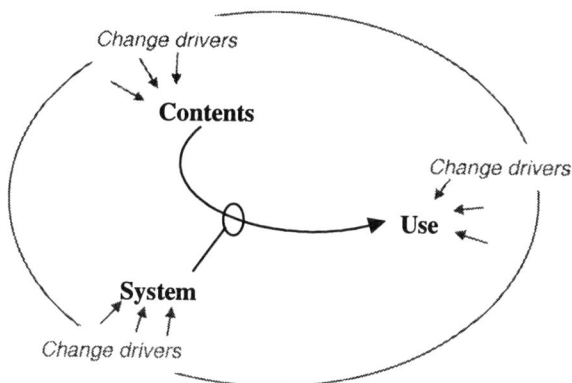

Figure 1: Three dimensions of change. The main purpose of any computer system is to make contents available for use. Systems should be designed to accommodate changes in the *system*, *contents* and *use* dimensions, without unnecessary cross-interference.

The *system* dimension refers to the information technology (IT) infrastructure of the solution. It covers all hardware and general systems software, as well as software architecture, tools, generic components and functionality, database servers, and runtime platform – everything needed for an operational system basis. The *contents* dimension covers all domain-related data, information, knowledge and behaviour that the system is capable of handling, i.e. what can be stored, worked on and retrieved from the system. The *use* dimension addresses how contents in the system support the business processes of the organization, i.e. how information, knowledge and functionality is made available to support the work of agents inside and outside the organization. By agents we refer not only to people filling various roles, but increasingly also to intelligent computer systems.

Ideally, changes in one dimension should not affect the other dimensions. Consider for example the redistribution of tasks and responsibilities among employees, which are typical consequences of reorganisations or revised business processes. Such *use* changes should be readily accommodated by IT applications, without fear of introducing disturbances to the *system* basis or domain knowledge *contents*. Similarly, taking advantage of new technological capabilities for the system should be possible without concern for disturbing application behaviour, i.e. how contents is used.

1.4 The JIT KS approach

As the rate of change in business knowledge increases, up-front training of personnel becomes correspondingly less cost-efficient. Increasing mobility of personnel aggravates this problem. To meet this, the JIT KS approach aims to enable personnel with sound basic training to carry out their work with sustained on-line knowledge support during the work execution, learning in real time.

Furthermore, with most of the world's information only a mouse click away, expectations are rising that everybody should be aware of and make use of the most relevant information in any given problem situation. In practice, this is not so. JIT KS therefore also uses active central knowledge bases to ensure that operational behaviour automatically conforms to new regulations from day one.

The JIT KS approach does this by taking the *use* dimension as its primary frame of reference. A JIT KS application is structured around the work processes of the organization, with explicit models representing the organization and its tasks. All user interaction is structured around this model. Domain knowledge is represented in separate modules, each adapted to the appropriate set of change drivers. On this basis, the application can direct the user through his tasks in an intelligent way, while actively volunteering knowledge to the user relative to the task at hand, as and when the work is performed.

2. Tracing the Origins of JIT KS

The following project reviews show how the Just-in-Time Knowledge Support approach has evolved from a series of applied research and development projects in Computas from the mid-eighties to the mid-nineties. More than fifty knowledge-based systems projects were carried out in Computas in this period, bringing together research results from the fields of Artificial Intelligence, Object Technology and User Interaction in real-world applications.

2.1 Visual knowledge objects: KIPS

KIPS (Knowledge-based Interface to Process Simulation) was the first significant project in Computas with a JIT KS focus, carried out in 1984-87 for Statoil [4]. KIPS researched how an intelligent front-end to a mainframe process simulation package could improve simulation performance, by assisting chemical process engineers in the tedious and costly task of preparing valid textual input files and interpreting result files. Process engineers spent too much time struggling with a complex command language and other software intricacies. KIPS aimed at eliminating the needs for software training, allowing the process engineers to work directly in the simulation domain with familiar terms and concepts.

KIPS took full advantage of the windows-based and mouse-driven graphical user interface of the Xerox AI workstations. Novel features in the experimental language Loops [5] added objects, rules, active values and knowledge bases to Lisp. To produce an intuitive and efficient user interface, KIPS explored object-oriented, hierarchical process diagrams as a user interaction metaphor. A generic, object-oriented knowledge modelling language was defined, supported by a graphical editor with CAD[1]-features for building and interacting with the process models. The editor provided immediate model validity feedback, eliminating the long delays imposed by batch simulation cycles.

2.2 Model-based meta-reasoning: XFRAC TUTOR

The XFRAC TUTOR project (1987-88) investigated the use of existing knowledge bases for training purposes [6]. The XFRAC expert system was developed on Xerox AI workstations and ported to standard hardware. XFRAC TUTOR demonstrated how the rule-based representation of expertise for diagnosing fractures could be reused for tutoring purposes, by adding user modelling and a meta-reasoning layer, inspired by the SOPHIE architecture [7]. Declarative knowledge representation separated from the reasoning engines, as opposed to algorithmic programming of behaviour, allow the tutor level introspectively to meta-reason about the domain reasoning progress in the expert system.

The XFRAC TUTOR research established an initial framework for understanding the applicability and limitations of just-in-time knowledge support in comparison with up-front awareness building and just-in-time skills training. A motivating factor behind XFRAC TUTOR was that the expert system itself was impractical to deploy in the computer-unfriendly work situation of field surveyors. This is about to change, as fully functional wearable computers are becoming feasible. Many work situations, which previously have required up-front training, will then be able to benefit from on-the-job performance support.

---

[1] CAD: Computer Aided Design

2.3 Integrated knowledge support architecture: STEAMEX

STEAMEX is a seminal system in the development of JIT KS. The STEAMEX research project, carried out in 1986-89 for the Norwegian Research Council and Norsk Hydro, explored and incorporated many advanced capabilities for real-time and model-based reasoning in an integrated knowledge support architecture [8].

STEAMEX provides on-line decision support to steam plant operators. The goal is to conserve energy and minimize costs, while staying capable of supplying the requested amount of steam. Temporal aspects are important. Starting an oil boiler takes an hour. Calculating a current optimal steam production distribution is not sufficient. STEAMEX continuously calculates and revises plans for when boilers should be started and stopped, based on predictions of future steam consumption, start-up costs, and many other factors.

Operator performance varies and skilled operators typically advance to better job positions. STEAMEX uses pedagogical means to make best practices available to less experienced operators. The main display shows an interactive diagram of all steam consumers and producers, colour-coded to indicate need for action. Key economic indicators, including an econometre dial, continuously show how the actual situation deviates from the calculated optimum. Operators can carry out what-if analyses for training purposes and for checking consequences of possible actions.

All domain knowledge, including definitions of plants, boilers, tubes and sensors, their operational characteristics, decision rules, etc, are represented declaratively. A rule language based on Prolog semantics was developed and integrated with the domain object model. The application is multi-threaded, with many reasoning processes running in parallel. All object model variables are temporal and multidimensional. This high degree of flexibility and integration in the knowledge representation architecture was required to allow background data acquisition, optimization and planning threads to interact with and be interrupted from foreground threads under control of the user.

Practical aspects are critical for user accept of JIT KS systems. The operators embraced STEAMEX. The user interface is simple and self-explanatory. Instructions fit on one sheet of paper. Several levels of automatic restart strategies allowed for frequent new releases to be deployed, without risks of disturbing operator patience. The system was in daily use in the steam plant for more than five years after the project ended. As the plants evolved, local process engineers updated the reader-friendly knowledge declaration files.

STEAMEX was developed and deployed on PC ATs with 2.5 MB memory, using Digitalk Smalltalk/V286. Many general mechanisms were developed to support multi-threaded, real-time reasoning and user-interaction, making extensive use of Smalltalk's meta-programming capabilities. These extensions were released in 1989 as Computas Multiprocessing Extension Kit to Smalltalk V/286.

STEAMEX effectively proved that expressive knowledge architectures could be realized on inexpensive PCs, in a programming environment resembling the efficiency and capabilities of expensive AI workstations.

2.4 Model-based application integration: OPA

A comprehensive intelligent operator support system, OPA, was developed for the Norwegian Power Board in 1987-90, to assist operators in supervising the power distribution network in Norway [9]. OPA integrates data from many sources in a single view, and serves as an intelligent front-end to several analysis and calculation programs. OPA was originally developed on Symbolics Lisp machines using Intellicorp KEE. It was ported to Gensym G2 in the mid-nineties, and is still in active operational use. Several research projects have been conducted in the context of OPA, taking advantage of the comprehensive ontology and live, model-based representation of the power distribution network in Norway.

2.5 Graphical procedure support: HALDRIS

HALDRIS is a real-time decision support system developed for Norsk Hydro in 1989-92 to assist operators in optimising the product/energy yield ratio of a large aluminium production plant [10]. Building on previous operator support research in Computas, HALDRIS introduces a work-procedure dimension. A textual operator handbook existed before the project started. The handbook contained much useful information, but as usual with textual handbooks, it was not often used. It was therefore decided to refine and formalize the handbook and use procedures as the main organizational entities in the knowledge base.

The main innovation of HALDRIS in relation to other Procedure Management Systems (PMS) research [11] is the explicit representation of both analytic and heuristic knowledge, using specific knowledge representation tools. HALDRIS makes extensive use of the rich real-time knowledge representation and reasoning facilities in Gensym G2. For the first version of HALDRIS, a knowledge representation based on rules was chosen. Each step of each procedure was translated into a corresponding G2 rule. While the rule format was easy to understand and apply, the structure of the procedure was lost in the translation. The next version therefore introduced a schema-based knowledge representation based on the object-oriented features of G2. Procedures were visualized graphically, with steps as textual nodes arranged in directed graphs. These graphs were introduced to structure the user interaction and to visualize execution. A process editor was developed to create and maintain the work procedures.

The procedural representation and reasoning mechanisms explored in HALDRIS are direct forerunners to the process representation formalisms and process engine used in FrameSolutions.

2.6 Executable procedures and checklists: KRM

The Nordic joint research project KRM (Knowledge-based Risk Management), carried out in 1989-1992, combined the procedural approach in HALDRIS with the low-cost Smalltalk-basis of STEAMEX [12]. KRM addressed the application of risk assessment in industrial operations, researching how design-time plant knowledge could be captured and made available to operators in the operations phase as work procedures and executable checklists.

The procedural reasoning capabilities in HALDRIS were further refined in KRM. Practical experience with several KRM prototypes in different industries made it clear that the declarative representation of procedures as graphs of executable steps was a promising knowledge support approach with broad applicability.

2.7 Knowledge-based application of statutes: KOMPANS

KOMPANS was developed in 1989-93 for the Norwegian Ministry of Education, Research and Church Affairs. KOMPANS assists municipal schools in the teacher recruitment process. It streamlines administrative paper work and uses rule bases for assessing teacher competency, seniority and salary according to prevailing statutes and regulations. KOMPANS' diligent application of rules significantly reduced the number of salary disputes and appeal cases. A contributing factor in this was the union's sanctioning of the system, after having ascertained that KOMPANS resolved several yearlong disputes in the teachers' favour.

KOMPANS users enter data for each applicant via registration forms into a database. The rule system then makes assessments based on each applicant's education and experience records. The reasoning engine asks for missing information and required human judgements. When finished, the system generates draft reply letters, with assessments included. An explanation facility constructs a plain-text basis for the assessments, with (hypertext) references to applied statutes and regulations.

KOMPANS rule bases are implemented in the expert systems tool Nexpert Object from Neuron Data. The user interface is developed in Asymetrix ToolBook, a multimedia authoring tool. While this unusual combination of tools for administrative applications was effective in demonstrating advanced and user-friendly behaviour, it also clearly showed the disadvantages of not having an integrated architectural basis, especially with regard to software reuse and maintenance.

School offices using KOMPANS have generally been very satisfied. Its intuitive user interface, along with the integrated tutoring system, made it possible even for users without Windows experience to use the application without prior training. KOMPANS was a supported product until the end of 1999.

2.8 Enterprise level business process task support: HELENE

HELENE (1994-95) marks a new milestone in the JIT KS ancestry. It is the first mission-critical application to apply the emerging JIT KS principles, and the first application to include all essential elements of the subsequent FrameSolutions architecture, until then only explored in industrial applications. At first glance, the work processes of industrial process control may not seem to have much in common with administrative work processes. But when abstracted to a sufficiently generic level, the similarities are striking. Both operators and executive officers handle cases. Confronted with a new case, their task is to collect, interpret and analyse information from relevant sources, make decisions and perform actions based on experience and in compliance with relevant instructions, regulations and procedures, and communicate and log results in documents, journals and databases.

Since May 1995, officials in the Norwegian National Insurance Institution have been using HELENE as their full-time tool for collecting child support debts from divorced parents that do not have custody [13]. The officials may, when required, use enforced debt collection. Their work is governed by a complex set of statutes and regulations set up to protect the rights of both children and parents. Officials interact with HELENE through computerized work procedures, based on the executable checklist approach pioneered in KRM, as shown in figure 2 below. These checklists ensure that every official works in compliance with the relevant statutory requirements, depending on the characteristics of each case. An explicit organization model and embedded workflow capabilities ensure that work is allocated to employees in accordance with their roles and authority.

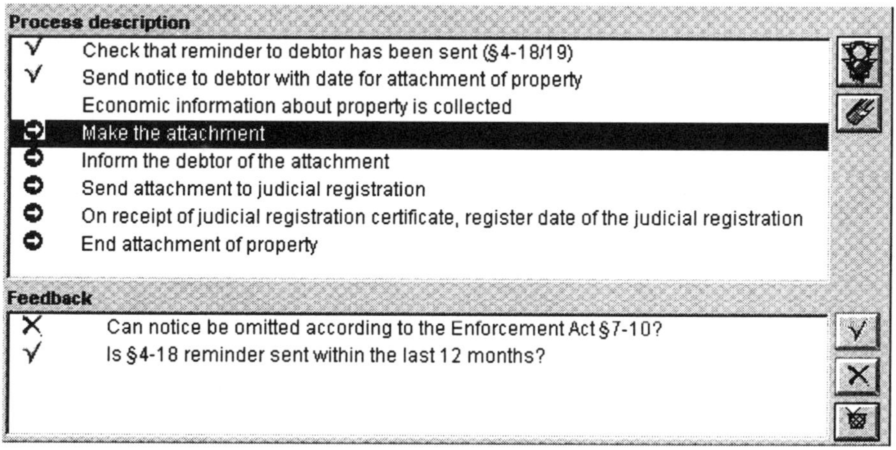

Figure 2: Basic checklist interaction principles in HELENE. A parent has not paid after repeated reminders, and the official has started a procedure to access the debtor's bank account. The top pane shows the work process and the activities involved. The activities marked √ have been executed. Activities marked ➊ are mandatory.

HELENE is implemented in Smalltalk as a client application, with data in a relational database system on a server. It communicates with a back-end accounting system on a mainframe. A standard word processor is integrated for handling documents, and a hypertext tool shows detailed instructions for each step in the procedures. The instructions also contain hypertext links to relevant legal rules and regulations.

2.9 The JIT KS principles line of descent

The object-oriented knowledge modelling approach in KIPS proved very efficient, in terms of letting users benefit from component knowledge precisely when they were dealing with given problems. A weakness was that there was no aid for structuring the knowledge in accordance with the overall work situation, and the just-in-time knowledge support aspect remained only an implicit potential of what was basically an intelligent configuration tool for knowledgeable users. XFRAC TUTOR, likewise, was not in itself capable of supporting an overall work setting with just-in-time knowledge. It did, however, demonstrate that declarative knowledge representations for automatic reasoning could also be actively brought to the mind of a user during the problem solving process, by giving the system a capacity for meta-reasoning about the work process of the user.

It is with STEAMEX and OPA that we start to see the contours of a comprehensive JIT KS supported work situation based on declarative knowledge representations. The multi-threaded STEAMEX architecture, with its many parallel processes, shows how the background reasoning of the system can be set up to interact with the foreground work of the user. OPA furthermore demonstrates that this approach can sustain an underlying diversity of information sources and tools, presenting these through a unified intelligent front-end. Although both STEAMEX and OPA provide active knowledge support in a broad work context, they still lack the explicit representation of work processes for structuring the work of the user and activating knowledge accordingly.

While also incorporating the features of the earlier systems, HALDRIS' interactive procedure graphs let system and user share an up-to-date perception of the overall work progress in its context, instead of merely co-operating in brief interludes triggered by ad-hoc events. This paves the way for giving the system a more active role in monitoring and reasoning about the needs of the user, and supplying knowledge accordingly. The executable checklists of KRM let the user explore and redesign his task space, further adding to the flexibility required to support the plethora of semi-structured work contexts with knowledge support.

The major JIT KS challenge encountered in the reviewed projects is ensuring that the knowledge support offered is not only correct, but consistent and to the point. Users have limited forgiveness with a system that disturbs work with irrelevant or unwanted interruptions, or that takes charge unduly. Correspondingly, when the user has come to expect a certain level of support, the system is obliged to meet these expectations reliably. It is mandatory to identify a level of knowledge support that is appropriately tuned for the work context, and that can be sustained technically, if a JIT KS application is to be accepted at all in a business critical context.

HELENE was the first project to negotiate this holistic challenge, taking the emerging JIT KS principles into an extensive, mission-critical work process support setting. Because Norwegian public administration authorities had previously seen the successful incorporation of active statutes and regulations in KOMPANS, there was considerable motivation to see the same facilities exploited in a more comprehensive work context. HELENE did this, while also projecting the general JIT KS principles of work process support from the industrial domain into the administrative domain, ultimately giving birth to FrameSolutions as an integrated reuse knowledge architecture product.

3. Realizing JIT KS: Computas FrameSolutions

Following HELENE, a growing number of enterprise JIT KS applications, ranging in cost from less than one million to tens of million euro, have successfully been built and deployed, all based on the common architectural basis and methodology, called FrameSolutions. The architecture has continually been generalized and refined, and made available for the currently leading technology platforms. FrameSolutions achieved notable market acceptance in Norway in 1999, after winning several large tenders by generous margins in terms of both functionality and cost. At the turn of the millennium, Computas FrameSolutions stands out as a family of efficient and flexible tools for building scalable Internet-hosted business solutions, positioned for entry into a broader international market.

3.1 Model-based architecture

FrameSolutions applications are structured around four interacting, executable models: organization, domain, processes and rules. The models are built with knowledge editors, and organized in local knowledge bases managed by agents. This distributed, agent-based architecture ensures that FrameSolutions applications can easily be extended with new knowledge base formats and reasoning engines. It also means that FrameSolutions applications are enabled for cooperating with other agent-based applications "speaking" KQML[2] or FIPA ACL[3].

The *organization model* represents users, roles and organizational structures, along with relevant relationships and attributes, including permissions, competencies and authority. Once a basic structure has been established, end users with the required authority can easily perform daily changes to the organization model.

The *domain model* is a generic object-oriented model of all the "things" and "concepts" that the system is capable of handling. Domain objects, relationships and attributes are typically modelled in UML[4]. Applications may use both object-oriented databases and relational databases. FrameSolutions efficiently maps domain objects to persistent storage, while ensuring transactional data integrity and execution performance in large distributed systems.

The *process model* represents the business processes supported by the system. Processes are modelled from a *use* perspective, as sets of declarative representations of pertinent actions and decisions, arranged in dynamic, context-sensitive, checklists. At runtime, a process engine executes the process definitions. Co-operative work between users is modelled with reference to a built-in workflow engine. Routing takes place implicitly, based on definitions in the organization model and the rule base. Users may also explicitly route work to other users.

---

[2] KQML: Knowledge Query and Manipulation Language
[3] FIPA ACL: Foundation for Intelligent Physical Agents - Agent Communication Language
[4] UML: Unified Modelling Language

The *rule base* represents conditions and constraints between representations in the other models. Rules can be used for both backward and forward chaining, and are expressed using first-order predicate calculus statements, i.e. statements built up using the familiar connectives and, or, and not as well as bounded versions of the quantifiers forall and exists. Rules interact with processes and objects in the domain and organization models by means of an innovative *object expression* integration mechanism, for which several patent applications have been filed.

3.2 Operational modelling approach

Experience from many real-world applications has taught us that useful just-in-time knowledge support is best achieved through an iterative and incremental development and deployment strategy. FrameSolutions provides frameworks, components and knowledge editors that allow project teams to address the *use*, *contents* and *system* dimensions in parallel. With FrameSolutions' flexible and configurable system infrastructure, a running system basis is established early on in the project, allowing developers to focus on building and refining knowledge models, rather than on traditional design and programming.

Most approaches to process modelling take a view on modelling as a vehicle for understanding and analysis, with less focus on operational requirements. Analysis models typically have a graphical representation, depicting objects, relationships and flows from a bird's-eye view. The models are refined using top-down decomposition. In our experience, such models become complex as they approach the operational level, when all sorts of exceptions and oddities must be accounted for. Business Process Reengineering models and Workflow models are similar in this respect.

In contrast, operational process models, as required for JIT KS, are built bottom-up. The focus is on one task at a time, from the point of view of an individual performing specific operations at a specific point in time. The goal of modelling, in this context, is to build executable models that are able to represent the required business objects and business logic, for supporting users in their work. All actions and decisions to be supported by the system are modelled, in terms of preconditions, post-conditions, required information and tools, and relevant knowledge support. The resulting models are engineering constructs, not intended to be depicted graphically or comprehended holistically.

Logical dependencies and constraints in business processes are more easily grasped when modelled as pre- and post-conditions applicable to a specific local context. Many complexities in top-down modelling that arise from viewing deviations as exceptions disappear when modelling is approached in this way.

3.3 Native technology platforms

FrameSolutions is natively available for three technology platforms. These platforms have different strengths and weaknesses, both technologically and in terms of market acceptance. Currently, COM+ and EJB are the leading component model architectures. These component models differ in several important ways, requiring slightly different implementations of FrameSolutions for optimal performance. Native implementations have other advantages as well, including improved runtime performance and reduced learning curve for application developers.

FrameSolutions/SARA is an implementation based on IBM VisualAge for Smalltalk and GemStone object-oriented application and database server. Being a direct continuation of early Smalltalk research on JIT KS, this platform excels in flexible support for prototyping of new ideas and concepts. FrameSolutions/BRIX is a Microsoft COM+ implementation, making full use of Microsoft's component-based architecture. FrameSolutions/BEANS is a vendor-independent Java implementation, available for all compliant Enterprise JavaBeans servers.

All FrameSolutions implementations share common representation formats. Knowledge models and information contents can be exchanged and communicated between any FrameSolutions systems and implementations using XML and appropriately specified DTDs[5], based on industry standards where applicable.

3.4 Diverse application domains

FrameSolutions is applicable in many contexts. In addition to serving as a general application integration environment, it is suited for intelligent eBusiness and intranet solutions, business-to-business process automation, ERP[6] functionality, engineering design, case management, project management, quality management, etc.

Many JIT KS applications have targeted Norwegian public administrations, including several for the judicial sector. BL96 assists more than 6000 police officers in handling administrative work processes related to criminal proceedings. A system for management of court cases in the Supreme Court of Justice was deployed in 1999. Judicial systems under development include JIT KS to immigration officers for handling applications from immigrants and refugees, and a system for the foreign services managing visa applications according to the Schengen convention. A comprehensive application for the Labour Market Administration will soon help 3000 officers improve their services to job seekers and employers.

Other significant JIT KS applications include PAGA, a payroll and human resources product, and DNV NAUTICUS, a world-wide, distributed application covering all technical and administrative processes related to ship classification, for the full lifecycle of ships, from pre-contracting, design, building, and operation through to condemnation. A related design-oriented system, focussing on knowledge based engineering aspects, is under development for house building. Other JIT KS systems are being developed for eBusiness, transport management and banking.

This diversity of applications illustrates the broad applicability of the JIT KS approach and the FrameSolutions architecture.

---

[5] XML DTD: eXtensible Markup Language - Document Type Definition
[6] ERP: Enterprise Resource Planning

4. JIT KS Perspectives

Rather than simply listing definitions, we shall show what the term JIT KS denotes by showing a few scenarios in which the salient features are pointed out. As in JIT KS application design, we shall adopt a *use* frame of reference. How does a JIT KS system differ from other systems, by way of supporting, reshaping and integrating the daily work of the individual user, and how is the sustained business redesign process of the organization supported?

The following accounts represent ideal JIT KS from the user point of view, as we currently see it. In some respects, this goes beyond the systems so far delivered, indicating areas for further research and development.

4.1 JIT KS in daily work

Let us imagine a typical end user of a JIT KS application. Ann is a senior official, responsible for a portfolio of some hundred cases in various stages of processing, which may last from days to months, involving the coordinated efforts of a team.

When logging in, her electronic desktop contains an intelligently sorted overview of her cases, reflecting their relative urgencies in accordance with the business logic of the organization. Ann selects the case that she wishes to work with, and is presented with a graphical image of the internal task structure of the case, broken down into activities. This task structure is generated and dynamically maintained by the system, using its rule bases to reflect necessary, recommended and voluntary aspects of how she can work with the case. Her graphical image is optimized to show at a glance what she has done, which dependencies exist and her further progress options. For any activity that she has already performed, a mouse-click is enough to retrieve the statutory basis, regulation or other knowledge that played a part in completing the activity.

4.2 Guided emergent workflow

Pete, another JIT KS user, is assigned responsibility for a new product development project. As each project is unique, Pete will not find predefined JIT KS processes directly supporting execution of his project in the process library. Instead, he has access to meta-level JIT KS support. Through this support, Pete is able to rapidly instantiate a draft project plan, with all important company practises and trade regulations pertaining to his project included. Pete is happy to find that the generated plan is operational, i.e. it exists as a set of instance-level JIT KS processes, with executable checklists for Pete and his team-mates to follow, ready to be further detailed and modified.

Pete focuses his effort on refining and delegating the tasks that must be initiated right away. Let us assume that he has a hunch for extensive up-front market surveys. He elaborates the definitions of these tasks, taking care to add some of his own knowledge and experience in this area, to ensure that his team-mates carry out the work to his standards. He leaves refinement of other tasks to later, as he knows that new external conditions require running revisions to the plan.

Throughout the project, the project team continually creates new knowledge. JIT KS contributes to organizational learning, by inviting team members to extract and contribute lessons learned as part of their normal work, with minimum effort required from the team members. The JIT KS system also ensures that contributed knowledge is approved by authorized personnel before being made generally available. This way, knowledge in the system is kept live, current and trustworthy.

Support for dynamic and ad-hoc work has been extensively researched in the AIS project [14], led by SINTEF. Their *Emergent Workflow* approach addresses unstructured and partly structured representations of work, and allows dynamic modification of workflows during performance. Concepts from AIS and Frame-Solutions were recently brought together in a Master Thesis, as *Guided Emergent Workflow* [15]. This work further explores the use of JIT KS in meta-processes, as an aid to users in defining and modifying instance-level work processes.

4.3 JIT training

While watching the future-situated action movie the Matrix, Tom, being responsible for his company's training department, takes notice of a scene at the end, which eloquently demonstrates the ultimate solution to just-in-time skills training: Trinity urgently needs to fly a helicopter to get out of a dangerous situation. The problem is, she has never flown a helicopter before. "Luckily", being connected to the Matrix, she is able to get a helicopter training module downloaded for direct implantation in her brain. A few seconds later, her brain has acquired the required skills, and she is ready to take off.

Direct brain implantation of skills is of course out of reach, at least for the time being. What caught Tom's attention was the interesting analogy between Trinity's way of acquiring new skills, and the just-in-time training capabilities of his company's new JIT KS system. After installation of the system, classroom training is rapidly being replaced by just-in-time training sessions, embedded in the JIT KS application. Users invoke training sessions on an as-needed basis, either when they are approaching new subject matters, or when they need skills to carry out tasks faster than what can be achieved through real-time JIT KS learning.

Tom's pedagogical competence has found an exciting new outlet. He now spends most of his time designing training modules, exploring novel ways of combining guided walkthroughs with supervised tutoring and simulations, based on the live knowledge models in the underlying JIT KS system.

Many of the early JIT KS systems referred to in this paper combine JIT KS with simulation and tutoring capabilities. As was pointed out with XFRAC, a model-based approach allows the tutoring level to meta-reason about the reasoning progress in the operational system, thus allowing for a far richer tutoring and training context than can be achieved with canned presentations.

4.4 Ubiquitous JIT KS

Our final JIT KS user, Susan, is heading her company's IT Department.

We find ourselves in a situation where advances in broadband wireless communications and wearable computing devices have once again turned the world's existing software base into legacy applications, almost overnight. Fortunately, when the previous component-based revolution occurred, Susan was foresighted, and decided to move all critical business applications to a model-based FrameSolutions platform. When she learns about the release of FrameSolutions/NEW, with native support for all the bells and whistles of the new technology platform, Susan decides to lift the entire system infrastructure of her company to the new technology platform. With system aspects clearly separated from contents and use, she is confident that the conversion risks and upgrade costs are under control.

FrameSolutions/NEW comes with meta-process support, which assists users in upgrading existing applications to the NEW platform, while taking advantage of new features and opportunities. Susan uses the JIT KS support to establish an operational plan for her team to smoothly carry out the conversion process.

A few weeks later, all *contents* in the old systems have been transferred to the NEW platform. No significant glitches occurred. All previously accumulated knowledge and business logic is available, automatically adapted to fit with the new interaction paradigms. The first groups of employees have already taken advantage of the new "invisible" wearable computing clients, with the latest in speech and animation technology, providing JIT KS anywhere and anytime.

5. Conclusions

Building complex software will never become easy. There is no panacea on the horizon. The shift in focus from information systems to knowledge systems will only make development more demanding.

Having the right tools and conceptual frameworks can make a big difference, though. Simply by changing the perspective from cartesian to polar coordinates, appearantly "intractable" mathematical problems become straight-forward calculations. In many respects, JIT KS and FrameSolutions represent a similar change in perspective. The model-based foundation lets developers transform complex requirements into operational behaviour in a natural way, leaving much of the work of resolving complexities to the built-in reasoning engines.

Acknowledgements and trademark notice

This paper focuses on research conducted in Computas. Though we are aware of much related research, comparing and contrasting our approach with research elsewhere is outside the scope of the paper. Apart from the references included, other research has only contributed indirectly to the development of JIT KS.

The authors wish to acknowledge the contributions of the Computas staff and research partners that have played parts in realizing JIT KS. Special thanks go to Roar A. Fjellheim and Thomas B. Pettersen for short-notice proof-reading and valuable feedback.

Computas is a registered trademark and FrameSolutions, FrameSolutions/ SARA, FrameSolutions/BRIX and FrameSolutions/BEANS are trademarks of Computas. All other brand and product names referred to are trademarks or registered trademarks of their respective holders.

References

[1] Dretske, F. I., 1981, *Knowledge and the Flow of Information*, Blackwell, Oxford.
[2] Nonaka I and H. Takeuchi, 1995, *The Knowledge-Creating Company*, Oxford University Press, New York.
[3] Edvinsson L. and M. S. Malone, 1997, *Intellectual Capital: Realizing Your Company's True Value by Finding Its Hidden Brainpower*, Harperbusiness.
[4] Fjellheim, R. A., 1985, *KIPS - Knowledge Based Interface to Process Simulation*, AI Simulation Conference Proceedings, Ghent.
[5] Bobrow D.G. and M. Stefik, 1993, *The LOOPS Manual*, Xerox Corporation.
[6] Fjellheim, R. A., G. J. Coll and B. Johansson, 1988, *Modularity and User Initiative in an Expert System for Fracture Analysis*, in Pham, D.T. (editor), "Artificial Intelligence in Industry: Expert Systems in Engineering", IFS Publications, Springer-Verlag, Berlin.
[7] Brown, J. S., R. R. Burton and A. G. Bell, 1974, *SOPHIE: A sophisticated instructional environment for environment for electronic troubleshooting (an example of AI in CAI) Final Report*, Report 2790 (AI Report 12). Bolt, Beranek and Newman.
[8] Dehli, E., T. B. Pettersen and O-W Rahlff, 1989, *STEAMEX: A Real-Time Expert System for Energy Optimization*, AIENG-89, Applications of Artificial Intelligence in Engineering, Cambridge, UK.
[9] Nordgard H., O. Gjerde, L. Holten, A. Mæland, E. Dehli and P. Støa, 1988, *OPA - Knowledge Based Systems for Power System Operation*, 1988 CIGRE-session, Paris
[10] Fjellheim, R. A., 1992, *HALDRIS - Process monitoring and operator procedure support for improved efficiency and regularity of aluminium production*, FAIST '92, Fielded Applications of Intelligent Software Technologies, Toulouse.
[11] Øwre, F., S. Nilsen, 1986, *Use of PROLOG in a Prototype System for Computerization of Procedures*, OECD Halden Project Workshop on Expert Systems in Process Control, Storefjell, Norway.
[12] Heino, P., I. Karvonen, T. Pettersen, R. Wennersten and T. Andersen, 1994, *Monitoring and analysis of hazards using HAZOP-based plant safety model*, in Reliability Engineering and Systems Safety, 44, pp 335-343.
[13] Johansson, S. E., B. H. Kallåk, T. B. Pettersen and J. E. Ressem, 1997, *Expert Workflow – Building Knowledge Based Workflow Systems with Object Technology*, Addendum to Proceedings, OOPSLA '97, Atlanta, Georgia, USA.
[14] Jørgensen, H. D. and S. Carlsen, 1999, *Emergent Workflow: Integrated Planning and Performance of Process Instances*, Workflow Management '99, Münster, Germany.
[15] Mong, E., 2000, *Active Process and Task Support in Knowledge Intensive Firms*, Master Thesis, FIM/IDI, Norwegian University of Science and Technology.

STEP PC: a Generic Tool for Design Knowledge Capture and Reuse

Nicolas PRAT

ESSEC Business School
avenue Bernard Hirsch - BP 105 - 95021 Cergy-Pontoise cédex - FRANCE
prat@essec.fr

Abstract: Knowledge management is currently considered as a key issue within organisations. Knowledge management is crucial in all design activities, since a designer often relies on his or others' experience to solve a new problem. This paper presents a tool for design knowledge management, focused on knowledge capture and reuse. Design knowledge is stored in process traces, which represent the successive steps leading to the solution (as opposed to just the solution). The tool is generic: it is independent both from the design method and from the application domain, thanks to the combined use of meta-modelling and case-based reasoning. We describe the tool and its application to a real Business Process Reengineering case.

1. Introduction

Knowledge management can be defined as the acquisition, storage, diffusion and use of knowledge within organisations. Knowledge management thus involves two fundamental activities, which are frequently interleaved: knowledge capture (capitalisation of experience) and knowledge reuse.

Knowledge management is currently considered as a key issue by researchers and practitioners from various disciplines, among which artificial intelligence, information systems and management science. Knowledge is an asset, which has to be capitalised i.e. formalised and stored for subsequent reuse. The capitalised knowledge should benefit not only the original producer, but also other members of the company, hence the concept of "corporate memory" or "organisational memory".

Knowledge management is of paramount importance in design activities: since design problems are typically complex and ill-structured [1], a designer often solves a new problem based on his or others' experience i.e. by analogy with one or several previously solved design problems .

In this paper, we present a generic tool for design knowledge capture and reuse, called STEP PC (*Système de Traçage et d'Exécution de Processus à Partir de Cas* i.e. system for process tracing and case-based replay). Design knowledge is stored in so-called process traces, which represent the set of successive steps leading to the solution as opposed to the solution itself i.e. the design artefact (product). This reflects our *process* view of design knowledge representation. While the information systems and artificial intelligence communities have traditionally focused on the representation of the design artefact (*product* view), we believe that the representation of the process is even more crucial since the latter incorporates most of the designer's expertise, in particular the design heuristics (alternatives considered, arguments, decisions...).

STEP PC is a generic tool in that it is independent both from the design method used (the method of Business Process Reengineering [2], a particular method of information systems development like OMT [3]...) and from the application domain (organisation, information systems, manufacturing...). Independence from the method is achieved by the use of meta-modelling, while case-based reasoning ensures independence from the domain.

The paper is organised as follows. Sections 2 and 3 set the terminology and background for the rest of the article, by clarifying the concepts of meta-modelling and case-based reasoning. Section 4 presents the tool along with an example session. Section 5 concludes and develops further research work.

2. Meta-modelling

In the domain of information systems, [4] has proposed a standard framework to represent the product at different levels of abstraction. In figure 1, this framework is adapted to represent not only the product (i.e. the result of design), but also the process (the path to the final result [5]).

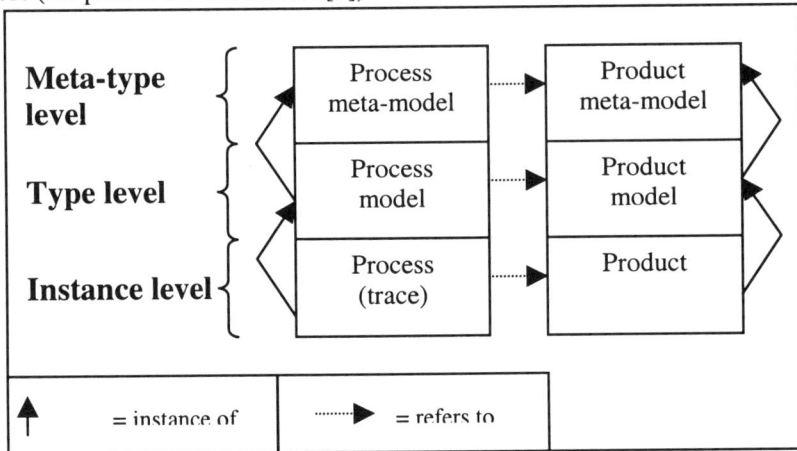

Figure 1: Product and process modelling levels (adapted from [6])

The product and process are represented at three levels of abstraction, each level being an instance of the level above it and a type for the level below:

- The intermediate level is the *type level*. This is the level of design methods. A method, e.g. the information system design method based on the entity-relationship formalism [7], consists in a product model and the associated process model. For example, the entity-relationship process model comprises the action-type "Create the entity-types", which references the concept of entity-type in the product model.

- The lower level is the *instance level*. It consists in the design processes (stored in traces) and in the resulting design products (artefacts). A process is an instance of a process model. For example, the action "Create entity-type Customer" is an instance of the action-type "Create the entity-types" in the entity-relationship process model; the product "Entity-type Customer" is itself an instance of the concept "Entity-type" in the entity-relationship product model.

- The upper level is the *meta-type level*. It comprises the process meta-model and the product meta-model, which shall be instantiated into methods i.e. into process models and product models respectively. For example, the concept of action-type in the process meta-model shall be instantiated into the action-type "Create the entity-types" in the entity-relationship process model.

The framework depicted in figure 1 underlies the repository of such process-oriented software environments or tools as MENTOR [8], PRO-ART [9] and STEP PC. Thanks to meta-modelling, these tools are independent of any particular design method and may support several methods. Typically, a method is defined by instantiating the product and process meta-models (top-down approach [10]) and/or by abstracting from traces (bottom-up approach [11]). Once a method has been defined, it is then "enacted" i.e. instantiated to guide the designer in the creation of a design product. However, design methods and more specifically design process models are often incomplete and/or hard to define a priori. In such cases, a design session can be guided following a case-based reasoning approach, i.e. by *adapting* process traces of similar design problems (instead of *instantiating* a predefined process model).

3. Case-based Reasoning

Case-based reasoning is a form of learning from experience, which consists in solving new problems by reusing the solutions of known, similar problems [12].

Case-based reasoning is a cyclic process, composed of four phases [13]:

1. *REtrieve*: the objective of this phase is the selection of one (or several) case(s) which solve a problem similar to that of the new case (also called the target).
2. *REuse* (adaptation): the target and the retrieved case (source) are combined to reach a solution. The solution of the source is adapted to account for the differences between the target and the source.
3. *REvise*: the purpose of this phase is to make sure that the proposed solution is correct and shall lead to success if it is applied.
4. *REtain*: the new case and its solution are stored into the case base. Thanks to this learning phase, the system acquires new knowledge at each reasoning cycle.

Case-based reasoning has been widely applied to design tasks, for which it is particularly appropriate [1] [14] [15] [16].

4. The STEP PC Tool

STEP PC is a generic tool for design knowledge capture and reuse. It has been implemented with Microsoft ACCESS. The current version totals more than 125 tables and over 150 pages of Visual Basic code.

STEP PC is independent from any design method: it is a meta-tool, whose repository is based on the three level framework depicted in figure 1. STEP PC focuses on a particular type of design methods, namely the methods for which the product model is known a priori but the process model is not known. There exist many examples of such design methods, for instance the method of Business Process Reengineering (BPR): the product model of the BPR method may easily be defined a priori (with such concepts as the business process, the objective or the actor), but it is hazardous to define the BPR process model a priori, due to the uncertainty, complexity and incompleteness of the domain [17].

Since the process model is assumed as unknown a priori, it cannot be used to guide the designer. Instead, STEP PC applies case-based reasoning techniques to guide the designer with process traces of similar, previously solved design problems. Process traces are thus the primary source of knowledge. Therefore, we detail how these traces are represented (trace model), before the static and dynamic presentation of the tool. Finally, an example session is presented. This session corresponds to the BPR application described in the appendix. This application and its process trace (partially represented in the appendix) shall be used for illustrative purposes throughout the paper.

4.1. Representing design knowledge as process traces

Figure 2 represents the trace model, i.e. the formalism used in STEP PC to represent and store process traces. We use the notation described in [18], based on the entity-relationship model [7]. Rectangles represent entities (concepts), which are linked by oriented relationships. Entities may have attributes. Relationships have cardinalities. For example, in figure 2, a Chosen Context is justified by a Justification. The cardinalities show that a Chosen Context is justified by at least 0 and at most 1 Justification, while a Justification justifies at least 1 and may justify several (N) Chosen Contexts. The Justification has an attribute (text). The notation also represents inheritance links, with bolded arrows. The # associated with inheritance links indicates a partition constraint. In figure 2, the combination of inheritance links and # indicates that there exist three and only three distinct types of steps: plan steps, choice steps and executable steps.

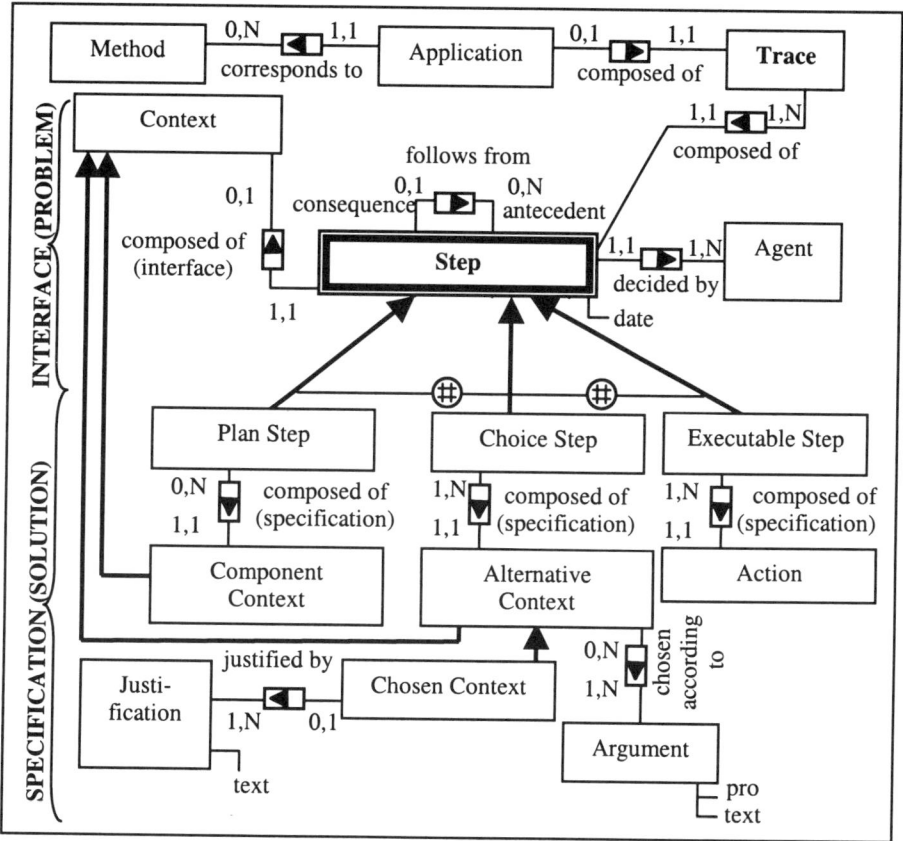

Figure 2: Trace model

The main concepts of the trace model are the concepts of trace and step.

A *trace* stores the execution of a process. There is one trace for each application (e.g. the application of redesign of the hospital described in the appendix). An application corresponds to a method (BPR method...).

A trace is decomposed into *steps*. A step corresponds to a goal which the deciding agent (designer) is trying to achieve, coupled with the situation in which the agent is trying to achieve this goal. A step is itself decomposed into an interface part (the problem i.e. the what) and a specification part (the solution i.e. the how).

The main constituent of the *interface* of a step is its interface context (represented by the "composed of (interface)" relationship in the trace model). The interface context represents the intention (the goal) which the agent is trying to achieve, as well as the situation (the initial state) in which he is trying to achieve this intention. The interface also contains the link with the precedent steps of the trace (relationship "follows from"), the deciding agent and the creation date of the step.

A step is *specified* depending on its type. Following the distinction made by [19], three types of steps are defined: executable steps, choice steps and plan steps.

If the intention of the interface context of a step may be achieved directly by one or several actions on the product (creation/modification/deletion), this step is defined as an *executable step*. The actions constitute the specification of the step.

If the intention of the interface context of a step may be achieved by several alternative intentions, this step is defined as a *choice step*. Every alternative intention, coupled with its situation, forms an alternative context.

In addition to the alternative contexts, the specification of a choice step contains the chosen context, the arguments and the justification.

An argument is associated with an alternative context. The argument opposes or suggests the choice of the alternative context. The "pro" attribute of the argument is true if the argument suggests the choice of the alternative context, otherwise it is false. The text of the argument is expressed in natural language. For example, in step 1 of the trace represented in the appendix, three alternatives are considered to redesign the business processes of the hospital. The first alternative (context CTX002) consists in redesigning the business processes based on the objectives. The advantage of this solution is that it "takes strategic considerations into account".

The choice of a context among the alternatives may be explained by a justification. In step 1, the designer justifies his choice of the first alternative by the fact that "the strategy of the hospital should be guided by objectives".

If the intention of the interface context of a step may be achieved by a set of intentions, this step is defined as a *plan step*. All of these intentions, coupled with their situation, form a component context. The component contexts constitute the specification of the plan step.

Figure 3 complements the trace model by detailing the formalisation of contexts. The concept of context plays a major role in the trace model. The modelling of this concept enables us to represent design processes at a micro level of detail.

A *context* associates an intention of the agent with the situation in which this intention appears. It is represented as a <situation; intention> couple, for example <Enterprise Schema "Enterprise Schema of the hospital"; implement(the chosen solution)Res>.

A *situation* is built on (refers to) one or several product part(s), each product part being an instance of a product part type i.e. an instance of a concept of the product model. For example, in the context above, the situation (Enterprise Schema "Enterprise Schema of the hospital") is built on the product part "Enterprise Schema of the hospital", which is an instance of the concept of Enterprise Schema in the product model of the BPR method.

Following our goal representation formalism [20], an *intention* is composed of a verb and of one or several parameters.

The *verb of* an *intention* is characterised by its text and by the goal verb that it references in the linguistic dictionary.

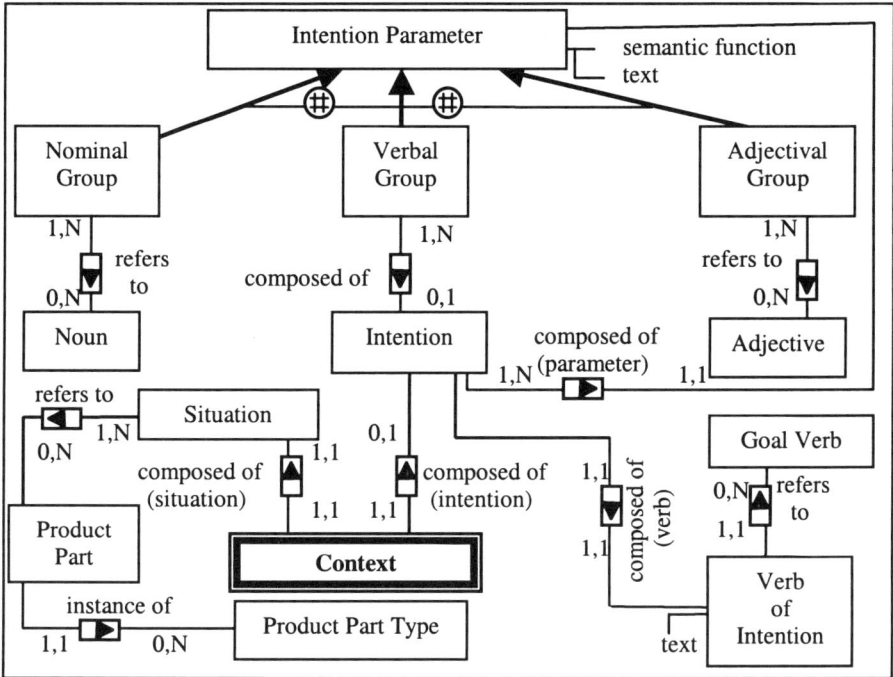

Figure 3: Modelling contexts

The text of a verb of intention is generally a verb in the infinitive form (for example, "redesign" for the intention of the interface context of step 1); it may take another form when the intention is itself the parameter of another intention. For example, in step 1, the manner ("Man") in context CTX003 is the parameter "by analysing(interdependencies between actors)Obj". The text of the verb of this intention is "by analysing".

The verb of an intention *refers to* a goal verb in the linguistic dictionary. For example, "by analysing" refers to the goal verb "to analyse".

An *intention parameter* is characterised by its semantic function and its text.

The *semantic function* [20] indicates the role played by the parameter with respect to the verb of the intention. The possible semantic functions are Object, Result, Source, Direction, Reference, Manner and Means. For example, the Object represents the entity or entities *a*ffected by the intention, whereas the Result represents the entity or entities *e*ffected by the intention.

The *text* of a parameter is the natural language expression of this parameter (for example, "the Business Processes of the hospital" for the Object parameter of the interface context of step 1).

Intention parameters are specialised depending on their grammatical nature: nominal group, verbal group or adjectival group.

A *nominal group* refers to one or several nouns of the linguistic dictionary. For example, the Object parameter of the interface context of step 1 refers to the noun "business process", which is a concept of the BPR product model.

A *verbal group* is composed of one or several intentions, represented recursively. For example, in step 1, the <u>Man</u>ner parameter of the context CTX003 is a verbal group, which consists in the intention "by analysing(interdependencies between actors)Obj".

An *adjectival group* refers to one or several adjectives of the linguistic dictionary.

4.2 Static view of STEP PC

The architecture of STEP PC, represented in figure 4, is composed of the data repository and the different modules (tools).

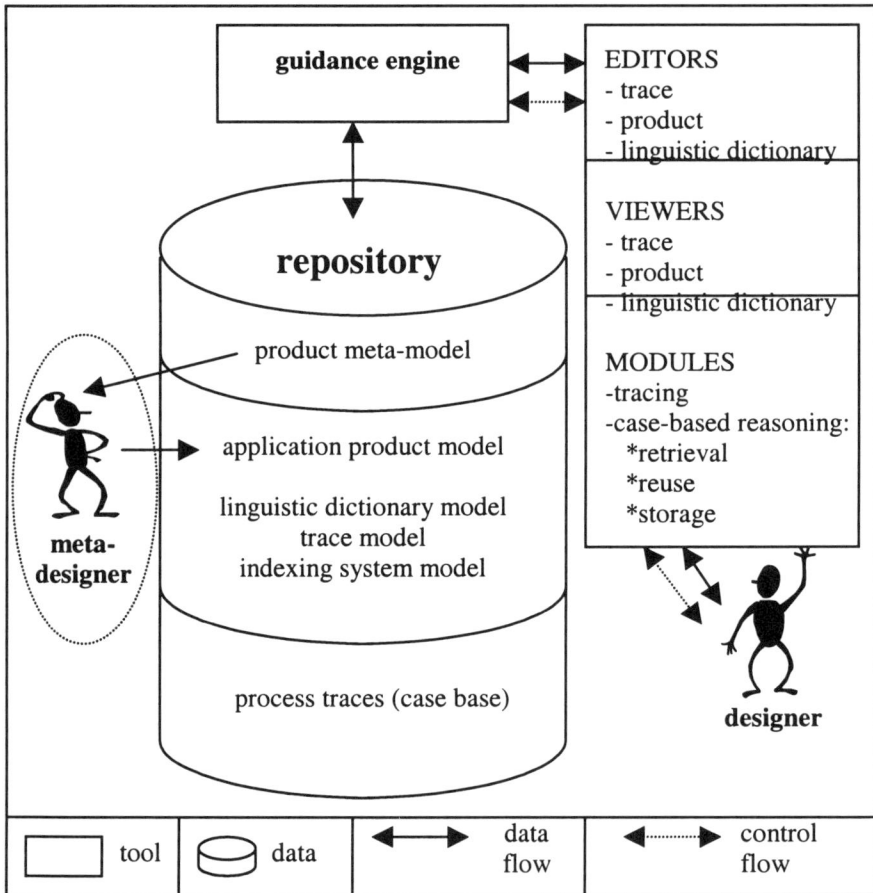

Figure 4: STEP PC: static view

4.2.1 The repository

The repository is organised according to the meta-modelling framework depicted in figure 1 (meta-type level, type level and instance level).

The meta-type level

This level makes STEP PC independent of any particular design method (the BPR method as illustrated in this paper, information systems design methods...).

Remember that STEP PC focuses on methods for which the process model is not predefined i.e. is unknown a priori. Therefore, the *process meta-model,* which is usually used to define process models by instantiation, is not necessary and is not represented in the repository.

Since we have assumed that the design methods supported by STEP PC have a predefined product model, the product model of a particular method needs to be defined before this method may be used (in the BPR case, the BPR product model has been defined before running STEP PC for the application described in the appendix). The product model is defined by the meta-designer by instantiating the *product meta-model.*

The type level

The *application product model* is, in our example, the BPR product model. There is one application product model for every instantiation of the product meta-model by the meta-designer i.e. for every design method supported by STEP PC. Once a product model has been defined, the corresponding actions (creation, modification and deletion actions on product parts) are created automatically. For example, after the definition of the BPR product model by the meta-designer, STEP PC automatically defines the action of creation of a factor, the action of deletion of a business process...

Apart from the application product model, the repository comprises three other models at the type level: the linguistic dictionary model, the trace model and the indexing system model. These models also result from the instantiation of the product meta-model but they have been defined once and for all and are independent of the application domain.

The *linguistic dictionary model* is the structure for storing linguistic knowledge. It represents nouns, verbs and adjectives, along with their synonymy relationships and the hierarchy of nouns and verbs. All this information is used by the case-based reasoning module for similarity computation. The linguistic dictionary is enriched by adding new nouns, verbs and adjectives as the tool is used.

The *trace model* is the model depicted in section 4.1. The trace model is generic: it is independent of any particular design method.

The *indexing system model* describes how traces are organised in memory. Traces are not stored sequentially but in a discrimination network [16]. This discrimination network plays a key role in the case-based reasoning process during the retrieval phase. The indexing system model is detailed in [21].

The instance level

This is the level of process traces, which constitute the case base. Process traces are the primary source of design knowledge.

4.2.2 The modules

The guidance engine

The guidance engine is a central module. It is the execution mechanism that enables STEP PC to function. Its plays the same role as an inference engine in an expert system.

[22] distinguishes two types of guidance: *flow* guidance and *step* guidance. Flow guidance helps to choose the next interface context i.e. the next step to specify. Once the step has been chosen, step guidance helps the designer to specify this step i.e. to determine if it is a plan, choice or executable step and to define its component contexts (for a plan step), its alternative contexts with their arguments as well as the chosen context and its justification (for a choice step), or its actions (for an executable step).

In STEP PC, flow guidance uses no heuristics and is based on an elementary strategy. We focus on step guidance, which is provided by the case-based reasoning module: a new step may be specified by reusing the specification of a previous, similar step.

The tracing and case-based reasoning modules

The *tracing module* helps the designer to define the process at hand (a process is defined by defining its trace). A trace is built by defining its steps.

The *case-based reasoning module* is seamlessly integrated with the tracing module. The retrieval and storage phases are automated. In the current version of the tool, adaptation is not automated and is performed by the designer.

The editors and viewers

The *editors* are used to define or modify the traces, the product and the linguistic dictionary. The edition of the product consists in the execution of actions following the definition of executable contexts. Actions are what makes the design product evolve. Following this evolution of the product, new situations can be recognised, and therefore new contexts and steps can be defined.

Similarly to the editors, the *viewers* are used to view the traces, the product and the linguistic dictionary. The excerpt of trace represented in the appendix was obtained with the trace viewer.

4.3. Dynamic view of STEP PC

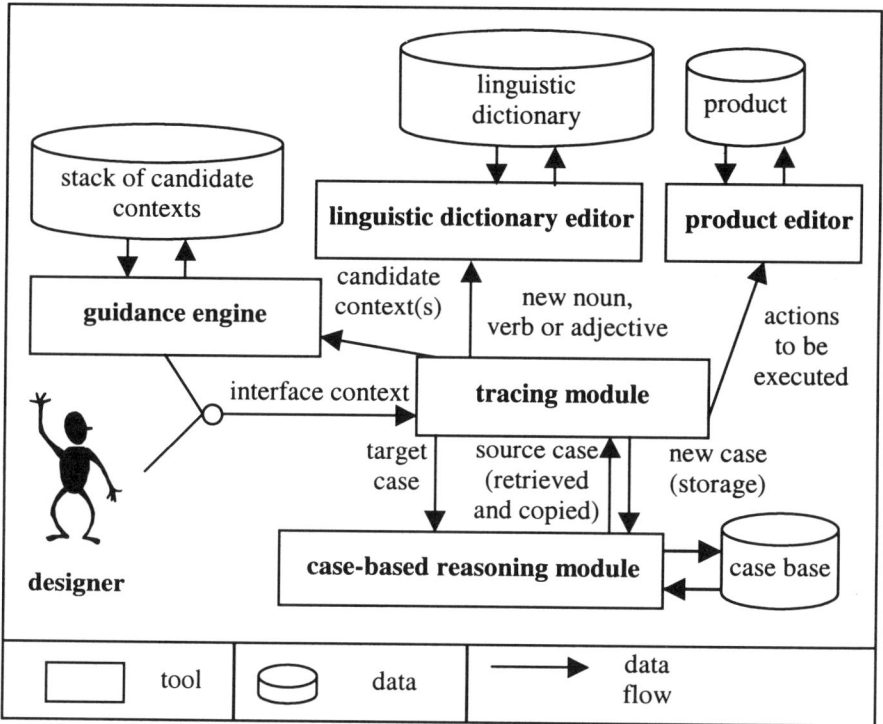

Figure 5: STEP PC: dynamic view

Figure 5 above shows how STEP PC works, by pointing out the key modules and the data flows.

A STEP PC session is a series of successive definitions of trace steps. A step is defined by defining its interface and its specification.

4.3.1 Defining the interface of a step

This definition primarily consists in defining the interface context (i.e. the goal which the designer pursues together with the situation in which he pursues this goal).

An interface context may be obtained in two alternative ways, which correspond to the two options of the Progression menu in STEP PC:

- The first option consists in choosing a *candidate context* (a context which results from the specification of a previous plan or choice step of the trace). Candidate contexts are managed in a stack. The tool removes the upper context in the stack and suggests it as default candidate context; this context forms the interface context of the new step.

- The second option consists in defining a *new* interface *context*, by specifying its situation and its intention. This possibility adds flexibility to flow guidance. For example, the designer may enter a context that does not follow from previous steps of the trace.

Apart from the interface context, the other elements of the interface of a step (agent, date of creation of the step, link with the previous steps) are defined.

4.3.2 Defining the specification of a step

The case-based reasoning module plays a key role at this stage. In STEP PC, *a case is a trace step*.

Once the interface of a new step has been defined, the case-based reasoning module uses this interface (target case) to retrieve a similar step (source case) from the case base. If STEP PC finds a similar step and the designer accepts it, the specification of this step is copied and proposed as default specification for the new step (the tool assumes that the new step is of the same type – plan, choice or executable – as the retrieved step). When there is no sufficiently similar step in the case base, the designer specifies the step from scratch.

During the definition of a step, when the designer specifies an intention parameter, the tracing module calls the linguistic dictionary editor to add any referenced noun, adjective of verb which is not already in the dictionary.

The end of the specification of a step triggers the following operations:

1. *The actions are executed* (in case the step is an executable step). The tracing module delegates this task to the product editor. The actions are executed sequentially.

2. *The trace is updated* by the tracing module, by adding the new step.

3. Simultaneously to the trace update, the case-based reasoning module indexes the new step, by *updating the indexing system* (discrimination network). The new step is thus made available for immediate reuse. By considering that a case is not a trace (like in PRODIGY/ANALOGY [23]) but is a fine-grained element of a trace (the step), we maximise the reuse potential of traces. It is even possible, within a trace, to reuse previous steps of this trace (internal analogy [24]).

4. *The stack of candidate contexts is updated* depending on the type of the new step. If it is an executable step, nothing is changed in the stack. If it is a plan step, the component contexts are stacked according to their order in the plan step. If the step is a choice step, the chosen context is stacked.

4.4. Example session

To illustrate how STEP PC works, we consider the example of the BPR application described in the appendix.

In the beginning, the case base is empty: the BPR product model has been defined by the meta-designer, but no BPR application has been defined yet.

Since the stack of candidate contexts is also empty, the designer chooses to define step 1 by defining himself the interface context of this step. This interface context is defined using the screen represented in figure 6.

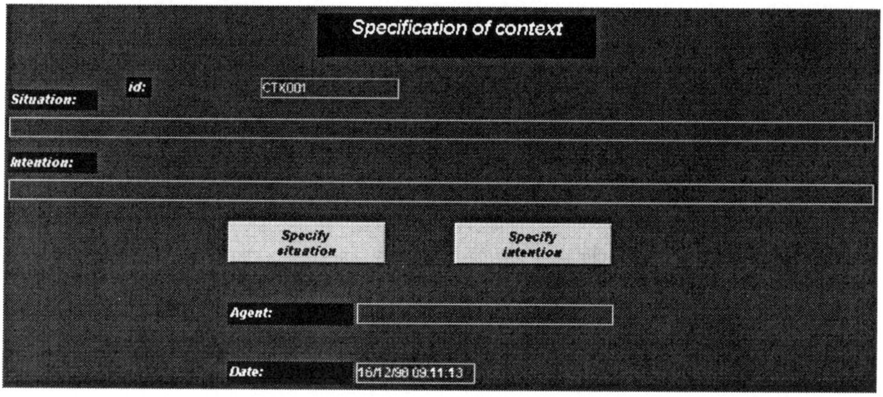

Figure 6: Defining the interface of a step

The designer specifies the situation and the intention of the interface context and enters the agent (i.e. himself and, if appropriate, other agents taking part in the design process).

Once the interface of step 1 has been defined, the case-based reasoning module uses this interface to search for a similar step in the case base. The search fails since no trace has been defined yet for the BPR method. Therefore, the designer specifies step 1 from scratch, by specifying the type of the step (choice step) and the alternative contexts, the arguments, the chosen context and the justification. Even though step 1 is defined from scratch, STEP PC guides the designer's decision process. In particular, the tool suggests him to consider and specify the alternatives and associated arguments before he makes a choice.

Once step 1 has been specified, it is immediately added to memory by the tracing and case-based reasoning modules. The guidance engine updates the stack of candidate contexts by adding the chosen context (the context <Initial Situation INI001; redesign(the Business Processes of the hospital)Obj(based on the Objectives)So>).

To define the interface of step 2, the designer asks STEP PC for a candidate context in the stack of candidate contexts. The guidance engine returns the context <Initial Situation INI001; redesign(the Business Processes of the hospital)Obj(based on the Objectives)So> which is the only context in the stack. This context forms the interface context of step 2. The other elements of the interface of step 2 are defined with default values. In particular, the agent is assumed to be the same as in the previous step.

After completion of the interface of step 2, STEP PC searches for a similar step in the case base i.e. a step with an interface similar to that of step 2. The retrieval algorithm, described in [21], is completely automated. It consists in a top-down search in the discrimination network. Here, the system retrieves the previously defined step (step 1) which is not surprising since the interface contexts of steps 1 and 2 (<Initial Situation INI001; redesign(the Business Processes of the hospital)Obj> and <Initial Situation INI001; redesign(the Business Processes of the hospital)Obj(based on the Objectives)So> respectively) differ only in the Source parameter. However, the designer refuses to reuse step 1 because STEP PC indicates that this step is a choice step and, although step 2 is not specified yet, the designer already knows that it is going to be a plan step.

The process continues, alternating the specification of steps from scratch and by reuse. In step 6, STEP PC suggests the designer to reuse step 5 after determining that this is the most similar step to step 6. The designer accepts to reuse step 5 and obtains the screen shown in figure 7: the system proposes the specification of step 5 (an executable step) as a default specification of step 6. By clicking on "Show interface context of source", the designer displays the interface context of the source case (here the context CTX008). In the proposed default specification, the designer only needs to change the first parameter of the first action (in the second action, the second parameter designates the result of the first action and shall be replaced at runtime by the identifier of the weakness generated as a result of this creation action).

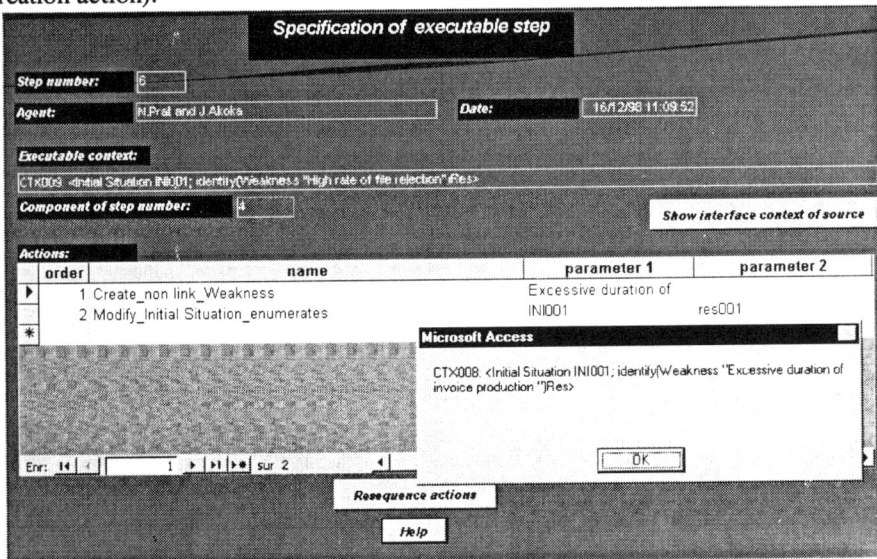

Figure 7: Reusing an executable step

On the whole, the use of STEP PC to solve the BPR case of the hospital has shown a high reuse rate: over the 60 steps of the trace, more than 35 steps have been specified by reusing previously specified steps.

5. Conclusion

In this paper, we have presented and illustrated a tool for design knowledge capture and reuse. Design knowledge is stored in so-called process traces, reflecting our *process* view of design: instead of reusing design solutions, we reuse design problem solving episodes, stored in process traces. The process incorporates much more knowledge than the product considered independently. In particular, it incorporates the design heuristics (alternatives considered, arguments, decisions...). Therefore, efficient design guidance may only be achieved by adopting a process view.

The trace model used to formalise and store process traces enables us to represent design knowledge at a micro level of detail.

The main originality of our work with respect to other approaches lies in the combined use of meta-modelling and case-based reasoning techniques. Together, they ensure independence both from the design method and the application domain.

Apart from BPR, STEP PC has also been applied successfully to a scenario-based information systems design method, in the context of the CREWS (Co-operative Requirements Engineering With Scenarios) ESPRIT project. Further work includes application to other design methods as well as partial automation of the adaptation phase in the case-based reasoning module.

Acknowledgement

The author wishes to thank the Research Centre of ESSEC (CERESSEC) for its support, as well as Professor Jacky AKOKA for the BPR case developed in this article.

References

[1] Hinrichs T.R., 1991, *"Problem solving in open worlds: a case study in design"*, PhD thesis, Georgia Institute of Technology, Atlanta, Georgia, September 1991
[2] Hammer M., Champy J., 1993, *"Reengineering the corporation"*, Harper Business, New York
[3] Rumbaugh J., Blaha M., Premerlani W. et al, 1991, *"Object-Oriented Modelling and Design"*, Prentice Hall International
[4] ISO-IEC 10027, 1990, *"Information Technology - Information resource Dictionary System (IRDS) Framework"*, ISO-IEC International Standard
[5] Olle T.W., Hagelstein J., McDonald I. et al, 1991, *"Information Systems Methodologies: a Framework for Understanding"*, Addison Wesley
[6] Rolland C., Nurcan S., Grosz G., 1999, *"Enterprise knowledge development: the process view"*, Information & Management, volume 36, number 3
[7] Chen P.P., 1976, *"The Entity-Relationship Model – toward a Unified View of Data"*, ACM Transactions on Database Systems, volume 1, number 1, March 1976
[8] Si-Said S., Ben Achour C., 1995, *"A tool for guiding the requirements engineering process"*, 6th Workshop on the Next Generation of CASE tools (NGCT'95), Jyvaskyla, Finland, June 1995
[9] Pohl K., 1995, *"A Process Centred Requirements Engineering Environment"*, PhD thesis, RWTH Aachen, Germany, February 1995

[10] Plihon V., 1996, *"Un environnement pour l'ingénierie des méthodes"*, PhD thesis, University of Panthéon-Sorbonne, January 1996

[11] Prat N., 1996, *" Using machine learning techniques to improve information systems development methods "*, 2nd Americas Conference on Information Systems (AIS '96), Phoenix, Arizona, August 1996

[12] Riesbeck C., Schank R., 1989, *"Inside Case-Based Reasoning"*, Erlbaum, Northvale, NJ

[13] Aamodt A., Plaza E., 1994, *"Case-Based Reasoning: Foundational Issues, Methodological Variations, and System Approaches"*, AI Communications, volume 7, number 1, March 1994

[14] Hennessy D.H., Hinkle D., 1992, *"Applying Case-Based Reasoning to autoclave loading"*, IEEE Expert 7(5), pages 21-26

[15] Smyth B., Cunningham P., 1992, *"Déjà Vu: a hierarchical Case-Based Reasoning System for Software Design"*, 10th European Conference on Artificial Intelligence (ECAI'92), Vienna, Austria, August 1992

[16] Kolodner J., 1993, *" Case-Based Reasoning "*, Morgan Kaufmann

[17] Bareiss E.R., Slator B.M., 1992, *" From Protos to ORCA: reflections on a unified approach to knowledge representation, categorisation, and learning"*, technical report N°20, Institute for the Learning Sciences, Northwestern University, Illinois, January 1992

[18] TEMPORA Esprit project, 1994, final report

[19] Rolland C., Souveyet C., Moreno M., 1995, *"An approach for defining ways of working"*, Information Systems Journal, volume 20, number 4

[20] Prat N., 1997, *"Goal formalisation and classification for requirements engineering"*, 3rd International Workshop on Requirements Engineering: Foundation for Software Quality (REFQS '97), Barcelona, Spain, June 1997

[21] Prat N., 1999, *"Réutilisation de la trace par apprentissage dans un environnement pour l'ingénierie des processus"*, PhD thesis, University of Paris-Dauphine, February 1999

[22] Si-Said S., Rolland C., 1997, *"Guidance for requirements engineering processes"*, 8th International Conference and Workshop on Database and Expert Systems Applications (DEXA)

[23] Veloso M., 1994, *"Planning and learning by analogical reasoning"*, LNAI N°886, Springer Verlag

[24] Hickman A.K., Larkin J.H., 1990, *"Internal analogy: a model of transfer within problems"*, 12th Annual Conference of the Cognitive Science Society, Norhvale, NJ, Erlbaum

[25] Loucopoulos P., Rolland C., Kavakli V. et al, 1997, *"Using the EKD approach. The modelling component"*, ELEKTRA project deliverable, version 1.2, April 1997.

Appendix

This appendix presents the application of STEP PC to the BPR (Business Process Reengineering) method.

The *BPR product model* is predefined (as opposed to the BPR process model). This product model consists in the objectives submodel, the actors and roles submodel and the objects submodel [25]. The concept of enterprise schema is common to these three submodels. The objectives submodel comprises such concepts as the initial situation, the factor and the objective. The initial situation enumerates factors which may be internal (strengths and weaknesses) or external (opportunities and threats). These factors motivate the identification of objectives.

Business processes are then derived from these objectives. The complete BPR product model is described in [21].

The *BPR application* considered here is a real case. The aim of this application was the redesign of the business processes of a large French hospital. The first six steps of the trace are represented below.

Step 1 /* redesign of the business processes of the hospital */
Interface CTX001: <Initial Situation INI001; redesign(the Business Processes of the hospital)Obj>
Agent N.Prat and J.Akoka *Date:*16/12/98 10:12:23 *Type of step:* choice step
Alternative contexts:
- CTX002: <Initial Situation INI001; redesign(the Business Processes of the hospital)Obj(based on the Objectives)So>
    PRO: Takes strategic considerations into account
- CTX003: <Initial Situation INI001; redesign(the Business Processes of the hospital)Obj(by analysing(interdependencies between actors)Obj)Man(and with interviews)Mea>
    PRO: Takes the human factor into account
- CTX004: <Initial Situation INI001; redesign(the Business Processes of the hospital)Obj(with a bottom-up approach)Man>
    PRO: Takes existing processes into account
Chosen context:
- CTX002: <Initial Situation INI001; redesign(the Business Processes of the hospital)Obj(based on the Objectives)So>
    JUSTIFICATION: The strategy of the hospital should be guided by objectives.

Step 2 /* redesign of the business processes of the hospital based on the objectives*/
Interface CTX002: <Initial Situation INI001; redesign(the Business Processes of the hospital)Obj(based on the Objectives)So>
Agent N.Prat and J.Akoka *Date:*16/12/98 10:16:32 *Chosen at:* step 1
Type of step: plan step
Component contexts:
- CTX005: <Initial Situation INI001; build(the Enterprise Schema "Enterprise Schema of the hospital")Res>
- CTX006: <Enterprise Schema "Enterprise Schema of the hospital"; implement(the chosen solution)Res>

Step 3 /* building of the enterprise schema of the hospital */
Interface CTX005: <Initial Situation INI001; build(the Enterprise Schema "Enterprise Schema of the hospital")Res>
Agent N.Prat and J.Akoka *Date:* 16/12/98 10:45:42 *Component of:* step 2
Type of step: plan step
Component contexts:
- CTX007: <Initial Situation INI001; identify(the Factors)Res>

Step 4 /* identification of the factors */

Interface CTX007: <Initial Situation INI001; identify(the Factors)Res>
Agent N.Prat and J.Akoka *Date:* 16/12/98 10:51:43 *Component of:* step 3
Type of step: plan step
Component contexts:
- CTX008: <Initial Situation INI001; identify(Weakness "Excessive duration of invoice production ")Res>

Step 5 /* identification of the weakness "Excessive duration of invoice production " */
Interface CTX008: <Initial Situation INI001; identify(Weakness "Excessive duration of invoice production ")Res>
Agent N.Prat and J.Akoka *Date:* 16/12/98 11:05:39 *Component of:* step 4
Type of step: executable step
Actions: 1. Create_non link_Weakness(Excessive duration of invoice production)
        2. Modify_Initial Situation_enumerates(INI001;res001)

Step 6 /* identification of the weakness "High rate of file rejection" */
Interface CTX009: <Initial Situation INI001; identify(Weakness "High rate of file rejection")Res>
Agent N.Prat and J.Akoka *Date:* 16/12/98 11:09:52 *Component of:* step 4
Type of step: executable step
Actions: 1. Create_non link_Weakness(High rate of file rejection)
        2. Modify_Initial Situation_enumerates(INI001;res001)

WebGrid: Knowledge Elicitation and Modeling on the World Wide Web

Mildred L. G. Shaw and Brian R. Gaines

Knowledge Science Institute
University of Calgary
Alberta, Canada T2N 1N4
{mildred, gaines}@cpsc.ucalgary.ca

Abstract: WebGrid-II is a knowledge acquisition and inference server on the World Wide Web that uses an extended repertory grid system for knowledge acquisition, inductive inference for knowledge modeling, and an integrated expert system shell for inference. This article describes WebGrid-II, and illustrates its use in eliciting and modeling knowledge.

1. Introduction

The development of knowledge-based systems involves knowledge acquisition from a diversity of sources often geographically distributed. The sources include books, papers, manuals, videos of expert performance, transcripts of protocols and interviews, and human and computer interaction with experts. Expert time is usually a scarce resource and experts are often only accessible at different sites, particularly in international projects. Knowledge acquisition methodologies and tools have developed to take account of these issues by using hypermedia systems to manage a large volume of heterogeneous data, interactive graphic interfaces to present knowledge models in a form understandable to experts, rapid prototyping systems to test the models in operation, and model comparison systems to draw attention to anomalous variations between experts.

However, existing knowledge acquisition tools are largely based on personal computers and graphic workstations, and their use in a distributed community involves moving software and often computers from site to site. The process of building and testing knowledge models across a distributed community would be greatly expedited if wide-area networks could be used to coordinate the activities at different sites. The initial objective of the work reported in this paper has been to use the World Wide Web to support distributed knowledge acquisition by porting existing knowledge acquisition tools to operate through the web. A further objective has been to use the web on a continuing basis to support distributed knowledge management in which the acquisition, representation and application of knowledge become an integral part of the activities of an organization.

Repertory grids based on personal construct psychology [1] have been used for knowledge acquisition since the early years of knowledge-based system development [2, 3] and have been refined over the years to support increasingly complex knowledge structures [4-6]. Since grid elicitation tools are used directly by experts it would be very useful to have them accessible through any personal computer or workstation with access to the web. This is feasible because the primary data input format is through rating scales, and this can be done effectively by using popup menus which are available through the HTML graphic user interface, and is the basis of WebGrid [7]. WebGrid-II extends WebGrid with new data types and analysis techniques to support expert system development.

The next section gives an overview of the web architecture and protocols, and following sections illustrate how knowledge acquisition systems have been ported to the web to operate in a distributed client-server environment.

2. WebGrid-II: Repertory Grid Client-Server Implementation

WebGrid was developed in 1994 to offer repertory grid elicitation and analysis through the World Wide Web. The server is based on our RepGrid/KSS0 program [8] which operates on an Apple Macintosh and offers a wide range of elicitation and analysis capabilities for single and multiple repertory grids and has been widely used in expert system development [9-11]. Offering access to RepGrid through the web made much of its functionality available to users throughout the world through standard web browsers operating on any computer.

WebGrid was intended primarily to introduce repertory grid techniques to new users and focused on basic capabilities. However, it rapidly came into widespread use for major research projects and interest grew in having a full range of knowledge acquisition tools available on the web. WebGrid-II extends WebGrid to provide integration with multimedia, additional data types in the grid, entailment analysis to induce rules from grid data, and an expert system inference engine to allow the rules to be tested on new cases as part of the elicitation process.

2.1 Elicitation

Figure 1 shows the initial screen of WebGrid-II. The HTML form requests the usual data required to initiate grid elicitation: user name; domain and context; terms for elements and constructs; default rating scale; data types allowed; and a list of initial elements. It also allows the subsequent screens to be customized with an HTML specification of a header and trailer—this capability to include links to multimedia web data is also used to allow annotation, text and pictures, to be attached to elements.

The problem domain is that of "selecting among graduate program applicants" and the intention of the user is to develop a set of rules defining the criteria for selection. The system will incorporate the text typed in by the user into the later dialog, and the user has stated that the elements in the grid are to be called "students" and the constructs "qualities." The user has set the types of constructs to be "Ratings+Names+Categories+Numbers"—indicating that the user wants to use more advanced facilities for knowledge modeling. She has left the default rating scale to be 1-5 and has entered the initials of 4 applicants typifying the range of selection decisions normally made.

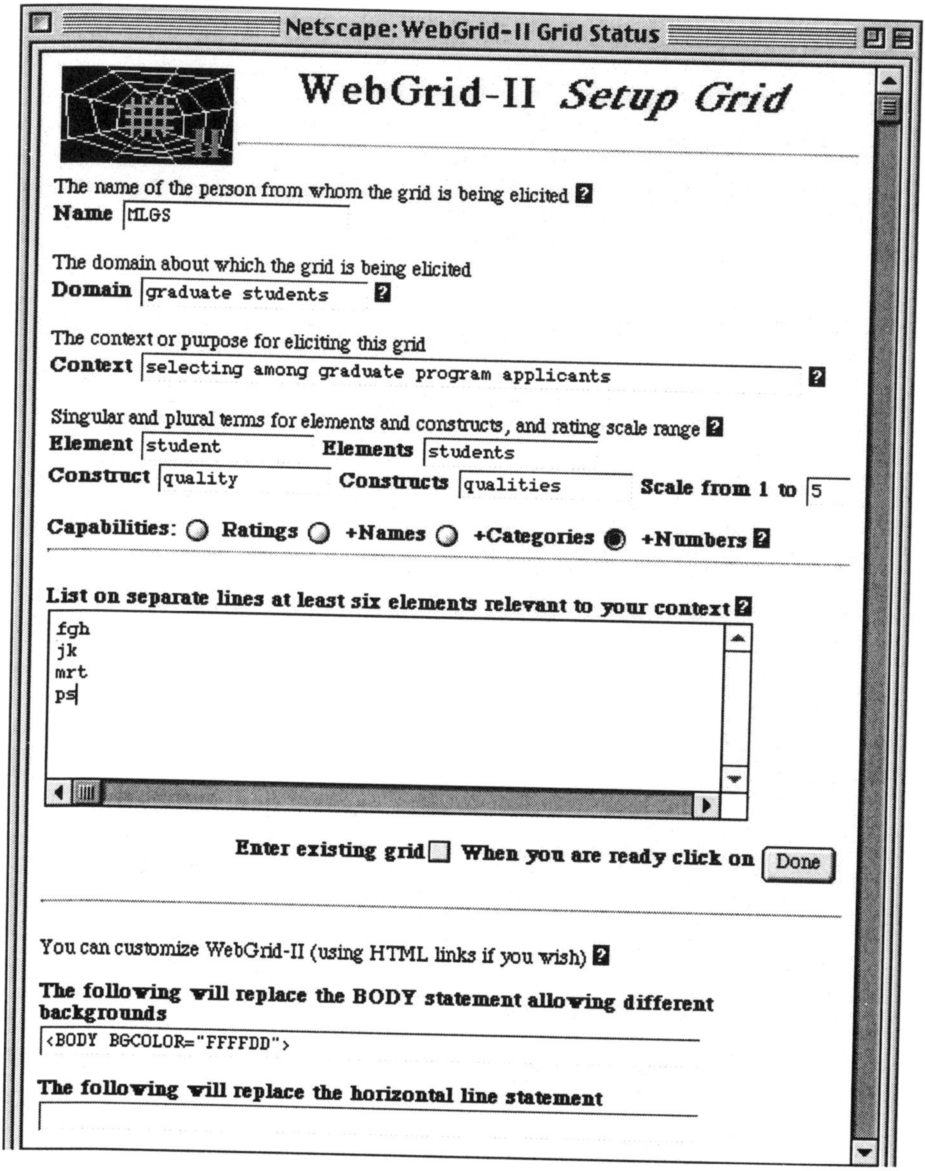

Figure 1: Repertory grid elicitation initial screen

When the user clicks on the "Done" button at the bottom, the browser transmits the data entered to the remote server which passes it through its common gateway interface to a specialist knowledge acquisition auxiliary server. The server processes the data and generates an HTML document that it returns to the browser resulting in the screen shown in Figure 2 eliciting a construct from a triad of elements.

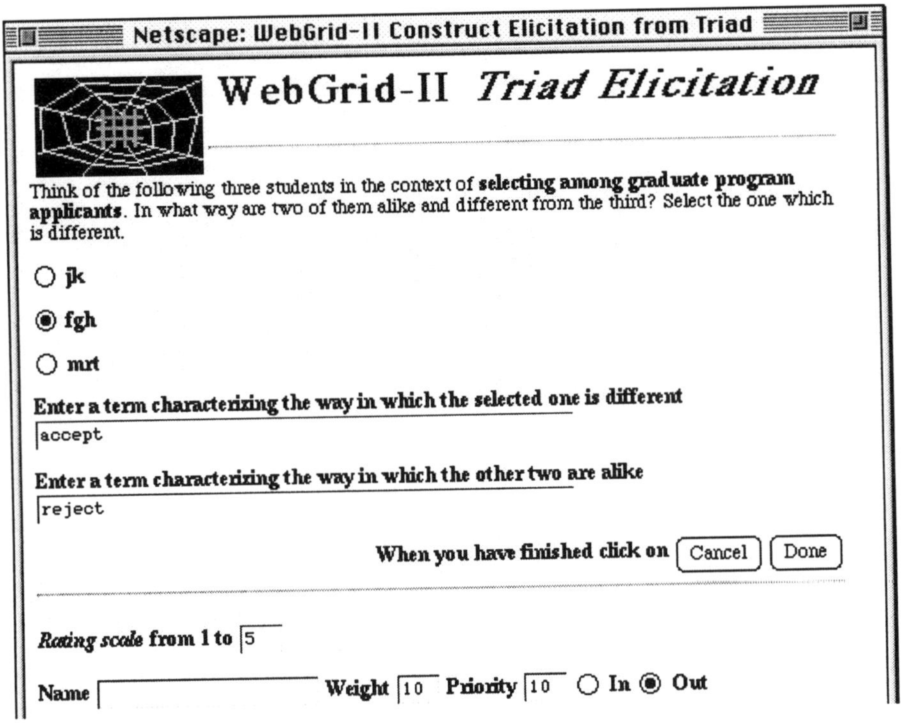

Figure 2: Construct elicitation from a triad

The user clicks on a radio button to select an element which she construes as different from the other two and enters terms characterizing the construct. The lower part of the screen is generated because advanced features were. It allows the user to vary the rating scale for an individual construct, to name the construct, to give it a weight in clustering, a priority in asking questions in an expert system application, and to specify that the construct is an input for decision making purposes or an output to be anticipated. Since the construct entered is the main decision to be made, to accept or reject an applicant, and the user wants the system to anticipate the correct decision from the way the applicant is construed she has made it an output.

When the user clicks on "Done" the server generates the screen shown on the left of Figure 3 which places a popup menu rating scale alongside each element enabling the user to rate each one along the new construct as shown on the right. She is also able to change the terms used if they seem inappropriate in the context of all the elements, and to change the ratings of the already entered elements if appropriate.

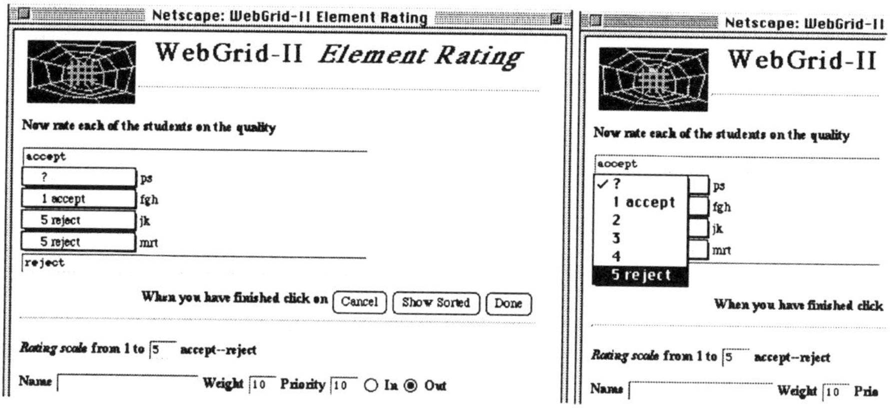

Figure 3: Rating elements on constructs

Clicking on the "Done" button in Figure 3 sends the ratings back to the server which generates the status screen shown in Figure 4.

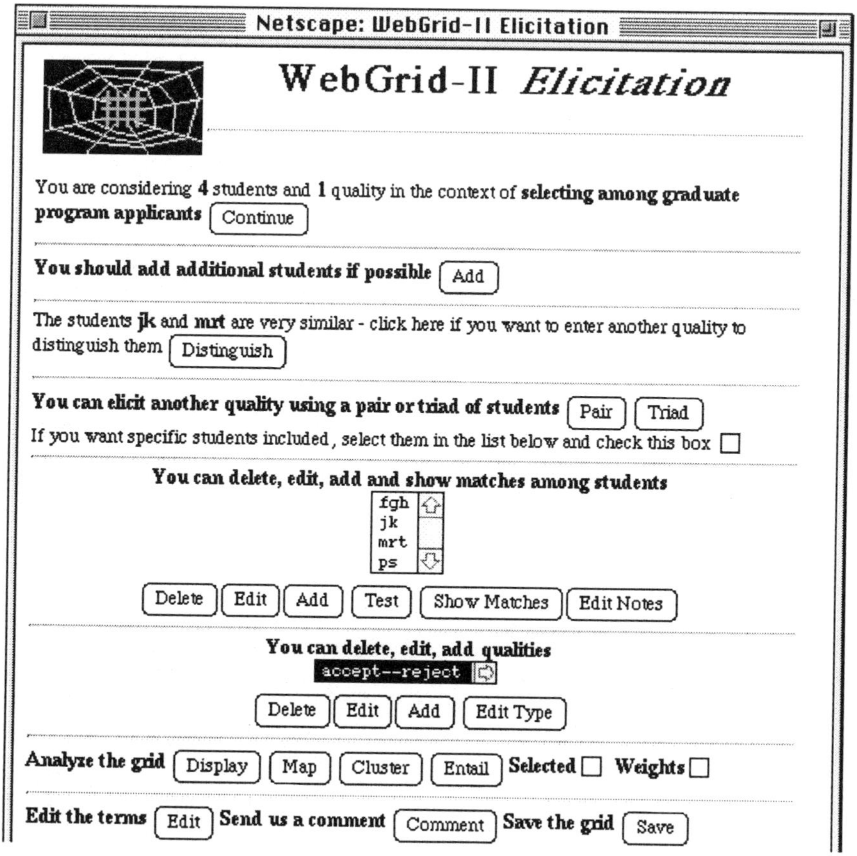

Figure 4: Status screen

This shows the elements and constructs entered, allowing them to be selected for deletion, editing and so on. It also offers various suggestions as how to continue the elicitation based on the data entered so far, facilities for analysis, saving the grid, and so on.

2.2 New data types in grids

The user decides to develop another two constructs from triads of elements and enters "high gpa—inadequate gpa" and "good references—inadequate references" since she knows that these are critical to acceptance. The applicant's gpa is a numeric rating between 0.0 and 4.0 and hence cannot be entered on a normal grid rating scale. It requires one of the new data types that we have used to extend repertory grids for knowledge acquisition.

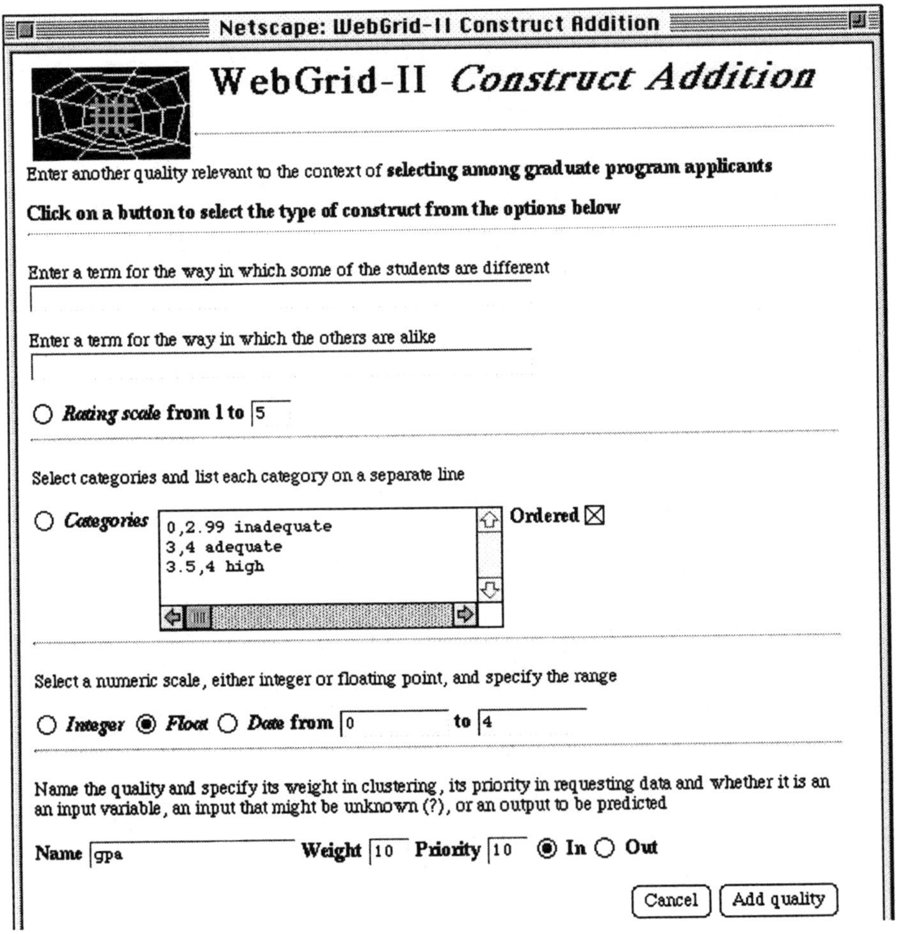

```
Netscape: WebGrid-II Construct Addition

WebGrid-II Construct Addition

Enter another quality relevant to the context of selecting among graduate program applicants

Click on a button to select the type of construct from the options below

Enter a term for the way in which some of the students are different
[                    ]

Enter a term for the way in which the others are alike
[                    ]

○ Rating scale from 1 to [5]

Select categories and list each category on a separate line

○ Categories  [0,2.99 inadequate        ]  Ordered ☒
              [3,4 adequate             ]
              [3.5,4 high               ]

Select a numeric scale, either integer or floating point, and specify the range

○ Integer  ● Float  ○ Date  from [0]        to [4]

Name the quality and specify its weight in clustering, its priority in requesting data and whether it is an
an input variable, an input that might be unknown (?), or an output to be predicted

Name [gpa]        Weight [10]  Priority [10]  ● In  ○ Out

                                        [Cancel] [Add quality]
```

Figure 5: Adding a numeric construct with categories

Figure 5 shows the gpa categories being added as a new construct. WebGrid-II makes provision not only for rating scales but also for a set of categories, ordered or unordered, and for numerical rating scales of the type integer, float or date with prescribed ranges. The user has entered "inadequate", "adequate" and "high" as possible categories for gpa, has entered "gpa" as the name òf the construct, and has noted that the value is a floating point number. The only difference the name entry makes is to shorten the process since WebGrid will generate the names "gpa = inadequate", etc, rather than have the user enter "gpa is inadequate", etc, as category names. The elements are rated on a purely categorical construct through popup menus and on a numeric construct through a numeric entry box.

2.3 Entailment analysis through rule induction

The status screen now shows three constructs and the user clicks on the "Entail" button near the bottom of Figure 4 hoping the data is sufficient to generate a model of the way in which the constructs entered entail the "accept—reject" decision. The system develops a model using the Induct algorithm [12, 13] for inductive derivation of entailments and generates the screen shown in Figure 6.

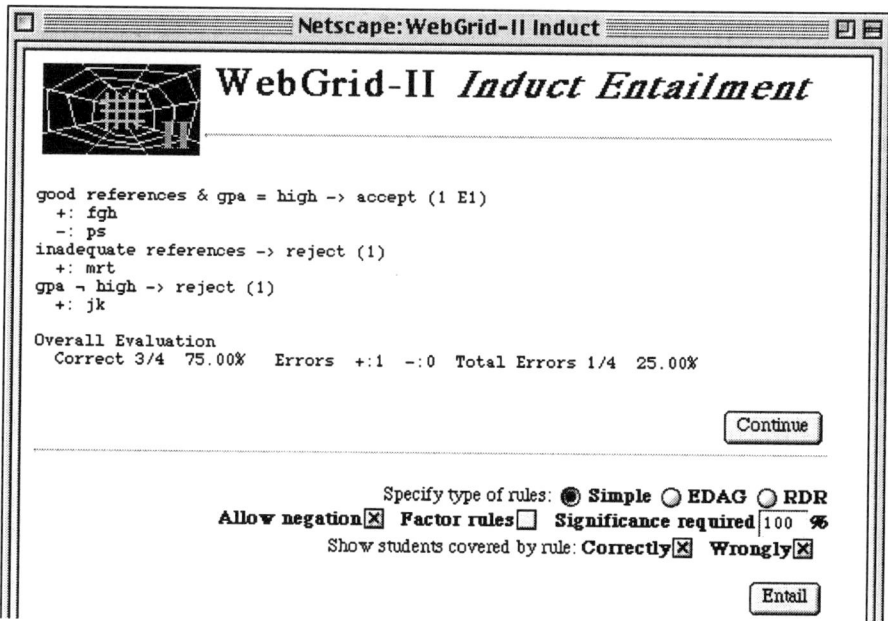

Figure 6: Rules entailed by the grid after 4 elements and 3 constructs

She is disappointed to receive the entailments shown which look sensible but work for only 3 cases in 4. The output consists of a set of rules defining the anticipatory entailments and each rule is followed by the positive elements where the correct decision is anticipated and the negative ones where that anticipated is incorrect. On thinking about it she realizes that the applicant "ps" has a high gpa and good references but was rejected because there was no-one willing to supervise him. She returns to the status screen, adds a construct "willing supervisor—no supervisor" and clicks on "Entail" again. This time the rules entailed by the grid returned as shown in Figure 7 correctly anticipate the decisions made for all 4 cases.

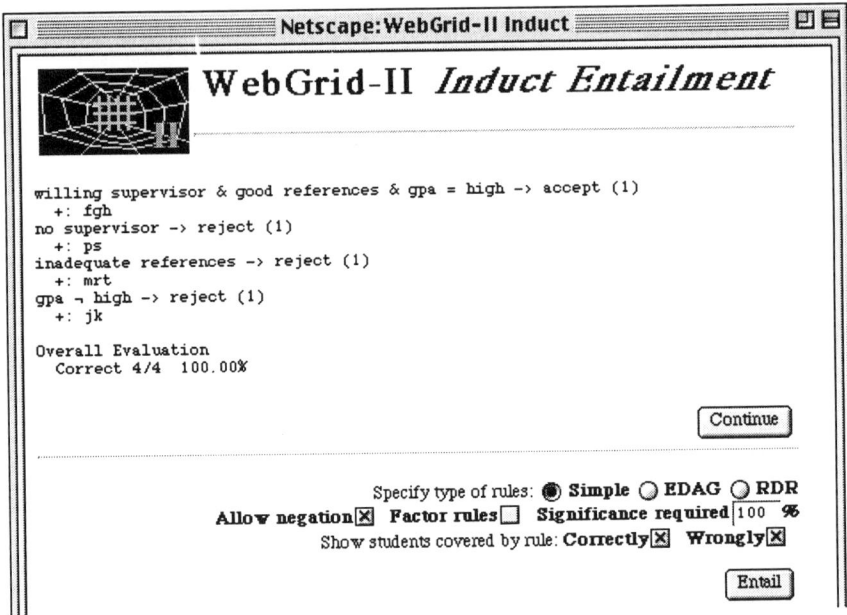

Figure 7: Rules entailed by the grid after 4 elements and 4 constructs

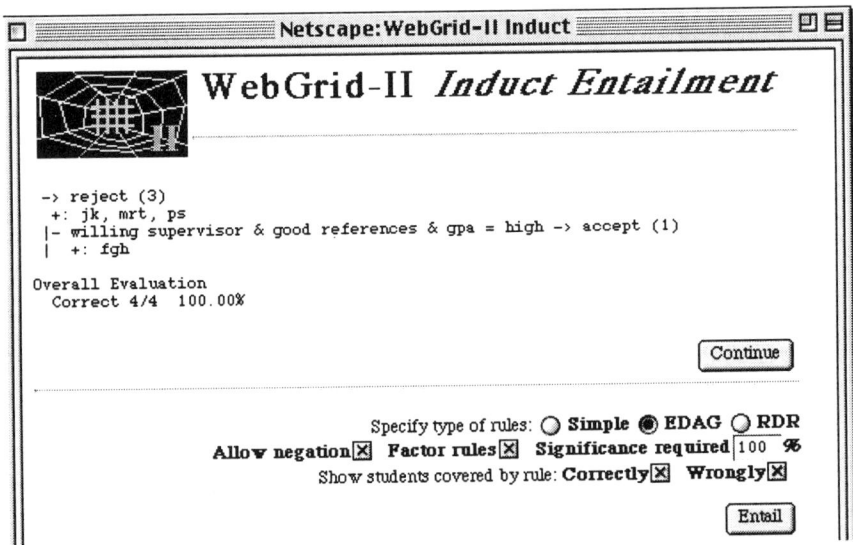

Figure 8: Rules with exceptions entailed by the grid after 4 elements and 4 constructs

The user knows that Induct is also capable of producing default rules with exceptions [14, 15] that are often more natural in their correspondence to psychological models, and runs Entail again with "EDAG" selected in Figure 7. The system generates the screen of Figure 8 which has the default rule to reject all applicants unless they have a high gpa, good references and a willing supervisor— a good initial characterization of the selection process.

2.4 Inference through an embedded expert system shell

The user decides to test the entailed rules on further cases and returns to the status screen of Figure 4 which now shows four constructs. She clicks on the "Test" button under the list of elements and the system returns the screen shown in Figure 9 which allows her to specify a test case and click on "Infer" to see what decision is anticipated.

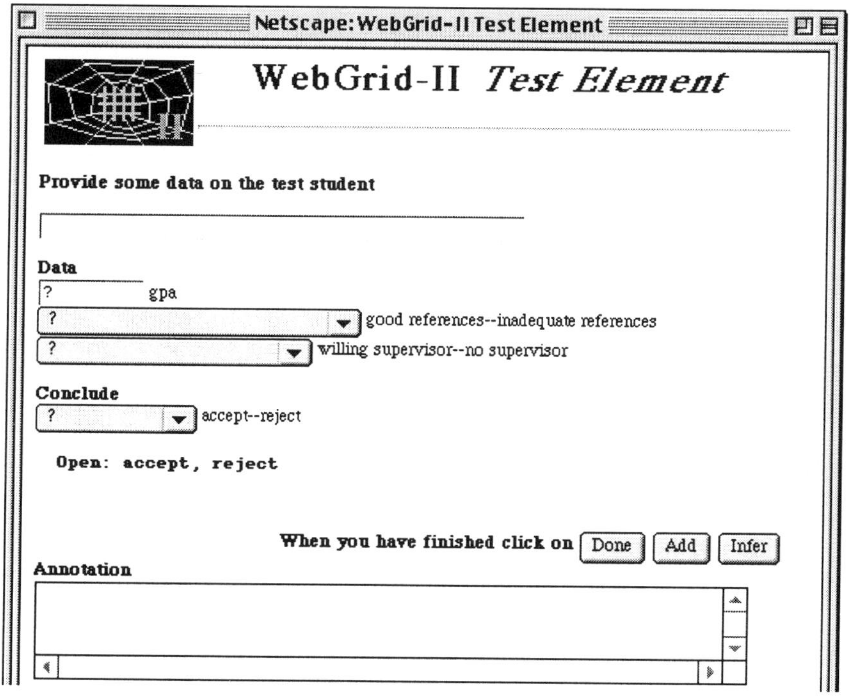

Figure 9: Entering a test case into the entailed expert system

The user does this for several similar cases and the conclusion returned is correct but when she does so for the applicant "lemp" as shown in Figure 10 the conclusion is "reject" when she knows the applicant has been accepted.

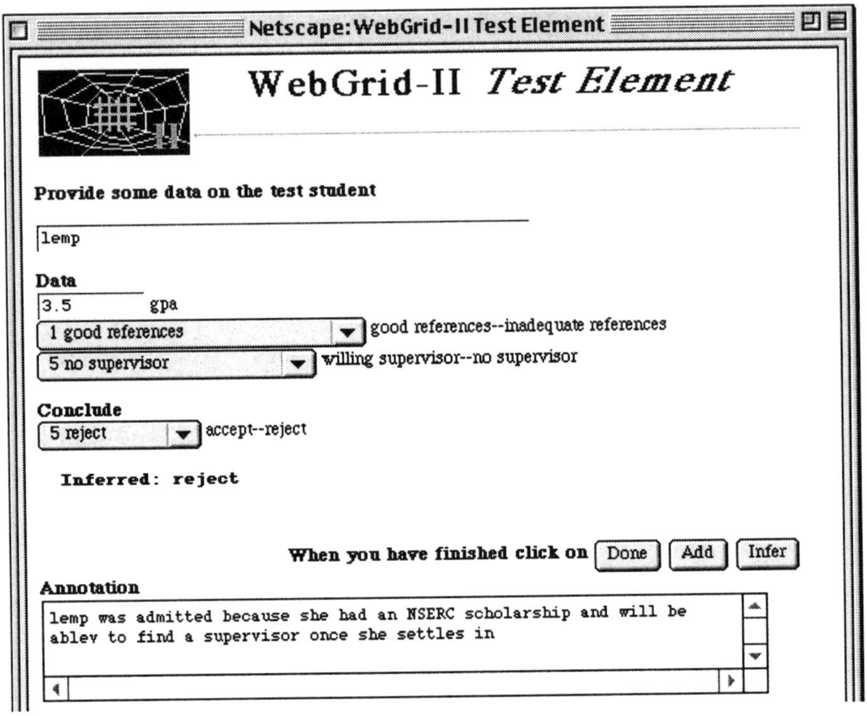

Figure 10: Conclusion anticipating the decision for a test case

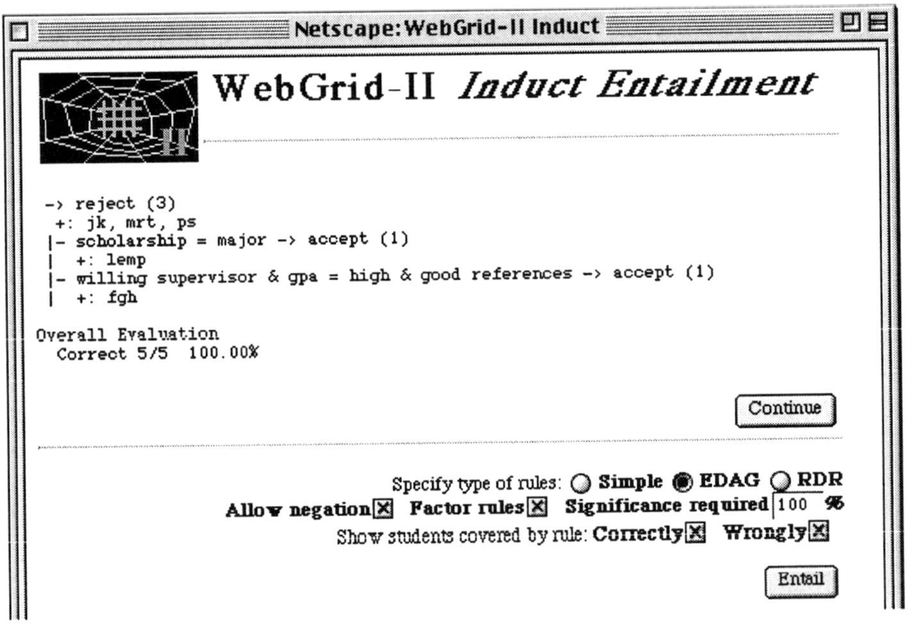

Figure 11: Rules with exceptions entailed by the grid after 5 elements and 5 constructs

The user realizes that applicant "lemp" was admitted because she had a major scholarship, corrects the conclusion to "admit", and clicks on "Add" to add this applicant as a new element in the grid. The system returns to the status screen of Figure 4 with "lemp" added and the user clicks on "Add" under the construct list to enter the categorical "scholarship" is "major" or "none" She then clicks on "Entail" and the system generates the entailed rules with exceptions as shown in Figure 11 which now account for "lemp" also. The default rule is to reject all applicants unless they have a major scholarship or a willing supervisor, high gpa and good references when they should be accepted unless they have no supervisor and no scholarship—a more refined characterization of the selection process. However, she also realizes there are applicants with major scholarships who have not been admitted because they have a poor gpa or inadequate references.

The user enters some relevant examples, returns to test case entry again and this time come across a applicant who was admitted to a new program, an industry-based software engineering degree with the requirement that the applicant have substantial industry background. She enters some relevant cases and constructs leading to the entailed decision rule shown in Figure 12, that applicants should be rejected by default unless they have good references and in addition either have a high gpa but not no scholarship and no supervisor, or are a software engineering application with industry experience and an adequate gpa.

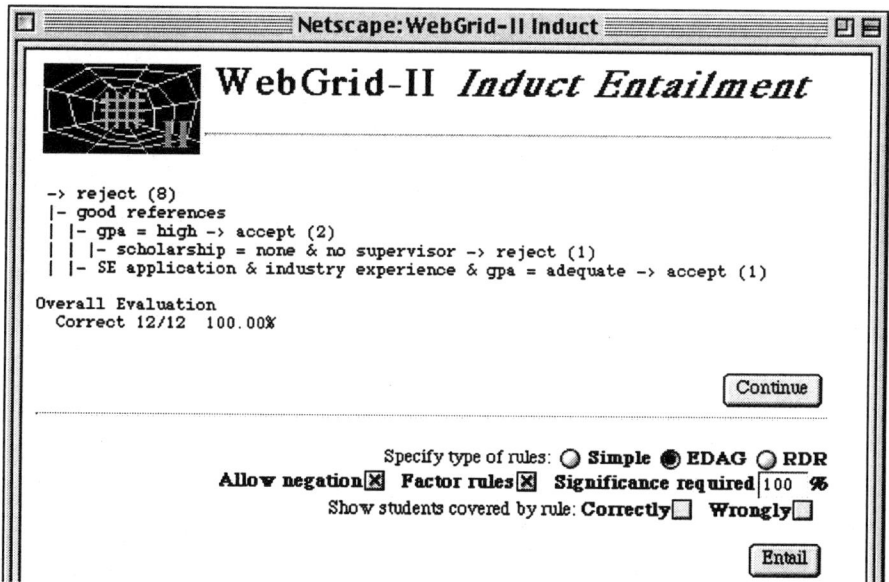

```
-> reject (8)
|- good references
|  |- gpa = high -> accept (2)
|  |  |- scholarship = none & no supervisor -> reject (1)
|  |- SE application & industry experience & gpa = adequate -> accept (1)
Overall Evaluation
   Correct 12/12  100.00%
```

Figure 12: Rules with exceptions entailed by the grid after 12 elements and 7 constructs

Note how the decision rule entailed by the grid is becoming more refined as it encompasses more cases and also how the Induct algorithm generating it changes its structure to be as comprehensible as possible as it grows in complexity. This attempt to generate an anticipatory structure which is natural for people is an important feature of EDAG generation [15]. It is also noteworthy that a fairly small grid of only 12 elements and 7 constructs can be used to generate such a complex decision rule. This is because the repertory grid elicitation technique with PEGASUS-style [16] feedback leads to the elicitation of typical elements and relevant constructs, allowing knowledge to be elicited highly efficiently compared with rule induction from a database where thousands of examples might be required to derive the same knowledge [12].

3. Conclusions

The technologies, tools and protocols developed for the World Wide Web make it possible to provide knowledge acquisition, representation and inference systems internationally to anyone with access to the Internet. Such knowledge-based technologies integrate naturally with other applications and hypermedia systems to provide a distributed knowledge medium capable of supporting the knowledge processes of professional communities.

This paper has illustrated what is possible through examples of the development of knowledge modeling tools operating through the web to provide knowledge acquisition, representation and inference through repertory grids. These systems were developed primarily for locally managed projects, and their availability has only been advertised through specialist list servers for the knowledge acquisition and personal construct psychology communities. However, in the period June 1995 to June 1997 WebGrid was accessed from over 5,000 different sites in some 60 countries, delivering over 200Mbytes of grid data in some 70,000 transactions. The web is essentially an anonymous medium and we do not track the activities of outside users. However, occasionally users contact us to discuss their applications and, if the local network goes down, we receive mail from serious users reporting the problem and requesting notification of when the knowledge acquisition programs will be available again. An interesting example of serendipitous use was a masters student in Holland studying operators' models of nuclear reactors who came across the grid elicitation system on the web in September 1996 used it with his experts to elicit their conceptual models, and included these in an additional chapter in his thesis for examination in October 1996—the web certainly accelerates research processes.

In conclusion, we note that the results presented in this paper demonstrate that the World Wide Web client server model operating through the Internet can effectively support the porting of existing knowledge acquisition, modeling and representation tools to operate in a distributed, open architecture environment. This opens up new possibilities for collaborative research in which the methodologies and tools developed by one group are made available on the network for use by other research groups world-wide. It enables past studies of the integration of hypermedia, knowledge acquisition and inference tools within a local environment [17, 18] to be generalized to integration on a distributed basis across wide area networks. A major advantage of this approach is that tools can be operated at the site of the originators responsible for them, and their maintenance, upgrading and monitoring can be done very much more effectively than when the software is distributed to other sites.

We envision a collaborative research program internationally in which the first objective is to make tools available across the Internet in interoperable forms, and the second objective is to develop some integrated knowledge management systems that use these tools in a variety of configurations to provide a new generation of knowledge-based systems

Acknowledgments

Financial assistance for this work has been made available by the Natural Sciences and Engineering Research Council of Canada.

URLs

WebGrid can be accessed at http://tiger.cpsc.ucalgary.ca/WebGrid/
Related papers on WebGrid and World Wide Web can be accessed through http://www.RepGrid.com/articles

References

1. Kelly, G.A., The Psychology of Personal Constructs. 1955, New York: Norton.

2. Shaw, M.L.G. and B.R. Gaines, A computer aid to knowledge engineering, in Proceedings of British Computer Society Conference on Expert Systems. 1983, British Computer Society: Cambridge. p. 263-271.

3. Boose, J.H., Personal construct theory and the transfer of human expertise, in Proceedings AAAI-84. 1984, American Association for Artificial Intelligence: California. p. 27-33.

4. Boose, J.H., et al., Knowledge acquisition techniques for• group decisions support. Knowledge Acquisition, 1993. 5(4): p. 405-447.

5. Gaines, B.R. and M.L.G. Shaw, Basing knowledge acquisition tools in personal construct psychology. Knowledge Engineering Review, 1993. 8(1): p. 49-85.

6. Ford, K.M., et al., Knowledge acquisition as a constructive modeling activity. International Journal of Intelligent Systems, 1993. 8(1): p. 9-32.

7. Shaw, M.L.G. and B.R. Gaines, Comparing constructions through the web, in Proceedings of CSCL95: Computer Support for Collaborative Learning, J.L. Schnase and E.L. Cunnius, Editors. 1995, Lawrence Erlbaum: Mahwah, New Jersey. p. 300-307.

8. CPCS, RepGrid 2 Manual. 1993: Centre for Person Computer Studies, PO Box 280, Cobble Hill, BC, Canada V0R 1L0.

9. Shaw, M.L.G., B.R. Gaines, and M. Linster, Supporting the knowledge engineering life cycle, in Research and Development in Expert Systems XI, M.A. Bramer and A.L. Macintosh, Editors. 1994, SGES Publications: Oxford. p. 73-86.

10. Shaw, M.L.G. and B.R. Gaines, Group knowledge elicitation over networks, in Research and Development in Expert Systems X. Proceedings of British Computer Society Expert Systems Conference, M.A. Bramer and A.L. Macintosh, Editors. 1993, Cambridge University Press: Cambridge, UK. p. 43-62.

11. Gaines, B.R. and M.L.G. Shaw, Eliciting knowledge and transferring it effectively to a knowledge-based systems. IEEE Transactions on Knowledge and Data Engineering, 1993. 5(1): p. 4-14.

12. Gaines, B.R., An ounce of knowledge is worth a ton of data: quantitative studies of the trade-off between expertise and data based on statistically well-founded empirical induction, in Proceedings of the Sixth International Workshop on Machine Learning. 1989, Morgan Kaufmann: San Mateo, California. p. 156-159.

13. Gaines, B.R., Refining induction into knowledge, in Proceedings of the AAAI Workshop on Knowledge Discovery in Databases. 1991, AAAI: Menlo Park, California. p. 1-10.

14. Gaines, B.R., Exception DAGs as knowledge structures, in AAAI-94 Workshop: KDD'94 Knowledge Discovery in Databases. 1994, AAAI: Menlo Park, California.

15. Gaines, B.R., Transforming rules and trees into comprehensible knowledge structures, in Knowledge Discovery in Databases II, U.M. Fayyad, et al., Editors. 1995, AAAI/MIT Press: Cambridge, Massachusetts. p. 205-226.

16. Shaw, M.L.G., On Becoming A Personal Scientist: Interactive Computer Elicitation of Personal Models Of The World. 1980, London: Academic Press.

17. Gaines, B.R. and M. Linster, Integrating a knowledge acquisition tool, an expert system shell and a hypermedia system. International Journal of Expert Systems Research and Applications, 1990. 3(2): p. 105-129.

18. Gaines, B.R., A. Rappaport, and M.L.G. Shaw, Combining paradigms in knowledge engineering. Data and Knowledge Engineering, 1992. 9: p. 1-18.

Mappings for Reuse in Knowledge-based Systems

John Y. Park and Mark A. Musen

Stanford Medical Informatics, Stanford University School of Medicine
Stanford, CA 94305-5479, USA
park@smi.stanford.edu, musen@smi.stanford.edu

Abstract: By dividing their contents into *domain knowledge* and *problem-solving methods* that reason over that knowledge, knowledge-based systems seek to promote reusability and shareability of their components. However, to produce a working system, the components must be connected within a global structuring scheme. We present the design for one such structuring scheme: the *virtual knowledge base constructor* (VKBC). VKBC enables developers to compose working applications from reusable knowledge-based components. The design is based on the concept of declarative *mapping relations*, which are explicit specifications of the conversions necessary for the syntactic and semantic connections between entities in the domain knowledge and the problem-solving method components. The mapping relation types supported by the VKBC enable a broad range of compositions and transformations of the modeled domain knowledge to match the input expectations of the selected problem-solving methods. This chapter covers the conceptual design and functional implementation of VKBC as a component of the Protégé system. It also describes several evaluation studies applying VKBC to component-based reuse projects.

1. Reusable Components and Mappings in Knowledge-based Systems

Knowledge-based applications often make the distinction between declarative domain knowledge and the problem-solving methods that reason over that knowledge [1]. This dichotomy promotes a degree of independence among the given components, thus offering hope of sharing and reusing these components. Work over many years in knowledge-based systems has resulted in large numbers of knowledge bases (KBs) for many diverse domains, many in standardized representational forms that encourage reuse. Related work has also led to a variety of implemented and refined problem-solving methods (PSMs)—algorithms for computing problem solutions—to reason over the KBs [2]. Given a new task, there is a good chance that the relevant domain KBs and an appropriate, implemented PSM already exist. Significant savings in development time would be realized if we could bind the instantiated KBs and PSM, using limited syntactic and semantic glue, into an application to accomplish our task.

As part of the Protégé project, we are exploring the issues of knowledge-based system design, sharing, and reuse [3,4]. We are also addressing the background issues of creation, formal description, and selection of PSMs [5]. The research presented in this chapter assumes that the candidate components have been selected and analyzed. We address the problem of binding the chosen elements into a working application entity, and providing the infrastructure for the reuse task. Our work is concerned with the theoretical and engineering-related issues of how to connect these components, and the nature of translating objects and concepts between components, using sets of declarative *mapping relations* that define the translations. We demonstrate a Protégé subsystem—called the *virtual knowledge base constructor* (VKBC)—that implements this approach to solve three diverse problems through component-based reuse. VKBC allows the user to define a structure for a new, PSM-accessible KB to be *constructed* from a set of source KBs, with the new KB's elements being created dynamically at runtime: thus, the *virtual KB* part of the system's name.

2. Component-based Reuse

Our research focuses on synthesizing knowledge-based applications by combining discrete, independent components. Knowledge-based systems can be divided into components in several ways. A simple, natural decomposition common to many frameworks, and that used within the Protégé system, is to divide knowledge-based systems into two fundamental classes: domain-specific KBs and the domain-independent PSMs that reason over these KBs [3]. The KBs encode knowledge about the task domain: the subset of the knowledge in a problem domain (e.g., medicine) that is relevant to the particular task of interest (e.g., diagnosis). It can be modeled with two entities: a structured definition of the types of objects and relations that can exist in the domain (the *domain ontology*) and the set of instances of members of that structure (the *KB instances*) [5]. In such a model, knowledge can be represented either in the structure of the ontology or in the instances themselves; the decision of where to encode what knowledge is often neither uniquely correct nor optimal.

The second major component of knowledge-based systems is the PSM. It can be thought of as a knowledge-processing algorithm or computational system. For example, there are PSMs that can compute solutions to constraint satisfaction problems, or identify possible faults in diagnosis problems. Researchers have done much prior work on reusable, domain-independent tasks and PSMs [1,2,6]. We can also describe the input and output requirements of the PSM using a *method ontology* [7]. Systems created with Protégé use a formal method ontology, which not only describes accurately the conceptual expectations of the PSM, but also specifies the exact expected representational format.

In the component-oriented model, an underlying design goal for the various entities is to support and encourage shareability and reusability. Therefore, the domain ontology, although oriented to a given task, is not intended to be matched directly to any one of the specific PSMs that can achieve that task. Similarly, the method ontology is neutral with respect to application domains. Furthermore, even for simple domains, any particular design for the domain ontology usually will reflect the personal abstractions of its designer, and there will be variability in structure across designers. (For example, for the disease domain, one possible ontology might group diseases under affected organ systems, while another might organize diseases by cause.) Therefore, incompatibilities are bound to occur between the domain ontology and the method ontology, no matter how well matched one is to the other. To bridge the resulting conceptual mismatch between the two ontologies, we need to define a syntactic and semantic "glue" to bind the components together.

Given a specific task, different knowledge-based system development methodologies use a range of solutions for how the knowledge and method should be brought together [8]. Some methodologies treat the problem as one of model refinement, starting from a model of a generic task and its knowledge roles, and evolving the model until it becomes domain-relevant [9]. However, these evolving-model schemes do not directly address the situation where both the KB and PSM components already exist. Another approach is to model the domain knowledge and method components independently, fusing the two together with an explicit conversion layer. We believe that this model more closely approximates the situation in most practical reuse scenarios, and we build our system upon such a model.

The framework for our model is to conceptualize the binding of the domain-KB and PSM components as generating mappings between concepts in the domain and analogous concepts in the method's universe of discourse (i.e., its ontology). We therefore need to define what mappings are in our framework.

2.1 Definition of mappings

In its general sense, *mappings* are defined as whatever mechanisms are used to convert between structures existing in one component and analogous structures expected by another. In the research described in this chapter, we focus specifically on mappings between structures in the domain KB and structures expected by the PSM. However, mappings would also be relevant to conversions of structures passed among component submethods in a composite PSM, or to any other intercomponent interface. We classify mappings into two principal types: *implicit* and *explicit*, with the latter further specialized into a *declarative* mapping type. Each is defined and described in the following sections.

2.2 Implicit mappings

Implicit mappings are conversions someone performs by specializing one or the other component (or both) to make the object definitions in one fit the requirements of the other to achieve the task at hand. This technique is sometimes also called *component adaptation*. For example, modifying public-domain software to work on one's own data structures would be a form of implicit mapping. In general, implicit mappings can take the form of either a modification of a pre-existing artifact or a customization of the overarching design in anticipation of creation of a specialized component. In our dichotomy of KB and PSM, either the KB can be modified to match the PSM's expectations, or the PSM can be altered to work with the native KB objects. In the first case, we have to modify the domain ontology, and either perform a one-time conversion of the instances in the KB, or work with the domain expert to reengineer the KB. In the latter case, both lexical and structural changes to the PSM implementation might be necessary.

The advantages of the implicit-mapping approach are that it is conceptually direct and straightforward. It also has problems, mainly in the areas of specialization and maintenance. Because the resulting artifacts are application task-specific, multiple instances of reuse of a given component lead to parallel sets of KBs or PSM implementations, each of which must be maintained individually. Another problem is that the specialization of working components can lead to the introduction of errors in the modified versions of both the KB and the PSM implementation. Also, the modifications are often neither principled nor well specified.

2.3 Explicit mappings

The alternative to implicit mappings are external, or *explicit*, mappings. Instead of modifying the original components, we create a third, separate type of component that moderates between the original components. In the database world, there is precedent for this in the work on *mediators* [10]. We have also experimented with such mediators in Protégé [11].

A simple approach is to create a monolithic procedural component from scratch for each new application task that involves reusing a new combination of components. The downside to this approach is that there is significant redundancy across reuse applications. Therefore, a promising expansion on this approach would be to abstract out the core mapping machinery that is common across mapping projects, and to build it into a reusable mapping component. Just as important, by making the interface to this generic mapping functionality logical and intuitive, we would be in essence creating a language for descriptively specifying mappings. This language would allow mappings to be defined declaratively.

Declarative mappings specify the conversions between the types of entities stored in the KB and those computed over by the PSM [7]. The run-time environment consists of two modules: a set of declarative *mapping instances* (*mappings* for short) and a *mapping interpreter*. In our work, mapping instances are explicit specifications, in a frame-based syntax, that define conversions for translating objects between the KB and the PSM. The mapping interpreter is a translation engine that parses the mappings and performs the run-time conversion to provide input to the PSM. For example, Figure 1 shows a declarative mapping that specifies that all instances of the domain class *domain-widget-A* are to be converted to instances of method class *method-widget-B* for input to the PSM, by scaling the *X*, *Y*, and *Z* slot values by 2.54 (to convert units from inches to centimeters), and by dropping all other slots (since these are of no use to the PSM). The mapping interpreter would parse these specifications, then search for all instances of the class *domain-widget-A* and perform the conversions to produce proper *method-widget-B* instances. We give a detailed description of declarative mappings in Section 3. There are several advantages to the declarative approach. First, a large part of the work—the mapping-interpreter implementation—is a reusable, generic engine common to all application tasks. Second, the declarative mapping specifications convey the *intent* of the mappings, and are not encumbered with the *mechanisms* of the mapping process, which are embedded in the mapping interpreter engine. By abstracting out the *what* from the *how* of the conversion specification, declarative mappings convey the intent of the designer more efficiently and clearly than procedural mappings would; this is important for future analysis of the mappings. The mappings are designed to meet the dual goals of being transparent to humans and of being efficient to implement.

3. Mappings in Protégé

The core purpose of Protégé is to support component-based development, sharing, and reuse of knowledge-based systems [4]. Protégé comprises both a framework and a tool kit, and supports end-to-end development of such systems. It provides an ontology editor that aids the user in structuring the domain knowledge to be acquired. It can then use this domain ontology to generate domain-specific knowledge acquisition tools for filling a KB with knowledge instances. Protégé also supports the structuring of the knowledge needs of PSMs into a method ontology. Through VKBC, Protégé also supports the use of declarative mappings to couple the KB and PSM components into complete working systems. The concept of ontologies is central to all of these tasks.

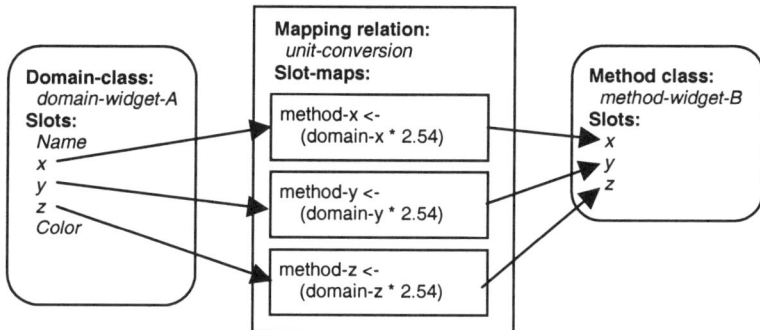

Figure 1: A diagrammatic example of a mapping that scales some slots numerically and drops others that do not correspond to PSM inputs.

3.1 The central role of ontologies in Protégé

Protégé uses explicit ontologies to organize information into regular, well structured hierarchies. There are three primary ontologies involved in any given application task: (1) the domain ontology, (2) the method ontology, and (3) the mapping ontology. For a given task, the *content* of the knowledge outlined in the domain ontology should meet the input requirements of the PSM, but the *form* of that knowledge might need to be restructured before the PSM can operate on it. The domain ontology is designed to be meaningful to domain experts—for example, in a medical domain, it can refer to entities such as "drugs" and "symptoms." In contrast, the method ontology [12] is domain-independent, and its elements are described in domain-neutral terms; for example, instead of "diseases," it might refer to "faults." The mapping ontology [7] describes the range of declarative mapping types that are supported by Protégé for bridging between the classes in the domain ontology and method ontology. The mapping ontology is used to guide the acquisition of instances of mappings and, by design, circumscribes the set of conversion types that are supported by the Protégé mapping engine.

It is important to note that the mapping ontology is an integral part of Protégé/VKBC, and is static across Protégé-based applications, whereas the domain and method ontologies are generally specific to each application.

3.2 Mapping classes and instances

Mapping instances in Protégé are formal, concrete entities belonging to a prescribed set of types, which are meant to be common across all Protégé-based applications. The relationships among the mapping ontology's member classes, the mapping relations created from these classes, and the larger application context are diagrammed in Figure 2. At the top of the figure is the set of general classes of mappings defined in the mapping ontology, which cover the mapping functionalities that might be needed in any given component-based integration application. At design time, the application engineer first matches each required input object class in the method ontology with the best semantically corresponding class in the domain ontology. For each pairing, she selects the most appropriate mapping class based on the type of conversion necessary. She then instantiates the chosen mapping class, filling the instance slots with the specifics of the required conversion. (This is represented by the middle layer in the figure.) Subsequently, at execution time, each mapping instance drives the conversion by the mapping interpreter of instances from one or more domain classes into instances of the corresponding method class. (This is the bottom layer in the figure.) Thus, design of the mappings for an application involves working with classes in the domain and method ontologies, while execution of the mappings involves converting instances from the KB into input instances for the PSM.

As an example of this process, we briefly describe our mapping of an elevator-configuration KB to the propose-and-revise constraint satisfaction PSM [13]—a heuristic algorithm that uses local repair knowledge to propose, evaluate, and update solutions. The entities in the elevator configuration domain—system components, such as doors, motors, and cables, and operational rules, such as operating limits on the components and permissible component combinations— must be mapped to the state variables, constraint expressions, and other inputs expected by the propose-and-revise PSM. The system designer examines each class of method inputs needed by propose-and-revise, determines which class (or classes) in the elevator domain corresponds to these method inputs, and then decides how to convert the slots in the latter into slots in the former. The designer then determines which of the prescribed set of mapping classes will suffice for this conversion. She might choose a simple renaming mapping for translating elevator component classes to propose-and-revise state variables, since the renamed slots' contents do not need any modification. On the other hand, she might need complicated expression manipulations for translating elevator component operating limits into constraint expressions. Once she has determined the extent of the necessary mappings, she then explicitly specifies the conversion by instantiating the mapping class with the particulars of this class-to-class mapping. This design process is repeated for all of the classes required by the PSM.

At run-time, the mapping interpreter reads in each mapping instance and parses it, finds the corresponding domain class—elevator doors, for example—and

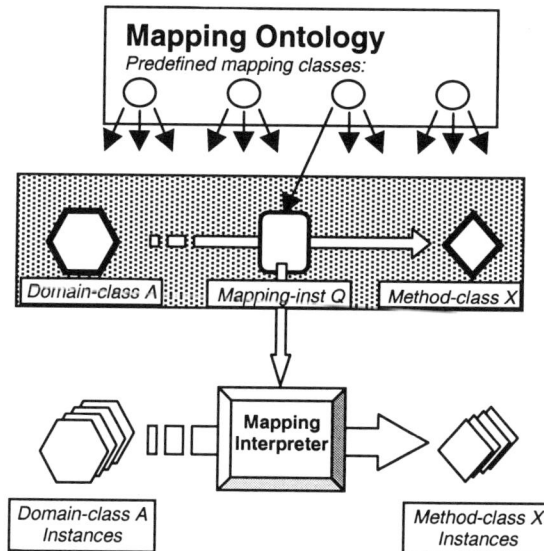

Figure 2: The mapping process. The path from mapping ontology to domain knowledge conversion fans out at several levels. The mapping ontology defines the allowable classes of mappings, each of which can be instantiated into many mapping instances. In turn, each mapping instance specifies the conversion of a source domain class to a target method class. The Mapping Interpreter uses each mapping instance to direct the conversion of multiple instances from the domain class to the corresponding method-class instances.

gathers all the instances of that domain class. Then, for each domain instance it has found—each elevator door type, to continue our example—it applies the mapping specifications in the mapping instance to create an appropriate method instance and fill its slot values. This process loops until all the mapping instances in the mapping declaration file are processed. At that point, the propose-and-revise PSM is passed a set of instances defined by the classes of its own ontology, which it can easily process.

4. The Mapping Ontology

In this section, we will outline the design process that led to the creation of the current ontology of mappings covered here.

4.1 Six desiderata for a mapping ontology

Initial experience with an early prototype for a mapping ontology [7] led us to conduct a principled, ground-up redesign of the mapping system, starting with a declaration of a set of desiderata for the new system:

1. *Expressiveness*: The set of mapping types should be fairly powerful, allowing almost arbitrary mappings if desired; in other words, while we are not obliged to optimize the mapping tool for every possible contingency, we should at least allow for many possibilities of the kinds of mappings users might want.
2. *Ease of use*: The mapping ontology should have a broad range of classes, from simple to complex, to make the designing of mapping relations easy and straightforward, given the task at hand.
3. *Clarity*: The created mapping relations should be easy to peruse and comprehend, and the designer's motivations should be readily apparent from the structure.
4. *Parsimony*: The set of mapping relations needed for a given project should be minimized, and the number of mapping classes in the ontology should also be kept small.
5. *Efficiency*: Mappings should be efficient to implement and execute.
6. *Principled design/natural distinctions*: The mapping ontology should be based on careful analysis of theoretical and practical requirements of mapping tasks, and should embody any natural distinctions common to these tasks.

Following naturally from this last point, we will now present a conceptualization of the mapping task that should shed some light on what an appropriate design should embody.

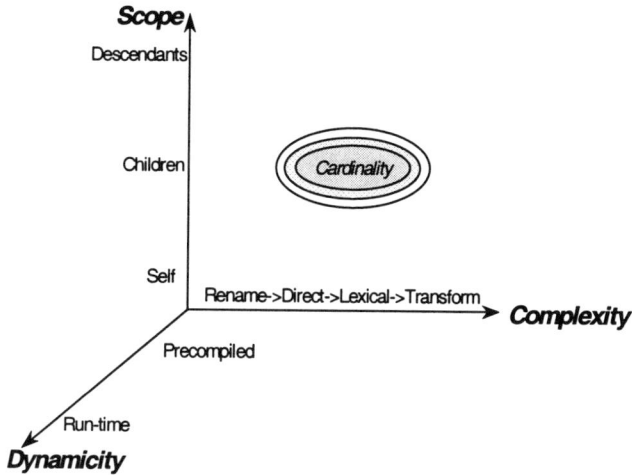

Figure 3: The four-dimensional space of mapping features.

4.2 A four-dimensional conceptualization of mapping properties

In formulating the design for the mapping ontology, we derived four orthogonal dimensions for describing attributes of the mapping task: *power/complexity*, *scope*, *dynamicity*, and *cardinality* (Figure 3), described here:

- *power/complexity:* This dimension deals with the spectrum of allowed expressiveness of the transformation of the datum or data into a new mapped value. Transformations range from the simple renaming of a slot all the way to arbitrary, functional transformations of multiple input slots into a single composite target slot. The principle here, and one of the recurring themes in our design, is that the mapping functionality should not be the limiting step in reuse, although it should reflect the cost of the complexity of the mapping.

- *scope:* The second dimension describes the range of domain classes to which the mapping should apply. We should be able to restrict the mapping to specific classes, or allow it to be inheritable to controllable depths.

- *dynamicity:* This dimension controls when and how the mappings should be invoked: whether all the instances should be mapped prior to PSM invocation, or mapped on-demand at run-time.

- *cardinality:* The last dimension specifies whether the mapping should be a simple one-to-one mapping, converting single domain instances into corresponding method instances, one-to-many, generating a set of related instances for each input instance, or even many-to-one, compositing objects from several instances across multiple classes into a single method class instance.

By organizing our properties as points in this four-dimensional space, we gain a distinct perspective on the mapping process, centered more on what mappings should accomplish than on how they should do it. Our experience from previous mapping experiments implies that any particular instance of a desired mapping can fall nearly anywhere in this four-dimensional space, and that the four axes are mostly independent: for any given application, the set of mappings may range over the entire space. Thus, as well as being a useful conceptualization scheme, preserving the independent nature of these four axes would be useful for any actual working design for a mapping system.

4.3 The structure of the mapping ontology

Structuring our design to cover the four axes of our multidimensional conceptualization, and guided by our six functional desiderata, we created an ontology of mappings, shown diagrammatically in Figure 4. We will start our analysis of the mapping ontology by studying the set of possible mappings it defines for individual slot values. We will then see how these slot mappings are composed into mapping classes in the ontology. Lastly, we will examine the ancillary features that implement properties such as slot composition and recursion.

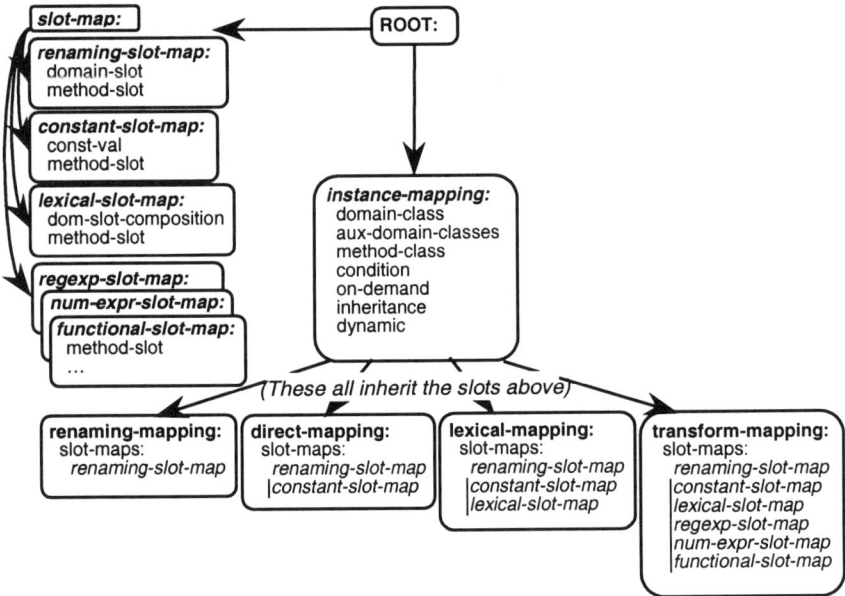

Figure 4: The new Protégé ontology of mappings. The *-mapping classes along the bottom represent the set of basic mapping classes supported by Protégé, while the classes arrayed along the left-hand side are used to define the individual slot-mapping elements of the mapping classes.

4.3.1 The spectrum of slot mappings

A mapping relation is composed of a set of component slot mappings, each of which specifies the mapping instructions for creating a single slot value in the method class instance. Since the primary component of any given mapping relation is the set of slot mapping specifications, understanding these slot mappings is essential to understanding what a mapping does.

The set of slot mapping types covers a spectrum of mapping power and complexity. (Functionality that covers the other dimensions of our space of mapping features is addressed later.) The supported mapping types are:

- *renaming-slot-map:* This slot mapping type is used where a slot simply needs to be copied from the domain class instance to the method class instance, with only a possible change in slot name.

- *constant-slot-map:* This slot mapping type is used to make constant value assignments—usually default values—to the method class slots that don't have corresponding slots in the domain class instance.

- *lexical-slot-map:* This slot mapping type allows method slots to be composed from lexical concatenations of multiple domain slots. It is most often used for minor syntactic variations on domain slot values, or for composing expressions from multiple fields.

- *regular-expression-slot-map:* This is an enhanced version of the lexical-slot-map; regular-expression-slot-maps allow arbitrary regular expression editing of a composed field, which allows the contents of slot values to be modified, unlike lexical-slot-maps.

- *numerical-expression-slot-map:* This slot mapping type allows arithmetic processing of one or more domain instance slots into a numerical method instance slot.

- *functional-slot-map:* This type is the fallback mapping for arbitrarily complex transformations of slot values. It allows user-supplied functions to compose the method slot value from any subset of the domain class slots, using any transformational logic. The interface supports multiple well-known interpreters.

Using these six slot mapping types, the developer defines a class mapping into a method class one slot at a time. Examples of some of these slot-maps are shown in Figure 5.

4.3.2 The spectrum of mapping classes

The set of slot mapping classes described in the preceding section are used as slot elements in four *mapping classes—renaming*, *direct*, *lexical*, and *transformational*—which are instantiated to create mappings. These classes cover a spectrum of mapping capabilities of increasing complexity. There is a clear parallelism between the spectrum of slot mappings and the spectrum of mapping classes. The higher-complexity mapping classes are implemented by monotonically adding permissible slot mapping classes, creating a strictly subsuming hierarchy. This strict ranking means that more complex mapping classes can accomplish everything that the relatively simpler mapping classes can. The motivation for using the simplest sufficing mapping class is that the choice of mapping class then conveys information about the level of expressive power needed for the mapping, and therefore gives cues about the level of mismatch between the domain and method classes being paired. For example, the choice of a *direct-mapping* class for a given mapping implies that only slot renamings and constant value assignments are necessary, and indicates that this is a relatively straightforward mapping between similar classes.

4.3.3 Cardinality, substructures and composite instances

In addition to the spectrum of complexity in the slot and class mappings, the mapping ontology defines a simple, intuitive way to specify the other functional dimensions outlined in Section 4.2 (and shown in Figure 3).

We will first address the issue of cardinality. During the mapping task, it is sometimes necessary to compose multiple domain instances into a single method instance. Conversely, it is also sometimes necessary to create multiple method instances from a single domain instance. A common cause of both of these scenarios are domain or method ontologies that contain nested, recursive structures. This is often used in ontologies to convey information about common substructurings in classes. For example, a slot called "position" might be common to several classes in the domain, and the slot would probably be implemented as a nested substructure instance with x, y, and z slots. VKBC handles this kind of structural remapping via an extension to the slot specification syntax that allows a

```
;; This mapping relation is of type lexical-mapping, and so can
use renaming, constant,
;; and lexical slot mappings to fill the method's slots.
([instance_1] of lexical-mapping
    (mapping-name "test-map")    ;; unique identifier
    (condition "t")              ;; unconditionally apply mapping
    (domain-class "my-domain")   ;; the domain class to be mapped
    (method-class "your-method") ;; method class to be mapped into
    (slot-maps [instance_2] [instance_3] [instance_4]))
;; Rename the domain slot "label" into the method slot "name",
leaving the value intact
([instance_2] of renaming-slot-map
    (method-slot "name")
    (domain-slot "label"))
;; The method slot "cost" has no equivalent in the domain, so just
assign a default 0
([instance_3] of constant-slot-map
    (method-slot "cost")
    (const-val "0"))
;; Use a lexical mapping to composite a method slot from multiple
domain slots
([instance_4] of lexical-slot-map
    (method-slot "stop-condition")
    (dom-slot-composition "(< *<.param>* *<.max-value>*)")))
```

Figure 5: The mapping relation "1" is composed of the three slot-mapping relations 2, 3, and 4, which are instances of a renaming slot map, a constant slot map, and a lexical slot map, respectively.

wide range of structural reorganizations.

Another common case of higher-order cardinality occurs when the domain ontology is structured hierarchically, partly into classes whose instances are individual domain elements, and partly into classes whose instances encapsulate descriptions about *sets* of other domain instances. For example, we might have an ontology of vehicles, where the class *cars* has an instance for each car model, and the class *body-types* has instances covering whole subclasses of cars (e.g., one instance describing minivans, another pickups). We would run into problems if the PSM expects all of the information for any vehicle to come in as slots in the individual car instances. However, the member instances are not likely to have pointers to the group instances from which they are conceptually inheriting, so our substructuring scheme will not work here. VKBC handles this by allowing multiple domain classes to be composed into *virtual instances*, much like the JOIN-with-condition operator of relational databases.

4.3.4 Scope and dynamicity

VKBC allows the user to control the *scope* of a mapping definition, by letting her explicitly specify the inheritance and recursion properties for each mapping. It also provides fine-grained control of the mapping process itself. In addition to *static mappings*, where all mappings are processed by the Mapping Interpreter before the PSM begins execution, VKBC also supports several modes of *dynamic mappings*, which are executed incrementally on demand in the course of the execution of the PSM. These modes can be useful for cases where space or computational load issues are a factor, or for numerous other reasons.

5. Mapping Evaluation Experiments

We have applied VKBC and its ontology of mappings in three reuse experiments that we developed in prior Protégé mapping research. The first of the three experiments is a recreation of the Protégé implementation of the Sisyphus-II elevator-configuration project [13], and involves the composition of a KB concerning elevator configurations with a generic implementation of the propose-and-revise PSM, introduced in Section 3.2. We map instances of elevator components, specifications, performance bounds and component upgrade rules to the propose-and-revise inputs of state variables, constraint specifications, and violation fixes. A promising result is that the new mapping relations are not only more terse, but also more understandable, compared to those in the original prototype mapping experiments; usually, parsimony and perspicuity would be traded off against each other.

The second experiment is the reuse of propose-and-revise in a completely different application domain: ribosomal conformation prediction [11]. The task's goal is to posit plausible three-dimensional conformations for the set of ribosomal subunits. Again, we map domain class instances like protein clusters and RNA helices, along with geometric positioning constraints and biological evidence, to propose-and-revise's state variables and constraints. In addition to the new spectrum-based range of class and slot mappings, the inline specification feature of domain class substructure slot accesses in the domain-slot specification were particularly useful in this experiment. This experiment also benefits from the explicit control of recursion, since substructure instances must be left intact.

The third experiment is applying a different PSM—the Protean n-way yoking method [14]—to the ribosomal structure domain. The results are very similar to the second experiment. Since Protean is a constraint satisfaction problem solver, the ontology for this PSM is very similar to that for the propose-and-revise PSM. Therefore, we were able to reuse much of the experience from the ribosome/propose-and-revise mapping to help design the mappings for this application. The original ribosome/propose-and-revise experiment is already a reuse experiment, and we are conceptually reusing the artifacts of reuse—the mapping relations—so this experiment is as much a lesson in reusing mappings (meta-reuse) as in creating mappings.

6. Summary

In the work presented here, we have investigated the nature of mappings between knowledge-based components, and designed an ontology of mapping classes to facilitate creating mappings between ontologies for reuse. Our mapping ontology was designed to accommodate a broad spectrum of functionalities along multiple dimensions, reflected in both the range of mapping classes and the range of slot-mapping types from which the classes are composed. The mappings permit a wide range of transformations, from simple slot renamings and constant value assignments, to arbitrary functional transformations. The mapping ontology also supports explicit control of cardinality, run-time context, recursion, inheritance, and conditional interpretation.

We have presented evidence that component-based reuse through declarative mappings is a viable concept; here, and in much previously published work on Protégé (e.g., [15]), we have demonstrated many examples of reuse, covering the gamut of domains from elevator configuration to ribosomal conformation prediction to medical diagnosis.

Our last point concerns meta-reuse: in this case, conceptually reusing mapping relations from previous projects. When reusing components, whether they are pre-existing domain KBs or legacy PSMs, a good start for the task of defining the mappings is to examine prior mappings involving either of the components. In fact, one might start with the original mapping relations as a basis for evolving a new mapping project. Thus, the clarity of intent and implementation of the mapping relations is important. Our claim is that our declarative ontology of mappings enhances the prospects for this level of meta-reuse.

References

[1] Breuker, J.A. and van de Velde, W. (Eds.) (1994). *The CommonKADS Library for Expertise Modelling.* Amsterdam: IOS Press.

[2] Mcdermott, J. (1988). Preliminary steps toward a taxonomy of problem-solving methods. In S. Marcus (Ed.), *Automating Knowledge Acquisition for Expert Systems,* Boston, MA: Kluwer Academic.

[3] Eriksson, H., Shahar, Y., Tu, S.W. et al. (1995). Task modeling with reusable problem-solving methods. *Artificial Intelligence,* **79**(2), 293-326.

[4] Musen, M.A, Gennari, J.H., Eriksson, H. et al. (1995). Protégé-II: Computer support for development of intelligent systems from libraries of components. *Proceedings of Medinfo '95,* pp. 766-770, Vancouver, BC.

[5] Grosso, W., Gennari, J.H., Fergerson, R. et al. (1998). When knowledge models collide. *11ᵗʰ Workshop on Knowledge Acquisition, Modeling, and Management 1998,* Banff, Alberta, Canada.

[6] Chandrasekaran, B. (1986). Generic tasks in knowledge-based reasoning: high-level building blocks for expert system design. *IEEE Expert,* **1**(3).

[7] Gennari, J.H., Tu, S.W., Rothenfluh, T.E. et al. (1994). Mapping Domains to Methods in Support of Reuse. *International Journal of Human-Computer Studies,* **41**, 399-424.

[8] Fensel, D. (1994). A comparison of languages which operationalize and formalize KADS models of expertise. *The Knowledge Engineering Review,* **9**(2), 105-146.

[9] Gil, Y., and Melz, E. (1996). Explicit representations of problem-solving strategies to support knowledge acquisition. *Proceedings of the Thirteenth National Conference on Artificial Intelligence,* pp. 469-476, Menlo Park, CA: AAAI Press/MIT Press.

[10] Wiederhold, G. (1992). Mediators in the architecture of future information systems. *Computer,* **25**(3), 38-49.

[11] Gennari, J.H., Cheng, H., Altman, R.B. et al. (1998). Reuse, CORBA, and knowledge-based systems. *International Journal of Human-Computer Studies,* **49**(4), 523-546.

[12] Gennari, J.H., Grosso, W., and Musen, M.A. (1998). A method-description language: An initial ontology with examples. *11ᵗʰ Workshop on Knowledge Acquisition, Modeling, and Management 1998,* Banff, Alberta, Canada.

[13] Rothenfluh, T.E., Gennari, J.H., Eriksson, H. et al. (1996). Reusable ontologies, knowledge-acquisition tools, and performance systems: PROTÉGÉ-II solutions to Sisyphus-2. *International Journal of Human-Computer Studies,* **44**(3-4), 303-332.

[14] Altman, R.B., Weiser, B., and Noller, H.F. (1994). Constraint satisfaction techniques for modeling large complexes: Application to central domain of the 16s ribosomal subunit. *Proceedings of the Second International Conference on Intelligent Systems for Molecular Biology,* pp. 10-18, Stanford, CA.

[15] Gennari, J.H., Altman, R.B., and Musen, M.A. (1995). Reuse with Protégé-II: From elevators to ribosomes. *Proceedings of the ACM-SigSoft 1995 Symposium on Software Reusability,* pp. 72-80, Seattle, WA.

Mining Very Large Databases to Support Knowledge Exploration

Neil Mackin

Decision Support Applications Manager,
WhiteCross Data Exploration, Waterside Park, Bracknell, RG12 1RB, UK.
Neil.Mackin@WhiteCross.com

Abstract: Exploitation of data mining may be increased by moving away from dependence upon statistically trained experts, and by making data mining, both in terms of its application and its results, more easily deployed within the business. Full volume data mining avoids the constraints of sample based approaches. A data mining framework, enabling software applications to be developed with simplified interfaces, increases the usability of these techniques. Encapsulating resultant predictive models as components enables easy deployment of the results within a business. A workshop style environment is appropriate where business managers need to work quickly with a powerful data mining capability.

1. Applying Data Mining to the Business

Data mining is routinely prioritised as a major force in garnering information in order to gain competitive advantage. However, the promise of delivering a deep understanding of the drivers behind customer behaviour and developing the predictive models that can be deployed within the organisation is often not realised. Three of the main reasons why data mining projects fail include:
1. There is a disconnect between the business managers and the quantitative experts in understanding the problem to be addressed.
2. Data mining tools are often unable to scale appropriately to the size of the data mining task.
3. Tools don't support deployment of data mining results into operational systems.

In a competitive environment successful businesses are driven forward by visionary and creative managers.

In data mining it must be remembered that the business manager is the problem owner – it is he or she who has to make decisions, set prices, and design customer propositions based upon the results of the data mining. It is he/she who has the responsibility for driving forward the business, and whose personal skill set is likely to be business administration or specific tasks such as marketing.

In contrast to the requirements of the business manager, most data mining packages available today are tools designed for advanced users with a strong statistical background. This bias is due to the skills needed to use the tools, and the background required to correctly translate a business problem into a format suitable for analysis through a data mining tool.

Commonly, businesses find themselves in a paradigm of relying upon highly skilled mathematical experts to perform Data Mining studies. This is unfortunate, and perhaps indicative of the relative immaturity of data mining, because the consequences of this mode of working include:

1. Delay in delivering the solution as the problem is transferred between the manager and the analyst
2. The risk that a language barrier between the manager and analysts leads to misinterpretation of the business problem and/or results.
3. Loss of creativity in exploring the possible solutions to the business problem.

This final point is perhaps most significant because data mining projects are not easily structured into distinct phases, such as a requirements specification, functional specification, analysis and results. They are inherently far more fluid and iterative. As initial results are found, the business problem is often changed as managers gain insight into their data and refine or redesign the goals of the exercise. This close interaction with the data and repeated redefinition of the business problem are key success factors in data mining work. It is akin to asking questions, and based upon the knowledge gained in the answers, asking further or new questions.

If the paradigm of transferring the analysis between the business manager and the quantitative expert is maintained, then the number of iterations may be reduced, the time expended fruitlessly exploring dead ends increased, and the overall quality of the results compromised. This is exacerbated by any urgency attached to the project – and most worthwhile projects in a competitive environment require urgent results.

A major area of technical complication in data mining is the use of statistical sampling to perform data consolidation. This arises because the volumes of data involved in business problems are often well in excess of those that can be handled on data mining platforms. A frequently taken compromise is to sample the original dataset in order to give one, two or more orders of magnitude reduction in the data volumes. Problems associated with this kind of sampling are discussed below. The disconnect between the business problem owners and the data mining analysis may be overcome by :

1. Simplifying the interface to data mining technologies – embedding data mining within vertical business applications rather than deploying only horizontal data mining workbenches
2. Scaling up the data mining to the full volume of data to avoid the complexities and risks introduced by sampling.
3. Using workshop style interaction sessions that enable business managers to exploit an extended data mining capability.

2. Data Mining Using Sampling

Data mining tools sometimes use sampling to get around the fundamental computational limitations of most hardware and software environments. Traditional hardware environments can't compute the necessary number of iterations within reasonable time periods. The use of client-side processing, where a data-cube is brought back from the database server onto the client workstation for processing, inherently constrains the data set size to that which can be handled by the workstation, typically only tens of megabytes in memory. Many of the algorithms involved in data mining are not scaleable with respect to the size of the source data set. As the size of the data set increases, the computation time of the algorithm rises exponentially.

While sampling has been necessary in order to make complex problems tractable, it must be noted that the there are costs associated with sampling which can be divided into technical and practical problems.

2.1 Technical sampling issues

The confidence with which results can be generalised is related to the sample size. As the sample size is reduced, so the ability to predict the behaviour of the population being modelled is reduced. This is a problem for all data mining algorithms, but it is particularly acute for decision trees; each new level of the tree decreases the number of rows in the terminal nodes exponentially.

As you go deeper into a decision tree, the number of rows represented by a node reduces until the point where there are insufficient rows to enable any more specific rules to be discovered. Taking a naïve mode, where each node represents the penetration of a Boolean objective and tree segments the population into regions of significantly high and low penetration, then a Binomial probability model may be used to predict the necessary node sizes, in order to generalize at a given confidence interval. For example, a 500 row node expressing a penetration at 20% would generalise at the 95% confidence interval to a penetration of [16.5%, 23.5%] in the population or a 7% interval size [1].

In order to maintain the ability to generalise confidently, it is usual to limit the sophistication of models; a simple method is to restrict the number of variable interactions that are considered.

A further problem with sample-based data mining is that the results of the project are only valid to the extent that the sample is representative of the population being considered. In terms of classification-based decision trees, where the implication is that once the tree has been generated and before a rule can be used, the data at every separate terminal node in the tree should be verified as being representative of the similarly constrained data in the whole population. In the realm of knowledge discovery, in order to be confident of a sample, you must verify the sample in light of the discovered knowledge. This necessarily introduces considerable delay and repetition into the data mining project.

2.2 Practical sampling issues

The use of samples as a basis for any decision model necessarily results in that decision model being merely an estimate rather than the absolute truth. Wherever practicable it is desirable to know the absolute truth rather than an estimate thereof. This is especially important when working in the financial/auditing arena.

Handling rare events is difficult with samples. Events such as fraudulent behaviour may only occur in a very small percentage of the base; taking a simple sample would be very likely to over-represent or under-represent this behaviour pattern in the data.

Data mining is only one part of exploring data in order to investigate a business question. Other common operations include "slicing and dicing" data; that is, repeatedly drilling down into the data making increasingly specific qualifications on the data. For example, in the travel industry one might initially view the data by year, then by year and resort, then by year, resort and week. This style of drilling into the data very quickly hits limits in samples. Examining the most common values, or TopN analysis, for example identifying the top 1000 telephone numbers called in a given month, is also severely restricted by sampling.

However carefully the initial sampling may be undertaken, and whatever scheme for selecting a sample is used, it is inevitable that discarding data introduces a risk that significant information about the business problem is also discarded.

Aggregation techniques may be employed as a data reduction technique. However, the loss of information through aggregation should be considered. By way of example, consider retail data for sales indicating the volume of sales per item per hour. This could be aggregated by summing all the data items on a weekly basis, which would significantly reduce the data volume. However, this method removes any pattern of sales with respect to the day of the week or the hour of the day.

Whilst sampling may be seen as a technique for accelerating the development of data mining models by generating them over a smaller dataset, it is important to note that the model generation is only one part of the overall data mining process [2]. Building a dataset for mining is a key preceding step, which is often time consuming when a sampling process is involved. This will have to be repeated many times if new fields, which should be brought into the dataset, are identified during the mining activity.

2.3 Business users not supported by sampling

Many of the issues raised above that relate to sampling can in some part be mitigated against by very careful handling of the analysis. Indeed, a skilled, statistically trained analyst may review the above list of sampling issues, and consider immediate work-arounds to each problem. However, a key business issue for deploying data mining is to move away from the dependence upon a small number of key statistical gurus in an organisation, and enable the business managers to interact with their data using intelligent tools.

The alternative to reliance upon highly skilled statistical analysts is to increase the ability of the business manager to engage with the data mining techniques directly.

3. In-Place Mining of Data with VLDB

Data mining applications tend to push the limits of the data warehouse. Problems with data integrity are often undiscovered until the Data Mining exercise.

A number of approaches to mining data in place in databases have been described by Freitas [3]. These can be summarised as :

1. Parallelising sequential code automatically – which is a sub-optimal because opportunities for speeding up the algorithm may be missed.
2. Manually parallelise existing algorithms – which may be impossible for some algorithmic approaches.
3. Re-design algorithms from scratch for parallel computing architectures – which involves considerable rework.

The computational requirements associated with data mining algorithms, when run over very large data sets, is high. Data sets commonly comprise millions of rows and several hundred variables or columns.

Data mining algorithms require multiple passes over the data set, each pass uses the information from the previous pass to target or narrow down the subsequent iterations. Each pass is potentially computationally intensive. For a telecommunications project, *each* run of a decision tree may require about 6,000 queries on a database consisting of many millions of records, each containing several hundred variables.

The application of data mining algorithms to very large data sets requires the algorithm itself to scale, and the computational environment to be of sufficient speed to provide the underlying processing power.   There are two important components of scalability here: (i) increases in the number of rows within the database; (ii) increases in the number of variables or columns. Given careful choices in algorithms, parallel computing environments make it feasible to apply data mining algorithms to these full volumes of data.

There is no one perfect data mining tool or technique. Frequently one needs to mix and match tools to achieve all the dimensions of information that are required to answer one question. One of the real challenges is how to use multiple data mining tools on the same data set, without restructuring the data set. There are two viable alternatives to this challenge. The first is to use a data mining tool kit, such as those provided by IBM, SGI or Thinking Machines. The second alternative is to select a group of tools that have a generic back end interface such as an ODBC connection. We present here the use of the second alternative, the power of a database supercomputer coupled with MiningSTUDIO, which is a data mining workbench developed by integrating the ANGOSS KnowledgeSTUDIO tool [4] with extensions to handle very large datasets using specific drivers for a WhiteCross computer.

The approach taken by the WhiteCross data mining tool MiningSTUDIO to generate decision trees is one of a classic client server tool. The client acts as a query generator, submitting SQL requests to the database server and interpreting the query result-sets. The algorithms employed are based upon the information theoretic approaches of ID3/C4.5 [5,6] and IT-Rule [7]. A major advantage of the decision tree is that we can easily take the generated rules and use them proactively.

The WhiteCross computer has been designed for very fast online analytical processing, rather than transactional processing. The server has a shared nothing, massively parallel processing (MPP) architecture. The very tight integration of the database server hardware and software enables considerable performance improvements over conventional portable database implementations. The server is easily connected to user applications that contain an ODBC (open database connectivity) back-end. [ODBC is a standard that was introduced by Microsoft in 1993. It provides an open interface based on Structured Query Language (SQL). ODBC consists of a standard, universal library of functions for building database-independent applications.]

A typical installation for a WhiteCross server may have 100 processors and approximately 50Gb of RAM for interactive analysis of data. This data can be queried as though every single column were pre-indexed – which in turn enables *ad hoc* queries against billions of records to be executed in a few seconds. Using this capability, MiningSTUDIO's algorithms understand the data by repeatedly drilling into it with increasing levels of granularity until a sufficiently accurate understanding is attained.

As part of the decision tree process, the user must create an objective or business goal. As MiningSTUDIO creates a decision tree, it finds records that match the goals and then present a breakdown of characteristics or profiles for the records. The business goals are expressed in terms of SQL clause fragments. A strong feature of this particular implementation is that multiple tables can be used to support both the SQL query and the later tree generation. The user does not have to join the relevant tables prior to the generation of the decision tree. The SQL query calculates how many rows match that the goal.

An interesting point to note about the client server implementation of data mining in WhiteCross In-Place Mining is that the time required to generate a decision tree is not significantly affected by the number of rows over which the tree is grown. However, the number of columns evaluated as possible splits at every node in the decision tree is a major factor. A user-settable option in WhiteCross' MiningSTUDIO controls the number of variables considered at every split. Pruning the search space enables even the largest datasets involving several hundreds of columns to be reduced to a tractable problem and executed within minutes.

There are a number of methods that can then be used to refine the model so that the user can obtain a better result. The main methods include; changing the decision tree parameters, performing statistical attribute analysis so that the default categorisations can be modified, adding or deleting business attributes, refining the business goal, filtering the data set, and using relational joins.

4. Integration of Multiple Data Mining Algorithms

Powerfully generating decision trees from data residing in the database is a useful technique in its own right. In the MiningSTUDIO tool it is integrated with additional exploration and interpretation tools as well as a number of further algorithms that together form a powerful toolkit.

- Data exploration. An interactive charting tool that allows the user to flexibly graph and explore the distributions of variables across the full volume of data.
- Interpreting results. A wide range of graphical views of the decision tree show the variable distributions for each node's sub-population; many text-based formats for rules such as English prose, Java or SQL code; analysis of the decision tree's efficacy in terms of a lift or ROI curve and scoring the tree against validation data sets.

- Clustering Algorithms. These unsupervised techniques (in that they are not specified a particular goal to search for but rather uncover hidden structure) that can improve the performance of supervised techniques. They work well with categorical, numeric, and textual data.
- Neural Network Algorithms. These are supervised techniques when used to generate predictive models. A number of architectures and algorithms are included in the MiningSTUDIO tool. (i) MLN - multi-layer perceptron with a single layer of hidden neurons, typically used to train a multi-dimensional data set with a continuous underlying model; (ii) PNN - probabilistic neural network using a Memory Based Reasoning technique to create a predictive model such that each training record may be considered as a neuron in the network; (iii) RBF - radial basis function predictive, an optimized form of the PNN which embeds clustering to predict based on cluster centres.

The MiningSTUDIO workbench also implements a three-tier client server architecture for data mining. Users may choose which computer to use to execute the computationally intensive algorithms. The users' own workstations can be used or alternatively a remote NT or Unix/Solaris server. DCOM is used as the protocol, or client/server middleware, to facilitate the communication between the client and the compute server.

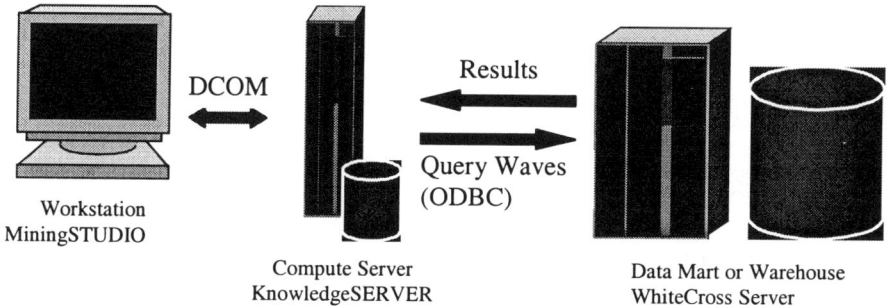

Figure 1: MiningSTUDIO's three tier client server architecture

5. Deployment of Data Mining Through Applications

The implementation of data mining against the full volume of data frees the business manager from wrestling with statistical issues relating to sampling. However, the increasing complexity of data mining, with workbenches such as MiningSTUDIO including multiple algorithms and considerable flexibility, works against the business manager. This can be overcome by providing simplified interfaces to data mining, which are specific to the needs of the business. Rather than describing data in technical terms and forcing the user to browse through SQL datasource catalogues, an application can simply present the user with the dataset for their specific needs.

This data mining application development is enabled through componentisation of the data mining workbench. This use of components is common place in business application software. When you have a spreadsheet or graph inserted into a document, there are two principle components, the document and the embedded object, using COM as the underlying mechanism to facilitate communication.

Every major element of MiningSTUDIO is implemented as a component with both a visible and a hidden interface. The visible interface provides the user interface, for example the decision tree browser from MiningSTUDIO could be hosted within a simple web page, if that is what a business application requires. The hidden interface enables complete embedding of the data mining technology within an application, so that the user needn't be aware of it's existence. This would be comparable to embedding bayesian algorithms to support the spell checking in word processors.

Data mining applications can be considered in terms of their intended life cycle. A distinction may be readily made between software developments that are intended to be long-lived – such as a Pay-Roll application - and those which are intended to have a short and limited life-span. Indeed, it is commonplace for software developers to undertake maintenance programming on systems there the code was written before they were even born. Within this context it is illuminating to compare and contrast the two modes of development and review the place of data mining applications.

Long-lived software is generally the result of a long design cycle, expending many many man-years of effort and an extensive period of testing. Such projects may fail due to their complexity. Their costs may overrun, risking cancellation or they may miss the business goals for which they were initiated because the business has moved on at such a pace that the application is obsolete at the time of delivery.

In contrast, short lived software is generally the result of a very short development cycle – indeed 'Rapid Prototyping' is a common expression in this regard. The design is often rigid and extension to the application may be very hard to undertake without a complete re-write.

MiningSTUDIO, and the data mining toolkit therein, supports both modes of development. The components with Studio have been crafted so that they can be incorporated within bullet-proof applications. Conversely, they integrate excellently with rapid application development environments such as Visual Basic. It is conceivable that the result of a short data mining engagement may no longer be just a static hard copy report – but it may be a data mining application that the user can interact with and explore the relationships in the data for themselves. This potential for "throw-away" applications means that their development must be very fast and efficient. Two examples data mining applications are described below.

Example 1 - Enhancing Account Management Reports

Consider a sales account manager working for a telecommunications company with a client base of small businesses. Before visiting or calling a client, the manager would want to review their past and present activity. This may well take the form of generating a standard report showing number of lines, total bill, average calls per day, or average calls per hour.

A data mining application could be used to enhance this report by interpreting the data, by analysing what was distinctive about the client's activity in contrast to the overall base and to other companies in the same business sector. Time series analysis of phone activity would support predictive modelling, looking for future usage trends. This knowledge would support the manager in better serving the client's telecommunication needs.

Example 2 - Early Warning Systems for Churn

Consider the same account manager, but this time his concern is to identify clients who might be about to leave for a competitor. The benefits of keeping a profitable customer are of course immediate, but the real value of a maintaining a high retention rate is only realized over time [8]. Customer attrition must be understood on an on-going basis. Many factors add up to make customer retention a long-term requirement and, perhaps, merit the effort of building a permanent churn application.

- The cost of customer acquisition is high. Many companies lose money on first year customers. This can be calculated by adding up the costs of *all* customer acquisition versus revenue earned from first year customers.
- The cost of maintaining experienced customers is less. They understand the services and are less likely to burden technical and customer assistance lines.
- Existing customers are targets for cross-selling opportunities and a source of referrals and therefore more profitable.

As new competitors enter the market, your own product offerings change and your competitor's offerings change, so the reasons for attrition are also likely to change. This reinforces the need to build permanent churn applications.

More specifically, the problem is one of identifying *valuable customers* who have a *high propensity* to switch to a competitor – the italicized words indicating the key concepts that must be modelled in a predictive application.

1. Valuable customers – by estimating the potential future spend and future cost to service from historical transaction records for each customer a measure of their future worth to the business may be derived.
2. Propensity to churn – or more specifically the probability that the customer will defect to a competitor within a specific time period.

This problem may be modelled by relating account information to call detail records and using a mechanism to identify the date of churn from a sequence of calling patterns over successive time periods.

6. Deployment of Data Mining Through Predictive Models

The predictive models that result from data mining activity are corporate assets. They can be produced as part of an application, as described above, or produced by analysts using a Data Mining tool. However, knowledge management should ensure their effective deployment.

One method of using a predictive model would be to undertake 'Bulk scoring'. This is the process of reading a set of records and writing the likelihood of an outcome, (e.g. response) for each record. Once this is obtained, the population can be sorted and the *best* candidates can be included in the marketing list. This would be undertaken using code generation – whereby the mining tool generates code to encapsulate the rules of the predictive model. Decision trees can be easily expressed in the form of code. MiningSTUDIO can generate a model in the form of a SQL Case statement. This allows the predictive model to be deployed directly into the database engine thus solving scalability issues. This is often the most efficient way to deploy a model and is relevant for both batch and real time model deployment.

Another alternative to bulk scoring that some application developers have embraced is to read the rules output and convert it into another form. This approach is applicable to virtually all data mining tools that can generate rules. This leads to real time model deployment, whereas batch deployment is only effective for campaign management, mail shots, etc., that are not time critical. Other applications require models to be run interactively, because batch scoring can be cumbersome with huge data volumes. Imagine a call centre application that requires a predictive model to indicate which tele-service to pitch a customer. Batch scoring could be used to write the results of the model for each customer record. When the customer phones the centre, the model results could be looked up. When the model changes, all the customers must be re-scored.

Clearly, it would be preferable to call the model on a real-time basis, for example, when the customer is on the phone. Using SQL Case rules is a natural way to deploy decision tree models on a real time environment.

An alternative to rule generation is to use the KnowledgeSERVER Interface to call predictive models on a real time basis. This has the advantage of abstracting the model type, so that an analyst could switch between a neural network and a decision tree without any changes to the client code, by simply saving a new model. The following code-segment shows how a client application can open a project and call a predictive model.

Another distinct advantage of real time scoring versus batch scoring is the ability to use dynamic inputs, such as time of day or events, during the conversation.

7. Exploration Workshops for Knowledge Discovery

A final way in which a business might exploit the advanced data mining approaches discussed above is through the use of Exploration Workshops. These workshop-style sessions for knowledge discovery bring together business managers and their business problems along with a powerful data exploration capability and data mining analysts to support them.

```
Dim proj as KSI.Project
Dim pm as KSI.PredictiveModel
Dim AttribInfo as KSI.AttributeInfo
Dim Arr As Variant
Dim Results As Variant

Set proj = New KSI.Project
proj.Initialize (False)

proj.Open ("n:\appfiles\CrossModel.ksp")
Set pm = proj.Items("Cross Model").Object

ksModel.ScoreRecord(AttribInfo, Arr, Results)

MsgBox pm.TrainAttributeInfo.DependentVariable.Name + " will be " _
       + CStr(Results(1)) + " (" + Format(Results(2), "0.00%") _
       + " Confidence)"
```

Figure 2: Example code for real time model deployment

As a precursor to an exploration workshop, a preliminary brainstorming session would be undertaken to identify:

1. The scope of business problem.
2. Initial ideas as to how it might be addressed.
3. The necessary data to support the analysis.

Ideally, the value of addressing the business problem should be specified – this may be a quantitative assessment (e.g. if we improve our fraud detection rate by 1% it will result in a $150m/year saving) or it may be more qualitative (e.g. if we improve the market segmentation then it will assist in campaign design).

Following on from the initial brain storming session, the necessary data for the analysis would be taken and loaded into the analytical environment. The data should be checked in two ways: verification and validation. Data verification checks that the data loaded is that which was intended. The data experts who understand the source systems would be needed to clarify the meaning of any columns and provide any additional meta-data necessary to understand the data. Validation checks the data with respect to the business problem to provide assurance that it is suitable for the analysis intended. Data may be unsuitable because it is of poor quality, having many missing or erroneous values.

The Exploration Workshop sessions would then follow, each being a facilitated session of 2 to 4 hours whereby one or more business managers engage with their problems and with analysts supporting their work using the data and the data mining toolkits. A very important aspect is the interactivity – allowing business managers to hypothesize about possible solutions to their problems and using the support of the analysts study these interactively.

This close coupling of data exploration power –very large databases avoiding the complexities of sampling, data mining tools along with analytical skills – provide the business manager an ideal environment to very quickly explore and resolve business problems.

References

[1] Mackin N., 1997, *Application of WhiteCross MPP Servers to Data Mining*, pp1-8, Proceedings of the First International Conference on the Practical Application of Knowledge Discovery and Data Mining,, Practical Application Company, 1997

[2] www.crisp-dm.org

[3] Freitas, A.A. 1998. *Mining Very Large Databases with Parallel Processing*, Kluwer Academic Publishers, London.

[4] www.angoss.com

[5] Quinlan, J. R., 1992. *C4.5 : Programs for Machine Learning*, Morgan Kaufmann Series in Machine Learning, 1992

[6] Quinlan, J.R., 1986. *The Induction of Decision Trees*. Machine Learning, 1 (1)

[7] Smyth, P. and Goodman, R., M., 1991. *Rule Induction Using Information Theory*, p. 159-170, Knowledge Discovery in Databases, Piatetsky-Shapiro, G. and Frawley, W., eds., AAAI Press.

[8] Reichheld, F.F., 1996. *The Loyalty Effect*. Harvard Business School.

Chapter 5

Industrial Case Study

Knowledge Management for the Strategic Design and Manufacture of Polymer
Composite Products
T. J. Lenz and J. K. McDowell

Cutting Tool Design Knowledge Capture
J. Bailey, R. Roy, R. Harris and A. Tanner

An Organic Architecture for Distributed Knowledge Management
G. R. R. Justo, T. Karran and M.J. Zemerly

Life-cycle knowledge management in the design of large made-to-order (MTO)
products
C. Laing, B. Florida-James and K. Chao

Xpat: A Tool for Manufacturing Knowledge Elicitation
B. Adesola, R. Roy and S. Thornton

CommonKADS and Use Cases for Industrial Requirements Gathering
A. D. Steele, G. Verschelling and R. Boerefijn

Knowledge Management for the Strategic Design and Manufacture of Polymer Composite Products

Timothy J. Lenz[1] and James K. McDowell[2]

[1]Michigan State University, East Lansing, MI, USA, lenz@cse.msu.edu
[2]Mad Dog Composites, Inc., Holt, MI, USA, maddog@grafix-net.com

Abstract: A corporate knowledge handling strategy should permeate an organization, but the design and manufacturing arenas are especially important. Considering the design and manufacturing activities holistically and conceptually promotes strategic application of knowledge. The holistic approach allows the consideration of issues from all dimensions and avoids decision making in a vacuum. A conceptual approach allows the suppression of detail, providing relevance and clarity. The idea is not to control technologists at the preliminary phases but to dispense focused and structured decision making knowledge. Such a knowledge management approach can facilitate successful product development without hindering creativity or needlessly restricting product descriptions. This discussion takes these themes and integrates them with knowledge systems philosophies into a vision for strategic knowledge management. Specifically, this vision is applied to the design and manufacturing of polymer composite products.

1. Introduction

The notion of a corporate aegis for knowledge is not novel. Such structured care of knowledge is a requisite part of being competitive in today's fast paced global economy. Beyond the typical corporate teams of CFOs, COOs, CTOs and CEOs, companies now employ Chief Knowledge Officers (CKOs). Executed properly, a corporate knowledge handling strategy should permeate every aspect of an organization, not just the board room and corporate office. Nowhere is knowledge management more important for product companies than at the front lines in the design groups and on the manufacturing shop floor. Design decisions must be made with the manufacturing consequences in mind; shop-floor modifications of design specifications should not be made without knowledge of their impact on the design's functionality. Too often, this interchange of knowledge is thwarted, one way or another: the entropic leanings of the workplace foster hermetically isolated patterns of behavior. Rather than hermetic isolation, a holistic approach to the general design and manufacturing activities must be emphasized. To ensure efficiency and effectiveness, a concerted effort at administering knowledge at this level must be made. This presentation focuses on the application of such knowledge management for polymer composite product design and manufacturing.

This discussion takes the themes of knowledge management and the systems development experiences of knowledge systems for design/manufacturing and integrates them into a vision for conceptual level knowledge management. Specifically, this vision is applied to the design and manufacturing of polymer composite products. Considering the design/manufacturing process both conceptually and holistically enables the strategic application of knowledge. A critical and important aspect of this knowledge management approach is that it can facilitate successful product development without hindering creativity or needlessly restricting product descriptions. The idea is not to control technologists at the conceptual phases but to support them with properly contextualized knowledge needed to make effective decisions. These intelligent decisions are made to streamline product development and consider the interacting issues of materials, design, and processing with the goal of supporting successful product realization.

1.1 Knowledge Management in Polymer Composites

Technologists working in expanding the use and application of polymer composites are often faced with several barriers to a successful product introduction. The end customer is often unfamiliar with the material technology, the structural design concept, and/or the manufacturing approach. The customer may have a background in sheet metal forming that has little in common with the new polymer composite product except, say, gross geometry and attachment requirements. It is important to guide the end customer through the critical issues and arrest any superfluous concerns. Multidisciplinary teams and design reviews, while well-intentioned, are other sources of difficulty. While an integrated, multidimensional approach is critical to the success of polymer composite products, improper considerations from a spurious dimension can sabotage the development process. Such improper issues and concerns are those that do not influence a product's functionality or final utility for the customer. Additionally, experts can have subconscious biases that do not apply to the product under development. It is important to recall the context of knowledge application and the actual consequences of a design decision. A principled knowledge management effort can help this by presenting focused and structured decision making knowledge. As a result, both customers and experts (in-house and external) can support the technologists involved in the development of a product.

Careful organization and handling of knowledge can do much to advance the development of polymer composite products. The success of a product using polymer composites depends largely on the application of critical knowledge (e.g., materials knowledge, structural design knowledge, and processing knowledge) at key points in the development cycle. Success hinges on the timely use of the proper knowledge in the correct product development context. Composite material product difficulties and failures often occur when knowledge is ill-timed or incorrectly applied. Processing issues, say, can dominate product decisions without addressing the implications of interacting issues and determining whether these issues are critical to the product functionality. Conversely, ignoring the manufacturing consequences of a design decision can needlessly increase the cost of the product or make it impossible to fabricate. Knowledge management can insure that critical information is available during the development cycle and simultaneously prevent expertise biases from leading the development along unfruitful paths. The ultimate goal is the affordable and effective use of polymer composites in new applications.

1.2 Holistic Perspective

The use of a holistic approach to the general design/manufacturing process dovetails well with the domain of polymer composites. A holistic approach allows considerations from many different dimensions and emphasizes the downstream consequences beyond the immediate result. The emphasis is on integration and functional confluence and not on isolation and unilateral decisions. Design and manufacturing, while sequential in the actual product development cycle, are considered as a whole.

For polymer composites the holistic approach is critical for covering the contribution of material issues, structural design issues and processing issues and the interactions among them. From the materials side the reinforcing fiber and polymer resin system must be effectively combined to produce the expected physical properties and mechanical behavior. The polymer matrix protects the fiber and allows load transfer across the fiber architecture. Fiber type and fiber architecture must be chosen with the wide range of loading types and magnitudes expected in the performance envelope. The structural design takes the capabilities of the composite material and begins addressing geometric constraints and requirements of the artifact. It also considers materials combinations and configurations that might achieve the necessary functionality. Processing issues must be resolved so that the final product is composed of the expected material, with controlled product quality and dimensionality. The manufacturing process must bring together the fiber and polymer resin, achieve a desired fiber volume fraction, cure-out the polymer (for thermosetting matrices) and maintain the desired fiber architecture. This must all happen while producing a part with the desired final geometry. For multicomponent assemblies, issues of bonding and fastening must also be addressed by the materials choices, the geometric features of the components and the selection of adhesives and mechanical fasteners.

1.3 Conceptual Perspective

Coupled with this theme of holism is a conceptual design approach for both the product and the manufacturing processes used for product fabrication. At the early stage of the design, issues are presented at a more abstract level, rather than delving into detailed descriptions of the product and processes. The conceptual phases of a development cycle offer the most opportunity for improving product affordability and effectiveness. Decisions at the conceptual phase, while made early in the cycle, dominate the costs found in the final product (e.g., [1]). Additionally, high quality decisions at the conceptual phase lead to fewer instances of design backtracking, product modifications, tooling rework and product retesting. This is highly desirable. Holistic and conceptual handling of knowledge can improve the quality of decisions and capture the context under which these early decisions are made. Such knowledge management can also help to identify interactions among critical polymer composites issues (especially for materials, structures and processing) and highlight the consequences of decisions made in these areas.

On the conceptual design side the goal is to present the technologist with design alternatives and provide explication, characterization and evaluation of the knowledge system results. This is done to impart understanding of the design landscape, the possibilities and the tradeoffs to the technologist. The knowledge system must exhibit a considerable amount of flexibility and allow the user to interact with partial requirements and partial design results as a possible starting point.

2. Knowledge System Technologies

Researchers in the field of artificial intelligence have come to realize that there are several different types of tasks for which AI can be used. These tasks range from the mundane (e.g., vision, robot control) to the formal (e.g., chess-playing, logical mathematics) and the expert (e.g., engineering design, medical diagnosis). While these tasks are all difficult and require knowledge, the expert tasks are the most dependent upon the nature and structure of the embedded knowledge. The dependence of these expert tasks on knowledge is so great that the approaches generated to solve them are collective known as knowledge system technologies.

Specific AI techniques for problem solving have been used as the bases for knowledge management systems for over a decade. Algorithms typically useful for this emphasize the use of common sense and discovered data. Examples of these AI techniques include ontologies and case-based reasoning (CBR), among others (e.g., machine learning, information extraction, and petri-nets). This section addresses four different types of knowledge system technologies that may be applicable for generating knowledge management systems: compiled approaches, ontologies, case-based reasoning, and knowledge level approaches.

2.1 Compiled Approaches

A compiled knowledge system approach to knowledge management implies the commingling of several different approaches. Generally, such combinations are not driven by sophisticated considerations; rather, ease of use and convenience dominate. This type of approach typically results in systems that, while useful, are quite brittle. Two particular compiled approaches to knowledge management, CALL and Égide, are presented here to convey a sense of the capabilities of such systems.

The Navy Combined Automated Lessons Learned Information Center (CALL) is a web-based repository of technical engineering experiences [2]. It stores knowledge about myriad issues (e.g., data management, design engineering, quality assurance, and training). CALL is a database that consists of over 5,000 learned lessons, and its keyword search feature allows quick and easy access to them. When consulted properly, these archived experiences can serve to engender informed decisions for a variety of technical engineering issues.

The Égide [3] approach to knowledge management is particularly evocative. The Égide system can both evaluate conflicts and reach design solutions while documenting the choices made and the reasons why. It is based on extensions made to the IBIS (issue-based information system) network approach [4]. These augmented IBIS networks contain a picture of an evolving design rationale expressed in terms of issues, positions, arguments, and conflict resolutions. Declarative use of an IBIS structure enables tracking of discussion issues, suggested alternatives, and the details of issue resolution. This helps not only to document the process of decision making thoroughly, but it also enables an individual awareness of group work. This, in turn, facilitates learning from previous experiences.

2.2 Ontologies

Ontologies generically encapsulate knowledge to enable a practical and ubiquitous understanding of a domain. A knowledge representation ontology such as FrameOntology [5] contain elements used to formally describe exemplars of knowledge, enabling other ontologies to be expressed in its representation style. The most famous of the general ontologies is the CYC ontology [6], which purports to mimic human common sense. Top-level ontologies (e.g., Sowa's boolean lattice [7]) embody a rigid structure under which all elements of the ontology are collected. Finally, domain ontologies contain concepts and relationships among them for a particular domain. Examples of domain ontologies include the Enterprise Ontology [8] for modeling businesses and the Plinius ontology for ceramics [9, 10]. There are many open issues in ontology research and Uschold discusses them later in this volume [11].

2.3 Case-Based Reasoning

Case-based reasoning is a problem solving paradigm that is arguably different from other major AI approaches. Instead of relying entirely on general knowledge of a problem domain, or establishing abstracted relationships between problems and solutions, CBR can use the specific knowledge of previous concrete problems. This is an experience-based method that attempts to deal with current problems through a comparison to past solutions. Experiences cached as cases allow incremental learning, which in turn allows the reasoner to become more efficient. This incremental learning occurs as a case-based reasoning system accumulates both specific and general knowledge. Potter, et al. discuss the potential application of CBR to knowledge management elsewhere in this volume [12].

2.4 Knowledge Level Approaches

At the root of all knowledge systems research is a desire to quantify knowledge intensive problem solving behavior. Newell's Knowledge Level [13] provides a way of understanding a problem solving agent apart from the implementation details. Although this allows a deeper understanding of problem solving than that possible through a symbolic level evaluation, it does not always allow the prediction of the behavior of an agent.

An organizational approach intended to facilitate interaction between different problem solving tasks has been proposed [14]. This hypothesis enables the specification of the behavior of a multitask system by explicitly representing the interactions between its agents. Other researchers have presented alternatives for the integration of multiple problem solving types with the Knowledge Level as a motivating factor. Goel has developed a technique for the combination of case-based reasoning and model-based reasoning in KRITIK [15]. Principles of Punch's TIPS architecture [16] allowed for the flexible integration of different problem solving types. The SOAR system [17, 18] also used a flexible integration architecture for its problem solving operators.

Several detailed methodologies for implementing systems at the knowledge level have been developed. These foundations simplify the knowledge acquisition process by focusing on the high level descriptions of problem solving and not on the specific low level implementation languages. The commonKADS approach [19, 20, 21], while not originally intended for use in knowledge management, has proven to be a successful support tool in practice. It presents guidelines for how knowledge analysis and knowledge-system development can be used as techniques within a general knowledge management approach. Task specific architectures [19, 22] and Generic Tasks [23 - 26] both support a long range goal of an engineering science style methodology for designing and developing intelligent decision support systems, and are readily applicable to knowledge management.

3. Knowledge Management for Polymer Composites Design and Manufacturing

The standard procedures used for designing a composite artifact are often heuristic in nature or the result of tedious and expensive trial-and-error testing. Although the composites designer has an overwhelming amount of information available, this information typically consists of experimental data and collections of unorganized heuristics, neither of which is conducive for use in design. Consequently, the design of polymer composites is essentially an artform guided by a combination of experience and scientific principles. A disciplined knowledge management system can help to impose structure on this chaos, enabling design and manufacturing to be done more effectively.

Recall, however, that this is knowledge management at a conceptual level. This means that all of the following discussion addresses knowledge management accordingly. A primary feature of this conceptual level consideration is the explicit generation of design alternatives.

Additionally, this is a methodology heavily influenced by a knowledge systems technology implementation bias. As such, it is driven by explicit assumptions about the practice of modeling expertise. Smithers has encapsulated these assumptions in five principles [27]: knowledge application (KA), knowledge level (KL), role limiting (RL), differentiated rationality (DR), and knowledge typing (KT). KA states that all actual design is the application of appropriate domain and task knowledge. KL extends Newell's notion of the Knowledge Level and demands a conceptual problem solving description that is independent of implementation details. RL insists that all embedded knowledge have identifiable and restricted roles in the design process. DR permits the use of role-limited knowledge in generating results by applying the appropriate knowledge in the appropriate way. Finally, KT specifies three different kinds of knowledge: domain knowledge, task knowledge, and inference knowledge. The application of these five principles will generate a comprehensive and consistent knowledge systems methodology.

There is a natural division between design and manufacturing for the methodology that follows. However, this division does not imply independence. Meticulous attention to the type and nature of the interactions between design and manufacturing enables them to be rigidly defined; only then can design and manufacturing be considered "separate." The locations and nature of these interactions are evident in the discussion that follows.

Beyond just the knowledge system methodologies for design and manufacturing, several important features are necessary for an effective knowledge management system. These include several contributions that add value beyond just the results generated by the design and manufacturing methodologies. This following discussion addresses these, identifying their importance to effective knowledge management.

3.1 Knowledge System Methodology for Design

A definition of the design of an artifact/product is a procedure that describes an object as a collection of components and features that, when considered collectively, satisfy the requirements. One important note is that there is no distinction between de novo design and redesign: all design is handled in the same way. The way in which the procedure is developed includes several different aspects, and this discussion presents a high level depiction of them.

The first step is to develop an understanding of the design problem. This is done by explaining the design requirements to the system in terms that it can understand. A specialized ontology serves as the language with which the requirements are conveyed. This ontology encompasses not only material and structural information, but also processing details. The ontology also serves as a common language for exchanging information between the design and manufacturing portions of the methodology. Once this information has been entered, the system "understands" the requirements in terms of functional and configuration knowledge models. These models are different perspective of the same description of the design requirements.

The system then examines these knowledge models for ways in which they can be modified. This potential modification looks to capitalize on known strengths of polymer composites by adapting functionality, behavior, geometry, or structural decomposition. Several different modifications are possible for each knowledge model, and the result is a set of plans that outline ways in which the artifact can be produced from composites.

The plans are used to generate a series of sparse conceptual composite assemblies [28]. These assemblies include top-level descriptions of the composite constructions (e.g., honeycomb panel with woven composite skin, surface laminated casting). Preliminary critique and analysis of these sparse designs identify ways in which improvements can be made. A secondary instantiation along the identified avenues of improvement leads to yet more detailed designs.

The result of the design process is a conceptual composite assembly. This entails a hierarchical depiction of the artifact (assembly/component/feature) with the initial requirements. The user can then identify which of the designs should be passed on for further consideration of its manufacturing plans.

3.2 Knowledge System Methodology for Manufacturing

The design of a manufacturing plan involves defining a sequence of physical processes that will produce components and features that, when assembled, provide the intended functionality of the artifact. A conceptual composite assembly, either built directly by the user or generated by the design portion above, is the beginning.

The first step in the generation of a manufacturing plan is the specification of a sequence of processing steps. These processing steps include the use of polymer composite fabrication technologies such as oven curing or autoclaving. This defines the paths from raw materials to individual components. Multiple alternatives for the fabrication of these components are possible. Next, the methods for adding features to the components are defined. Again, alternatives are possible. After that, the sequence for joining the components into the subassemblies or assemblies is decided. Methods for adding features to the assemblies are then specified. Finally, the manufacturing sequence for generating the artifact is compiled.

The result of the manufacturing planning process is set of organized plans that shows a sequence of states where components/features are generated over time, progressing from raw materials to assembled artifact. These states may be dependent upon access to or the existence of auxiliary equipment (e.g., molds, bonding jigs, machining fixtures, and machining operations).

3.3 Value Beyond Results

While providing design alternatives for product descriptions and manufacturing plans is useful, it is very easy to get lost in the multitude of results. Knowledge management requires a capability to visualize and manipulate the generated results. For the integrated design and manufacturing of polymer composite products, three forms of value-added interactions emerge: characterization, evaluation, and explication.

3.3.1 Characterization

The nature of the multitude of design and manufacturing alternatives implies the existence of significant data beyond just a naive presentation of the results of the system. Characterization is a means to examine that additional level of information. The most basic characterization is a summary of the design/planning results. This allows the user to comprehend the scope of the design results without having to examine each design and catalog similarities and differences.

For composite products, results might be summarized along common elements in descriptions or plans (e.g., designs that use carbon fiber, manufacturing plans with oven curing). Other functional, configurational or plan groupings might also be used to organize results: designs that use sandwich construction for stiffness, designs that use both composites and metals, or plans that machine all joining features.

3.3.2 Evaluation

An important issue to address is how to discern whether one design is significantly better than another. This is especially important considering the nature of the conceptual knowledge systems approach in which families of alternatives are generated. Another issue is how to decide whether a redesign should be modified rather than considered inferior to another redesign. These issues are typical of a divergent-convergent philosophy, where a design problem is expanded into many solutions before it is narrowed to one solution.

An evaluation may involve the generation of a score of a design/planning result based on absolute and relative criteria. User-assigned weights can be used to influence the final rankings and the results of the evaluation may be used to remove or segregate results. Ideally, one or a few results will be superior to others.

3.3.3 Explication

For the knowledge management methodology above to be successful, an implementation of it must include ways for users to easily interact with it. Two methods of interaction are typically used in expert systems: explanation of results and knowledge acquisition/modification. Of these two, an explanation facility is paramount for this approach to be accepted. Design engineers do not often accept results (especially from a knowledge system) on blind faith; there must be adequate justification of the results. Additionally, beyond just a simple explanation of results, there should be a capability to explicate any design failures. This adds to the system's capabilities in two ways: it guides the user to alternative requirements that may generate other results, and it helps to check errors in the embedded domain knowledge.

Two classic examples of expert systems that included these types of user interaction were the TEIRESIAS [29, 30] and SALT [31] systems. The TEIRESIAS system was probably the first program to support explanation and knowledge acquisition. It enabled a user to ask "Why?" and "How?" in response to the generated requests for information and results, respectively. The system could respond by generating a trace through its rule-base. The SALT system was a knowledge acquisition system used to build propose-and-revise expert systems. It could answer questions like "Why not?" and "What if?". These questions were used to locate incorrect or missing knowledge in the system.

Similar modes of operation are envisioned for the knowledge system methodology outlined above. Potential interactions with the system could be in any of several ways. One could be a simple explanation of generated results upon request. Two, the system could automatically notify the user of a potential error during runtime (e.g., "Honeycomb-stiffened panel is selected but I am unable to fully parameterize a design"). Additionally, the system could notify the user of near-misses and the reasoning that caused the miss (e.g., "I want to use carbon for the fiber type as weight savings is critical and load severity is high, but the need for elongation is blocking this choice"). This addresses an atypical side of explanation in design, i.e., explaining why a system cannot generate an answer.

Beyond these potential explanation options, there is yet another way in which the user could interact with the system. If the user is given the ability to enter the "design pipeline" at multiple points in the process with partial requirements and partial design results, a different type of explanation is necessary. For example, if the user says, "I want an airfoil leading edge made of carbon fiber," the system could respond with *"Carbon fiber has high tensile and stiffness properties that are not really required for your application, but it works well in weight critical situations like racing. However, carbon fiber is very poor under impact conditions and not recommended as a protective shield."* The system then could suggest alternatives.

4. Examples of Knowledge Management for Composites

While this discussion does not espouse a specific knowledge systems technology for a knowledge management system, it is important to point to and discuss relevant application work. Examples include the COMADE system, the DSSPreform system, and the Raven/Socharis suite. The following discussion addresses each of these tools, looking at how each approach handled knowledge management.

4.1 COMADE

A system for polymer composite material system design, COMADE [32, 33], emphasized the acquisition and organization of domain knowledge for polymer composite material design and the inclusion of that knowledge into an integrated Generic Task problem solving framework. COMADE considered the performance requirements (e.g., tensile strength, flexural modulus) and the environmental conditions (e.g., chemical environment, use temperature) an assembly may face and generated multiple material system designs. It specified polymer composite material systems as combinations of polymer matrix materials, chemical agents (curing, cocuring, reactive diluents), fiber materials, and fiber lengths. COMADE could generate over one thousand material system designs, ranging from simple polyesters to exotic thermoplastic systems.

COMADE provided a focus for composite material system design, while also presenting possibilities for families of composite material systems that may not have been immediately obvious. An integrated environment was built for composite material design that embodied COMADE, a case-based reasoning tool, and a design modification algorithm. This enabled a complete handling of design. However, COMADE did not consider fabrication issues, and therefore could not handle comprehensive knowledge management.

4.2 DSSPreform

The DSSPreform (Decision Support System for Preforming) system was developed at the University of Delaware with Lanxide Corporation [37]. This system was developed to aid in the design and manufacture of porous preforms which are subsequently infiltrated with a matrix material to form ceramic- and metal-matrix composites. The goal was to provide a comprehensive tool for realizing rapid, efficient manufacturing that could start with a material specification and finish with control of the manufacturing process, covering all points between.

The design philosophy upon which DSSPreform was based distinguished between material selection, process selection, and the parameterization of the manufacturing process. A linearized design process beginning with material selection, going to the process selection and ending with process parameterization was developed and implemented using a mixture of a knowledge-based system and model-based reasoning.

The approach to the design of the entire process from material selection to production was quite detailed and the interactions between the varying portions of the design were enumerated explicitly. While the framework for DSSPreform was quite detailed, the tool was never much more than a prototype. The portions of DSSPreform that were implemented, most notably the Materials Expert, were quite detailed and did provide part of the intended comprehensive tool. If the system had been completely implemented, it had the potential to be a very effective tool for knowledge management in ceramic composites.

4.3 Raven & Socharis

The Raven and Socharis are intelligent assistants for conceptual design and manufacturing in polymer composites. Although each is a standalone system, Raven and Socharis can cooperate to provide design assistance for both structural concepts in composites and concepts for process planning. Both systems are based on Generic Tasks, a knowledge-based systems technology.

The Raven system [34] works from the design requirements and an abstracted description of an existing metal assembly and produces conceptual descriptions of replacement polymer composite products. The generated polymer composite product designs can effectively replace the original assembly. The Socharis system [35, 36] takes a conceptual composite assembly (either generated by Raven, or built by the user/designer directly) as input and recommends manufacturing options. These options include component processing technologies, feature generation, and joining strategies for the assembly.

Both Raven and Socharis generate a large space of solutions that the user can explore. This enables an appreciation of the scope of the possible solutions and the critical features of the results. Both Raven and Socharis also have user-directed evaluation modules that help the designer to focus on a handful of the most plausible and effective designs or process plans. However, neither Raven nor Socharis could do disciplined material design. Both systems employ an *ad hoc* adaptation of COMADE for material design. Additionally, because Socharis does not emphasize a plan approach, it could not generate many potential aspects of the manufacturing plan (e.g., tool/mold count reduction, assembly/bonding fixture consideration, feature addition jigs/fixtures).

5. Discussion and Conclusions

Technology development requires that theory be brought into practice and knowledge management is no different. The issues described here are by-products of applied research and development efforts in knowledge systems for composites of the last several years. Implemented prototypes such as Raven and Socharis expose shortcomings in approach and deficiencies in knowledge. They also raise the bar for theory and expectation in the future. A strong central ontology, value beyond results (especially explication), and an ever increasing flexibility for use continue to challenge knowledge management technology.

While our current focus is the domain of polymer matrix composite products, there are natural extensions to this work. Composites remain an important and growing materials research area. Metal matrix composites and ceramic matrix composites share many complexities of materials, structures, and process faced in polymer composites. We also see elements of our knowledge level architecture that may be useful in other design and manufacturing activities. Specifically, these would be applicable when consequences of decisions have the potential to increase cost and disrupt functionality.

An important challenge for knowledge management in polymer composite product design and manufacturing is situational flexibility. A knowledge management system must not just have a single mode of interaction. Instead, it must allow considerable flexibility in knowledge processing. This does not mean applying knowledge in inappropriate contexts; rather it means if an impasse is reached the system allows the user to understand the impasse and change some aspect of the current situation to reroute the problem solving. Suppose, for example, the user wanted to construct the product using a single type of manufacturing process. The knowledge management system could enforce the requirements/ constraints could cause such a choice and allow the user to change manufacturing requirements and product description to achieve this result. Other scenarios might involve fully defining a product description and generating possible manufacturing plans or starting with partial descriptions of product or manufacturing details.

6. References

[1] Ullman, D. G. 1997. The Mechanical Design Process. 2nd edition. McGraw-Hill.

[2] http://www.nawcad.navy.mil/call

[3] Bañares-Alcántara, R., J.M.P. King, and G.H. Ballinger. (1995). Égide: A design support system for conceptual chemical process design. In Proceedings the 1995 Lancaster International Workshop on AI System Support for Conceptual Design. Ambleside, Lake District, England, 27-29 March.

[4] Rittel, H.W.J., and M.M. Webber. (1973). Dilemmas in a general theory of planning. Policy Sciences, 4:155-169.

[5] Gruber, T.R. (1993). A translation approach to portable ontology specifications. Knowledge Acquisition, 5:199-220.

[6] Lenat, D.G., and R.V. Guha. (1990). Building large knowledge-based systems. Representation and inference in the Cyc project. Addison-Wesley, Reading, Massachusetts.

[7] Sowa, J.F. (1999). Knowledge Representation - Logical, Philosophical, and Computational Foundations. Pacific Grove, CA: Brooks/Cole Pub Co.

[8] Uschold, M. (1996). Building ontologies: towards a unified methodology. In Expert Systems 96.

[9] van der Vet, P.E. and N.J.I. Mars. (1994). Concept Systems as an Aid for Sharing and Reuse of Knowledge Bases in Materials Science. Knowledge-Based Applications in Material Science and Engineering, J.K. McDowell and K.J. Meltsner, eds., The Minerals, Metals, and Materials Society: Warrendale, Pennsylvania 43-55.

[10] van der Vet, P.E. and N.J.I. Mars. (1998). Bottom-Up Construction of Ontologies. IEEE Transactions on Knowledge and Data Engineering, 10(4):513-525, July/August.

[11] Uschold, M. (2000). Ontologies: A vehicle for Knowledge Sharing, this volume.

[12] Potter, S., et al. (2000). The Development of Case-Based Reasoning for Design - Techniques and Issues, this volume.

[13] Newell, A. (1982). The Knowledge Level. Artificial Intelligence, 18, 87-127.

[14] Sticklen, J. (1989). Problem Solving Architectures at the Knowledge Level. Journal of Experimental and Theoretical Artificial Intelligence, 1(1), 1-52.

[15] Goel, A. K. (1989). Integration of Case-Based Reasoning and Model-Based Reasoning for Adaptive Design Problem Solving. Ph.D. dissertation, The Ohio State University.

[16] Punch, W. F. (1989). A Diagnostic System Using A Task Integrated Problem Solver Architecture (TIPS). Ph.D. dissertation, The Ohio State University.

[17] Laird, J. E., A. Newell, & P. S. Rosenbloom. (1987). SOAR: An Architecture for General Intelligence. Artificial Intelligence, 33, 1-64.

[18] Steier, D. M., R. L. Lewis, J. F. Lehman, et al. (1993). Combining Multiple Knowledge Sources in an Integrated Intelligent System. IEEE Expert, 8(3), 35-44.

[19] Breuker, J. and B. Wielinga (1989). Models of Expertise in Knowledge Acquisition. Department of Social Science Informatics. Amsterdam, The Netherlands, University of Amsterdam.

[20] http://www.commonkads.uva.nl

[21] Steele, A.D., et al. (2000). CommonKADS and Use Cases for Industrial Requirements Gathering, this volume.

[22] Steels, L. (1990). Components of Expertise. AI Magazine.

[23] Chandrasekaran, B. (1983). Towards Taxonomy of Problem-Solving Types. AI Magazine. 7: 66-77.

[24] Clancey, W.J. (1985). Heuristic Classification. Artificial Intelligence XX(VII): 289-350.

[25] Chandrasekaran, B. (1986). Generic Tasks in Knowledge-based Reasoning: High-Level Building Blocks for Expert System Design. IEEE Expert (Fall 1996): 23-30.

[26] Brown, D. (1987). Routine Design Problem Solving. Knowledge Based Systems in Engineering and Architecture. J. Gero, Addison-Wesley.

[27] Smithers, T. (1996). On Knowledge Level Theories of Design Process. Artificial Intelligence in Design '96, Palo Alto, CA. 561-579.

[28] Lenz, T.J., M.C. Hawley, J. Sticklen, et al. (1998). Virtual Prototyping in Polymer Composites. Journal of Thermoplastic Composite Materials, 11(5), 394-416.

[29] Davis, R. (1977). Interactive transfer of expertise: Acquisition of new inference rules. In Proceedings IJCAI-77.

[30] Davis, R. (1982). Applications of meta level knowledge to the construction, maintenance and use of large knowledge bases. In Knowledge-Based Systems in Artificial Intelligence, Eds. R. Davis and D.B. Lenat. New York: McGraw-Hill.

[31] Marcus, S. and J. McDermott. (1989). SALT: A knowledge acquisition language for propose-and-revise systems. Artificial Intelligence 39(1).

[32] Lenz, T., J.K. McDowell, B. Moy, et al. (1994). Intelligent Decision Support for Polymer Composite Material Design in an Integrated Design Environment. Proceedings of the American Society of Composites 9th Technical Conference, pp. 685-691.

[33] Lenz, T.J. (1997). Designing Polymer Composite Material Systems Using Generic Tasks and Case Based Reasoning. Ph.D. Dissertation, Michigan State University.

[34] Zhou, K., T.J. Lenz, C. Radcliffe, et al. (1999). A Problem Solving Architecture for Virtual Prototyping in Metal to Polymer Composite Redesign, American Society of Mechanical Engineers Design Engineering Technical Conferences: Design Automation Conference (ASME DETC99/DAC), paper 8593.

[35] Martinez, I., O. Lukibanov, T.J. Lenz, et al. (1999). Augmenting Conceptual Design with Manufacturing: an Integrated Generic Task Approach. American Society of Mechanical Engineers Design Engineering Technical Conferences: Design for Manufacturing Conference (ASME DETC'99/DFM), paper 8948.

[36] Lukibanov, O., I. Martinez, T.J. Lenz, et al. (2000). Socharis: The Instantiation of a Strategy for Conceptual Manufacturing Planning. AIEDAM Special Issue on AI in Manufacturing. In press.

[37] Pitchumani, R., P. A. Schwenk, V. M. Karbhari, et al. (1993). A Knowledge-based Decision Support System for the Manufacture of Composite Preforms. Proceedings of the 25th SAMPE Technical Conference, Philadelphia, PA.

Cutting Tool Design Knowledge Capture

J. Bailey[1], R. Roy[1], R. Harris[2] and A Tanner[2]

[1]Department of Enterprise Integration, School of Industrial & Manufacturing Science, Building 53, Cranfield University, Cranfield, Bedford MK43 0AL, UK, j.i.bailey@cranfield.ac.uk, r.roy@cranfield.ac.uk
[2]Widia Valenite, 12 Alston Drive, Bradwell Abbey, Milton Keynes, MK13 9HA, UK, Richard_Harris@milacron.com, Andrew_Tanner@milacron.com

Abstract: This paper presents the challenges and issues that are encountered when capturing design knowledge in an industrial environment. Identifying and finding a representation for the relevant design knowledge are seen as the key activities in modelling design knowledge. Identification of design knowledge is shown through two case studies undertaken by the first author. In these cases, knowledge is considered to be the difference between the expert and the novice (Knowledge = Expert - Novice). Cutting Tool Design knowledge consists of design, manufacturing, external, internal and technical knowledge and designers consider many of these factors when designing. This paper presents preliminary results of ongoing research carried out at Cranfield University.

1. Introduction

Design in all engineering domains is a complex and knowledge intensive process. At present, most designs are performed by experts using their previous experience. With the demographic change in the labour market, it is becoming increasingly important to capture knowledge in a computer-based system so that it can be reused. In the cutting tool industry there is a shortage of new designers and this problem is worsened as it takes several years for a designer to become knowledgeable about the cutting tool domain. It has been highlighted that current designers come through an extensive apprenticeship, which provides the theoretical and practical knowledge of actually cutting metal. After this stage the designing actually begins. It is difficult for an expert to pass on his/her knowledge to the novice designer and the ongoing research at Cranfield University aims to provide a computer based learning tool or a decision support system to aid both the expert and the novice designer. The paper discusses some issues and initial results in capturing cutting tool design knowledge.

Throughout each stage of the design process the designer can access various items of knowledge and information in order to achieve the final design. Design does not take place in a vacuum; the designer has to consider many aspects in the product design. This paper looks at the information and knowledge needs of the cutting tool designer.

The design of cutting tools is complex and knowledge intensive. In addition to this the market is changing and with the high average age of designers and the lack of recruitment into engineering and, especially the cutting tool industry, results in the loss of this expertise, and according to current Knowledge Management literature this is likely to affect the commercial performance of the company. One answer is to collate and store this information and knowledge in a computer-based system or manual to ensure that the expertise stays within the company. This research is looking into a typical design process of the cutting tool industry in order to establish the information and knowledge used at each stage of the design process. The design process has been modelled using the $IDEF_0$ modelling technique. It begins with the initiation of the designer by a proposal via a salesperson. With this initial specification the designer can then search for the closest few designs which are similar in nature to the new proposal. This can become a case of redesign, modifying an existing design to provide a full manufacturing drawing. This paper aims to identify the types of knowledge and information that the designer requires to complete a design. In the future, the research will move on to capture knowledge from the whole design process in an attempt to develop the decision support system.

2. Models and Methods of Design

Design is an integral part of any product or process. Designers go through a number of processes to achieve the final specification from an initial list of requirements known as a design brief. The designer will solve problems through the design search space by a process of divergence and convergence to the eventual solution. Several iterations can be undertaken to find a solution that is acceptable. Final communication of a design is often in the form of drawings and depending on the complexity of the design, a full scale model of the artefact could be made.

The design process has received the attention of the design community for many years and many authors have attempted to provide maps or models of the process of design. These either describe the activities involved in the design process (descriptive models) or prescribe (prescriptive models) showing how to perform the activities in a better way. A more recent addition to design models have been the computational models, which emphasise the use of numerical and qualitative computational techniques, artificial intelligence techniques in conjunction with computing technologies. Design methods can be regarded as any procedures, techniques, aids, or 'tools' for designing [1]. They represent a number of distinct kinds of activities that the designer might use and combine into an overall design process.

2.1 The design models

Design models are the representations of philosophies or strategies proposed to show how design is and may be done [2]. Three classes of models can be seen to emerge – prescriptive, descriptive and computational models.

2.1.1 Prescriptive models of design

Prescriptive models of design are associated with the syntactics school of thought and tend to look at the design process from a global perspective, covering the procedural steps (that is suggesting the best way something should be done). These models tend to encourage designers to improve their ways of working. They usually offer a more algorithmic, systematic procedure to follow, and are often regarded as providing a particular design methodology [1]. They emphasise the need to understand the problem fully without overlooking any part of it and the 'real' problem is the one identified. They tend to structure the design process in three phases – analysis, synthesis and evaluation. An example of a prescriptive design process can be found in Hubka [3, 4, 5].

2.1.2 Descriptive models of design

Descriptive models are concerned with the designers' actions and activities during the design process (that is what is involved in designing and/or how it is done). These models emanated both from experience of individual designers and from studies carried out on how designs were created, that is what processes, strategies and problem solving methods designers used. These models usually emphasise the importance of generating one solution concept early in the process, thus reflecting the 'solution-focused' nature of design thinking [1]. The original solution goes through a process of analysis, evaluation, refinement (patching and repair) and development [2]. Finger & Dixon [6] further suggest that these models build models of the cognitive process – a cognitive model is a model that describes, simulates, or emulates the mental processes used by a designer while creating a design.

2.1.3 Computational models of design

A computer-based model expresses a method by which a computer may accomplish a specified task [6]. A computer-based model may in part be derived from observation of how humans think about the task, but this does not have to be the case. Often computer-based models are concerned with how computers can design or assist in designing. The former include those that make decisions and those that assist in the design process provide some kind of analysis (provide information on which design evaluations and decisions may be based). Finger & Dixon [6] suggest that computer-based models are specific to a well-defined class of design problems. These are parametric, configuration and conceptual design problem types.

> *Parametric* – the structure or attributes of the artefact are known at the outset of the design process. It then becomes the problem of assigning values to attributes which are called the parametric design variables.
> *Configuration* – or structure design, a physical concept is transformed into a configuration with a defined set of attributes, but with no particular values assigned.

Conceptual – functional requirements are transformed into a physical embodiment or configuration.

Computational methods focus on mapping function into structure and investigate which are intended for computer implementation. Within these models design is considered to be a process that maps an explicit set of requirements into a description of a physically realisable product which would satisfy these requirements plus implicit requirements imposed by the domain/environment [2].

2.2 Design methods

Design aids, tools and support systems are used in order to arrive at a realisable product and/or process. Design methods generally help to formalise and systemise activities within the design process and externalise design thinking, that is they try to get the designer's thoughts and thinking processes out of the head into charts and diagrams. There are several techniques which enable the designer to explore design situations (literature searching), search for ideas (brainstorming), explore the problem space (interaction matrix) and evaluate designs (ranking and weighting). A fuller descriptions of these and 35 other methods can be found in [1, 7].

2.3 Summary

The models presented here and in the literature do provide a logical approach to the design process which encourages designers to articulate the decisions, strategies that they undertake to achieve a design or artefact. However, many do try to overcomplicate the design activity by providing too detailed a description of the processes in the models. The argument stems that if a designer is constrained to a particular model, then the creativity that is inherent in any type design (engineering, industrial etc.) is lost. Most of designing is a mental process, that is the design is often done in the head. The models enable designers to provide a visual record of the processes that they undertake to achieve a particular design, along with the sketches and drawings that are also produced. This provides a series of rationales of why particular routes were taken in order to produce the artefact.

3. Design Knowledge Capture

Design knowledge, is in general, comprised of descriptive information, facts and rules. These are mainly derived from training, experience and general practice. Most design knowledge is vague and lacks order and is therefore difficult to capture, store and disseminate [8]. Furthermore, the knowledge is often accumulated over a number of years [9] and that most of the knowledge exists as separate 'islands of knowledge' [3]. The problem then becomes one of how to capture this knowledge and is often a difficult and time consuming process. Artificial Intelligence (AI), and in particular micro-level knowledge capture techniques have helped to solve design problems through modelling designer activity, the representation of designer knowledge, and the construction of either systems that produce designs or systems that assist designers [10].

This section presents an overview of the challenges faced while identifying, eliciting, recording and organising the design and manufacturing knowledge. It identifies current practices in the knowledge capture and reuse. It is observed that categorising the type of knowledge and matching a suitable representation for it are the major challenges faced in this area. The knowledge capture requires a good understanding of the design and manufacturing processes, and the development of a detailed process model helps in this understanding.

3.1 Capturing the knowledge

Knowledge capture is the eliciting, recording, and organising of knowledge [11]. The task is extremely difficult to achieve successfully and it requires an understanding of what kinds of knowledge to capture, how to represent it, and how it can be used in the future. The elicitation of knowledge is crucial and has been underestimated in the past [12]. The difficulty arises from identifying the right type of knowledge to achieve the particular functions required by support systems [13] and the knowledge users. Secondly, and probably the most difficult, is the actual elicitation from the domain experts who will have, their own language to categorise the knowledge that is being represented.

A process model of the design has been utilised by the authors to develop an $IDEF_0$ (Integrated Definition Method) model to understand the functional and structural relationships [14] in the process, and highlight where the designers find information and knowledge. The aim has been to describe in detail each of the inputs, outputs, constraints and resources that go into the design process [15], thus highlighting the categories of information and knowledge which need to be captured and represented.

3.2 Knowledge acquisition

Knowledge acquisition is the process of acquiring knowledge from a human expert (or group of experts) and using this knowledge to build a knowledge based system [16]. Knowledge acquisition and elicitation represents a large amount of the development time in the knowledge capture process. Two types of knowledge may be elicited from the domain expert, explicit and tacit knowledge. Eliciting tacit knowledge is the more difficult of the two, as this knowledge rests in the head of the expert. The problem is that often the expert doesn't know how to express the knowledge he or she uses on a day to day basis – it has become second nature to them.

There are various techniques of eliciting the knowledge from the expert, but it is mainly performed by interacting with the expert. These can be broken down into four main areas adapted from [17], shown in Table 1. An example of an indirect approach (used several times by the author) is to use a case study to show the difficulties faced by a novice in the domain when designing a component. Both the expert and knowledge engineer undertake the same design work from the design brief stage to the final design.

Table 1 – The main approaches to knowledge acquisition.

Direct Approaches	The knowledge engineer interacts directly with the expert to obtain an explanation of the knowledge that the expert applies in the design work
Observational Approaches	The knowledge engineer observes the expert in the performance of the design task
Indirect Approaches	The expert is not encouraged to try and verbalise his/her knowledge and the knowledge engineer uses other methods to elicit the information
Machine-based Approaches	Elicit knowledge through use of either knowledge-engineering languages or through induction from databases of domain examples

This method has highlighted, however, that the novice with basic CAD experience can design the basic component, but without the expert knowledge there are many areas of the component that need further revision, e.g. in a cutting tool design task - the angle at which an insert sits in the shank to achieve a particular cutting condition.

Machine-based knowledge elicitation tools use a computer to elicit and capture knowledge from the designer. There are several systems developed through research [11], such as IDE 1. 5 & 2.0, a hypermedia tool incorporating semiformal models of the domain and design of the component. Designers are required to perform a task on a component, after that period of work they must stop and record on the system, the decisions that led to the resulting design in that period of work.

Machine learning is a branch of AI concerned with the study and computer modelling of learning processes. It offers the potential, not only to alleviate the problem of knowledge acquisition, but also to enhance a system's problem-solving performance [18]. The design of cutting tools is based on experience and influenced by past designs. Machine learning provides a support tool to learn from this experience and past designs by obtaining, using, and maintaining this knowledge [10].

A further tool to capture knowledge includes the well-developed KADS (Knowledge Acquisition and Document Structuring) system [17], which provides an appropriate conceptual model of the domain, which can be translated into a design for the final system. It works by providing a 'template' knowledge structure to which the designer can add the elements for the particular application. This has now been taken over by the comprehensive CommonKADS methodology which develops knowledge based applications by constructing a set of engineering models of problem solving behaviour in its concrete organisation and application context [19]. This modelling concerns not only expert knowledge but also the various characteristics of how that knowledge is embedded and used in the organisational environment.

XPat [20] is a process driven elicitation technique that engages the experts in mapping the process and the knowledge themselves. Knowledge is elicited through inputs, process and outputs based around the $IDEF_0$ technique. It is primarily a paper based technique allowing representation of tasks graphically therefore avoiding the need for lengthy descriptive text. XPat is based upon three stages: pre-analysis, problem identification and collecting and interpreting the knowledge. The technique requires a direct elicitation approach with the domain experts at all stages of the process.

3.3 Knowledge representation

The activity of knowledge representation is the means of organising, portraying, and storing knowledge in a computer program which, leads to knowledgeable behaviour [21] using several techniques. These techniques include formal mathematical logic, state-space-speech, semantic nets, production systems and frames & objects [17]. There are four types of knowledge that need to be represented in a computer about a particular domain, declarative knowledge, procedural knowledge, heuristic knowledge and descriptive knowledge. Procedural knowledge refers to how to perform a task, whereas declarative knowledge is factual information and knowing what to do [3].

Heuristic knowledge refers to problem solving methods that are utilised by experts which have no formal basis or can be regarded as a 'rule-of-thumb'. This heuristic knowledge is often regarded as 'shallow knowledge' as the heuristics often ignore the formal laws and theories of a problem. Thus the level of knowledge an expert can have about a particular domain can be either 'deep' or 'shallow', shallow knowledge occurs when an expert has a superficial surface knowledge of the problem, whereas with deep knowledge and expert has full thorough grasp of the basic fundamentals of a problem.

Descriptive knowledge is the formulation of the heuristics used by an expert highlighting concepts in the problem domain which are central to the problem solving process which in effect represent a distilled version of the expert's background knowledge [22]. It provides a description of the problem domain by highlighting the important features and characteristics of the domain.

A knowledge representation can be considered by the roles it plays: as a surrogate or a substitute for the thing itself, an ontological commitment, a fragmentary theory of intelligence, a medium for practically efficient computation and a medium of human expression [22]. There are many tools and techniques that are available to facilitate knowledge representation. Most of these tools are based around the most commonly used representation techniques such as specialised languages, logic, objects, semantic nets, frames, procedural representations and production rules [23].

Ontologies are content theories about the sorts of objects, properties of objects, and relations between objects that are possible in a specified domain of knowledge [24]. The knowledge representation languages and techniques described are used represent ontologies. Ontologies facilitate the construction of a domain model providing a vocabulary of terms and relations with which to model the domain [25]. The benefit of using ontologies is the sharing and reuse capability which promotes a shared and common understanding of a domain that can be communicated across people and computers [25]. Noy and Hafner [26] review several prominent ontology design projects comparing purposes of the ontology, its coverage, its size, and the formalism used.

3.4 Summary

This literature survey has identified the processes involved in design and manufacturing knowledge capture. The objective of the knowledge capture is to reuse the knowledge through a computer based system. There is a need to develop a generic framework of knowledge capture for a category of products. The design and manufacturing knowledge need to be represented within one framework so that the designer can have access to manufacturing experience and knowledge and vice versa. Future work will involve the development of a representation technique that can handle both design and manufacturing knowledge. Identification of cutting tool design knowledge is the first stage in the knowledge capture process. The following section describe the first authors indirect approach to knowledge capture illustrated with two case studies.

4. Design Knowledge in an Industry based setting: A Case Study

Metal cutting is a mature technology, involving several disciplines of science. It is continually changing in line with strategies, material developments throughout the manufacturing industry worldwide, and also the developments within the metal cutting industry. The competitive challenge here is the continual provision of improvements to metal cutting production, thus leading to a race to provide better tool materials, cutting edge geometries and methods of toolholding. Metal working know-how and skill can be traced back many centuries. However, metal cutting known in today's industry began with the industrial revolution of the eighteenth and nineteenth centuries and accelerated during the twentieth [27].

Cutting tools range in complexity from the simple single point tools such as turning tools to multi point tools such as milling cutters. The purpose is to remove material from a component or surface to achieve a required geometry e.g. the machining of a casting. For the turning tool shown in Figure 1, a 'P-system' clamping mechanism is used, and the type of clamping/holding mechanism depends on the application.

The tool shown in Figure 1 is a turning tool which is designed and manufactured to machine an undercut on a cam-shaft. The backend of the tool is a standard DIN69893 fitting (the dimensions are standard and are available on microfiche) and as such is explicit. The actual design is carried out on the insert pocket and the shape of flank to meet the dimensions of the component avoiding any interference with the component. Special attention is paid to the design of the insert pocket to ensure dimensional accuracy.

4.1 Types of knowledge in cutting tool design

In order to capture relevant knowledge of the domain, it has been necessary to design actual cutting tools for customers from the proposal stage through to detail work. This was necessary, as the first author had very limited theoretical knowledge about cutting tool design. He is considered as a typical novice for the domain.

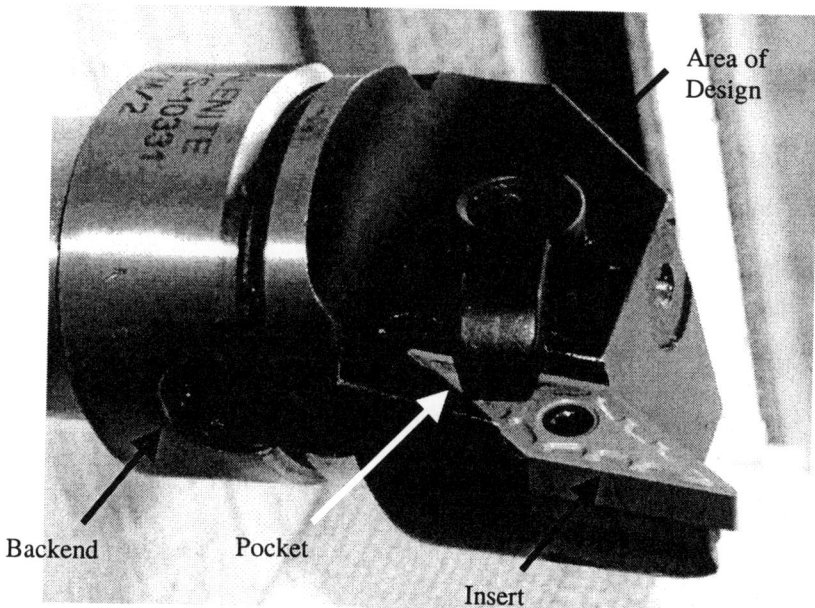

Figure 1: A turning tool.

The purpose of this design work is to understand the principles involved in designing cutting tools and to interact with the experts to identify the types of information and knowledge they utilise throughout the design process, including where they go to find this information and knowledge.

Many cutting tools have been designed by the first author over a period of 5 months, within the organisation, from single point turning tools to more complex milling tools. It has illustrated the difference between the novice designer (the first author), and an expert, who has around 20 years experience designing cutting tools.

The design knowledge is defined as:

$$Knowledge = Expert - Novice$$

This suggests that knowledge within the cutting tool domain is the difference between an expert designing a tool and a novice (the first author). Although it is possible to use textbooks to teach the rudimentary tool design to a novice, there are many instances when specific knowledge of the expert incorporates specialist knowledge of the designed products, which has been gained over the years.

So far only the technical knowledge that is needed when designing a cutting tool has been mentioned. It is certain that a designer would have to have many more types of information and knowledge at their fingertips in order to fully understand the issues that a complete design would have associated with it. Typically, this would involve the customer, economic and supplier knowledge that the designer holds on a personal basis and at a organisational level that has been built up over a number of years. In the cutting tool industry, reuse of past designs is high. Often it will be the designer remembering what they have done before, (see the tool in Figure 1). Here the designer was able to recall the design task carried out and the application with little or no trouble. Therefore, if this extra non-technical knowledge can be captured as well then one would have useful design rationales, taking into account the a global view of the design process. For instance, if in five years time a solution to a design problem points to a design undertaken five years ago, then the designer would have a complete picture of the state of the environment of that designed element. It is expected that this would improve the decision making to design the later product. Knowledge and information types required in cutting tool design are discussed below.

4.1.1 Knowledge in cutting tool design & manufacture

All design has some kind of formal process, in which the designers will follow a set of procedures to produce a final product, a design process. The design process as described within Widia Valenite is shown in Figure 2, which represents one model of design. As with knowledge, getting the designers to articulate this process and make it explicit was found to be difficult as they have been undertaking this process for many years and it has become second nature to them. It was felt important that this process be modelled as to understand the routines that a designer would undertake and also it provided the means to examine how the designers carried out their work.

The process in Figure 2 shows a conceptual stage (proposal) and a description of the technical system which is the checked detailed drawing. This provides a proposal via a salesmen which contains information about the cutting conditions, existing tools used in the operation, speeds and feed rates and dimensions of the component to be machined. With this information the designer will produce a proposal drawing, which is then cost estimated. The proposal and estimate are then sent back to the customer for approval from which an approval drawing is sent out, which returns to be detailed to a full manufacturing specification. After this, the drawing is checked and verified for functionality and manufacturability. If no modification is required then it is sent to contractors for manufacturing. Modifications are made by the detailing engineer which are then checked. At each of these stages the designer uses information and knowledge retrieved from many sources, and in these situations designers spend a lot of valuable design time. It has been observed that knowledge of the manufacturing process is utilised at the proposal stage, as it is here that the designer needs to make sure that the product can be manufactured before the proposal is sent to the customer.

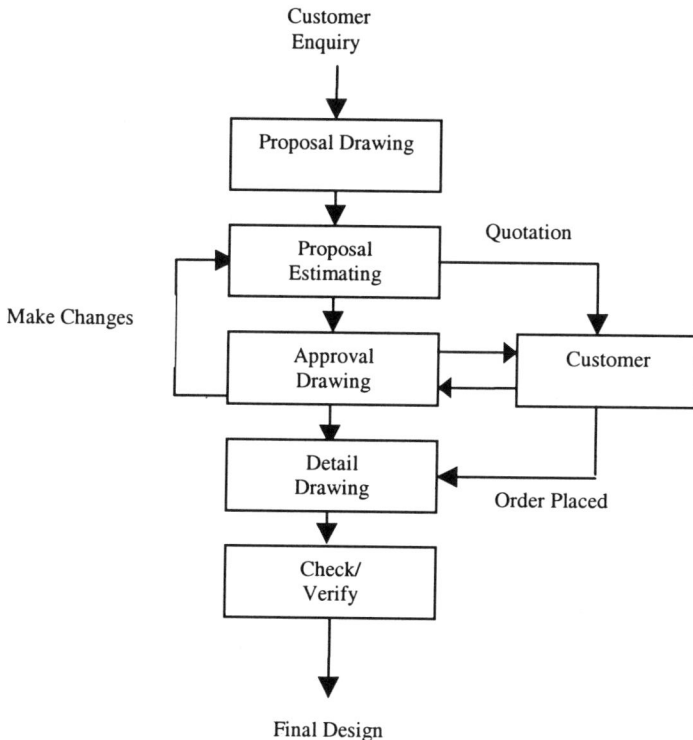

Figure 2: The design process at WIDIA VALENITE [28]

4.1.2 Knowledge in the design

Designers need knowledge and information of past designs, materials, manufacturing information and standards among others to be captured and represented in a computer-based tool. In the process described above the majority of the design work is to modify and reuse past designs. In this situation the designer retrieves only the drawing information from the CAD system without any knowledge of the intent in the design which would describe the decisions undertaken to achieve the design from the customer requirements. Design rationale is an explanation of how and why an artefact, or some portion of it, is designed the way it is [29] to meet the specifications set out by the customer. Capturing this rationale would be useful as the author has experienced several instances when the designer of a product is unavailable, and it has been too difficult to ascertain as to why the decisions were undertaken to design the product in a particular fashion.

4.1.3 Knowledge in manufacturing

Knowledge in manufacturing differs from that in design. For instance the types of knowledge of interest on the shop floor [30] are operation efficiency of plant and machinery, maintenance, control and raw material procurement, etc. The problem here is that the knowledge is located in many places, is of many differing topics and lends itself to different levels of precision [30]. What is important to the designer of cutting tools is the set-up costs for tooling in order to manufacture the design artefact, reliability and quality of the work carried out in previous cases. Also the capability of the manufacturer is an important factor i.e. whether the manufacturers have the expertise and equipment to undertake the task. At the conceptual and proposal stages during the design process the manufacturability of the artefact would be taken into account by the designer. Because of the background and experience of the designers are mostly based on an apprenticeship, the appreciation of what can be manufactured and the processes that are needed for manufacture is borne through trial and error. During observation of this small group of designers, it is worth highlighting that the designs are not 3D modelled and analysed by advanced computer methods but analysed through picturing in the head of the designer and then a through checking/verification procedure of what 'feels right'. Both design and manufacturing knowledge, however different in their categories and context of knowledge, still use the same methods of capturing and representing knowledge including the rationale as to how the process is planned the way it is [31].

4.1.4 Non-technical knowledge in cutting tool design

The pace of political, economic and technological change means that the design environment has become more challenging. It is now important that a designer considers a wide range of issues when designing the artefact in order to continuously innovate and to keep up with the increasing competition. As described above, not only is the technological knowledge important but also external (outside the organisation) and internal (inside the organisation) knowledge. These types of knowledge are shown in Figure 3 (not exhaustive). A designer needs a basic knowledge of each of the items mentioned in Figure 3. Each of the knowledge types can be broken down into more specific knowledge types as shown with customers, organisational culture and manufacturing possibilities. In the knowledge type 'customers', the designer would know what the current trends are for cutting tool materials (which the customer would know of through advertising), typical capabilities of the customer's machinery and the costs that the customer is likely to be satisfied with.

4.2 Identification of design knowledge

During the design training and observation case study it was possible to identify knowledge of cutting tool design. Most knowledge is implicit and in these cases it is necessary to converse with the experts to find a solution to a specific problem i.e. the location and angle of an insert on a shank.

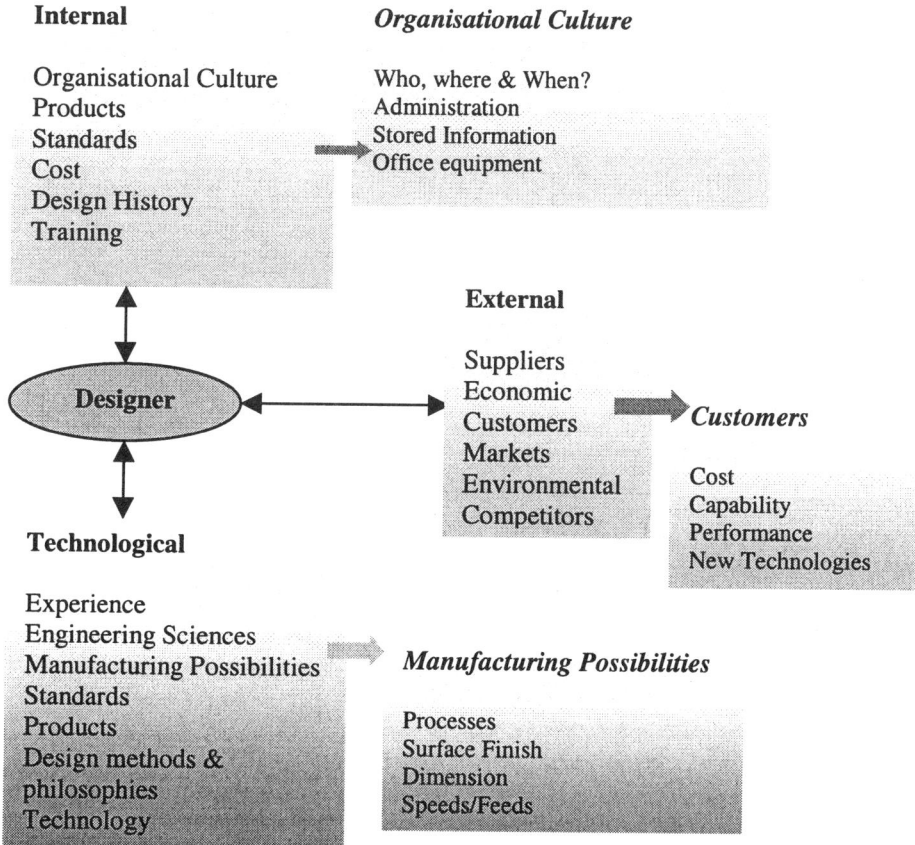

Internal   *Organisational Culture*

Organisational Culture   Who, where & When?
Products   Administration
Standards   Stored Information
Cost   Office equipment
Design History
Training

Designer

External

Suppliers
Economic
Customers   *Customers*
Markets
Environmental   Cost
Competitors   Capability
                 Performance
                 New Technologies

Technological

Experience
Engineering Sciences
Manufacturing Possibilities   *Manufacturing Possibilities*
Standards
Products   Processes
Design methods &   Surface Finish
philosophies   Dimension
Technology   Speeds/Feeds

Figure 3: Types of knowledge in cutting tool design.

There are a few examples of explicit knowledge, although poorly documented, which show typical procedures for designing particular types of tool. Examples of design knowledge in this domain include the application of inserts on an end mill shown in Figure 4, and the location of the tooling hole in the placement of the E-Z set unit as shown in Figure 5.

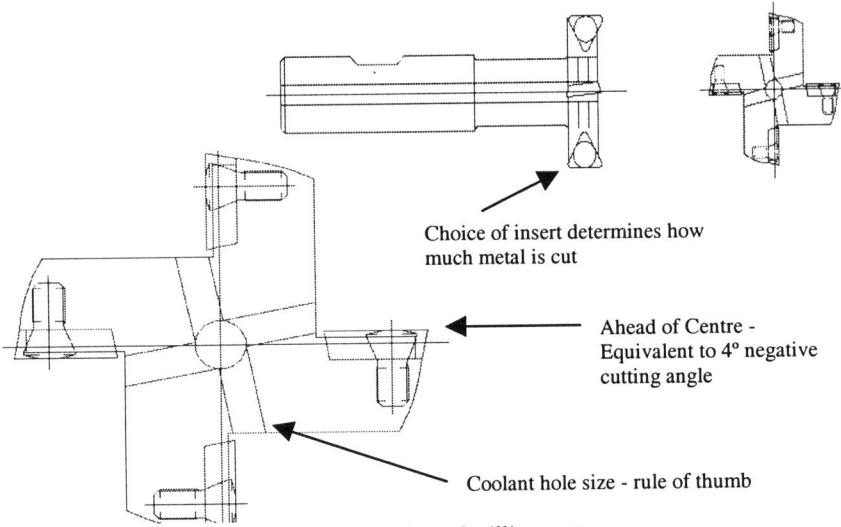

Choice of insert determines how
much metal is cut

Ahead of Centre -
Equivalent to 4° negative
cutting angle

Coolant hole size - rule of thumb

Figure 4: Design of milling cutter.

4.2.1 Case one

The design begins with the sketching of the machined component as shown by the hatched area in Figure 6 from the customer's drawing. Setting of the cutting diameter of 68mm is set, paying attention to 21.50mm cut. With this set, the diameter of the shaft can set with the difference of the overall diameter and the two 21.50mm cut areas. The shank is a template that is stored on the computer system and is inserted at the point required.

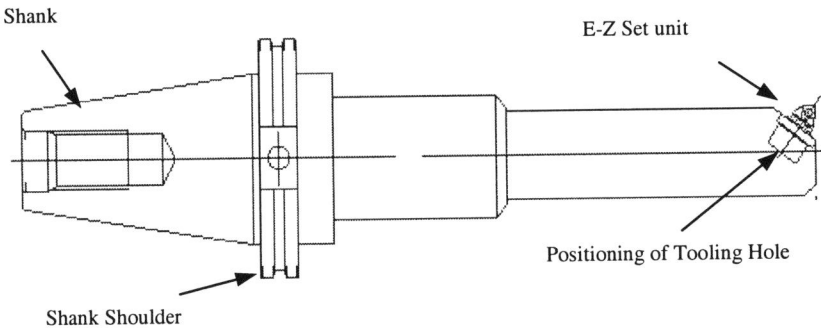

Shank

E-Z Set unit

Positioning of Tooling Hole

Shank Shoulder

Figure 5: E-Z Set unit.

The questions to arise from the first author when designing this component was the placement of the inserts and the types of inserts to be used. The constraint in this case is the size of insert that can be used due to the small size of the cutter. The size of the coolant holes and then the amount of material that is required around the inserts.

Designing the end mill for the component shown in Figure 6, required the machining of a 5mm pocket on the backside of the component. The motion of the tool is as follows:

- the tool comes in on centre with a 1mm clearance on the internal walls of the component bore,
- the tool then moves into position and makes the 5mm depth of cut.

Due to the size of the head of the end mill being only 9mm in section, only a particular type of insert could be used. The designer (the first author) found out by trial and error that the parallelogram insert shown would not do the job, after trying a number of other insert shapes the triangular shaped insert was chosen as shown in Figure 4. It was found, that with parallelogram insert not all the metal would be cut, but with careful positioning of the triangular insert this would not be a problem. The types of knowledge identified in Figure 4 are:

- Placing the insert ahead of the cutting centre line is the equivalent to having 4° negative cutting angle. This information is needed so that it can be machined at 90° to the spindle.
- The size and position of the coolant holes are performed manually/heuristically with the help of the expert relying on his experience of placing coolant holes successfully in previous designs. This ensures the right amount of cooling to contact areas during machining.

The knowledge here is heuristic or a 'rule-of-thumb' and is not written down and can only be accessed by asking the expert the right questions.

4.2.2 Case two

In order to design the cutting tool in Figure 5, product catalogues contain the E-Z Set unit information giving the critical dimensions that are needed in the calculations that are stored in a file. The calculations depend on the corner radius of the insert used. The calculations give the dimensions of the bar, the bar diameter around the E-Z Set unit and the tooling hole dimension. The shank is obtained as an object stored on the system.

Figure 5 shows the knowledge identified on a special E-Z SET bar which requires a set of calculations to be made in order to ascertain critical dimensions for it to be manufactured. The E-Z SET unit is the feature of the drawing that contains the insert, and the special feature of this type of unit is the micro adjustment which is achieved through the application of a micrometer gauge contained within the unit. The distance that is critical is the distance from the back-face of the shank shoulder to the 'tooling hole' which allows the pocket of the unit to be machined in at 53°8' to the centre line. The description of the calculations are given in Table 2 which very much depends on the insert corner radius and the diameter of the bar into which the unit is to be placed.

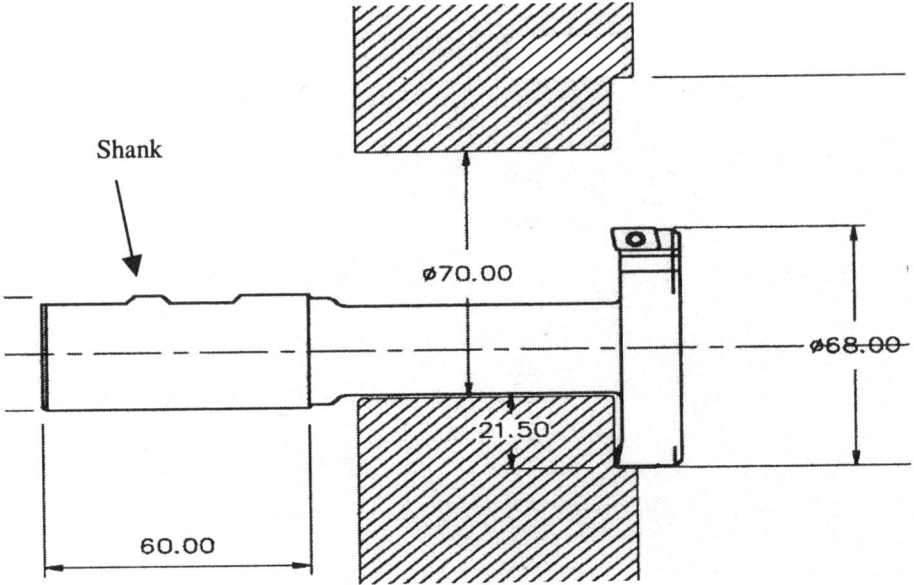

Figure 6: Component to be machined with preliminary design of cutting tool.

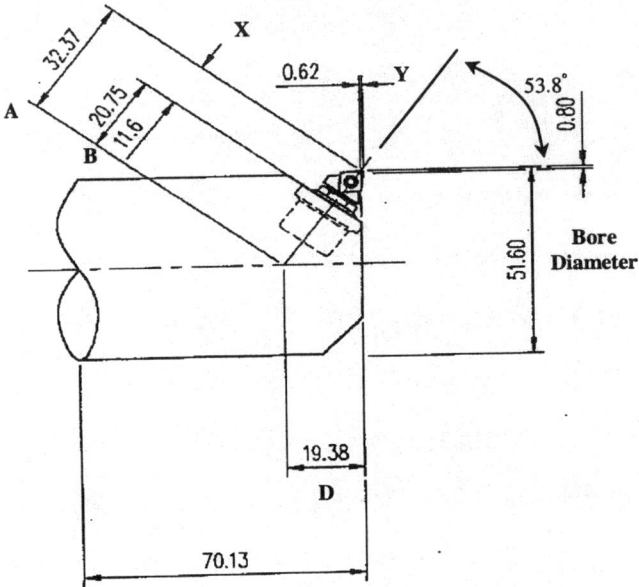

Figure 7: E-Z Set dimensions for calculations.

There is a very small adjustment made to one of the dimensions, as a result of experience, by the experts, and this is not written down. The resulting dimensions of the calculations are shown in Figure 7. Also, there is a diagrammatic method of achieving the correct front end dimensions of the tooling bar which is written down, but the experience of the expert comes in again, and by intuition they achieve the correct chamfer on the face opposite the E-Z SET unit.

Table 2: Typical E-Z Set calculation

1. Check your insert radius to correspond with the "X" dimension given in catalogue.
2. The "X" - dimension must have one third micro adjustment added to it.
3. To find "A" - divide bore diameter by 2, multiply that by the Cosec of $53.8°$. From that answer subtract "X" dimension, which will then give the answer "B".
4. To find "D" - multiply bore radius by Cotan of $53.8°$, then subtract "Y" dimension which will then equal "D".

5. Summary

The paper has outlined the different methods and models of design that have been developed over decades of research. The authors would argue that designers do conform to some features of the models and methods outlined, but in an industry situation such as the cutting tool industry it is not strictly followed. The design process highlighted in Figure 2 is based on the designer's experience of how best to perform the design within the organisation. The authors have also presented the issues that are relevant to capturing knowledge including systems that facilitate knowledge capture. The authors have identified the types of knowledge that are inherent in cutting tool design and have emphasised the need for a global approach to design by considering aspects other than just the technological factors of a design. The case study highlights mainly technical knowledge as this has been experienced first hand. The authors would argue that capturing knowledge about the economy, customer and suppliers as examples would lead to better and informed decisions during design in future years. Hence, it is important to capture design rationales to support the designs. The major challenges in capturing design knowledge and rationale are to organise and categorise the knowledge, and to find a suitable representation to complement it. The types of knowledge identified require a good understanding of the processes involved in the design, and thus it is useful to develop the detail process level model of the cutting tool design process. The research so far has identified cutting tool design knowledge as: Knowledge = Expert - Novice. The knowledge identification is based on an indirect approach, where the first author performed several designs as a novice and identified the knowledge required from experts (this is the difference between an expert and a novice).

6. Future research

The design rationale of the artefact are key to improving decision making and reducing design time. The short life cycle of cutting tool design offers the opportunity to study the whole design process and capture the information and knowledge that leads to successful and non-successful design. The next stage is to capture and represent this knowledge by providing frameworks that are feasible in an industrial environment using further knowledge elicitation techniques. The development of a computer-based interface that can provide support to the computer aided design (CAD) environment to aid the designers in both decision support and training are sought.

Acknowledgements

This is an Engineering Doctorate (EngD) research ongoing at Cranfield University and is supported by EPSRC and Widia Valenite UK. The first author would like to thank the employees of Widia Valenite UK for their help and support when undertaking the case studies.

References

[1] Cross, N., 1994, *Engineering Design Methods – Strategies for Product Design*, John Wiley & Sons, West Sussex, UK.
[2] Evboumwan, N. F. O., Sivaloganathan, S., Jebb, A., 1996, *A Survey of Design philosophies, models, methods and systems*, Proceedings of the Institution of Mechanical Engineers, Journal of Engineering Manufacture Part B, vol. 210 B4, pp301-320.
[3] Hubka, V., Eder, E. W., 1995, *Design Science*, Springer-Verlag, London (UK).
[4] Hubka, V., 1982, *Principles of Engineering Design*, Butterworth & Co Ltd, UK.
[5] Hubka, V., Eder, W. E., 1988, *Theory of Technical Systems*, Springer-Verlag, Germany.
[6] Finger, S., Dixon, J.R., 1989, *A Review of Research in Mechanical Engineering Design. Part 1: Descriptive, Prescriptive, and Computer-Based Models of Design Processes*, Research in Engineering Design, vol. 1, part 1, pp51-67.
[7] Jones, J. C., 1981, *Design Methods*, Wiley, Chichester, UK.
[8] Edwards, K., Murdoch, T., 1993, *Modelling Engineering Design Principles*, ICED-93, International Conference on Engineering Design, The Hague, August 17-19.
[9] Rodgers, P. A., Clarkson, P. J., 1998, *An Investigation and Review of the Knowledge Needs of Designers in SMEs*, The Design Journal, vol. 1, no 3, 1998.
[10] Brown, D. C., Birmingham, W. P., 1997, *Understanding the Nature of Design*, IEEE Expert, March/April 1997, pp.14-16.
[11] Gruber, T. R., Russell, D. M., 1991, *Design Knowledge and Design Rationale: A Framework for Representation, Capture, and Use*, Knowledge Systems Laboratory Technical Report KSL 90-45, Stanford University.
[12] Vergison, E., 1999, *Knowledge Management: a breakthrough or the remake of an old story*, PAKeM 99 - The Second International Conference on The Practical Application of Knowledge Management, 21-23 April 1999, London (UK), pp1-5.
[13] Khan, T. M., Brown, K., Leitch, R., 1999, *Managing Organisational Memory with a Methodology based on Multiple Domain Models*, PAKeM 99 - The Second International

Conference on The Practical Application of Knowledge Management, 21-23 April 1999, London (UK), pp57-76.

[14] Ranky, P. G., 1990, *Manufacturing Database Management and Knowledge Based Expert Systems*, CIMware Ltd, Guildford, Surrey, UK.

[15] Macleod, I. A., McGregor, D. R., Hutton, G. H., 1995, *Accessing Information for Engineering Design*, Design Studies, vol 15, no 3, pp 260-269.

[16] Smith, P., 1996, *An Introduction to Knowledge Engineering*, International Thomson Computer Press, UK.

[17] Winstanley, G (Ed.), 1991, *Artificial Intelligence in Engineering*, John Wiley & Sons, West Sussex (UK).

[18] Duffy, A. H. B., 1997, *The "What" and "How" of Learning in Design*, IEEE Expert: Intelligent Systems and their Applications, May/June 1997, pp71-76.

[19] Schreiber, G., Akkermans, H., Anjewierden, A. et al., 2000, *Knowledge Engineering and Management: The CommonKADS Methodology*, MIT Press, Cambridge, USA.

[20] Adesola, B., Roy, R., Thornton, S., 2000, *Xpat: A tool for Manufacturing Knowledge Elicitation* in Roy, R. (ed.), Industrial Knowledge Management, Springer-Verlag, (to appear).

[21] James-Gordon, Y., 1992, *Integrating a Knowledge-Based Design System into an Existing Computer Aided Design Environment*, MSc. Thesis, Cranfield University.

[22] Davis, R., Shrobe, H., Szolovits, P., 1993, *What is a Knowledge Representation?*, AI Magazine, vol.14, no.1, Spring 1993.

[23] Boy G. A., 1991, *Intelligent Assistant Systems in Knowledge Based Systems*, Academic Press Ltd, London (UK), Vol. 6.

[24] Chandrasekaran, B., Josephson, J. R., Benjamins, V. R., 1999, *What are Ontologies, and why do we need them?*, IEEE Expert: Intelligent Systems and their Applications, Jan/Feb 1999, pp20-26.

[25] Studer, R., Benjamins, V. R., Fensel, D., 1998, *Knowledge Engineering: Principles and methods*, Data & Knowledge Engineering, vol. 25, 1998, pp161-197.

[26] Noy, N. F., Hafner, C. D., 1997, *The State of the Art in Ontology Design - A Survey and Comparative Review*, AI Magazine, Vol. 18, No. 3, pp53-74, Fall 1997.

[27] Sandvik Coromant, 1994, *Modern Metal Cutting – A Practical Handbook*, Sandvik Coromant Technical Editorial Department, 1994.

[28] Bailey, J. I., Roy, R., Harris, R., Tanner, A., *Knowledge in Design and Manufacturing*, NCMR 99, 15th National Conference on Manufacturing Research, 6-8th September 1999, Bath, UK.

[29] Gruber, T., Baudin, C., Boose, J., & Weber, J., 1991, *Design Rationale Capture as Knowledge Acquisition Tradeoffs in the Design of Interactive Tools* in Birnhaum, L., and Collins, G. (eds), Machine Learning: Proceedings of the Eighth International Workshop, San Mateo, CA: Morgan Kaufmann, pp.3-12.

[30] Ravindranathan M, Khan T. M., 1999, *A Methodology for Structuring Shop Floor Knowledge*, PAKeM 99 - The Second International Conference on The Practical Application of Knowledge Management, 21-23 April 1999, London (UK), pp83-91.

[31] Roy R., Williams G., 1999, *Capturing the Assembly Process Planning Rationale within an Aerospace Industry*, ISATP '99, July 21-24, Porto, Portugal.

An Organic Architecture for Distributed Knowledge Management

G. R. Ribeiro Justo, T. Karran, M. J. Zemerly

Cavendish School of Computer Science, University of Westminster, London, UK
E-mail: {justog,karrant,zemerlm}@wmin.ac.uk

Abstract: The new breed of enterprise information systems is very complex in scope. They comprise multiple-component interactions on a variety of data models. New components have to be added to existing operational systems, in order for the enterprise to stay competitive by exploiting new sources for data capture. If new sources of data involve new Internet transaction components, for example, these should be integrated to the existing logistics and market components. This integration cannot be managed in an ad hoc way but requires a holistic architecture. This paper proposes a complex organic distributed architecture (CODA) to manage the integration of additional components to large distributed knowledge management systems. CODA achieves this through a layered architecture, dynamic adaptability, filtered information and role-based security. The layered architecture levels data knowledge according to cognitive complexity. The dynamic adaptability allows components to meet a variety of objectives by using critical success factors in each component. Filters are placed between the layers to ensure that each layer only receives relevant data. Finally, a security system based on roles and layers manages access to data. In this paper, we give an overview of CODA and present a simple case study, which illustrates the use of CODA.

1. Introduction

An enterprise information system needs to provide information for all levels of the enterprise. Enterprise information includes knowledge about resources, trends, potential initiatives, new markets and potential competitors. Singh [1] suggests that these systems be of "great value to enterprises seeking to empower their knowledge workers in order to achieve competitive advantage". As well as providing information, the knowledge management system should be able to prompt users when enterprise objectives are not being met and, in some cases, suggest courses of action. As an enterprise develops the information system needs to evolve. Organisations need to add functionality to existing information systems.

The resulting enterprise systems should allow organisations to evolve new initiatives. Fundamental to every modern competitive enterpise in this new century is the need to get involved in electronic commerce. This usually requires adding

Internet components to an existing information system. However, adding Internet components to provide for home delivery of products, for example, would require the addition of a monitoring component to show how well this new initiative is doing. Some controlling component is also needed to ensure that delivery logistics are effectively managed and do not interfere with existing capability. Furthermore, some extra monitoring component is needed to ensure that the expected growth is met and that the organisation is maintaining its competitive edge.

Organisations developing enterprise systems face, however, inhibitors when launching new initiatives. These include:

- Legacy systems: These consist of stable transactional database systems centred on a single static monolithic data model. Such systems effectively perform the existing operational functions of the organisation.
- The need for additional monitoring and controlling capabilities: New initiatives will require components for monitoring and controlling activities at every level. Such components are needed to ensure that the organisation can rapidly respond to environmental influences.

However, such additional functionality can prove incompatible with the existing operational systems for the following reasons:

- They may need to dynamically change operational data.
- They may need to dynamically modify operational processes.
- They may need to inhibit or modify operations in an unpredictable fashion.

This paper proposes a new architecture, which separates the information system into five functionally distinct layers, each supported by a data warehouse component. Each data warehouse component is separated by filter components, which restructure data into formats suitable for the enterprise functions being performed. This approach is based on the way that organic systems manage complex and adaptive behaviour [2]. The advantages of CODA are that it is intuitively easy to manage, and supports complex evolution by using data intelligently. At the same time CODA minimises information flow between the system components and provides sufficient component autonomy. This is illustrated with a small case study showing how a home delivery system can be managed semi-autonomously and at the same time adapt to new circumstances.

2. Requirements for Distributed Knowledge Management Systems Architectures

In most cases, enterprises have sufficient data at their disposal to provide knowledge. However, the data is stored in different formats and locations. More importantly, existing data needs to be analysed and processed into information suitable for decision-makers. Knowledge management systems should be able to integrate the disparate data held in the enterprise. Integrated information should support business analysis and decision-making functions as well as day to day operations.

The key to solving these problems is the formulation of clear architectures because they allow designers to better understand the components of the systems and their interactions [3], bringing numerous advantages such as flexibility and evolution. A number of such architectures have recently been proposed to support knowledge management.

Firestone [4,5] proposes a new architecture, called distributed knowledge management architecture (DKMA). DKMA is an object-oriented/component-based architecture applicable to multiple processing styles such as DSS (Decision Support Systems), OLAP (On-line Analytical Processing) and batch processing. It can also be applied to data warehouse systems. The main contribution of the new architecture is the addition of an object layer to provide dynamic integration through automated change capture and management. The object layer contains an Active Knowledge Manager (AKM) component, which provides process control/distribution services, an in-memory active object model accompanied by a persistent object store and connectivity to a variety of data store and application types.

Architectures for the centralised enterprise warehouse, such as typical DKMAs, draw data from multiple sources into one central structure using predefined metadata templates [6]. This is done through a centralised filtering process. This centralised filtering process cannot automatically limit data access to certain users. If data is changed as a result of a profile analysis, the filter process does not automatically know where the changed information should go. This may have serious security implications. In addition, the centralised filtering layer creates a system bottleneck in the case of complex large-scale ERP (Enterprise Resource Planning) systems. The reason being the enormous amount of information that needs to be processed and filtered by a single central component layer.

Singh [1] suggests that data warehouse architecture represents the overall structure of data, communication, processing and presentation of the end-user within the enterprise. He then proposes a data warehouse architectural model based on a typical layered architecture [3]. The main idea of his architecture is to separate platform-dependent component from high-level management components.

Concepts such as an active DKMA provide solutions to key problems of data warehouse and knowledge management but the current proposed architectures still present various limitations, as described in the next section.

3. Problems with Current DKMAs

Distributed knowledge management models present a new direction for enterprise architectures but the existing DKMAs suffer from using a centralised filtering process, as previously stated. This approach is impractical and is not organic in the sense that the filtering process does not know where the changed information should go and this may have serious security implications. In addition, the centralised filtering layer creates a system bottleneck in the case of complex large-scale ERP systems. The reason being the enormous amount of information that needs to be processed, filtered and passed to the correct layer in the system. The organic model described in the next section can solve these problems by efficiently decomposing and distributing the filtering process into a novel *dynamic component-based layered architecture* [3], using role-based access control [7]. More specifically, centralised architectures suffer from the following drawbacks:

• Legacy components can be added to existing systems but in most cases they cannot be successfully integrated because they usually make use of different data structures.

- DKMAs such as the one proposed by Firestone [5] filter the information flowing between operational, monitoring and controlling components. However, the centralised filtering process proposed means that an enormous amount of information needs to be processed, filtered and passed to the correct component of the system (operational, monitoring, or controlling).
- Active support for the restructuring of data is required for an enterprise information system to manage operational, monitoring and controlling components. The filtering process cannot be managed effectively unless there is a directed flow of information between the generic operational, monitoring and controlling components. The DKMA cannot automatically know where to send any changed information and any alert condition would require reprogramming. If the architecture can automatically differentiate between types of components, then it is possible to add multiple alert mechanisms and allow dynamic management of complex conditions.
- Secure access to data is problematic because there may exist many copies of data all requiring different accessing protocols. This may lead to serious problems of data contention.

4. CODA: A Complex Organic Distributed Architecture

New generation DKMAs need to include a means of monitoring and controlling objectives to allow the enterprise to evolve dynamically with a certain degree of autonomy [8]. This requires a theoretical foundation for modelling evolutionary enterprises that can be provided by (organic) principles.

4.1 The organic paradigm as a source for a reference architecture

An organic model is capable of providing a layered filtered architecture for managing information in a distributed context. CODA achieves this by refining and adapting Beer's Viable Systems Model [2] with reference to distributed information systems. Although originally applied to help define management principles, the model is further adapted to the control of data flow in a complex information system.

4.2 Layered architecture

The Viable Systems Model is based on the way a biological organism, such as the human nervous system, processes data in terms of objectives. Incoming data is levelled according to the type of activity performed and filtered so that only the relevant information is presented when decisions are made.

According to the cybernetic model of any viable system, there are five 'necessary and sufficient' subsystems involved in any viable organism and organisation [2]. To be viable a DKMA should therefore be organised according to those levels. We define these in CODA, focusing mainly on data complexity and business functions, as follows:

1. Operations: This layer deals with simple linear data, which usually corresponds to typical transaction processing and business operations. The operational data warehouse usually links together data from database from several locations.

2. Monitor operations: In this layer, the data is often dimensional and aggregated. For instance, data is organised by time or group. This layer is responsible for monitoring business operations.
3. Monitor the monitors: This layer deals with multidimensional data and provides capability for analysing trend behaviour. At this level, business operations are monitored in terms of external trends.
4. Control: This level should be able to "learn" about simple emergent behaviour, trends and forecast and be able to run predictions and simulations automatically.
5. Command: This is the highest level, which should be able to deal with any variety not treated by the lower layers [9]. This means to recognise new threats and opportunities.

In Table 1, we illustrate the CODA layers with the respective processing types, which correspond to typical data warehouse and business intelligence components, and also give an example of a business model for an accounting system.

Table 1: Cybernetic layers of an information system.

CODA Layers	Component Type	A Model for an Accounting System
Command	AI Systems OLCP type 2	Devise strategies for increasing account activity
Control	Online Complex processing (OLCP)	Aggregate accounts by profitability and produce trend control mechanisms
Monitor the monitors	Decision Systems OLAP type 2	Calculate profits on types and ranges of accounts and assess formulated trends
Monitor operations	Analytical Processing OLAP type 1	Aggregate activities by type, location, time and monitor for failure or success
Operations	Online Transaction Processing (OLTP)	Transact accounts

4.3 Dynamic adaptability

Above all, CODA provides evolutionary capability for complex information via the notion of *feedback loop* [2]. This involves applying Ashby's Law of Requisite Variety [9,10] using *critical success factors* (CSFs), in place of coenetic variables defined in [9,10], as a means of determining goals for each component and dealing with the complexity of the environment [11]. The structured breakdown into activities allows systems to evolve new behaviours as follows:

- Meeting objectives: The organic model allows system components to meet a variety of objectives by using CSFs within each component.
- Critical success factors: CSFs allow the information system to respond to change in data in a dynamic and complex way.
- Component autonomy: CSFs operating within a layered architecture allow components to be semi-autonomous. A component only reports unusual failure or success in meeting CSFs within a pre-set range of tolerances. Therefore components act relatively independently, unless an alerting condition is triggered. For example, a customer wishes to pay for an Internet transaction using a credit card. The payment component sends the details of the card to the credit card

authorisation component. It does not know what the clearance parameters are. It does not need to know how this information will be used.

4.4 Filtered information

Each component should only have the required information travelling through it. Any other information produces a huge excess baggage on the successful management of the distributed system. Moreover, the data reaching each component should be of the appropriate data type.

A filter separates each level and presents data in the required format for use by that layer. Thus:

- Information is safely filtered.
- Levels above are not swamped by detail unless an alert condition is triggered.
- Layered architecture.

Filters present information in a structured way. This is similar to the way biological systems use *homoeostats* [2]. (Homoeostats are the elements within the organism, which store variable information about the environment. The information used by a system may not be complete and therefore co-operation between systems is required. "An individual operation would be depicted as homeostatically balanced with its own management on one side and its market on the other [2].) These co-operate, filtering information and alerting higher levels only if necessary.

The higher levels of the human brain, for example, are not engaged in the everyday activities of the nervous system such as monitoring breathing. The higher brain is only notified of events, which cannot be handled by the normal routine. Functions filter information and pass information on to each other only when necessary.

In a business system the monitoring components monitor the rate of sale of various items. If the predefined rate of sale is not achieved (such as 100 items per day), then the monitoring component can put the item on special offer. Typical offers are: three for the price of two if the rate of sale is almost satisfactory, and two for the price of one if stocks are piling up.

4.5 Role-based security

Security is based on a reference model for role-based access control (RBAC) [7]. The RBAC defines user profiles, which grants access to components of a primary layer. However, access to further layers may be given depending on other roles played by the user. This means that access to data is securely controlled based on the organic architecture. In addition, changes to data can be traced back to roles.

5. The CODA Structure

5.1 Key architectural principles

Information processes (mapped to architectural components) are distributed and levelled. Each component can be upgraded and implemented incrementally. In a fully implemented organic structure, components are semi-independent, as required by any viable system [2,11]. They can manage themselves using critical success factors but can refer to components above. Each component has a part in fulfilling enterprise objectives. The combined interactions of the business processes demonstrate the emergent behaviour of the organisation. Each component should also have enough information to make the right decisions. Only a subset of the information available to the whole organisation is relevant to each component. They manipulate information and send results on for further processing if relevant. They must also perform some self-checking.

In the VSM, incoming data is levelled according to the type of activity performed and filtered so that only the relevant information is presented when decisions are made [2,11]. CODA achieves this by refining and adapting the VSM with reference to distributed information systems. Information processed (mapped to architectural components) are distributed and levelled. Each component can be upgraded and implemented incrementally. They can manage themselves semi-autonomously using critical success factors but can refer to components above, if necessary. Each component has a part in fulfilling enterprise objectives. The combined interactions of the business processes demonstrate the emergent behaviour of the organisation. Each component should also have enough information to make the right decisions. Only a subset of the information available to the whole organisation is relevant to each component.

5.2 Architectural elements

The general structure of CODA is shown in Figure 1. It corresponds to a typical object-oriented (OO) layered architecture [8,3,12], where each layer denotes a level of the VSM. Data sources, which can be data marts, data warehouses and legacy processing, must be attached to the correct layer. The criteria to allocate a particular element to a certain layer (level) in the architecture are defined according to their processing type as in the VSM, as described above. Observe that other data warehouse architectures [1] may suggest the separation of data in two levels according to the summarisation process; that is, lightly summarised data is a data that is distilled from low level of detail whilst highly summarised data is compact data. Unlike CODA, however, this levelling does not take into account the user of the data and their roles.

CODA provides a distributed OO architecture for knowledge management with global data warehousing capability. Therefore most components in the architecture correspond to a class or object, or are wrapped to provide an OO interface. For example, a business task is modelled as a method, if it is a simple task, or a class, if it is a complex task. More details can be seen in [13].

Interaction between the layers is achieved by two key elements of CODA, namely, the filters and the feedback loop. As previously stated, each layer is separated by filters, which ensure that only the necessary information reaches it. The feedback loop allows components in a layer to only report unusual failure or success in meeting their CSF to the above layers. This means that tasks in the above layer(s) must respond and possibly take action. Basically the feedback loop corresponds to an *event-based, implicit invocation style* [3], where one layer generates *alert events*, related to failures or successes, and the above layer registers the task(s) to be invoked. Each layer is structured in a similar way, providing filtering, feedback and security capabilities.

6. Architectural Properties of CODA

A DKMA (and its associated data warehouses) should make data readily accessible to decision-makers without interrupting on-line operational workloads [1]. This requires fast access to large volume of data. So, performance and scalability are key architectural properties. Since the DKMA deals with strategic information, security is also an important issue. Finally, key to any DKMA is its capability to integrate data from difference sources [14] and consequently allow new data sources to be easily added to the system. Below we summarise how CODA deals with these architectural properties:

- Performance: The serious bottleneck caused by the central management of monitoring and controlling processes is avoided since the data is correctly filtered and leveled. This results in two major performance advantages. Firstly, only data, which has changed, is refreshed. Secondly, if there is a failure, then it is limited to the level at which it occurs. Users at each layer 'buy' write access and this means locking procedures can be localised. The effects of a layer failure are minimised.
- Scalability: The organic architecture is more flexible than a centralised monolithic system because it does not rely on a single component. Existing processing components, which do not fit the scale, can be added to the model as external information components feeding information to processes at multiple layers. For example, existing management information system components with multiple functionality can be added as external feeders to the model. Their internal structure is therefore unaffected but the information flows they produce can be effectively monitored and controlled. The resulting architecture can cope with increases in the number of users and amount of processing performed by components (such as data marts, data warehouses or transaction processing).
- Security: The components must have a degree of independence for the system to be effectively adaptive. Components must have the power to make some decisions based on the state of critical success factors. Where the condition of the critical success factors merits it, the components can send alert events to the appropriate layers and in extreme cases halt a process, or suspend it pending clearance from another named source. Security is devolved to the user profiles and to the component objects, which automatically check the security status of the user. This is not a performance overhead because the system is fully distributed. It is particularly important if data is coming from external sources such as Internet service providers. Security controls and user profiles can be assigned to the different layers of the architecture. This provides a natural layered security architecture.
- Adaptability and upgradability: The model can be implemented incrementally leaving existing systems untouched. The component objects, including filters, are added in a structured way. It is possible to identify where each component belongs in the overall model. Evolution is managed within the structure. Adding Internet based components can be handled by using the appropriate filters.

The next section describes a case study where most of the above architectural properties of CODA are illustrated.

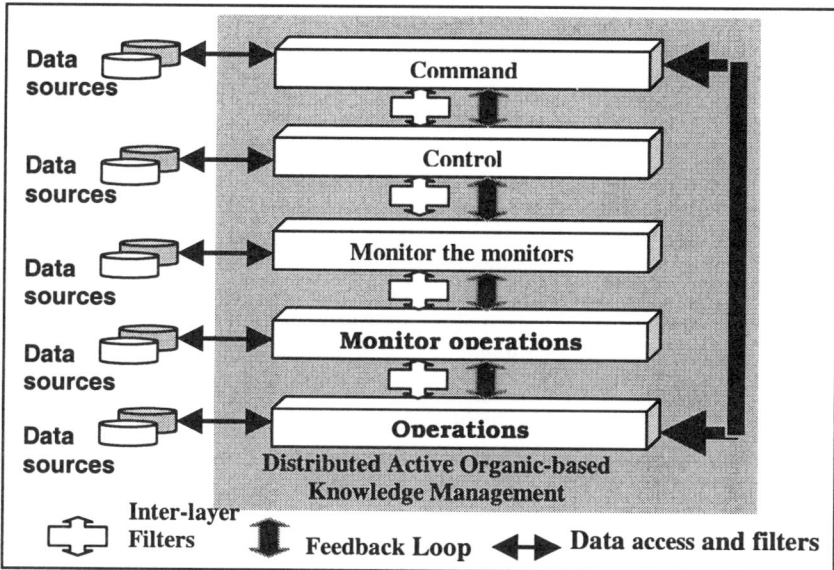

Figure 1: Overview of CODA.

7. Case Study: Adding a Home Delivery Internet Component to a Supermarket

7.1 Description of the original supermarket system

This case study demonstrates a number of the problems facing organisations implementing enterprise systems with internet and Intranet components. It considers a typical enterprise solution for a supermarket faced with problems in this very fast moving retail sector. In this scenario, an early eighties real estate policy involved expanding high street presence instead of moving out to cheaper premises with large car parking facilities. This led to serious erosion of market share during the nineties. These problems may now prove an advantage providing the supermarket can exploit the Internet and Intranet opportunities.

An enterprise analysis of customer base and logistics possibilities has shown a number of unique opportunities in the logistics and customer relationship management areas. The supermarket is well placed to promote personal services and delivery from its high street location. Firstly, it is possible to arrange efficient home delivery systems to customers within a two-mile radius from the store without a serious logistics overhead. This means that the supermarket is able to undercut its rivals who impose a £5 delivery charge for Internet orders while this chain is able to offer free home delivery to any customer living within two miles of any store.

Secondly, the supermarket could, at a later stage, exploit its personal service image by providing customers a browse facility using the store Intranet ordering component. This unique opportunity would, in the long term, allow customers to visit the local store for food tasting and information about products providing that the store can identify its high spending target groups.

The supermarket has existing systems comprising stock control and transaction components for stock moving through its existing retail outlets. The new system will interact with these in multiple ways and the architecture must cope with all possible evolutionary paths. For example, if a customer comes into the shop and orders products then the stock control system can cope fairly simply using the existing reordering system. However, if the customer orders directly from home, then it is possible to eliminate the storefront to a large extent. This depends on how successful the new components prove to be.

The additional Internet and Intranet components need a more sophisticated architecture because they will have to be adaptive. The business needs to have a very effective monitoring system to analyse the behaviour of its customer base and extrapolate trend behaviour so that it can evolve new selling strategies.

7.2 Adding components at the operational layer

At the operational layer, the store requires a new customer database. This will hold information about its home delivery customers. It requires a sophisticated control system, which supports two systems running concurrently, namely the remote ordering system and the store ordering system. Both of these will make calls on the stock data and the system must be sufficiently responsive to flag potential over-ordering situations and predict possible future buying trends. This can be achieved by adding an operational data warehouse (DW) to the system using the CODA schema as shown in Figure 2.

The initial operation system running the Internet components will manage information about the customer base. Some of this information will be dimensional. For example, age range, income range, employer type, family size and order list. This data is obtained about customers on-line. It is managed through the operational data warehouse as a class, called the **Customer Details**. Note that CODA is an OO architecture. The overall structure of this class is depicted in Figure 2. Some of the databases may be relational or legacy systems. This class is actually composed of data drawn from three databases, as illustrated in Figure 2. Observe that following the principles of CODA, the data drawn from external sources must be filtered, to ensure that the data is clean and secured.

Since CODA needs to replicate the behaviour of the enterprise, both the filter components and the classes need to express the business rules and responses. This is done by adding critical success factors to filters and classes (see Figure 2). The critical success factors act in two ways, firstly, by acting as guards to ensure the integrity and validity of the data, and secondly, by flagging an alert condition if the data has certain values.

As illustrated in Figure 2, the transaction database is located at the store nearest to the customer. The stock data relates both to the store and to the warehouse (depot) database. For example, if the store does not process/control the stock ordered by the customer, it should be possible to get information on ordering times from the stock control database located at the distribution depot.

The customer specific data is held on the customer database. This may be located at head office or at the store. It is unlikely to be located at the depot. The operational data warehouse should be located at head office. It should be logically distinct from the other databases.

The operational class allows customers to place orders quickly and efficiently. It is connected to the existing stock control database and to the transaction database through the operational DW. Customer specific business rules can be expressed at the class level which ensure that the customer cannot over-order, or order too many items. The 'validate order' operation takes items from the virtual shopping trolley and loads them into the item list. If the value of the order exceeds a certain predefined figure then the order will be queried. Further business rules, expressed as critical success factors, will be added to the operation as needed.

The filter component builds the classes from the three databases as needed. Once the customer has placed an order, the class will reside in the operational DW for a specified length of time. If the customer places no further orders then the critical success factors in the customer class structure will automatically send the data back to the transactional databases so that they do not affect the performance of the internet and intranet components.

All CSFs, whether at class level or at filter component level, are capable of being changed and added to. It is possible to change the CSFs at the class or filter component level, thereby supporting evolution. It is also possible to change derived attributes but it is more difficult to add new attributes, as this requires a modification of the underlying databases.

The information gained about the customers may change the transaction and stock data in several ways and the filter components must ensure that these are consistent and secure. For example, the customer's name, address details and telephone number should not be accessible from any but the operational layer. They should therefore be filtered out, when building monitoring classes. The customer access code and credit card number should be private to the class itself and not accessible from the operational layer.

7.3 Adding components at the monitor operations layer

Information obtained from the operational customer class will be stored as part of a customer profile at the monitor operations layer. This customer profile is not the same as the operational customer class. In CODA, it is stored at the monitor operations layer and is separated by a filter component located above the operational layer, as shown in Figure 3.

Monitoring information will be used by different user roles with specified access rights. The information serves a different business function. Some of the detail information must be filtered out and new aggregated information must be derived and added to the customer profile class. As a monitoring class, it will be subject to different business rules. These will be expressed as different critical success factors to those at the operational layer.

In the case of the new Internet component, the customer profile will be used in two ways, to monitor the customer and to obtain trend behaviour for groups of similar customers. This information is used to create special offers for specified groups of customers and to offer special shopping events.

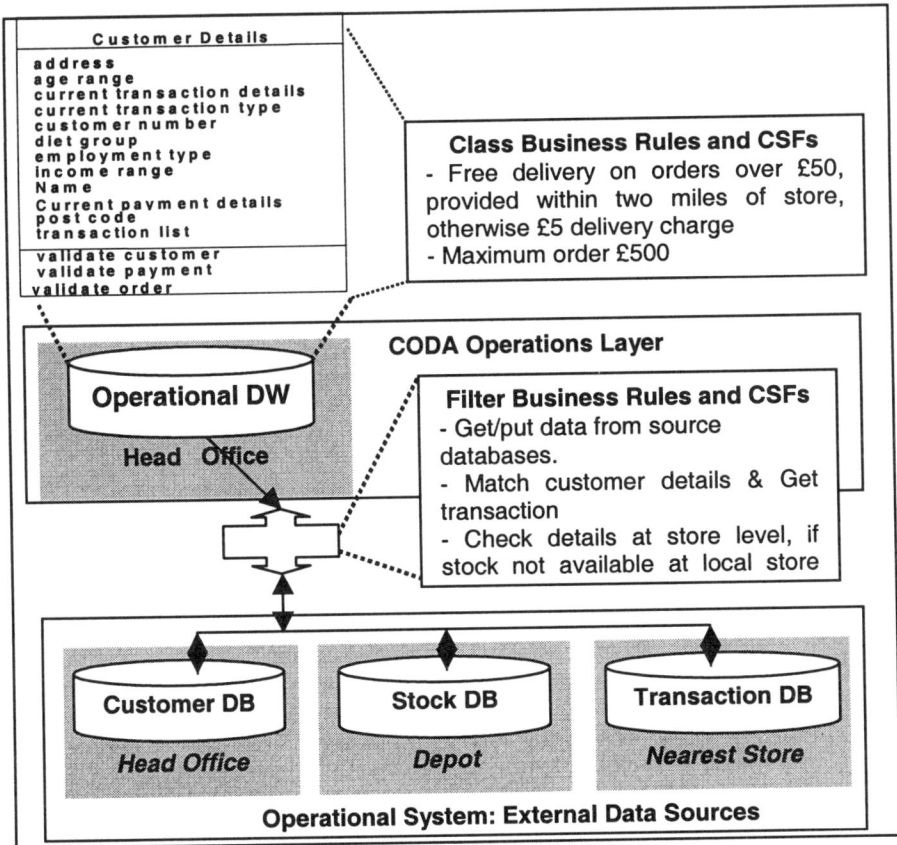

Figure 2: Operational layer of the Home Delivery System.

The monitoring class contains additional information about customers drawn from the stock and transaction databases, as illustrated in Figure 3. For example, it allows the business analyst to identify figures such as *spend per month per group* and *profit per customer or per customer group.*

The monitoring class will also contain a different range of business rules based on different critical success factors. For example, it may be possible to aggregate a number of complaints and to calculate the cost of complaints against the profits made from the customer.

The class may be modified in the same way as the operational layer class. For example, it is possible to add new critical success factors, such as complaint rate, this being number of complaints as a percentage of total number of deliveries.

Once profile types are established, it is possible to add critical success factors to the average spend and profit attributes. If a customer profile is not meeting predefined limits, the CODA feedback loop is able to make decisions on which role to alert. For example, the monitor operations layer may include have a set of parameters relating to food types. The 'aggregate food types purchased' operation may show a buying anomaly in a 'customer profile type'. It may be that the customer profile type being monitored is 'high-income customers under 25' (stored as HIC25). Analysis at the monitor operations layer may find that Internet orders from this group are predominantly for beverages. This feedback loop is depicted in Figure 3 by the event/action block.

This anomaly may be turned into an advantage. Initially, the monitor the monitors layer is alerted. This level is able to create new special offers in the store. Special offers may relate to the customer profile class at this layer. Trend analysis at this layer shows that a modification of the HCI25 behaviour might be achieved by introducing a free party catering service for orders over £150. New critical success factors have to be created to test this hypothesis and added to the customer profile stored at the monitor operations layer.

7.4 Adding components to higher CODA layers

By applying CODA to the Home Delivery system, it becomes clear that the operational and monitor operations layers are actually part of a more complex architecture with more than two layers. The CODA higher layers will offer business advantage by providing intelligent response to threats and opportunities presented to the enterprise from within and from the environment. To make the system truly effective further components need to be added to allow for the system to make predictions about the future behaviour of target customer groups. These are defined in the monitor the monitors and control layers of CODA. These layers are complex and adaptive. They use critical success factors derived from the business mission.

At the monitor the monitors layer the supermarket requires constant monitoring of sophisticated customer profiles. New profiles used to frame new marketing objectives and special offers may be obtained using clusters and associations.

All business objectives must map down to the operational layer of the enterprise. Since each layer takes care of different enterprise functions, information reaching each layer must be carefully filtered so that the user roles at the layer are not swamped by unnecessary detail. Information not required in the decision making process can be reached by a process of drill down provided the user role has the required access rights. All layers above the immediate monitor operations layer need to match the filtered data with data from the environment. This may be in the form of marketing data, credit references, stock market performance and analysis, etc. This supplementary information, added to internal data, allows the enterprise to make intelligent decisions based on a sophisticated information system. In a full CODA implementation, large parts of the decision process can be fully automated.

7.5 Evaluation of the case study

In the case study, for example, the business objective at monitor the monitors layer requires that the Internet and Intranet components grow by 10% per net value of aggregated transactions per week initially. If they do not manage this growth figure then an alert condition is automatically triggered in the stores failing to meet this growth. New actions will have to be taken at the control layer. A possible line of action is to transfer the Internet components to an area with more home computer ownership. This information is available from marketing information providers and held at the control layer.

426

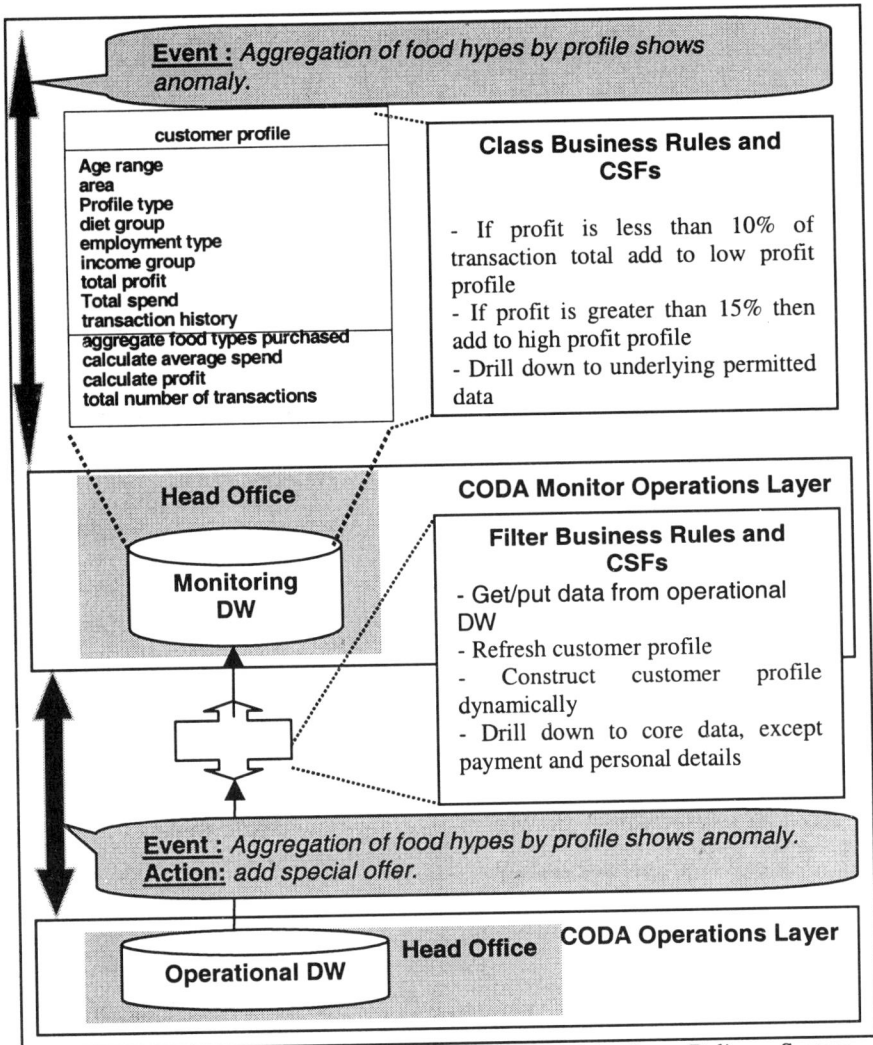

Figure 3: Operational and monitor operations layers of the Home Delivery System.

This type of adaptability would place a static and monolithic system under a great deal of strain and can only be managed effectively using a full layered architecture, as proposed by CODA.

8. Conclusions and Future Work

Distributed information systems architectures have evolved paradigmatically. The first generation of centralised monolithic systems gave way to distributed systems gracefully. In many instances both systems coexist without major performance or security overheads. The advantage of CODA is that it provides a structure for the design and implementation of distributed knowledge management systems using existing hardware and software components. This evolution is achieved by building filter components into the layers, and by adding critical success factors to components. The information system can evolve using responses from the CODA feedback loop and request action where necessary. However, such a system of carefully constructed components can be most effective only when a standard component structure is widely adopted. CODA provides the foundation for such a standard architecture.

The proposed CODA model has been applied to the development of several ERP and data warehousing applications. We are now developing a new complete component-based architecture. The architecture is currently being validated by developing some standardised components for the monitoring layers. There are two strands to this. Firstly to develop robust filter components. Secondly, to develop enterprise layer monitoring components using multidimensional models [14,15]. The key challenge in developing an organic-based architecture is the need for specifying CSF for application components and defining the feedback loop.

References

[1] Singh H. S., 1998, *Data Warehousing: Concepts, Technologies, Implementation and Management*, PTR.
[2] Beer S., 1984, The Viable System Model: Its Provenance, Development, Methodology and Pathology. *Journal of Operational Research Society, 35(1):7-25.*
[3] Shaw M. and Garlan D., 1996, *Software Architecture: Perspectives on an Emerging Discipline.* PTR.
[4] Firestone J. M., 1998, Basic Concepts of Knowledge Management. White Paper No. Nine, Executive Information Systems, Inc.
[5] Firestone J. M., 1998, Architectural Evolution in Data Warehousing and Distributed Knowledge Management Architecture. White Paper 11, Executive Information Systems, Inc.
[6] Introducing Sybase Adaptive Server™ Enterprise for Unix, Document ID 34976-01-1150, Sybase, Inc., 1997.
[7] Nyanchama M. and Osborn S, 1999, The Role Graph Model and Conflict of Interest. *ACM Transactions on Information and Systems Security,* Vol. 2, No. 1.
[8] Ribeiro Justo G. R. and Cunha, P. R. F., 1999, An Architectural Application Framework for Evolving Distribute Systems. *Journal of Systems Architecture.* Vol. 45, pages 1375-1384.
[9] Ashby W.R., 1965, *Introduction to Cybernetics*, Chapman Hall, London.
[10] Waelchi F., 1996, The VSM and Ashby's Law as Illuminants of historical Management Thought, *In The Viable Systems Model: Interpretations and Applications of Stafford Beer's VSM* Eds. R. Espejo and R. Harnden.
[11] Espejo R., Schuhmsnn W. and Schwaninger M., 1996, *Organizational Transformation and Learning, Cybernetic Approach to Management,* Wiley.
[12] Digre T., 1998, Business Object Component Architecture, *IEEE Software,* September/October.
[13] Karran, T., Ribeiro-Justo G. R. and Zemerly J., 1999, An Organic Distributed Knowledge Information Management Architecture, IASTED International Conference on Intelligent Systems and Control, October 28-30, Santa Barbara, CA, USA.
[14] Sen A. and Jacob V. S., 1998, Industrial-Strength Data Warehousing: Introduction to Special Issue on Data Warehousing, *Communications of the ACM,* Vol. 41(9), pages 29-31.
[15] Codd E.F., Codd S.B. and Salley C.T., 1993, Providing OLAP to User-Analysts: An IT Mandate. Technical Report.

Life-cycle Knowledge Management in the Design of Large Made-To-Order (MTO) Products

Chris Laing[1], Barry Florida-James[1] and Kuo-Ming Chao[1]

[1] Engineering Design Centre, Armstrong Building, The University, Newcastle-upon-Tyne, UK.

Abstract: Large made to order products consist of many complex functions and it is expected that no single designer will have all the necessary and relevant design skills. This results in specialized design teams, which operate with different design priorities. They often operate at different locations. In such operations, different models of the design are represented in different databases. Essential elements of the design process will emerge from these different databases. Since such elements will be a function of dfferent design priorities, different locations and possibly different temporal zones, a model for version control is proposed. Capturing the design process in this manner (rather than according to a pre-defined method), allows the recording of a much *richer* representation of the knowledge states in design models. These states are a representative explanation of the actual design processes rather than the perceived design processes. The version control is used to capture the decisions when moving from '*knowledge in design*' (design decisions) to '*knowledge in service*' *(*maintenance decisions). This version control forms part of a *coherent memory*, through which knowledge exchange can occur within and between design teams. This exchange allows stakeholders to develop questions that support the activities of exploration, assessment, evaluation and assimilation of a design domain. Questions that need to be asked in order to identify incubating failure conditions, to manage those conditions and to identify new opportunities that may arise from those conditions. It is demonstrated that a question driven strategic framework would help to avoid the '*we have always done it this way*' mentality and could encourage individuals to develop a selection of diverse beliefs to counter the "*phenomena of defective reality testing*" [1].

1. Introduction

The way in which engineering design is conducted is changing significantly with the adoption of concurrent engineering principles and the increase in international collaboration. Strategic alliances and partnerships often require design activity to be carried out simultaneously between widely distributed design agents. This is particularly the case when the products concerned are large and complex such as those in the aerospace and marine industries. These changes have demanded significant progress in the supporting information systems, towards the goal of an integrated engineering environment.

An environment should be produced where information is shared at formally defined levels not as ad hoc data transactions. In this way designers in traditional engineering disciplines should be aided in understanding product data from their collaborative partners, leading to a reduced product development cycle. In particular a system is required to address the need for complete product configuration and version control.

A key aspect of version control is that of grouping and monitoring derivative solutions and the recording of the design rationale. The version controller presented in this paper demonstrates this ability. It will be shown that the applied system, identifies solutions and keeps track of alternative solutions using a suitable labelling scheme. These labels are used to group similar derivative solutions. The lexical ordering of the label is used by the system (rather than the individual designers), resulting in a more meaningful label at the user interface. This information is clearly important in any knowledge acquisition process. Capturing the design process in this manner (rather than according to a pre-defined method), allows the recording of a much *richer* representation of the knowledge states in design models. These states are a representative explanation of the actual design processes rather than the perceived design processes. These states are active in the management of changing details and the organization of version sets within an engineering environment. The presented system is used to capture the decisions when moving from '*knowledge in design*' (design decisions) to '*knowledge in service*' (maintenance decisions), offering designers a consistent view throughout the complete design life-cycle.

As knowledge is derived from the data, the labelling mechanism and their consistency play a significant role in achieving knowledge management. Thus, the understanding of the version management is the essential first step to follow the proposed knowledge management system. The organisation of this paper is as follows; the following section starts with the introduction of the overall architecture of the version management and its components. The detailed operations of version management are also illustrated in the section. In order to assist the reader in better understanding the purpose of the version management, a comprehensive industrial case study is used to demonstrate the effectiveness and consistency of data that can be managed. Finally we demonstrate with examples from the case study how a unique approach to knowledge capture is achieved.

1.1 Related research

Knowledge management is the formal management of knowledge for facilitating creation, access, and reuse of knowledge, typically using advanced technology [3]. The concept of ontology has been used by a number of research groups to construct the domain knowledge, to characterise that knowledge, and to provide a library for knowledge reuse and sharing among corporate stakeholders. Interested readers are directed to [4], [5] & [6].

Recent research in work flow management [6] & [7] and enterprise modelling [8],[3] & [9] has made significant contributions to knowledge management (KM). The shortcomings of business processes can be realised by the analysis of the knowledge required for a specific task. This leads to the identifcation of potential needs of information technology support.

Currently, the knowledge in the knowledge-based systems is structured by formal knowledge representation languages. However, the knowledge in KM tools could be stored as informal text documentation or represented as semi-formal data structures (e.g. HTML and XML etc). The data mining and data warehouse methods have been introduced to retrieve the appropriate information [10]. Furthermore, maintaining knowledge is also an important activity in KM, which includes correcting, updating, adding, deleting knowledge in order to refine the knowledge [4]. In this research, we propose a data version management to ensure the data consistence and groupings as a step towards knowledge management.

2. Version Management

The scheme for version management is agent based. The goal of our agent architecture is global consistency of data and the ability to reflect change in all models of a disparate design process. An agent communication language is implemented on top of CORBA (The Common Object Request Broker Architecture [11]).

The version mechanism is session-based, that is updates are made at the end of a user session. The mechanism is intended to be used in full lifecycle support and also to record a history of the design process. In order to achieve this an agent architecture has been developed. This architecture consists of three layers which may be considered as the physical layer, the logical layer and the knowledge layer. Each layer is represented by a separate type of agent - resource, behavioural and global. A detailed description of these agents can be found in [12].

2.1 Version model

The version model which we present here has a number of key concepts which separate it from other schemes. Firstly, we make no unrealistic requirements on the underlying design tools. An ability to export data is the only assumption. Secondly, in a more natural representation, we describe the tools as communicating agents with asynchronous messaging capabilities rather than as remote objects or functions which may be invoked or called. Finally we allow design agents to collaborate without relaxing any autonomy constraints on the individual design tools.

We will proceed with a description of our model by defining the main terms used in it.

An **Entity** is an item which is considered as a design object in any participating model at some stage in the product lifecycle.

A **Configuration** is a unique set of entities which when logically related describe the complete product model at a given point in time.

A **Version** is a specific instance of a given entity, which may be derived from any previous version through a series of change operations.

Change operations are defined as **create**, **delete** and **modify**.

Version levels are described as **private**, **declared** and **recorded**.

2.2 Entity version management

Local models are managed by the local database and we assume no control or access to this structure. This allows designers to continue working with the tools they are familiar with and also to introduce the system to legacy applications. We use the commonly described wrapper method [13] to access these system but our wrapper is contained within the resource agents.

The proposed scheme is a forward deltas scheme [14] where deltas are stored as a list of entities on which primitive operations have been performed. It is assumed that the three primitive operations we describe, **modify, create** and **delete,** can be used to represent all design actions across all domains.

The labelling scheme applied by the resource agent is external to any local scheme the design tool or database may have. It is of course possible to utilise any available local scheme as long as in the global context we get a unique version set identifier. The scheme is adapted from that of Keller & Ullman [15] but at present we do not implement their optimisation processes. Entities are initially **created** within a version set with a unique identifier corresponding to the local model entity name. The entity's storage is physically within the resource agent but rules concerning the creation and deletion of entities and their relationships at the global level are controlled by the global agent. An advantage of our scheme of resource agents is that it is reactive in the sense that changes made in another model are automatically reflected in all local models by the application of a new label. This label is actually designated by the resource agent and it is required to be unique in the local context and to tell us which agent caused the change.

As stated earlier we do not distinguish between complex and simple entities in our versioning system. Entities are related in terms of hierarchy by the global agent. The system does not choose to ignore this and indeed this structure is available within the local models where it is very useful. We use a less rigid definition of an entity for global version control. As stated this allows entities to exist at different levels of detail in different models or domains. The reason this is useful is that models based on mismatched domains can be related and merged using hierarchies containing generalisations and specialisations of existing terms [12]. The difficulty is that somewhere a consistent representation of the knowledge of these relations needs to be stored somewhere. This is one role of the global agent.

3. Configuration Management

The global and behavioural agents combine to give an overall configuration management system for the complete product life cycle. In this system the behavioural agents may be described as the logical layer and the global agent as the knowledge layer. The behavioural agents define the rules for co-operation and change management whereas the global agent has knowledge about design entities and their relationships in the global context of the total product.

The first aspect of the configuration management scheme is the labelling of design models. Versions are caused by change and hence the process of change is represented within our labelling scheme. When a requirement for a change is issued a conversation on this change is started between the behavioural agents and the currently declared versions of each model are updated with a new label as shown in Figure 1.1. If this change is agreed the label on each version then becomes the timestamp and the other labels are removed. The change issue and delta storage is handled by the resource agent. In Figure 1.2 a change is issued but this conflicts with constraints in another model so this agent produces a set of alternatives which compromise both constraints and the label is now composed by adding the agent's alternative. At this point the behavioural agents are in a state of conflict resolution and all subsequent alternatives are labelled in the same way. When the conflicts are resolved the label reverts to the timestamp as in the previous scenario.

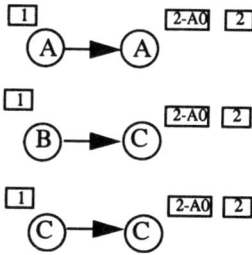
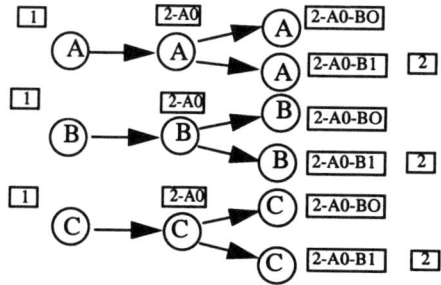

Fig 1.1   **Fig 1.2**

Figure 1: Sample Labelling Scheme

In order to cope with the situation where the above resolution fails at this point in time, that is, where parallel design exist we simply clone [16] the global agent and give it another label based on the same scheme. In our experience these *clones* very rarely stay active for long periods.

4. Case Study: Fast Ferry Concept Design

In this section a case study supplied from a ship design consultancy Armstrong Technology Associates is presented [17]. The Made To Order product under consideration is a mono hull fast ferry and it is the concept design phase of the process which is examined. The case study begins with what the company calls a *basis* ship from which design solutions are evolved. The basis ship is actually a design specification of an existing sea-going ship with similar design criteria. as the ship to be built. Firstly the basis ship is described. Then two consequent revisions to the specification are presented. The effects of these revisions on the design are described and the effects on the concept design process are detailed. Finally, it is shown how the version model supports and aids this process.

4.1 Basis ship

The business rationale for the basis ship is stated as follows. The basis ship will operate on an existing route and run in parallel with an existing monohull displacement ferry service. Therefore improving service to attract mainly non-freight cargo from other routes and modes of transport.

From this rationale the following conceptual design requirements are derived:

Crew 15	Vtrial (Vt) 40 Knots
Passengers 600	Vservice (Vs) 37 Knots
Lane Length 685m	Vcruising (Vc) 25 Knots
Number of Cars 136	Endurance (Es) 400 Nautical miles
Number of Coaches 6	End. Days1
Articulated Vehicles 5	

Figure 2 shows a profile of the vessel concerned.

Figure 2: Profile of Mono Hull FastFerry

4.2 Route data for the basis ship

This section describes what is called the *route data* for the basis design ship. As illustrated this covers the parameters of the potential operating route of the ship which impact on the ship design.

Distance	45 miles
Sailing Time	45/37 = 1 hr 13 mins
Berthing =	15 mins
Loading/Unloading = 30 mins	
Total = 1 hr 58 mins	
Sailing Intervals	2hrs
Sailings per day	8

From this data the following timetable can be derived which leads to a time in service of 16 hours per day and therefore 8 hours down time.

PortA		PortB	
Arrive	Depart	Arrive	Depart
-	07:00	08:30	9:00
10:30	11:00	12:30	13:00
14:30	15:00	16:30	17:00
18:30	19:00	20:30	21:00
22:30	-	-	-

Figure 3: Timetable for the Basis Ship Design

4.3 Description of design changes

In this section two consecutive revisions to the basis design are presented. The business drivers for these changes are indicated. The consequent changes to the ships specification are then derived.

Revision 1

From experience with a similar route operated by the company, the introduction of a fast ferry service generated a greater demand than that accounted for in the basis ship design. As demand is not at the start or end of the day it was not thought that an extra sailing would alleviate the potential problem. Therefore the cargo (passengers and cars) capacity needed to be increased by 25%. The result of this change in capacity is reflected in an increased required lane length in the garage deck of 856 metres.

Revision 2

The company has recently taken the corporate decision to inaugurate a new route currently served by a conventional monohull ferry service (passengers and freight) and a fast ferry (passenger only). Therefore there is potential to develop a new modern service giving modern facilities and reduced sailing times. This specification requires a design to cope with these additional requirements giving the vessel flexibility and interchangeability over the companies routes.

4.4 New route data

Distance   73 miles
Sailings per day   8
Sailing Time   Must give a 2.5 hr turn around time
Berthing = 15 mins
Loading/Unloading = 30 mins

Hence sailing time is 1 hr 45 mins which means a speed of
Vs = 73/1.75 = 41.71 = 42 kts
Also Es = 73 *8 = 584 + (margin) 75 = 660. Hence Vt = 45 and Vs=42.

PortC		PortD	
Arrive	Depart	Arrive	Depart
-	06:00	08:00	8:30
10:30	11:00	13:00	13:30
16:00	16:30	18:30	19:00
21:00	21:30	23:30	00:00
02:00	-	-	-

Figure 4: Revised TimeTable for Ship Design

In the next section an overview of the conceptual ship design process is given.

4.5 Conceptual ship design

The conceptual ship design environment is described in Figure 5. This conceptual design environment is recreated within the Engineering Design Centre by a number of analysis tools, each of which manage their own input and output data. This interaction is represented in Figure 6. The data flow diagram represents the distributed design environment on top of which the agent structure previously described sits.

Figure 5 illustrates how a conceptual ship design environment is recreated at the Engineering Design Centre. It also shows the information flows between the isolated design tools. The Prime Contractor issues a specification to the design consultancy (Flow 1). The design consultancy perform some preliminary analysis producing output which is sent concurrently to the hydrostatic (Flow 2a), hydrodynamic (Flow 2b) and shipyards (Flow 2c). The hydrodynamic consultancy require that the hydrostatic consultancy have completed (Flow 3) before they can produce their design analysis. The final information flow is back to the prime contractor (Dashed Lines).

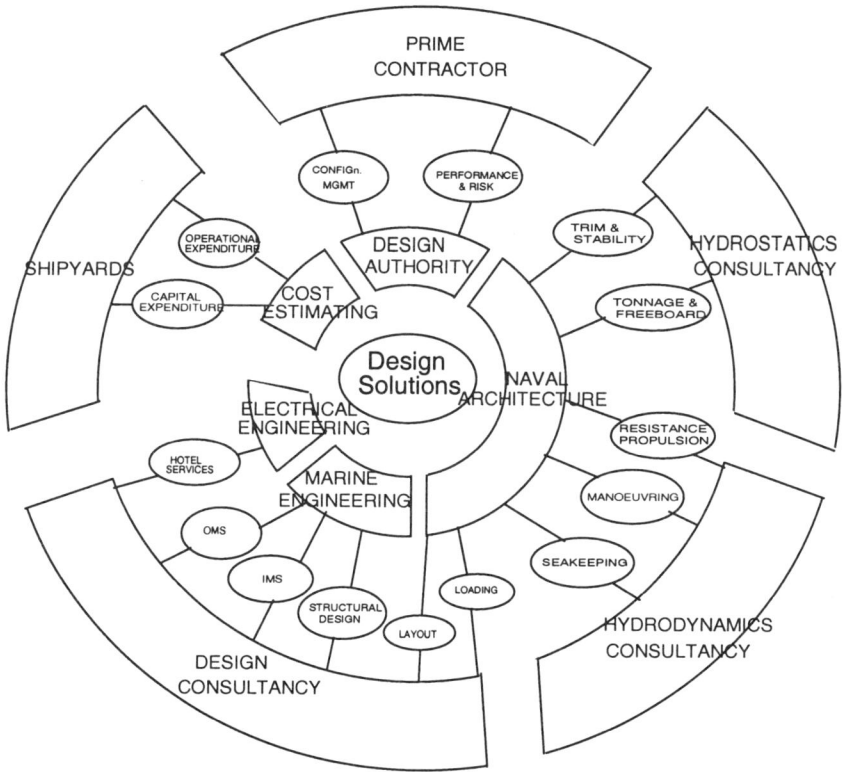

Figure 5: The Conceptual Ship Design Process

4.6 Case study details

In this case study the first revision to the *basis* design requires an increase in the number of passengers. To achieve this the prime contractor decides to increase the length of the ship. A range of three designs are initially evaluated which have an increased length of 100 metres (Design A), 105 metres (Design B) and 110 metres (Design C). These designs all cause a violation within the hydrostatics consultancy and the first two violate the cargo constraints at the design consultancy. At this point the hydrostatics consultancy suggest an increase in freeboard and two alternatives are now tried 2.5m (Design D) and 3 m (Design E) . The hydrostatics consultancy state that the second of these designs now produces the required stability. However, the design consultancy states that it feels that the Length over Breadth ratio is too high. Hence the overall breadth is increased to two new values 14.5 metres (Design F) and 15.5 metres (Design G). At 14.5 metres the power requirements of the ship are too high and again the design consultancy raises a conflict in its model. To compensate for this the design consultancy changes the shape of the ship by altering the midship coefficient (Cm). This is reduced from 0.68 to 0.65 (Design H). However this violates passenger comfort requirements (hydrodynamics), so Cm is increased slightly to 0.66 (Design I). At this stage we have 3 designs which are valid and it is chosen to progress only 2 of these, the two cheaper options, which we now refer to as version a (Design E) and version b (Design G). Table 1 shows the set of designs so far evaluated.

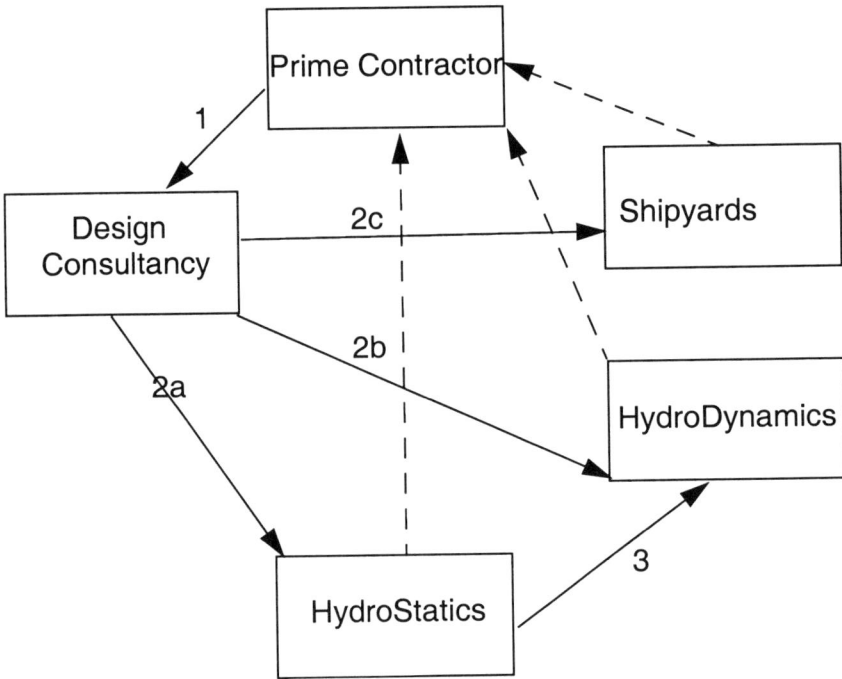

Figure 6: Data Flows Between Design Teams

Table 1: Set of Design Solutions

Design	Length	Breadth	T	Cm	Fb	Valid Design
A	100	13.7	3.3	0.68	2	no
B	105	13.7	3.3	0.68	2	no
C	110	13.7	3.3	0.68	2	no
D	110	13.7	3.3	0.68	2.5	no
E (a)	110	13.7	3.3	0.68	3	yes
F	110	14.5	3.3	0.68	3	no
G (b)	110	15.5	3.3	0.68	3	yes
H	110	14.5	3.3	0.65	3	no
I	110	14.5	3.3	0.66	3	yes

4.7 Impact of route change

From revision 1 there are now two live versions under consideration. In this section the impact of the second revision, that is the adaptation of the design to a new route, is described.

Version a

The design consultancy inform the prime contractor that the increase in speed requires a larger set of engines which can be accommodated but this reduces the number of cars which can be carried by the ship causing a design violation. To compensate for this the breadth of the ship is increased to 16.5 metres (Design J) which compensates for the bigger engines but now the hydrodynamics state this causes a conflict in their model. Two alternatives are now proposed- a further increase in breadth (Design K) or an increase in the draft (Design L); which is allowable on this route. The first alternative again causes a conflict in the hydrodynamics consultancy's model. The increase in draught causes no violations in any models

Version b

This design was originally wider than Version a and can accommodate the bigger engines but these again reduce the number of cars that can be carried. To compensate the hydrodynamics consultancy suggest an increase in draft and freeboard (Design M) but this causes another violation in the hydrodynamics model, and hence the draft is reduced and the shape of the ship is changed by altering the midship coefficient. This produces a stable design.

At this stage version b is chosen as the most suitable design to progress. The overall characteristics of this design are perceived by the design consultancy to give better performance.

Table 2: More Design Solutions

Design	Length	Breadth	T	Cm	Fb	Valid Design
E (a)	110	13.7	3.3	0.68	3	yes
J	110	16.5	3.3	0.68	3	no
K	110	17	3.3	0.68	3	no
L	110	16.5	4.3	0.68	3	yes
G (b)	110	15.5	3.3	0.66	3	yes
M	110	15.5	4.6	0.66	3.5	no
N	110	15.5	3.3	0.72	3	yes

Examining the complex set of design interactions and solutions occurring in the case study. The version history graph looks as shown in Figure 7 where revision one and two are clearly indicated.

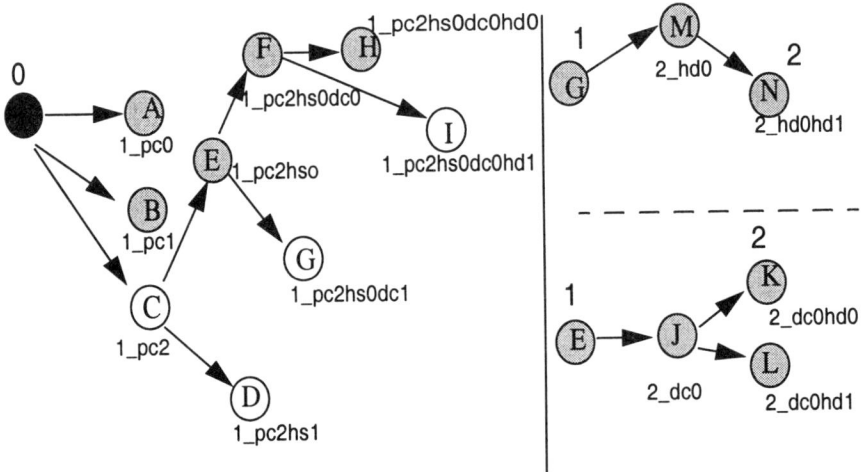

Figure 7: Graph of Version History

The key to the version model at this stage in the design process is really in the grouping and monitoring of the derivative solutions and the recording of the design rationale as the model is not too detailed. It can be seen from Figure 8 that the labelling scheme applied clearly identifies the solutions and keeps track on the alternative solutions and propositions chosen. Examining the labels derived from the *basis* design, which are 1_pc0, 1_pc1, 1_pc2. The labels indicate three alternative solutions, pc1, pc2, pc3 to the first design revision (1_) from the *basis* design.

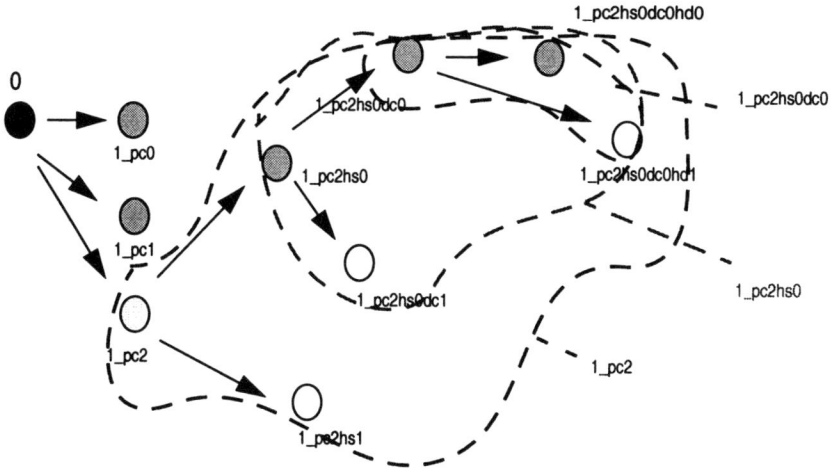

Figure 8: Version Grouping

Figure 8 shows how we use these labels to group similar derivative solutions. Whilst appearing cumbersome the lexical ordering of the label is useful [6]. This is used by the agent architecture rather than the human designers. A more meaningful label may then be applied at the user interface, such as low cost solutions. This information is clearly important in any knowledge acquisition process and also can be used at later stages in the design to indicate weaknesses that may have been avoided and could be avoided in the future. For example, while the current version control is able to identify and extract reusable knowledge and represent such quantative knowledge as associations and links (Figures 6, 7 & 8), it is unable use those links to enhance the qualitative understanding of the representation . In essence, it is unable to enhance the knowledge content of that captured knowledge [18].

5. Strategic Memory: Background

In this section, the discussion will focus on how to extend the version control proposed above into a formal *'strategic memory'*, through which knowledge exchange can allow designers to develop questions that support the activities of exploration, assessment, evaluation and assimilation of a design domain. It will be demonstrated through examples that a question driven strategic framework would help to avoid the *'we have always done it this way'* mentality, by, (i) providing the designer with relevant reflections and actions on how, when, where and why a previous design version occurred; (ii) pointing out to the designer which aspects have been altered and; (iii) suggesting to the designer similar derivative solutions, from which relevant questions that were needed to be asked (but were not during the initial construction of the design model) are identified. These questions are facilitated by the reflective aspects of a question driven strategic framework. The model behind these reflective aspects is one in which design building, (i) is represented as a reflective process that acts on the relationship between different versions of previous designs; (ii) an analysis of how those versions might perform and; (iii) the incorporation of the design updates. The nature of reflection in design making views design building as an activity that questions and tests the designers understanding of the external problem. Consequently, this interaction between the designer and the reflective version controller allows the reflective version controller to develop a better understanding of how to avoid weaknesses in the current design building process. This strategic memory of design building views learning as a questioning activity that encourages the designer to engage in an activity of hypothesis testing. When the current design version is found wanting then the reflective version controller displays alternative design versions. The use of alternatives encourages the designer to engage in reflective induction, in which the individual begins to question, (i) their understanding of the current design version; (ii) the way in which it relates to the current design problem and; (iii) the inferred outcome of that design process. This hypothesis testing and reflective induction may be viewed as a process of justification and could encourage individuals to develop a selection of diverse beliefs to counter the *"phenomena of defective reality testing"* [1].

5.1 A new and novel approach

The implementation of a reflective version control, requires a set of knowledge criteria for the evaluation of competing knowledge fragments generated from multiple design alternatives. These may be viewed as *interactive processes* and their *emergent properties*. Interactive processes are active agents in the construction of past and present experiences and would support *adaptive* learning [19]. That is, they would be concerned with adapting experience to help with an existing decision problem. Whereas, properties emerging from the testing of this interaction would support *generative* learning [19]. That is, they would be concerned with using testing to generate questions that should be asked at the time (*but were not*), and are instrumental in helping the design team learn and then decide about possible design options. In this respect a reflective version control is a function of a reflective loop between a process model explanation of design alternatives and the current design problem and may be viewed as strategic memory.

Firstly, this strategic memory may be characterized by the terms meaning and understanding. Meaning is a function of interactive processes, (i) develops from the dynamic interactions within the derivative solutions and can be organized by the *positive* interactions (successes) that a design achieves and the *negative* interactions (failures) that it avoids; (ii) interactive processes hold a body of tacit knowledge, shared by all agents within the design process (*global awareness*), some agents hold detailed, often technical knowledge (*specific awareness*), while the interaction represents knowledge of mathematical functions (*decision awareness*) all of which support the retrieval, modification and prediction activities of adaptive learning; (iii) the dependability of previous meaning values associated with previous interactive processes and the pre-conceptions of existing adaptive learning activities are constantly being tested. This can be seen in Figure 6, in which an increased length causes a violation within the hydrostatic model. In this example, the interaction between the design model and hydrostatic model represents a negative or failure. The hydrostatic model holds specific information on why this failure has occurred, while the interaction between the models represents a body of mathematical transformations on what has occurred, and can suggest ways of minimizing that violation. In suggesting alternatives, each of which can be reviewed against the initial basis ship, allows for the pre-conditions of that basis ship to be constantly tested. Understanding is generated by this attempt to falsify those assumptions, in which the domain exploration activities of situation assessment, evaluation and assimilation are pre-eminent. These activities (which form the reflective part of the version control), are composed of *resource awareness* (shared knowledge of who is an expert on what), *task awareness* (tension and conflict resolution), *chronological awareness* (instantaneous awareness that an agent has regarding the activities of other design models) and *activity awareness* (simulation of sequential, concurrent and selective decision activities). These activities facilitate generative learning in which the exploration and experimentation of solutions and the elaboration and re-definition of problems are undertaken, from which emergent *tactical* knowledge properties (the what and how questions that need to be asked) and emergent *strategic* knowledge properties (the when and why of using the answers to advantage) will be produced at many levels of definition. Consider once again Figure 6, each of the activities; design, hydrostatics, hydrodynamics, etc., represents the reflective value of the version control. Each of these activities must have a shared knowledge of who is an expert on what, since each activity passes relevant information to that particular expert model. Similarly, each activity must have an almost instantaneous awareness of the activities of other design models, since violations are made know within the version control. These model activities in creating alternatives are in this respect facilitating some form of generative learning. For example, in design H a reduction in the midship coefficient (Cm) from 0.68 to 0.65 violates passenger comfort requirements (hydrodynamics). A slight increase (design I) solves this problem. At this point, an emergent tactical question could be, "*what is the most appropriate midship coefficient?*" In future designs this becomes a strategic question to be asked at the beginning of the design process.

The adaptive interface between each of the design teams (Figure 6) generates the *predictive-intervention-states* warning of a *proneness-to-decision-failure*, whereas the generative interface between different versions within the design history (Figure 7) generate *evaluation-intervention-states* of *what-to-ask* and *when-to-use* system dialogues. Such dialogues will make available to the design team a rich set of attributes of similar processes, including the event sequences that have lead to previous failures or near misses. These attributes presented as interactive graphical descriptions will include the current and desired states, initiating, transforming, scenarios and consequence states. For example, the current and desired states of design E(a) (Table 1) may be the same. That is, the current state [<length:110> <breadth:13.7> <depth:3.3> <midshipCoefficient:0.68> <Fb:3>] could match the desired state. The initiating state for design E(a) is a [violation] within the hydrostatic model for design C. The transforming states will be the mathematical functions used to transform design C into design E(a). A scenario may be an [increase in speed], which has a consequence state [larger engines required results in reduced cargo space]

Secondly, the reflective version control would have to justify the *appropriateness* of those intervention conditions. That is, those attributes of the decision process that the reflective version control *believes* may assist the decision maker in understanding the *problem-decision process-solution* connection. This belief may be based on evidence of previous designs. The appropriateness of the evidence depends upon a *process memory* of previous design paths indexed by the positive and negative interactions. For example, the graph of the version history (Figure 7.) represents a design process flow, while the reported violation within the hydrostatic model is caused by a negative interaction between the models of the design and hydrostatic processes. The index will function as a *connectivity matrix*, discriminating between the positive and negative interactions between the design teams (Figure 6).

The reflective version (in identifying relevant questions and assigning evidence) uses qualitative process model descriptions with quantitative state transformations of the state variables. This has many advantages, (i) individuals appear to reason in a qualitative process fashion with the processes as patterns rather than numerical data representations of previous decision paths; (ii) the process models of the design path are built up from a deep knowledge of actions teased out using *grounded theory* [20] and therefore they contain the sub-processes ready for presentation at a higher level; (iii) representing the design path as a pattern means that the assessment of changes or failures can be tracked and consequently the initiating event sequence that caused a particular failure can be identified and isolated; (iv) constructing the process representation as hierarchically sub-processes allows for the evaluation of those changes to be conducted at many levels of process definition (from the local to the global); (v) pattern matching the represented design paths should improve the assimilation of information and; (vi) the process models will work with sparse quantitative data.

To conclude, Brechtel [21], has argued that solving problems and the way in which the problem solving task is performed appear to be dependent upon the nature of the problem and the way in which the individual understands the problem. This implies that the problem-solution process could be influenced both by the type of information given and the form of its presentation. Unfortunately, conventional decision support tools tend to present information as quantitative data rather than process flow. Clearly, the solution to a decision problem depends on the information models available to the decision maker. Quantitative data based decision tools are by their very nature limited since the models on which they are based are necessarily reduced versions of reality. Questions of the dependability, the usefulness and the sufficiency of these models in the situations in which they are actual used can be raised. To address these issues, it is suggested that models must help engineering practitioners articulate and reflect upon the processes and their interactions and the assumptions that can be derived from those interactions. Processes and their interactions, both within and between processes are the means through which individual and collective senses of reality are constructed. The representation of these processes are active agents in constructing process model explanations of the past and present and are instrumental in helping the user to decide about the future. The manipulation of these process and the interactions generated from that manipulation are the core activities of a reflective version control.

It is proposed that a reflective version control will help a decision maker identify incubating failure conditions that have occurred in previous case histories and the questions that need to be asked in order to manage those conditions away and to generate new opportunities. The key idea is that in any particular decision problem the questions that are raised by the stakeholders within the process are crucial to attaining success. By considering the interaction and exchange of design knowledge and how qualitative understanding emerges from this interaction may enable links to be made between technical and economic factors, environmental, cultural and social effects of design decision making. This is a very complex area of research about which methodologies and techniques are still embryonic. However, a Design Reuse project being undertaken at the Engineering Design Centre, University of Newcastle is attempting to address some of the issues by producing a comprehensive methodology and supporting software environment that help designers re-use emergent design knowledge.

6. Conclusions

The engineering of large made to order products is a process which creates large amounts of structured data. The structuring of this data is generally only optimised to the specialised domain where the data originated. Hence in a collaborative environment interpretation and domain knowledge are applied to the exchange of data at critical stages. These critical stages can de modeled as a directed acyclic graph. A version controller as described in this paper allows us to recreate this graph for a generalised design process.

It is hypothesized that a reflective version control aids the creation of an appropriate problem-solution environment between the users and the system by making available to the design teams a rich set of alternative and relevant design process explanations. These explanations will include the emergent processes that have led to previous failures or near misses.

Design model explanations that are selected could, (i) provide the users with relevant reflections and actions on how, when, where and why a problem has previously occurred; (ii) point out to the users which aspects are responsible for the problem and; (iii) suggest to the users relevant questions that were needed to be asked (but were not), during the initial design path.

References

[1] Turner, B. A., [1978:1997]. Man-made disaster, edited by B. A. Turner & N. F. Pidgeon. 2nd ed. Oxford: Butterworth-Heinemann, Previous ed. London: Wykeham, 1978. ISBN: 0-7506-2087-0, 1997.

[2] O'Leary D. E. "Using AI in knowledge management: knowledge bases and ontologies" *IEEE Intelligent Systems*, vol.13, no.3, May-June,1998, pp.34-9. Publisher: IEEE

[3] O'Leary D. E. "Enterprise knowledge management" *Computer*, vol.31, no.3, March 1998, pp.54-61. Publisher: IEEE

[4] Benjamins, V.R., Fensel, D., Decker, S., & Perez, A.G., "(KA)^2: Building Ontologies for

the Internet: a Mid-Term Report", *Journal of Human-Computer Studies*, Vol 51, 1999, pp 687-712.

[5] Abecker, A., Bernardi, A., Hinkelmann, K., Kuhn, O., & Sintek, M, "Toward a Technology for Organizational Memories", *IEEE Intelligent Systems*, Vol. 13, No. 3, pp 40-48, 1998.

[6] Kuhn , O, & Abecker, A. "Corporate Memories for Knowledge Management in Industrial Practice: Prospects and Challenges", *Journal of Universal Computer Science (Speical issue on Information Technology for Knowledge Management)*, Springer Science Online: URL: http://www.iicm.edu/jucs_8/coporate_memories_ for_knowledge 1997.

[7] Simone, C., & Divitini, M, "Ariadne: Supporting Coordination through a Flexible Use of the Knowledge on Work Process", *Journal of Universal Computer Science (Speical issue on Information Technology for Knowledge Management)*, Springer Science Online: URL: http://www.iicm.edu/jucs_8/aduadne_supporting_coordination_through, 1997.

[8] Uschold M, King M, Moralee S, Zorgios Y. "The Enterprise Ontology" *The Knowledge Engineering Review*, vol.13, no.1, March 1998pp.31-89. Publisher: Cambridge University Press, UK

[9] Gruninger M., "Integrated Ontologies for Enterprise Modelling" *Proceedings of ICEIMT '97, International Conference on Integration and Modeling Technology.* Springer-Veralg. 1997, pp.368-77. Berlin, Germany.

[10] Firestone, J. "Architectural Evolution in Data Warehousing" (White Paper No. Eleven, July 1, 1998). http://www.dkms.com/White_Papers.htm, 1998.

[11] Object Management Group (OMG), *"The Common Object Broker: Architecture and Specification* : Revision 2.1 ", OMG, August 1997.

[12] Florida-James, B., Hills, W. and Rossiter, B.N. , "Semantic Equivalence in Engineering Design Databases", *Proceedings, 4th International Workshop on Knowledge Representation meets Databases*, Athens, Greece, 1997.

[13] Roth, M.T. & Schwarz, P., "Don't Scrap It, Wrap It!" , *Proceedings of the Twenty Third international Conference on Very Large Databases*, Athens, Greece, 1997.

[14] Rochkind, M. , "The Source Code Control System", *IEEE Transactions on Software Engineering*, Vol SE- 1 No 4, pp 364-370, 1975.

[15] Keller A.M. & Ullman, J.D., " A version numbering scheme with a useful lexicographical order", *Proceedings of the IEEE Data Engineering Conference*, Taipei, Taiwan, 240-248, 1995.

[16] Dattola, A., "Collaborative Version Control in an Agent-Based Hypertext Environment", *Information Systems Journal*, Vol 20 No 4, pp 337-359, 1996.

[17] Hutchinson, K.W., Todd, D.S., Sen, P., "An Evolutionary Multiple Objective Strategy for the Optimisation of Made-To-Order Products with special reference to the conceptual design of high speed mono hull Roll-On/Roll-Off Passenger Ferries", *Proceedings International Conference of Royal Insitution of Naval Architects AUSMARINE '98*, Australia, 1998.

[18] Duffy, A. H. B., Smith, J. S., & Duffy, S. M., *Design reuse research: A computational perspective*-Keynote paper, Engineering Design Conference '98 (Brunel University, UK, 23-25 June 1998), edited by S. Sivaloganathan & T. M. M. Shahin, published by Professional Engineering Publishing Ltd. London. UK, ISBN: 1-86058-132-3, 1998.

[19] Senge, P. M., Kleiner, A., Roberts, C., Ross, R. B., & Smith, B. J., *The fifth discipline fieldbook: Strategies and tools for building a learning organization*, published by Nicholas Brealy Publishing Ltd, ISBN: 1-85788-060-9, 1994.

[20] Pidgeon, N. F., Turner, B. A., & Blockley, D. I., "The use of Grounded Theory for con-

ceptual analysis in kowledge elicitation", in Int. Journal Man-Machine Studies, Vol. 35, pp. 151-173, 1991.

[21] Brechtel, W., Connectionism & the philosophy of mind: An overview, in Mind & Cognition: A Reader, edited by W. G. Lycan, published by Basil Blackwell Ltd, ISBN: 0-631-16763-3, 1990.

XPat: A tool for Manufacturing Knowledge Elicitation

Benjamin Adesola[1], Rajkumar Roy[1] and Steve Thornton[2]

[1]Department of Enterprise Integration, School of Industrial and Manufacturing Science, Cranfield University, Bedford, MK43 0AL, U.K.
Email: {B. Adesola, r.roy}@cranfield.ac.uk
[2]Corus UK Ltd., Teesside Technology Centre, P.O. Box 11, Grangetown, Middlesbrough, Cleveland TS6 6UB, UK. Email: Steve.Thornton@corusgroup.com

Abstract: This paper presents **XPat** (e**X**pert **P**rocess Knowledge **A**nalysis **T**ool), a knowledge elicitation tool and describes an example of its application for the elicitation of knowledge from planning and scheduling experts in steelmaking. XPat approaches knowledge elicitation using a "process view" considering it as comprising of input, process and output. The approach serves to elicit knowledge with concurrent development of a process model to generate deeper understanding of the tasks involved. Direct involvement of the experts in mapping out their knowledge is a major strength of XPat.

1. Introduction

Knowledge elicitation is not "rocket science" yet it is difficult both in theory and in practice. As in software development, a large percentage of knowledge engineering costs are incurred during the earlier phases of knowledge systems development. Knowledge elicitation has been described as "the process of eliciting the expertise of authorities in an application area so that computer systems can render domain-specific advice." [1].

The problem of knowledge elicitation can be ascribed to difficulties in creating and agreeing on shared models of problem solving. The task of the knowledge engineer is to access and expose the mental processes employed by the expert to realise an explicit model, which can be shared and reused by others in some form. The challenge lies in achieving convergence of the mental process and the explicit model through a process of elicitation, validation and mutual agreement. Building a knowledge system from articulated experience of human expertise is not an easy task. Human knowledge tends to be highly tacit in nature; it is a network of several interconnecting reasoning patterns, which cannot be extracted without considerable effort. Therefore, it is often difficult to understand and interpret such expertise using one elicitation technique.

The task of knowledge elicitation is extremely cumbersome and requires understanding of what *type* of knowledge needs to be elicited from an expert and how best to *structure* the knowledge for reuse in the future. These lengthy and painful processes have been underestimated in the past [2]. However, the realisation that the performance of a knowledge system critically depends on the amount and quality of the knowledge embedded in the system [3], provided an incentive for the development of a variety of techniques to overcome the problems.

This paper describes a tool, which has been developed to facilitate the initial elicitation stage. The approach is termed **XPat** (e**X**pert **P**rocess Knowledge Analysis Tool). In Section 2, the context of the research is described outlining the characteristics of the oxygen steelmaking process, the functions of production planning and control, and the challenges posed for the research. Section 3, introduces the XPat knowledge elicitation technique, briefly describing the approach and the techniques utilised. Section 4 explains how XPat has been used to elicit knowledge of planning and scheduling for steelmaking. The Section describes the approach used for analysis of the process including template probe questions for generating rules. In Section 5, related work is discussed with emphasis on the problem of scheduling and knowledge elicitation, and the strengths and weaknesses of XPat are discussed. Finally, conclusions are drawn in Section 6.

2. The research context

2.1 Basic Oxygen Steelmaking (BOS) process

The oxygen steelmaking process presents significant challenges with respect to achievement of the manufacturing flexibility required to meet the market demands of today. Steel is made in batches of typically 250-300 tonnes, but is then continuously cast into solid product of convenient cross section and length via continuous casting machines. In the early 1980's, a very simple process route between the oxygen steelmaking (BOS) vessels and the continuous casters was the norm with just basic intermediate processing for alloy additions and stirring. The increasing requirement for more sophisticated and tighter specifications for chemical composition and cleanness has resulted in the addition of many additional intermediate process stages however, and consequent increase in operational complexity. This is illustrated in Figure 1, which shows some of these 'secondary steelmaking' developments.

Thus the process has become significantly more knowledge intensive, i.e. the knowledge required to construct feasible schedules has exploded, both from the static (typical constraints) and dynamic (current atypical constraints) aspects. In addition, the manufacturing instructions which need to be issued on the shop floor for execution of the schedule require a high degree of flexibility and tacit knowledge, on the part of manufacturing teams and coordinating functions.

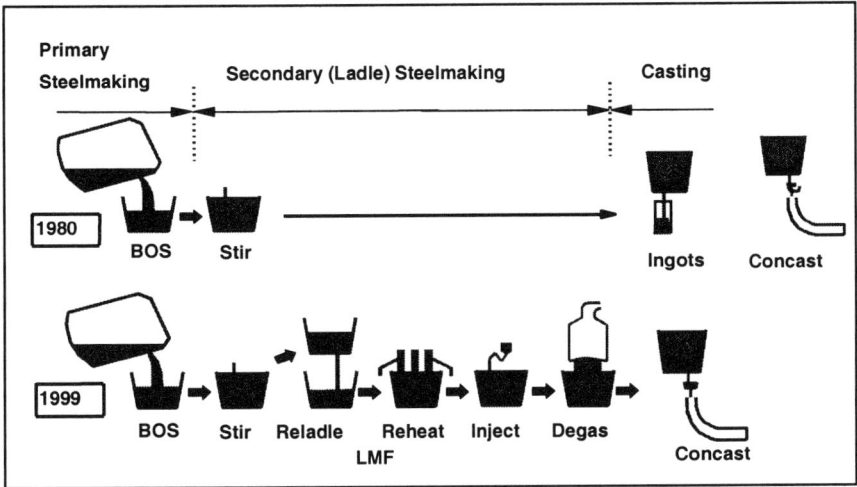

Figure 1: Development of Steelmaking Process Routes

2.2 Production planning and control functions

The three primary functions involved in production planning and control are planning, scheduling and manufacturing execution. Planning concerns the weekly allocation of orders to plant resources to meet customer due dates. Scheduling addresses the daily sequencing of orders to produce the best method for manufacturing the required orders according to raw materials constraints (principally hot metal supply for this process), known plant capability constraints, and economic factors. The function also monitors the progress of schedule execution in real-time and addresses the requirements for "rescheduling" when requests or triggers are invoked from agents in the application domain. Manufacturing execution concerns the actions required on the shop floor to realise the schedule. It is in the area of "scheduling" that there is a high requirement for manual manipulation and employment of tacit knowledge. Achievement of this function is mainly the responsibility of the "shift manager production control", who has been the prime focus for this research.

Figure 2 gives a high level illustration of the interactions between these three functional roles. Essentially, scheduling provides an interface function between planning and manufacturing, translating instructions into a form which can be used to make products in manufacturing, and interpreting feedback from manufacturing for use in planning. This feedback can be considered to be of three types:

'Proactive'   Provided for use at the medium term planning level in good time for consideration in the development of medium term plans.

'Anticipatory'   Provided where short term risk is recognised by the scheduling function and considered in the building of optimised short term schedules.

'Reactive'   Where schedule adjustments must be made in real-time because of some unanticipated event.

This feedback is provided through several channels including business mainframe computers, the plant process computers, and by verbal and visual means. In the case of proactive feedback this mechanisms are formalised and processes are in place to ensure that the planning function has timely access to this information. At the anticipatory and reactive levels however, the mechanism is more informal, relying on the human expert to detect problems or infer them from the available data. It is at this domain that this research is principally aimed.

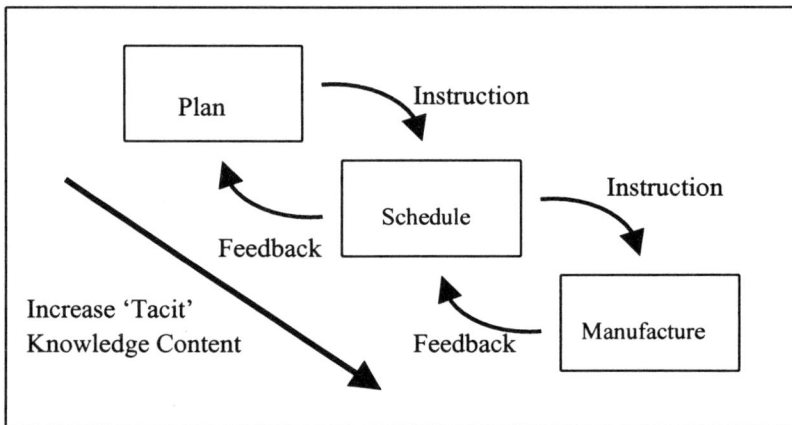

Figure 2: Interface between production planning and control functions.

2.3 Decision support for the scheduling role

As already described, the effective scheduling of Steelmaking is a demanding knowledge intensive task. Historically, this has been achieved primarily through the experience of the people close to the process, in combination with many informal communication models. As the complexity of the operations has increased, and the demands on cost, quality and reliability have heightened, computer based methods have been developed to provide decision support for devising and execution of schedules. These have greatly improved the effectiveness of the process, especially in terms of devising initial planned schedules, but measures to address the real-time requirements on the shop floor have typically been limited in scope and effectiveness.

One approach uses a "Scheduling Expert System" (SES) to address real-time rescheduling issues arising when unplanned deviations from the planned schedule occur [4]. The system has three components; information handling, problem detection and solution, and a user interface. The main driver for the system is receipt of "change requests" from the mainframe computer or other plant sources. When the SES receives such messages, analysis is made of problems, which may be arising, for example possible resource conflicts, impact on process synchronization etc. These problem areas are highlighted via the user interface. The user is offered three options, (a) to fix the problem himself (the system assists by confirming the feasibility of suggestions), (b) automatic problem resolution by the expert system itself, or (c) ignore the request for change. In reality, although the system provides extremely valuable assistance, the onus is usually on the user to effect necessary changes.

The limitations of the intelligent component centre mainly on the difficulties associated with furnishing the expert system with the knowledge required for the task (much of which is tacit in nature), and maintaining this over time as manufacturing tasks, requirements and constraints are added. Total reliance on the user to undertake this function may be a legitimate strategy but it presents significant potential drawbacks including:

- The level of tacit knowledge and experience varies from person to person and thus effectiveness of decisions can differ.
- Application and sharing of best practice is difficult to achieve.
- The risk element associated with decision making is assumed by the individual and not necessarily applied appropriately for the specific situation.
- There is high risk that knowledge can be lost due to retirements and job rotations.
- For new practitioners, the learning process is slow with increased risk of mistakes.
- Speed of decision-making is inhibited and so only limited options can be explored.

There is a strong motivation therefore to consider how the knowledge base component of such systems can be improved both from the perspective of content and management.

2.4 Research challenges.

The knowledge to be considered can be classified into two types; "domain" knowledge which concerns the processes, products and environment associated with the enterprise, and "task" knowledge which concerns the rules, methods and procedures which differentiate the expert from a novice to the field. Furthermore, the knowledge can be categorised under three main headings:

2.4.1 Declarative knowledge

Considered as "static" knowledge by the plant operators, comprising of explicit knowledge about standard operating procedures, recipes, specifications and orders. This typically is located in databases and plant systems.

2.4.2 Procedural knowledge

Existing explicit knowledge about business and plant processes, which can be applied to ascertain the status of real-time operations.

2.4.3 Tacit knowledge

This is the implicit knowledge of the process stored in the heads of people, which is applied outside of the computerised and written procedures, to generate information about changes which may need to be made to an in-progress schedule. The implication is that some of this at least can be converted to procedural knowledge.

The challenge for this research is to determine effective methods for capture of the knowledge which is employed in this process, to find appropriate ways to model such knowledge in a maintainable structure, and to establish deployment routes for effective integration in the application domain. This paper focuses on the first element of the challenge, i.e. effective capture of manufacturing knowledge.

3. XPat knowledge elicitation technique

As stated already, the task of knowledge elicitation is not "rocket science". The problem knowledge elicitation tries to address is "*how to capture process knowledge from the experts*". The purpose of knowledge elicitation is "*to provide raw material for knowledge modelling which will be used for building a knowledge system*". Through focussing on the problem and its purpose, the XPat approach is a deliberate attempt to deliver a complete solution to this problem. XPat is a process driven technique, the knowledge elicitation is structured within [Input], [Process], [Output] framework. Input and output, who or what is giving information, and who or what is receiving information, will define the "agents". Tasks and methods will come from the process analysis. The next section describes how XPat has been applied for capturing manufacturing knowledge.

3.1 The approach

3.1.1 Stage 1: pre-analysis

Generally the knowledge elicitation should start with pre-analysis. The aim of pre-analysis is to establish the scope, issues and challenges for knowledge elicitation in a project. Depending on the nature of the organisation and project scope, the content of this stage may vary the objective however remains the same. The output of this stage is an organisation chart, feasibility study incorporating alternative options, business case analysis and initial project definition.

3.1.2 Stage 2: identify the problem.

The aim of the problem identification stage is to achieve a deeper understanding of the application domain i.e. the tasks, the methods and the agents involved. The outcome of this stage will be used to guide further collection and interpretation of process knowledge. This stage can be split into three areas of interest:

(1) Knowledge intensive tasks - describe the task (e.g. planning, scheduling etc).
(2) Sources of knowledge - list all sources and locations (e.g. database, users, experts, artefacts, equipment, systems, tools).
(3) Types of knowledge - may include procedural, declarative, heuristic, structural.

The output of this stage is a list of documents describing the task, the sources of knowledge and the types of knowledge.

3.1.3 Stage 3: collect and interpret knowledge

The input to this stage is the list of documents describing the task, the sources of knowledge and the types of knowledge found in the application domain. This is where the bulk of the activities involved in knowledge elicitation are carried out. Once the problem has been identified the process begins with collection of knowledge, followed by interpretation.

The collection process is the task of interviewing the expert. It requires effective interpersonal communication skills to obtain the co-operation of the expert(s). Structured interviewing techniques (e.g. probe questions) and unstructured interview techniques should be applied for this stage. Interpretation involves the review of assembled data and information and the distillation of essential pieces of knowledge. The process requires the help of the expert to establish the problem goals, constraints and specific scope. It is recommended that the collection and interpretation process should be carried out in parallel. This will minimise risk of collecting irrelevant knowledge. Formal methods may be used later to further interpret and structure knowledge. The output of this stage is the expert's view of the process, it is recommended that "brown paper", "flip charts" and "post-it" notes are used as visual symbols to drive the process and to assist in achieving full engagement of the expert in the elicitation task. Formal methods may be used later to further interpret and structure knowledge arising from this stage.

3.1.4 Stage 4: analyse knowledge elicited

The process map provides the input to this analysis stage. Here, the knowledge engineer forms his opinion on how best to structure process knowledge elicited from an expert. At this stage, knowledge elements identified as input, process and output will be defined and structured in a graphical format. The knowledge engineer will need to be clear in his mind how an expert uses each of the pieces of knowledge identified. For process analysis the use of $IDEF_0$ is recommended for representing functional view of the process.

IDEF$_0$ is a well-established function modelling method, designed to model the decisions, actions, and activities of an organisation [5]. The output of this stage is the 'process model'.

3.1.5 Stage 5: design further elicitation techniques

The processes described in stages 2-4 are iterative following conventional requirements engineering approaches. The final stage is the design of further elicitation processes appropriate to the additional knowledge required, and for the particular expert involved. The design stage relies on the success of the analysis stage; assuming the result is favourable the outcome should provide guidance in designing techniques for gathering additional knowledge.

3.2 Techniques for knowledge elicitation

3.2.1 Interviewing

Interviewing is the most common knowledge elicitation mechanism. It requires higher order communication and interpersonal skills. The ability to conduct effective interviews is probably the most important skill for any knowledge elicitation technique, since the process serves to establish interaction and a relationship with the expert.

3.2.1.1 Unstructured interviews

The purpose of unstructured interviews is to explore heuristics (rules of thumb). The technique is concerned with trying to understand how the expert approaches the solving of problems, understanding their mindset, thoughts, feelings and formative experiences. There are no formal barriers or constraints as long as the knowledge engineer and the expert have a good relationship the technique can be used anytime anywhere. The lack of structure does however limit effectiveness.

3.2.1.2 Structured interviewing

This formal technique is particularly useful for identifying concepts and their relationships. Structured interviewing has the benefit of generating easy to analyse transcripts, and restricts the expert to abstraction from detailed principles of the application domain. In addition, as the expert generates scenarios, these are noted and transferred for inclusion in the detail level process view.

3.2.2 Protocol analysis

In addition to using interviewing techniques, protocol analysis [6] can contribute significantly to knowledge elicitation. Here the expert is observed in the performance of specific tasks while "thinking aloud." The technique captures both the actions performed, and the mental process followed in determining these actions. As a precondition however, the protocol analysis should take place after interviewing techniques have been exhausted.

4. Manufacturing knowledge elicitation: XPat for planning and scheduling steelmaking

The purpose of the section is to describe how XPat was used to capture manufacturing knowledge from planners and schedulers in Steelmaking. The technique used for collection and interpretation of process knowledge is a combination of structured and unstructured interviews and protocol analysis. XPat uses paper-based tools ("brown paper", "flip charts" and "post-it" notes). The brown paper provides a highly symbolic and tangible interface between the expert and the knowledge engineer, and provides a focal point, which is especially useful early in the process. All interviews were recorded on audiotape, text and graphical transcripts made. $IDEF_0$ was used to graphically represent the planning and scheduling process. The role of $IDEF_0$ is to facilitate the decomposition of task into subtasks. The technique provides a powerful means for analysis of the planning and scheduling function, which helps to minimise the need for elaborate descriptive text. Its graphical representation provided a clear representation of the complex aspects of manufacturing which, once familiar to the expert, provided a very effective medium for validation and verification of material.

4.1 Stage 1:pre-analysis

This provided initial input to the knowledge elicitation process. Pre-analysis was used to understand the overall organisation process. The result of this stage is an organisation chart, shown in Figure 3.

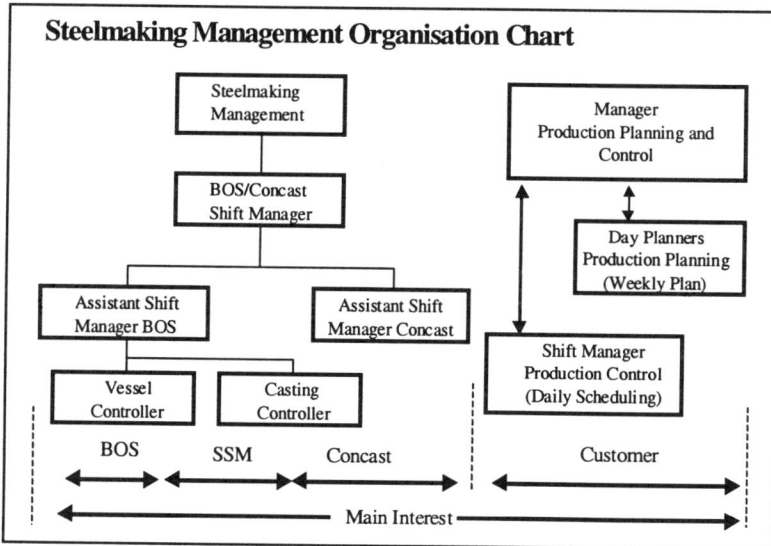

Figure 3: Steelmaking Management Organisation Chart

It is not possible to disclose other results (feasibility study, business case analysis, project definition) for reasons of confidentiality. The main technique used for organisation analysis was unstructured interviews, this was used to gain an understanding of the interaction between functions and people involved.

Essentially, production is controlled through collaboration of groups of people associated with the BOS, Secondary Steelmaking (SSM), Concast and daily scheduling (Figure 3). Whilst the management associated with BOS, SSM and Concast have prime responsibility for continuity and security of manufacturing operations, the shift manager production control is responsible for daily scheduling and thus represents the interests of the customers in the system. Through the provision of independent advice, this function ensures that whilst manufacturing effectiveness is maximised, the needs of the customer (required orders) are satisfied. As can be seen from the chart in Figure 3, this role provides shop floor representation for the planning function and it is this role in particular which provides the focus for this case study.

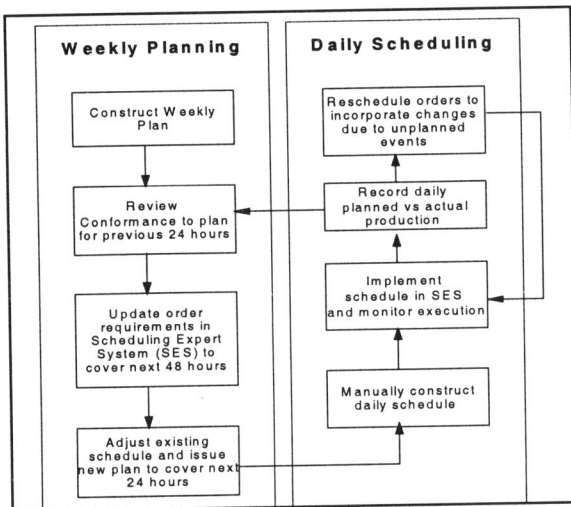

Figure 4: Knowledge Intensive Tasks in
Production Planning and Control

4.2 Stage 2: problem identification

Pre analysis is the main input to this stage, which considers three main areas of interest identified as production, scheduling and manufacturing execution. The focus for this research however is the scheduling element, which provides an interface between the other two areas as shown in Figure 2 above. As stated earlier the objective of this stage is to achieve a deeper understanding of problems faced by these functions. The technique used to identify the knowledge intensive tasks shown in Figure 4 was a combination of structured and unstructured interviews. The types of knowledge were briefly introduced in section 2.4.

Table 1: List of Domain Knowledge Sources and their Classification

Sources	Characteristics
Stakeholders, Experts and Users	1. Multiple experts with several years of experience e.g. 20-35 years. 2. A new user is fairly new to the scheduling, however with considerable knowledge of metallurgy. 3. Regular weekly meeting to discuss schedule changes 4. Research and Development personnel 5. Senior managers involved in decision making. 6. Knowledge Documents
Scheduling Expert System	An expert system for scheduling Steelmaking, Provides Gantt chart for graphical display of jobs, units, resources and constraints. Facility for messaging and dialogue screen for unsolicited messages
Information System	Mainframe computer terminals for information about orders, analysis information etc. Several screens are used for monitoring real time event in BOS and Concast plant.
Process Control Computer System	Real-time information and messages about plant status
Database	Stores the information on Steelmaking codes of practice, process routes, aim analysis.
Others	CCTV monitors for viewing incoming transport from the Blast furnace to the BOS / Concast plant.

The application domain is a technical domain, experts receive special training in metallurgy in addition to techniques for production planning and control. The sources of knowledge for planning and scheduling information are diverse in nature and a sample is presented in Table 1. It should be noted that some of the sources are available in real-time to directly influence the execution of the scheduling tasks whilst others are available only on an off-line reference basis. In addition, a typical domain expert will draw upon extensive practical experience.

4.3 Stage 3: collect and interpret knowledge

The collection process involves interviewing the expert(s). Structured interview technique was used. The probe questions are particularly useful for generating rules, describing concepts, attributes and their relationships. To explore the heuristics (rules of thumb) unstructured interviewing was used to understand the experts' approaches to problem solving. Interpretation was carried out in parallel with collection of knowledge. This involves review of data and information and the identification of essential pieces of knowledge. For this process the expert was asked to establish the problem goals, the constraints and specific scope. More formal interpretation and structuring took place immediately after the interactive sessions. The output of this stage represents the process view, which consist of the process map, the process model and the process glossary.

4.3.1 Mapping of top level process view

The aim of process mapping is to develop a shared understanding of the key processes and knowledge involved in planning and scheduling of steelmaking. The process analysis is a structured approach to identify and link inputs/outputs to process elements. The approach used to elicit knowledge was directed by the knowledge engineer, and the process map was constructed on brown paper.

The brown paper was divided into three sections, 'input', 'process' and 'output' as shown in Figure 5. XPat deliberately applies a process context for knowledge elicitation, as it has been discovered that this greatly assist experts in relating there experience to generate a 'process view'.

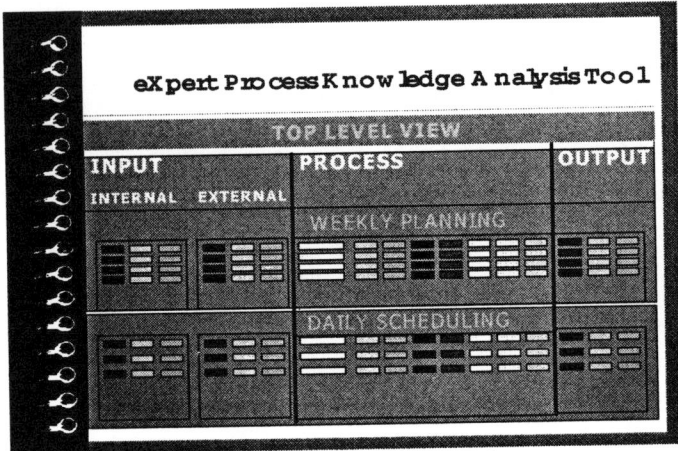

Figure 5: XPat -Top Level Process View

4.3.1.1 XPat template probe questions

Tables 2, 3, and 5 present the template-probe questions, which have been developed for the purpose of eliciting knowledge about manufacturing processes. These focus on the three areas of 'output', 'input ' and 'process'. A numbering regime is applied for ease of reference, together with the rationale for each question. These probe questions have been carefully selected based on previous experience of probing experts for knowledge to generate rules and problem solving methods. These questions are generic and can be applied in any other environment to elicit process knowledge, for example, requirements capture for business process improvement and reengineering. Using these template questions to elicit knowledge in this way greatly enhances subsequent modelling using $IDEF_0$.

4.3.2 Results: mapping of top level process view

The probe questions were used to reveal process knowledge in the form of rules, specific or generic to task and subtasks. To augment this approach, the expert(s) used flip charts to graphically represent complex problems. Each response to interview questions was captured on post-it notes, and mapped onto pre-allocated areas on the brown paper, (Figure 5). It should be noted that this process is physically demanding on both expert and knowledge engineer and so sessions must be designed to include regular breaks and efforts made to maintain a comfortable working environment.

4.3.2.1 Mapping "output" view of planning and scheduling functions

Figure 6 shows output view of functions at the top-level. The knowledge elicitation is done in the sequence: [Output], [Input], [Process]. It is observed that considering output first helps experts to focus onto tasks needed to deliver those outputs. The starting point is to ask the expert to list all relevant outputs using the probe questions in Table 2.

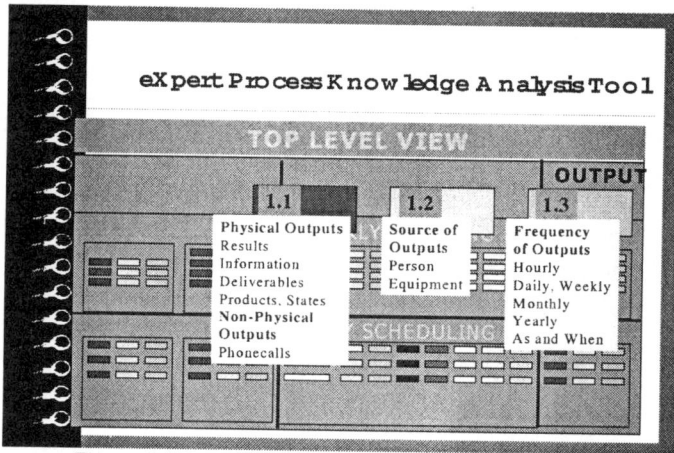

Figure 6: XPat -Top Level Process View of Output

Once a list of outputs is compiled on post-it notes, the next activity is to, ask the expert to list the destination/target of each output. Finally, ask the expert to identify frequency of all process outputs. For example, a top-level output for daily scheduling is information contained in daily production plans and reports, the destination is production control management office, BOS/Concast management office, process control department and test laboratory. The report is produced daily.

Table 2: List of structured probe questions for output view of functions at the Top-level.

Probe Identity	Probe Questions	Rationale for a probe question
O1	List all output from the process?	To identify specific outputs from the process in terms of information deliverables states product and results. To define types of output. To provide support for constructing IDEF$_0$ process model.
O2	Why would you need that output?	To generate rules for output information. IF <condition> THEN <action>
O3	How would you get that output?	To determine acquisition or reuse process.
O4	How would you use that output?	To generate detail level rules. IF <condition> THEN <action>
O5	What is the source of output?	To identify sources of output and interactions
O6	When would you generate this output?	To reveal specific or generic frequency of outputs To generate a detail level rule specific or generic output. IF <condition> THEN <action>
O7	What is the frequency of output?	To determine the dynamic nature of output (e.g. time relative to output – Hourly, Daily, Weekly, as and when required)

4.3.2.2 Mapping "input" view of planning and scheduling functions

Figure 7 presents the input view at the top-level; in this case the inputs need to be considered in terms of those which arise inside the domain (internal), and those which originate from outside (external). The starting point is to ask the expert to list all relevant internal inputs to the process. Next, ask the expert to list all sources of input. Finally, ask the expert to identify frequency of all process inputs. Repeat this activity for external input. The probe questions listed in Table 3 are used to facilitate the process. At this stage it is appropriate to introduce an additional probe question I8, which asks the expert about the relationship between inputs and the outputs already defined. An example of top-level input for daily scheduling is computerised list of sequences, the source is shift manager weekly planning and the frequency is weekly.

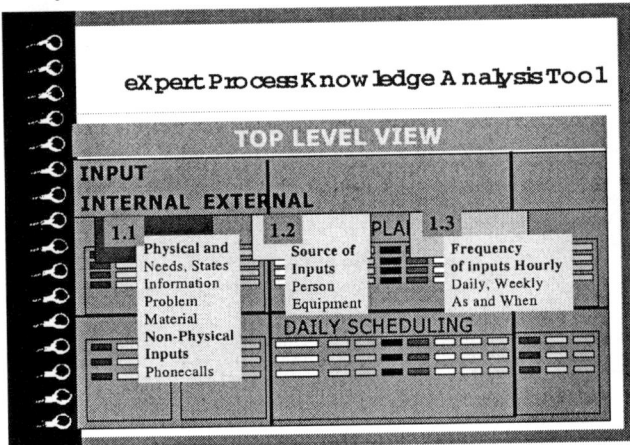

Figure 7: XPat – Top Level Process View of Input

Table 3: List of structured probe questions for input view of functions at the Top-level.

Probe Identity	Probe Questions	Rationale for a probe question
I1	List all input to the process?	To identify specific inputs to the process in terms of information needs, states, problem and material. To define types of input. To provide support for constructing IDEF$_0$ process model.
I2	Why would you need that input?	To generate rules for input information. IF <condition> THEN <action>
I3	How would you get that input?	To determine acquisition process
I4	How would you use that input?	To generate detail level rule. IF <condition> THEN <action>
I5	What is the source of input?	To identify sources of input and interactions
I6	What is the frequency of input?	To determine the dynamic nature of input (e.g. time relative to input – Hourly, Daily, Weekly, as and when required)
I7	When would you generate this input?	To reveal specific or generic frequency of inputs. To generate a detail level rule specific or generic input. IF <condition> THEN <action>
I8	What is the relationship between inputs and output elements?	To reveal the nature of relationships as either specific or generic.

4.3.2.3 Mapping "process" view of planning and scheduling functions

Mapping process elements in Figure 8, is where the bulk of process analysis is realised. The terminologies for describing process elements are shown in Table 4.

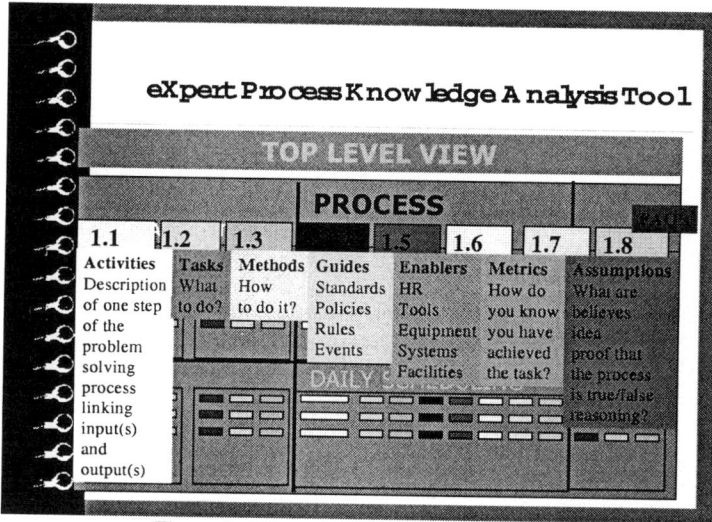

Figure 8: XPat – Top Level Process View

Table 4: XPAT-Process Terminology

Terms	Description
Activities	"an activity describes a step in a problem solving process" [8].
Task	"a task defines a reasoning goal in terms of an input-output pair" [7]
Methods	"a method describes how a task can be realised through a decomposition into subfunctions plus a control regimen over the execution of the subfunctions." [7]
Guides	a guide provides directions or applies a constraint to a problem solving process e.g. standards, policies, rules and events. The are two types of constraints. A hard constraint is a physical constraint that must not be violated. Soft constraint. A soft constraint is a policy constraint that can be relaxed.
Metrics	A metric describes how a process is measured in qualitative or quantitative terms.
Enablers	An enabler provides the means to solve problem e.g. human resources, tools, systems, equipment, and facilities.
Assumptions	An assumption describes beliefs, ideas and or proof that a process is true or false.
FAQ's	Frequently Asked Questions are illustrative examples, which can be used to enhance a future knowledge system.

The probe questions listed in Table 5 below are used to facilitate this process. Start by asking the expert to list all relevant activities. Next, ask the expert to list the tasks. Against each task, ask the expert to list methods used and associated guides and enablers, and use the probe questions to generate more knowledge elements.

Table 5: List of structured probe questions giving a process view at the Top-level.

Probe Identity	Probe Questions	Rationale for a probe question
P1	List all activities performed in a process?	To identify the steps in a problem solving process.
P2	In what context would you do that?	To identify matching input/output of an activity
P3	List all tasks specific to an activity?	To determine the type of task and subtasks of an activity. To decompose tasks into subtasks.
P4	List all methods specific to each task?	To determine what method is for specific tasks. To decompose method in subfunctions.
P5	List all guides specific to a task?	To identify guideline for specific task and sub tasks E.g. constraint/control related to a task such as policies, standards, rules, and events. To identify types of constraints (Hard constraints or Soft Constraints) To generate more rules for a task. To provide support for constructing IDEF$_0$ process model.
P6	When would you use these guides?	To reveal appropriate timing to use a guide.
P7	How would you use these guides?	To generate detail rules for a guide. IF <condition> THEN <action>
P8	What preferences can be made?	To reveal the choice in decision making. To generate rule for making decisions.
P9	List all enablers specific to a task?	To determine who does what. To determine what tool is used. To determine what system is used. To determine what equipment is used. To determine interaction between people and system To determine the rules for using the system or tool. To provide support for constructing IDEF$_0$ process model.
P10	When would you need that?	To reveal the frequency of participation.
P11	Why would you need that?	To generate rules for an enabler specific to a task. IF <condition> THEN <action>
P12	Who will need that?	To determine the rule for interaction.
P13	What are alternative enablers?	To generate more rules. IF <condition> is not available THEN <action>
P14	List all metrics the metrics for a specific task?	To generate rules for completion or state of a task.
P15	Why would you need that?	To generate rules for a measure specific to a task. IF <condition> THEN <action>
P16	How would you use this measure?	To generate detail level rules for metric. IF <condition> THEN <action>
P17	When would you need this metrics?	To reveal the generality of the rule and generate other rules.
P18	What are alternative measures?	To generate more rules. IF <condition> is not available THEN <action> IF <condition> is false THEN <action>
P19	List all assumptions for a task?	To identify the decision-making patterns. To provide additional information about a process
P20	List sources of Frequently asked questions?	To identify additional sources of knowledge.

Finally, ask to list assumptions, metrics and FAQ's. An example of how an activity is represented in IDEF$_0$ notation is shown later in Figure 9, "receive computerised list of sequences" (referenced as Node 21). The list of tasks includes A311, A312 and A313. The list of methods specific to each task describes how each task is performed. For example the method for task A312 states how several orders of incompatible quality codes can be combined to produce a longer sequence. Next, the expert can be asked to list guides' (controls/constraints) and enablers (mechanisms) taken from $IDEF_0$ notations.

Guides identify controls/constraints for an activity, they include standards, policies, rules and events. The focus here is to capture knowledge of how experts deal with hard or soft constraints that affect this process element. A hard constraint is a physical constraint that must not be violated e.g. process route for a particular steel grade. A soft constraint is a policy constraint that can be relaxed [9]. Each post-it note is numbered, the first response to each question about an output is labelled 1.1 for physical or non physical outputs, 1.2 for destination of output and 1.3 for frequency of output. The numbering regime will continue for the next question about output. This applies also to top-level inputs and processes.

4.3.3 Process analysis

The two deliverables from process analysis are the process model (Figure 9) and process glossary (Table 6). The process model is a graphical representation of the activities and tasks decomposed into a set of $IDEF_0$ diagrams. The process glossary describes in detail the terminology used in the knowledge domain. Process analysis provides much of the input for knowledge analysis and knowledge modelling. This activity results in the analysis of tasks performed by agents (e.g. systems, people) and the understanding of information flowing between agents.

The $IDEF_0$ diagram shown in Figure 9 has been customised to show a structured view of part of the Steelmaking scheduling task. The activities are shown in process blocks, input arrows go from left to right into the process blocks. Arrows emerging from the right side of the process blocks indicate outputs from the process. Guides and enablers are arrows entering the process blocks from the top and bottom respectively. "Sources" and "destinations" have been elicited for all inputs and outputs respectively. Inputs and Outputs are also annotated to indicate the frequency as elicited during the process mapping exercise, [D] for daily, [W] for weekly, [M] for monthly, [Q] for quarterly and [A] for "As and when". For methods, assumptions and metrics, a rectangle shape with letters [M] for methods, [m] for metrics and [A] for assumptions have been used to indicate that these knowledge elements have been elicited during process mapping.

For guides, circles with (H) for a hard constraint or (S) for a soft constraint are used. The presence of a letter in the circle shape indicates that knowledge elements have been captured during process mapping. These representations will support further analysis of process knowledge and description of tasks for knowledge model construction. Enablers and FAQ's have also been elicited during interviewing to further enhance understanding of the process. The $IDEF_0$ functional model has been used to develop a graphical representation of process inputs, outputs constraints and mechanisms. The process model developed is based on $IDEF_0$ notations with additional symbols to indicate the knowledge elements elicited during the process model development. XPat has thus enabled the knowledge engineer to elicit additional functionality to support description of input/process/output. The destination of output determines where result of any activity is stored in the process.

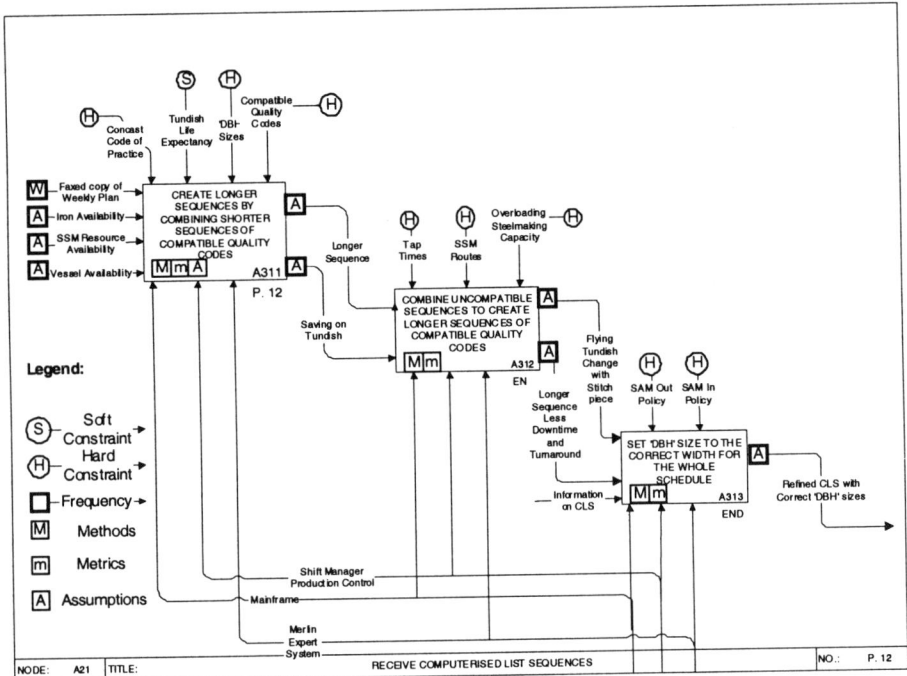

Figure 9: Sample of customised IDEF$_0$ showing Structured
Scheduling Process Model

4.3.4. Process glossary

A process or domain glossary is a very important output from the knowledge elicitation process. Essentially, this clarifies the language used within the knowledge domain, removing any ambiguity over terminology. This is especially important where there is impingement upon other knowledge domains, for example for the engineering function. Here a term like 'end of cast' can mean very different things to each function and removal of this potential ambiguity is essential if future cross-domain knowledge sharing is to be effective. Table 6 below presents a sample of the type of terms used within the planning and scheduling domain.

4.4 Stage 4: analysis of knowledge elicited

This is where the knowledge engineer forms his opinion on how best to structure and represent process knowledge elicited from an expert. The input to this stage is the process map, the process model and the process glossary. The activities involve penetrating the expert problem solving process further with sets of probe questions

Table 6: Sample Glossary of Terms for Steelmaking

Terminologies	Description
Acceptance Analysis	'Customer' acceptance limits for specific chemical elements for a quality code.
Aim Analysis	Target chemical range for the BOS plant to meet the customer's requirement.
Bloom Caster	Caster, which continuously cast steel into Blooms.
Cast	The solid product produced from continuously casting a ladle of liquid Steel.
Cast Analysis	Chemical analysis of a cast of Steel, whose samples are taken from the liquid Steel in the mould over the period of casting.
Casting Speed	Rate of withdrawal of the 'solid' Steel from strand of a machine in meter/min. Depends principally on quality codes, cross section of the mould and Tundish temperature of the liquid Steel
Concast Plant	Plant where casts of liquid Steel are continuously cast into solid products.
Due Date	Required delivery date for a semi order to a customer mill. This will be defined differently for customer mills to support their requirements.
End Cast	Finish of the casting process when all the Steel from the mould has been cast.
Hot metal	Liquid metal supplied from the Blast Furnace. In the Steelmaking domain this is often referred to as 'Iron'.
Misfit (Cast)	A cast whose actual analysis is outside the acceptance limit for the quality code it was made for.
Orders	Can be any of the following: customer orders, semi-finishing orders, schedule number orders.
Process Route	For planning and scheduling it is a database designed to store all acceptable process routes for a finished database product. For each quality code, there is one or more Process routes that define the BOS and Secondary Steelmaking processing required to make this quality.
Quality Code	Define quality information for an order. For example the required chemical analysis to meet the customer requirement, the processing required within secondary Steelmaking, casting practices and surface dressing.
Steel Type	Quality codes is split into generic groups based on aim analysis and called Steel types.
Weekly Plan	Plan generated each week by Production Control, summarising activities in the Steelmaking and semi-finishing are. This high level document is agreed with mill representatives and Steelmaking management.

designed to realise rules and methods for solving specific problems which occur due to disruption to operations. The output of this activity is a problem solving process in the form of a flow chart.

4.4.1 Detailed knowledge elicitation

The aim of this activity is to select a value adding activity from the process model for further analysis and observation. The concept of identifying activities where there is significant potential added value for the business is very important here, as a 'blanket' approach would be too time consuming and not cost effective. This aspect of involving the expert in deciding which elements of his work adds value to the business process is a critical success factor in this process.

This section describes the process of detailed knowledge elicitation to understand the tasks and logic behind problem solving methods used. The input to this activity is the process model. The expert was asked to select a value adding activity for further elicitation. Detailed knowledge elicitation was performed using the combination probe questions listed in Table 7, and unstructured interview technique to elicit detailed descriptions of problem solving processes.

Table 7: List of structured probe questions for detail knowledge elicitation.

Probe Identity	Probe Questions	Rationale for a probe question
P1	What do you do?	To establish a task description.
P2	Why would you do this task?	To convert a task description into a rule.
P3	When would you do this task?	To reveal the nature of a task as specific or generic to an activity. To generate more rules.
P4	How would you do that?	To reveal description of problem solving method and reasoning patterns for a specific task. To generate rules for a method.
P5	Why would you do that?	To convert method description into a rule for a task.
P6	What do you do when that happen?	To establish method for responding to unusual event.

The probe questions in Table 7 were used for detail knowledge elicitation. P1 - P3 can be used to elicit knowledge elements for tasks, whilst P4-P6 can be used for capturing problem solving methods. For the activity selected for further analysis ask the expert to briefly describe the tasks and relevant problem solving methods.

Probe questions were used to direct the elicitation process and to tease out complex reasoning processes. The knowledge elicitation process was recorded and transcribed. Protocol analysis technique was used to extract meaningful structure from the text of the interview transcripts. For example in Figure 9, activity node A21 "receive computerised list of sequences", the tasks identified were A311, A312 and A313. Detailed descriptions of each task problem solving process were captured.

4.4.2 Observation

The purpose of observation is twofold; (1) witness the expert performing the tasks specified in the process model; and (2) to validate the process model as complete, correct and consistent. The combination of structured and unstructured interview technique was used during analysis and observation. The responses to probe questions were used to generate rules and direction for further research. The observation periods served to clarify how the tasks were actually performed, and the decisions, which were taken. Whilst the top-level process view abstracts facts about the problem solving process, the detailed knowledge elicitation seeks to gain full understanding of the expert mental model of problem solving. The expert was observed while performing value-adding activities. The knowledge engineer asked the expert to describe the tasks in simple terms and the way in which the task solution or outcomes affect the choice of decisions. The expert's responses were later cross-referenced with corresponding tasks and methods specified during the detailed knowledge elicitation session. During the observation, an event resulting in a requirement to reschedule was witnessed. Rescheduling is the process of making "repair and improvement" to an original schedule which has become "flawed" due to an unplanned event. This was identified as a process where the potential added value was significant, and could be quantified. The activities were fully described and initially the expert sketched the problem solving process himself. This was later refined in conjunction with the knowledge engineer to produce a more accurate and complete representation. This was developed in the form of a process flow chart, a sample of which is given in Figure 10. For reasons of confidentiality, the full diagram, which extends further through the branches labelled (A), (B) (C), (D) and (E), is not reproduced in this paper.

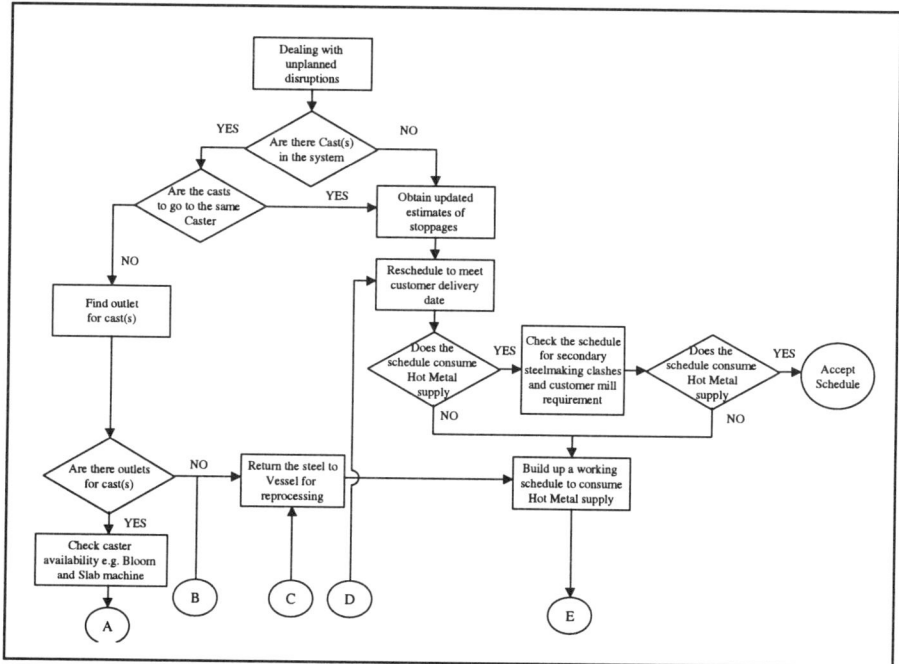

Figure 10: Sample Problem Solving Method for Scheduling Steelmaking

The availability of a flowchart of the problem solving process was extremely useful for validation of this process model. The flowchart was generated by a highly skilled domain expert [10], and validated by other experts working in the domain.

4.4.3 Rules from structured probing questions

The following section presents several examples of rule development from the product of the knowledge elicitation sessions. The extracts below are taken from transcript interviews with scheduling expert made during observation periods. The dialogue illustrates the type of probe question used e.g. P6. Knowledge Engineer and Experts are represented as KE and EX respectively.

Example: This dialogue is a description of method for dealing with out of specification casting.

EX		Here is an example where the Steel is cold and out of specification.
KE	[P6]	What do you do when that happens?
EX		Well this is quite rare but, basically, if the Steel is cold and out of spec we have to take a view on whether to try to cast it or not. Of course, it's possible that just one of these conditions will apply, i.e. it's either cold or out of spec but not both. If it's a bit cold then it could still be possible to cast it, but only if a higher casting speed can be used and the Tundish is not cold. There is a possibility also that we could return the Steel to the ladle furnace (for reheating). On the other hand, it might be a little too hot, in which case an option might be to cast it at a slower speed than normal. If the temperature is too hot though there would be an increased risk of breakout and so we should not delay casting. As for being out of spec, we need to consider whether we can take a "decant" option before considering a return to the vessel; this can also be used to address temperature problems in some cases as well Translates to rule format as follows:

IF	Steel is at right temperature
AND	Steel is in Spec
THEN	Cast Steel as Planned

Or going deeper into the knowledge as exposed using XPat:

IF	Steel temperature is > 5°C above Liquidus
AND	< 30°C above liquidus
AND	Chemical Analysis is within tolerance stated in code of practice
THEN	Allocate Steel to a Casting machine

In the above, the definitions of parameters such as liquidus and tolerance would be available from a data source such as the codes of practice database. A future opportunity however might be to introduce further rules and applications to calculate these. Similarly, the hard values used in the rule could be flexed according to operational circumstances. In reality the experts do apply some latitude in this area.

Examples of other rules following from this dialogue are as follows:

IF	the Steel temperature is < 5^0 Celsius of the Codes of Practice
AND	the chemical composition is out of specification
AND	Ladle Furnace is not available
THEN	decant (dilute) or as a last resort return the Steel to the Vessel

IF	the Steel temperature > 5° C of the codes of practice
AND	the Steel temperature < 15°C of the code of practice

AND   the Tundish temperature is $> 30^0$ Celsius above liquid Steel
AND   the chemical composition is in specification
THEN   cast the Steel at 10% slower than the Code of Practice speed.

Clearly there is a need for supporting rules, for example to determine whether "chemical composition is in specification" is "True", and these too could vary in sophistication in a future knowledge system, from simple data definitions to complex models which determine this on the basis of required physical properties.

4.5 Design of further elicitation techniques

Sections 4.1-4.4 described the stages in knowledge elicitation. The process described is iterative however, and it is frequently necessary to return to earlier stages, especially in the course of verification and validation. The analysis stage has been successful in eliciting product, process and environment knowledge.

4.6 Validation

The results of the process including those presented in this paper have been validated with the Experts and other potential users of any future knowledge systems, which might be developed from these results. The main method employed for validation of the process model was by repeated "walk through" testing with individuals. Where necessary, feedback from the walk-through tests was used to modify and extend the process model. Multiple experts and potential users have validated the flowcharts for problem solving process and rules. The results in this paper have all been validated for completeness, correctness and consistency.

5. Discussion and related work

5.1 Challenge of scheduling and role of artificial intelligence

Scheduling is a difficult task both in theory and in practice, [11]. In the past two decades several advanced scheduling methods have been developed [12], [13], [14], [15] but only a limited number have reached the stages of practical use [16], [17]. This is because highly automated methods are not necessarily the complete answer for the typically ill-defined problems which schedulers face in the real world. The highly interactive and knowledge intensive nature of the tasks involved, particularly in batch process industries, means that attempts at automation can result in compromising of the operation. Thus, the traditional expert system approach is not sufficient for this knowledge domain and the knowledge elicitation work to date has indicated that a combination of knowledge representation methods would be needed to even approximate the overall knowledge regime.

However, the work has clearly demonstrated the value of structured knowledge elicitation methods in exposing the tacit knowledge employed in the operation. Many components have been identified which could be represented successfully using artificial intelligence techniques, to provide effective support to ensure that the best decisions possible are made on a consistent basis. The key to achievement of real business benefits from this approach however lies in development of knowledge systems, which are maintainable, and in the effective integration of these into the complete business process.

5.2 Knowledge elicitation from process experts

Several research projects have attempted to tackle the problem of knowledge elicitation. The techniques involve direct and indirect methods namely interviewing, protocol analysis, laddering, concept sorting, repertory grid, rule induction, 20 questions, scenario analysis and many more. Many literature have proposed elicitation techniques [10], [18], [19], [20], [21], [22], [23], [24], [26], however there is no 'magic bullet' solution to the challenge of extracting knowledge from the expert.

In spite of much research, knowledge elicitation remains a difficult task. There is a need to understand and converse in the language of the experts and thus the initial development of a shared process view is essential. In the literature there are several tools available to support some knowledge elicitation techniques, for example, markup tools, rule editors and graphical support tools. In general, they impose artificial structure and require special skills and training for use. Although some are very useful for organizing and general management of the elicited knowledge, the methods used can impose limitations onto the process and do not address the requirement to fully involve the expert in the process.

The knowledge engineer must be careful when selecting the approach and tools he will use so as to avoid these pitfalls. For example, many techniques target the capture of production rules as their primary goal; these tend to be relatively shallow in nature and based on specific operating context. Maintenance of resulting knowledge systems can become difficult as the rule base grows, particularly with respect to interaction between individual rules and the affects of introducing new operational steps and methods.

The aim of interactive knowledge elicitation and encoding tools is to limit the workload for the expert. Instead they leave substantial role [27] for the knowledge engineers whose job is to structure and represent complex concepts, patterns and heuristics that best describe the expertise on reflection. This is not to say that software tools have no part to play in the art of knowledge elicitation, but that there is a stage during the process where software tools can begin to make a positive contribution. The precise point at which this applies depends on many factors related to the individuals involved, the nature of the knowledge domain, and the overall maturity with respect to the philosophy of organizational knowledge management.

5.3. The contribution from XPat

XPat aims to combine process modelling and knowledge elicitation, and to put the expert in charge of the knowledge elicitation process. This approach permits the 'fabric' of the domain knowledge to be teased out, and a shared understanding to be developed. Thus the complete requirements for a successful knowledge system can be identified and the likelihood increased of successful implementation and business benefits. The method also encourages system ownership by the experts within the domain and an acceptance of the responsibility for future maintenance of knowledge systems. Indeed, this can be extended further to encourage recognition of the concept of domain knowledge, and the responsibility for maintenance of this as 'part of the job'.

Probe questions form a major basis of the tool, as is the case with other established approaches such as CommonKADS [7]. The XPat probe questions however, differ from these in that they are designed specifically to guide knowledge elicitation following an [Input]-[Process]-[Output] template. This means that the approach elicits knowledge, but at the same time develops the process model to enhance understanding of the tasks involved. It actually gives a 'process view' to knowledge elicitation. The advantage of doing knowledge elicitation this way is that it facilitates the expert in articulating their experience, and serves to facilitate the achieving an alignment between the expert and the process.

Experience shows that there is a relationship between, elicitation techniques used to gather data and information, and the resulting model of expert knowledge. This suggests that there is a need for integration of existing techniques to help organizations to minimise risk associated with loss of vital knowledge. Much of the research in this area to date has focussed on developing new approaches and critiquing existing techniques. The research in this paper has focused on achieving reduction in the overall time needed for effective elicitation of knowledge from experts, and full mental and physical engagement of the expert in the process.

The following benefits are observed
- The technique is process led meaning that experts can relate their work to the process map quite easily.
- The technique provides adequate support for process analysis using probe question, and process modelling with tools such as $IDEF_0$.
- XPat is a structured approach hence support for verification and traceability
- Enhancement to $IDEF_0$, by providing additional notation to represent complex knowledge elements in a process: XPat has helped to structure and elicit the knowledge elements
- XPat can be applied to similar environment to elicit process knowledge. The tool is already being used in other projects at Cranfield University.
- The human expert once 'on board' can map the process himself. In fact the facilitator adopts an increasingly supporting role, as the expert becomes familiar with the process.

- The approach encourages recognition of the need for knowledge management within functional areas, and provides a basis on which to build this competence.

The following limitations are observed
- The process is time consuming.
- The process is physically demanding
- Significant effort is associated with transcription to generate rules

6. Conclusions

XPat has proved that knowledge elicitation is not "rocket science" by demonstrating the approach to elicit knowledge from experts in the domain of Steelmaking scheduling. A tool for knowledge elicitation has been developed which has proven useful for quick elicitation of abstract and detailed specific knowledge. The unique feature of the tool is the "*process view*"; this means the knowledge engineer and the expert can adopt the same language early in the elicitation process. The expert can relate the XPat questions better because they relate directly to the process, which he understands intimately. The research has also identified and represented new knowledge elements to enhance IDEF$_0$. Further research at Cranfield University is aimed at identification and application of a suitable modelling framework to the manufacturing knowledge elicitation outlined in this paper. Future research will map the output of this phase to a knowledge model that supports knowledge reuse.

Acknowledgements

The research is being carried out as part of a Doctoral study ongoing at Cranfield University and is supported by EPSRC and Corus UK Ltd. The first author would like to thank the employees of Corus UK Ltd. for their help and support during the knowledge elicitation.

References

1. Musen, M. (1988). *An Editor for the Conceptual Models of Interactive Knowledge Acquisition tools*. In the Foundation of Knowledge Acquisition. Knowledge Based System, 4, pp 135-160.
2. Vergison, E. (1999). Knowledge Management: A breakthrough or the remake of an old story. *PAKeM '99, The Second International Conference on The Practical Application Of Knowledge Management*, 21-23 April 1999, London (UK), pp.1-5.
3. Feigenbaum, E.A., (1977). *The Art of Artificial Intelligence: Themes and case studies of knowledge engineering*. Proceedings IJCAI 77, Cambridge, MA, pp.1014-1029.
4. Smith, A. W. and Caldwell W.G. (1993). Merlin - A Knowledge Based Real Time Rescheduling System. Technical Paper British Steel, General Steels, Redcar, Cleveland, UK.
5. Colquhoun, G.J. and Baines, R.W., (1991). Ageneric IDEF$_0$ model of process planning, International Journal of Production Research, Vol. 29, No. 11, pp. 2239-2257

6. Errickson, K.A. and Simon, H.A. (1993). *Protocol Analysis: Verbal Reports as Data, revised edition.* Cambridge, MA, MIT Press.

7. Schreiber, G., Akkermans, H., Anjewierden, de Hoog, R., Shadbolt, N., Van de Velde, W., and Wielinga, B. (1999), Knowledge Engineering and Management: The CommonKADS Methodology A Bradford Book, The MIT Press, Cambridge, Massachusetts.

8. Neubert (1993): Model Construction in MIKE (Model-Based and Incremental Knowledge Engineering). In Knowledge Acquisition for Knowledge Based Systems, Proceedings of the 7th European Workshop EKAW '93, Toulouse, France, september 6-10, Lecture Notes in AI, no 723, Springer-Verlag, Berlin, 1993, pp.

9. Helene, S., (1998). *Intelligent plant wide Steel scheduling and monitoring, vertically integrated with SAP and the production floor,* http://www.kbe.co.za/articles/ExpertSystems/SSP-Paper2.html

10. Chi, M.T.H., Glaser, R., and Farr, M. (1988), *The Nature of Expertise.* Hillsdale, NJ, Erlbaum.

11. Dorn, J., Kerr, R. and Thalhammer, G. (1995). Reactive Scheduling: improving the robustness of schedules and restricting the effects of shop floor disturbances by fuzzy reasoning, Int. J. Human-Computer Studies, 42, pp. 687-704.

12. Nemhauser, G.L. and Wolsey, L.A. (1988). Integer and Combinatorial Optimisation, Wiley, Interscience.

13. Davis, L. (1987), Schedule Optimisation with Probabilistic Search, Proceedings of 3rd Conference on Artificial Intelligence Applications, pp 231- 235.

14. Nilsson N.J. (1982) Principles of Artificial Intelligence, Springer- Verlag.

15. Bruno, G. (1986) A Rule-Based System to Schedule Production, IEEE Computers, pp. 32-39.

16. Nonaka, H. (1990) A support method for schedule revision using AI Techniques, Proceeding of IAEA Specialists' Meeting on Artificial Intelligence, Helsinki, VTT Symposium 110, vol.II, pp 219-231.

17. Numao, M and Morishita, S. (1988) Scheplan - A Scheduling Expert for Steelmaking Process, Proceedings of International Workshop in Artificial Intelligence for Industrial Application pp 467-472.

18. McGraw, K.L. and Harbinson-Briggs, K. (1989), *Knowledge Acquisition: Principles and Guidelines.* Prentice-Hall International.

19. Kidd, (1987) In Kidd, A.L., (ed.), *Knowledge Acquisition for Expert Systems: A practical Handbook,* Plenum Press.

20. Anjewierden, A. (1987), Knowledge Acquisition Tools. *AI Communications.* Vol 10 No. 1 pp 29-38.

21. Olson, J.R. and Reuter, H.H. (1987) Extracting expertise from experts: Methods for Knowledge Acquisition, *Expert Systems* 4(3).

22. Burton, A. M., Shadbolt, N. R., Hedgecock, A. P., and Rugg, G. A. (1987), *Formal evaluation of knowledge elicitation techniques for expert systems: domain .* In D.S. Moralee (ed), Research and Development in Expert Systems 4, Cambridge University Press.

23. Fox, J., Myers, C. D., Greaves, M.F. and Pegram, S. (1987), A systematic study of knowledge base refinement in diagnosis of leukemia. In Kidd, A.L., (ed.), *Knowledge Acquisition for Expert Systems:A practical Handbook,* chapter 4, pp. 73-90. Plenum Press.

24. Johnson, L. and Johnson, N.E. (1987), *Knowledge elicitation involving teachback interviewing.* In Kidd, A.L., (ed.), *Knowledge Acquisition for Expert Systems: A practical Handbook,* chapter 5, pp. 91-108. Plenum Press.

25. van Someren, M. W., Barnard, Y., and Sandberg, J. A. C., (1993), The Think-Aloud Method. London Academic Press.

26. Gaines, B.R and Shaw, M.L.G. (1995) Eliciting Knowledge and Transferring it Effectively to a Knowledge Based System Knowledge Science Institute, University of Calgary, Alberta, Canada.
http://ksi.cpsc.ucalgary.ca/articles/KBS/KSS0/

CommonKADS and Use Cases for Industrial Requirements Gathering

A. D. Steele[1], G. Verschelling[2] and R. Boerefijn[3]

[1] Unilever Research Port Sunlight, Bebington, Wirral, CH63 3JW, UK. (Andy.Steele@unilever.com)
[2] Unilever Research Port Sunlight, Bebington, Wirral, CH63 3JW, UK.
[3] Unilever Research Vlaardingen, Postbox 114, 3130 AC Vlaardingen, The Netherlands.

Abstract: This paper presents a novel approach to capturing the user requirements for support tools for consumer goods product development. Best practices from the fields of knowledge management and software requirements specification are combined. The CommonKADS methodology is used to analyse the organisational context and highlight bottlenecks in the product development tasks. Use Cases are used to capture the user requirements for a support system to overcome these bottlenecks and otherwise support product development. These methodologies are applied in a process of structured interviews followed by group workshops. This process has been applied successfully in 40 end user interviews and 3 group workshops involving product and process developers from a range of international Unilever companies. The main benefits found when applying this approach are:
- High degree of buy-in from end user community and senior stakeholders
- Understandable process and deliverables
- Effective use of end user time
- Process can be effectively applied by non KM specialists

1. Introduction

Unilever is one of the world's largest producers of branded consumer goods in the foods and home & personal care markets. The Corporate Purpose of Unilever states: "Our purpose in Unilever is to meet the everyday needs of people everywhere – to anticipate the aspirations of our consumers and customers and to respond creatively and competitively with branded products and services which raise the quality of life."

Knowledge management plays a significant role in enabling Unilever to fulfil its Corporate Purpose, by leveraging the vast amounts of knowledge embedded in its people, products, processes and repositories around the globe. Successful KM initiatives within Unilever have included the development and application of knowledge-based systems [1][2], knowledge workshops [3], knowledge mapping and structuring techniques [4] and knowledge strategies [5].

Unilever's desire to compete on the basis of innovative, consumer driven products is reflected by its significant expenditure on research and development. Unilever

Research recently embarked on a new project to support the development of innovative products. The goal of this research project is to develop a software support tool for product and process developers which enables the design of optimised products that better satisfy consumer needs. An important part of the project is to identify key areas where the support tool can add maximum value to the end users (product and process developers in Unilever Research and Innovation Centres).

The identification of these user requirements from a large number of international users in the Unilever business is itself a significant knowledge management activity. This paper details the methodologies and mapping and structuring techniques which were developed and applied during the course of these activities. Section 2 describes the need for a methodological approach to requirements gathering which combines elements of CommonKADS with Use Cases. Section 3 describes the process of applying this methodology in individual interviews and group workshops, emphasising the role of template and graphical knowledge structures. Section 4 details the deliverables of the process and how these add value. Finally, section 5 presents conclusions and main benefits.

2. A Methodological Approach to Requirements Gathering

The approach chosen to gather the user requirements for the support tool builds on best practices from the fields of knowledge management (CommonKADS methodology) and software requirements specification (Use Cases). These approaches were selected in order to mitigate the following perceived risks:

1. The project goal is very broad in scope, thus there is a possible risk that the 'real' needs of the business and the end users are not well met

2. The proposed end user community is large and varied, in terms of markets, consumers, expertise and resources. Therefore, there is a risk that the user requirements cover a very wide space of functionality and priority.

The CommonKADS Methodology

The first perceived risk prompted the use of the **CommonKADS** methodology [6] for knowledge engineering and management. CommonKADS has already been successfully used in the development of a number of knowledge systems within Unilever e.g. [7].

What is CommonKADS ?

CommonKADS provides two main types of guidance for a knowledge management activity. Firstly, it provides a set of standard templates and practical guidelines to enable an analyst to perform a detailed study of knowledge-intensive tasks and processes in an organisation. This allows the analyst to identify the opportunities and bottlenecks in how organizations develop, distribute and apply their knowledge resources. Thus, CommonKADS provides practical support for knowledge management. Secondly, CommonKADS provides a set of re-usable structures for modelling the domain and task knowledge of experts. This facilitates the design and development of knowledge systems software that support selected parts of the business processes.

The two main deliverables of a CommonKADS project are The CommonKADS Model Suite and Knowledge System software.

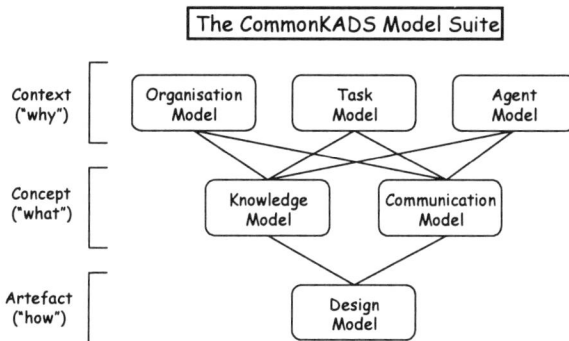

Figure 1: The CommonKADS Model Suite

The CommonKADS Model Suite is a set of 6 models, which are developed iteratively over the course of the project. Each individual model consists of a number of standardised documents and, where applicable, analysis and design structures.

Why CommonKADS ?

It is now well established that successful knowledge management activities must be properly situated in their wider organisational context. Indeed, a rule-of-thumb has developed at Unilever that 50% of the effort in successful KM initiatives is spent in establishing organisational enablers, as opposed to 30% for process and 20% for technology.

CommonKADS provides a valuable, practical set of tools for modelling the organisational context of a knowledge management solution. In particular, the Organisation Model provides templates and guidelines for capturing:
- problems and opportunities (with regard to vision, mission, strategy etc.)
- organisational structure
- people and resources
- business processes
- process breakdown into tasks (highlighting agents, knowledge assets and bottlenecks)

Typically, the above information is used to create a Feasibility Study document, which establishes the benefits, costs and feasibility of the identified opportunities. This type of document has been used successfully to judge the feasibility of a number of KM activities within Unilever e.g. [7].

Thus, the CommonKADS Organisational Model provided an excellent tool to analyse the global organisational structures and processes which form the context of the product development support tool being developed.

Use Cases

The second perceived risk led to the selection of **Use Cases** [8] as the method for specifying the end user requirements for the support tool.

What are Use Cases ?

Use Cases provide clear and simple descriptions of required system functionality (or services) from the user's point of view. Use Cases have been in use since 1987 and are now fast becoming the industry standard method for software requirements specification. They are extensively used in many software development methodologies, such as the Rational Unified Process [8].

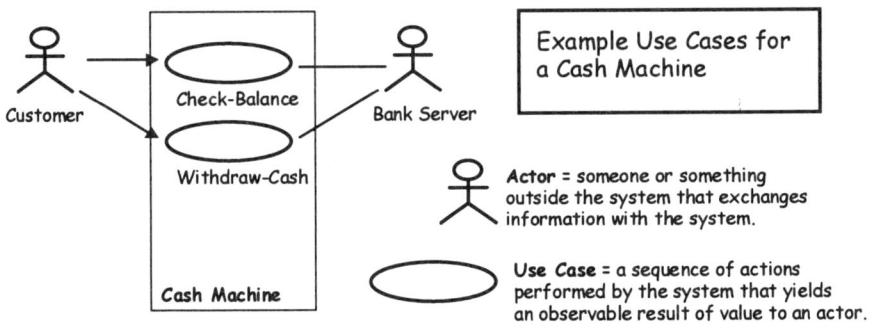

Figure 2: Example Use Case Model

Why Use Cases ?

Use Cases are always written in collaboration with the end user, using their terminology, making them easy to understand and analyse during the development cycle of the software tool. This makes them an ideal candidate for capturing the wide and varied range of requirements expected for the product development support tool. The use of end user terminology and the graphical component also helps in the discussion with end users and stakeholders about focusing the project resources on the most important areas of required functionality.

Combining CommonKADS with Use Cases

To gather the requirements for the software support system, it was decided to combine the Organisation Model from CommonKADS with the Use Cases approach from Rational Unified Process, as their strengths are complementary.

The Organisation Model provides excellent guidelines and templates for analysing the context of the proposed system. Software development methodologies such as Rational Unified Process also provide some guidelines for the analysis of the business context of a system. However, the emphasis of the Organisation Model on business processes, their breakdown into tasks, knowledge assets and bottlenecks, gives a particularly suitable way of analysing the broad and varying contexts of the international groups of end users.

The Feasibility Studies previously produced in Unilever using the CommonKADS Organisation Model have typically concluded with the high-level requirements for knowledge systems for selected opportunities. These were then further detailed into the Knowledge Model format through task knowledge elicitation from a small number of experts. This approach was not applicable to gathering the large scale and diverse functional requirements for the product development support tool. However, the Use Case approach is perfectly matched to this gathering of functional requirements from a large and diverse group of users.

It is proposed that the combination of CommonKADS Organisation Model with Use Cases for requirements gathering provides synergies which are very well suited to mitigating the major perceived risks in this project.

3. Process for Requirements Gathering

It was necessary to design a simple and effective process to apply the selected methodologies. The process must achieve the following benefits:
- Maximise buy-in from users and stakeholders – a highly participative, group-based process is desired in order to maximise buy-in from the globally dispersed groups of users
- Effective use of user and stakeholder time – product and process developers are very busy people and their time is a precious resource
- Applicable by non-KM experts – the requirements of a large number of users and stakeholders had to be gathered within a short period of time, requiring project team members with little KM experience to be involved in the process
- Usable internationally – the process should make use of clear graphical structures to capture requirements, in order to minimise any language barriers

A 2-step process of end user interviews followed by group workshops was proposed to achieve these benefits.

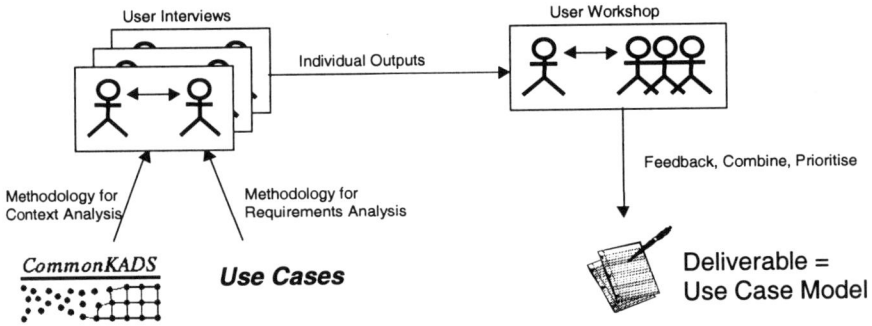

Figure 3: Requirements Gathering Process

User Interviews

Inputs

- CommonKADS Organisational Model Templates.

The standard CommonKADS templates OM-1 (Problems & Opportunities) and OM-2 (Organisational Aspects) were drafted from the contents of the project senior stakeholder interviews. These included graphical representations of the structure and business processes of the relevant organisations and provided useful context for the end user interviews. The actual template used to drive the interviews was a tailored combination of OM-3 (Process Breakdown) and OM-4 (Knowledge Assets):

Organisation Model : Adapted from Process Breakdown Worksheet OM-3 and Knowledge Assets Worksheet OM-4									
ID	Task Name	Performed by ?	Where ?	Task Significance ?	Knowledge Intensive ?	Knowledge Assets	Knowledge Availability ?	Current Support ing Tools ?	Bottleneck ?
0	Receive Marketing Brief (Ideas)	Product Developer	Innov. Centre	4	Yes	1. xxx 2. xxx	Xxx xxx	xxx	[type and description of any bottlenecks in this task]

Figure 4: Interview Template Based on OM-3 and OM-4

- Use Case Template.

A template was drafted to capture the Use Cases provided by the interviewees. The template consists of the essential parts of a Use Case i.e. ID & Name, Agents involved, Observed value to agents (summary of functionality provided) along with other information essential for this process i.e. which task & bottleneck this Use Case relates to:

Use Case Template	
ID / Origin / Name	**1 / Country / Get-Historical-Data-About-XXX**
Agents	Process Developer, Product Developer
Observed Value to the Agents	The system displays …
Related to Bottleneck & Task	Bottleneck ID & TaskID
Basic Flow of Events	− Xxx − Yyy − Zzz

Figure 5: Interview Template based on Use Cases

- Pre-reading pack for interviewees.

This contained two 1-pagers outlining the purpose and contents of the CommonKADS Organisational Model and Use Cases templates, and allowed interviewees to understand the deliverables and process of the interview.

Process

The **end user interviews** typically took 1 to 2 hours, with a User Requirements analyst from the project team facilitating the interview with the end user according to a two-stage process:

First, in the *Context Analysis* stage, the analyst and end user followed the CommonKADS Organisational Model templates and guidelines to breakdown the relevant business process (Product and Process Development to support product innovation, in this case) into the individual tasks, and to highlight the 'bottlenecks' in these tasks. Important here was the definition of bottleneck used – "anything which reduces the speed, quality or frequency of innovative product development" - which was wider than the traditional CommonKADS interpretation of a knowledge bottleneck (wrong knowledge form, place, time or quality). This reflects our emphasis on gathering requirements for a general support tool, and not specifically a knowledge system.

Secondly, in the *Requirements Analysis* stage, the identified bottlenecks were used as a springboard into a discussion about the requirements for a software tool which could help to reduce these bottlenecks, or could otherwise support the tasks in the business process. These requirements were recorded as Use Cases.

Output

The deliverables of the interviews were the completed Organisational Model template and a number of Use Case templates. A first attempt at prioritising the Use Cases was also made, to be refined in the group workshop.

User Workshop

Input

The outputs from the 12-15 individual interviews were amalgamated to form a starting point for the group workshop. Specifically, the following materials were created for the workshops:

- Graphical overview of the structure and business processes of the relevant organisations
- Graphical breakdown of business processes into actual product and process development tasks
- Graphical overview of the location and nature of bottlenecks in these tasks
- Graphical 1-page overview of use cases e.g.

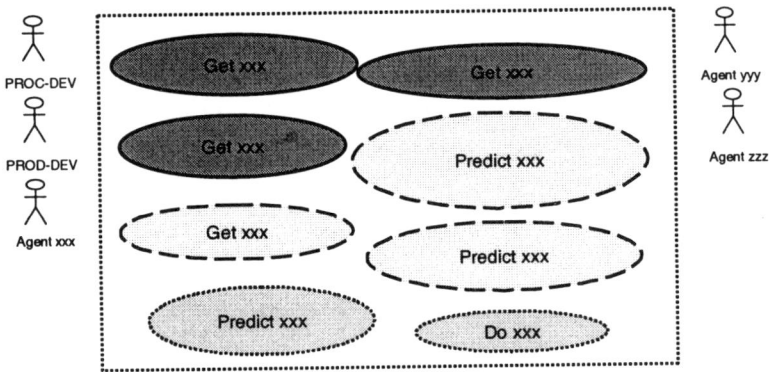

Figure 6: Graphical Overview of Use Cases

- Workshop Agenda clearly showing the goals, deliverables, and roles & responsibilities of those attending

Process

The **end user group workshops** typically took one afternoon (3 hours), with 2 User Requirements analysts from the project team facilitating a group of end users of up to 15 people. The main goals of these workshops were to:

- Feedback the outputs from the individual interviews to the group
- To prioritise the groups' identified bottlenecks in the Task breakdown
- To brainstorm and prioritise the groups' Use Cases

The workshop was run as an instance of a 'knowledge workshop', using predefined knowledge structures and graphical methods to map out the organisational analysis, task breakdown, knowledge bottleneck location and use case overview. Unilever has considerable successful experience in using this format of group workshop e.g. [4, 5]. This structured graphical approach allowed the group to keep the whole picture in mind and enabled them to more easily combine and prioritise the outputs from their individual interviews. Further, the extensive use of graphical representations helped to minimise language barriers when dealing with the international end user groups.

Output

The deliverables of the group workshop were an agreed and prioritised set of Use Cases representing the functional requirements of that user group for the product development support tool. These Use Cases were clearly linked to bottlenecks and tasks in the relevant business processes.

Another significant deliverable was a list of the bottlenecks which could not be addressed by the proposed product development support tool. These bottlenecks and their relation to the business processes are a valuable source of information to those planning future research & development and other improvement programmes.

4. Deliverables

A total of 3 group workshops were performed. The clear structures used to represent the requirements allowed the outputs from the 3 groups to be amalgamated into a single deliverable - a User Requirements document - structured according to the above process:

1. Context Analysis – Organisational Model templates and graphical overviews of organisation, business processes, breakdown into tasks, bottlenecks
2. Requirements Analysis – Use Case templates related to bottlenecks and tasks, with group prioritisation and ownership, and graphical overview

This User Requirements document is now being used to drive the develoment of functional specifications for the product development support tool, and forms the basis of ongoing discussions with stakeholders and users.

5. Conclusions

The above approach to requirements gathering has been applied successfully in 40 end user interviews and 3 group workshops involving product and process developers from a range of international Unilever companies. The following were perceived as the main benefits of the approach:

- High degree of buy-in from end user community and senior stakeholders

The structured approach of CommonKADS allowed the users to clearly see where the knowledge bottlenecks occurred in the tasks they perform, and how these related to higher level business processes and strategy. This made elicitation of requirements easier during the interviews, and led to a high degree of buy-in from both the end users and the senior project stakeholders.

- Understandable process and deliverables

The structured approach makes extensive use of templates and graphical representations of business processes, tasks, knowledge bottlenecks and use cases. This led to a set of clear and understandable deliverables from interviews and workshops, and helped to minimise any language barriers for the international groups of users.

- Effective use of end user time

The structured approach to the interviews and group workshops ensured that the precious time of the product and process developers was used very effectively.

- Process can be effectively applied by non KM specialists

A large number of interviews were conducted in a short period of time. This required that project team members with little experience of KM or requirements engineering techniques had to be used to conduct the majority of the interviews and to facilitate workshops. The high quality deliverables of the process and the positive feedback from end-users demonstrated that structured approach of CommonKADS for Context Analysis and Use Cases for Requirements Analysis can be applied effectively by non KM specialists.

References

[1] P. H. Speel, M. Aben; Preserving conceptual structures in design and implementation of industrial KBSs. *Special issue on Problem-Solving Methods for the International Journal of Human-Computer Studies*, Vol. 49, 547-575, 1998.

[2] K. W. Mathisen, A. Mjaavatten, R.A.G. Kooijmans; Industrial implementation and testing of model-based diagnosis in the PRIDE project. *Proceedings of SAFEPROCESS 98*, Lyon, France, June 1998.

[3] G. von Krogh; Care in Knowledge Creation. *California Management Review*, Vol. 40 (3), Spring 1998

[4] P. H. Speel, N. Shadbolt, W. de Vries, P. H. van Dam, K. O'Hara; Knowledge Mapping For Industrial Purposes. *Proceedings of the Knowledge Acquisition Workshop*, Banff, Canada, 16-21 October 1999.

[5] G. von Krogh, I. Nonaka, M. Aben; Strategizing in the Knowledge Economy. *Submitted to: Sloan Management Review,* 1999

[6] A. Th. Schreiber, J. M. Akkermans, A. A. Anjewierden, R. de Hoog, N. R. Shadbolt, W. Van de Velde, B. J. Wielinga; *Knowledge Engineering and Management : The CommonKADS Methodology,* to appear in The MIT Press 1999

[7] H. Akkermans, P.H. Speel, A. Ratcliffe; Problem, Opportunity and Feasibility Analysis for Knowledge Management: An Industrial Case Study. *Proceedings of the Knowledge Acquisition Workshop,* Banff, Canada, 16-21 October 1999.

[8] I. Jacobson, G. Booch, J. Rumbaugh; *The Unified Software Development Process,* Addison Wesley Longman, 1999.

Chapter 6

The Future

Model Framework for Sharing of Engineering Practice Knowledge through ubiquitous networks of Learning Communities
B. Bharadwaj

Managing Engineering Knowledge Federations
F. Mili

Languages for Knowledge Capture and their Use in Creation of Smart Models
B. Prasad

Ontologies for Knowledge Management
M. Uschold and R. Jasper

Model Framework for Sharing of Engineering Practice Knowledge Through Ubiquitous Networks of Learning Communities ·

Balaji Bharadwaj

Contact Address: Suite 3200 - Visteon Technical Center, Visteon Automotive Systems, 16630 Southfield Road, Allen Park, MI 48101, USA.

Abstract: Knowledge sharing of engineering practices between members of product development groups is important for improving the overall effectiveness of the product design process. In the recent past, various organizations have reported on the emergence of employee motivated learning communities that strive to share work related knowledge and experience among its members. Although at this point, these learning communities have been fairly successful in their endeavor to assist in organization wide knowledge management activities, yet a model framework has to be developed for sustaining such communities. In this chapter, a model framework incorporating appropriate support in the form of social, management and technological infrastructure is proposed along with its implementation. A review of some of the benefits to be accrued from the implementation of such frameworks is also highlighted along with the metrics to be used for assessing their influence on the performance of the product development process is presented in this chapter.

1. Introduction

Knowledge Management (KM) initiatives place emphasis on the importance of "sharing and dissemination" of engineering practice knowledge among members of product development (PD) groups within an enterprise. This applies both to the realization of PD tasks as well as to the improvement of such activities.

In the past, industrial organizations had focused on supporting fairly centralized knowledge sharing activities such as classroom training programs for improvement of technical skills, setting up of coaches/subject matter experts to educate personnel in certain core technical competencies, formation of quality improvement teams, centralized development of knowledge repositories and development of information technology based solutions to support collaborative work. However, these have had limited success in the creation and exchange of engineering practice information among practitioners. This has been due in part to the inherent bottleneck of centralized activities, as well as to a disconnection between KM and PD activities within organizations. In the past, companies were

---

· This paper reflects the personal views of the author and does not reflect the views or policies of Visteon Automotive Systems

PROCESS CONCEPT

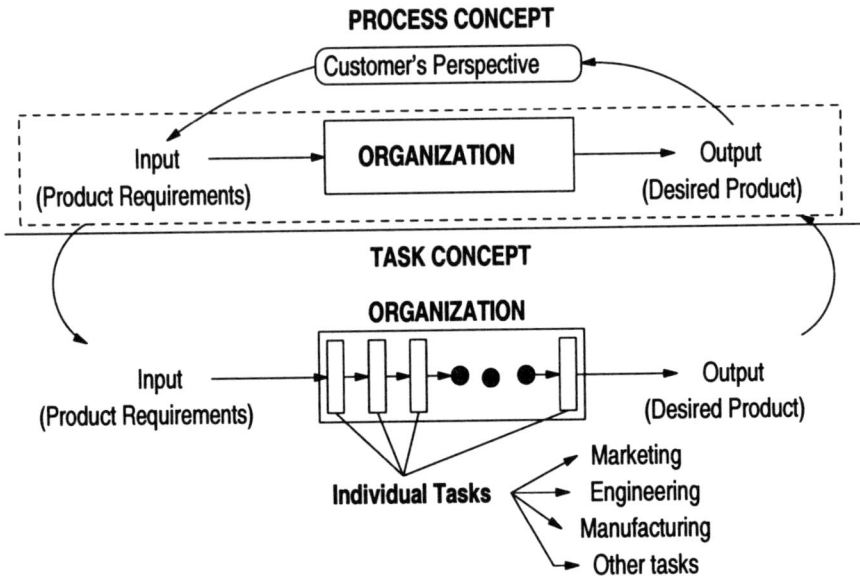

Figure 1: The concept of task and process centered organizations

often organized around tasks in restricted functional domains and technical competencies. This organizational structure offered good productivity within individual functions and competencies, but it had notable problems with matrix-management issues involving development of products across functions and technical competencies. This shortcoming of task-centered organizations resulted in the emergence of process-centered organizations (Figure 1) [1], with processes replacing tasks as the focus of activity. Thus, knowledge sharing as an activity was not only involved with supporting groups of individuals in specific technical competencies but also with fostering networks of communities involved in various tasks associated with a specific process in the PD cycle. As evident in various studies, process centered organizations recognized and supported process oriented cross-functional teams, which strove to improve the process that was under their responsibility. Such cross-functional teams formed the nucleus for employee motivated 'Learning Communities' (LC).

The opportunity for engineering practice improvement through more distributed means has prompted initiatives to support networks of employee motivated learning communities in various organizations. These learning communities are also known as Communities of Practice (CoP) and strive to share work-related knowledge and experience among members (practitioners) of that engineering practice community. The direction of this initiative has been influenced by various paradigm shifts such as that from technical to socio-technical solutions [2] for knowledge management, as well as by shifts in thinking from training to learning within CoP [3][4]. The above factors have prompted organizations to revisit their training programs and lay more emphasis on learning initiatives for groups of individuals with focus on enriching individual experiences from the activities of the group. Learning provides an ideal framework for KM activities from the perspective of the organization as well as the practitioner for the following reasons: (i) Organization – Generation of new knowledge/practices can help reduce time to market and product development cycle time as well as improve productivity, and (ii) Practitioner – Learning involves doing something 'new' which is a satisfying experience and revolves around the notion of 'rewards'. It is the above notions of learning that would be explored in greater depth through the rest of this chapter from the viewpoint of implementing KM activities in organizations.

Information ≠ Knowledge

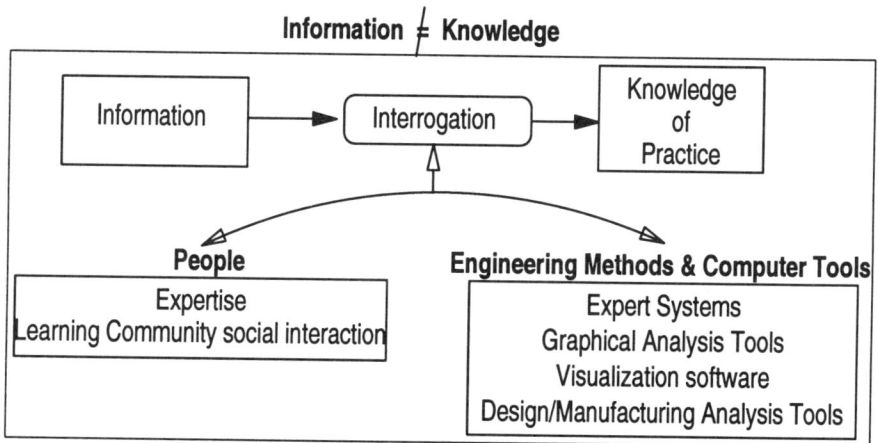

Figure 2: Knowledge creation - A socio-technical approach

2. Learning: Nucleus for Knowledge Management

Learning, put in the context of a LC would consist of each one of the following activities in sequence: (i) Acquisition and interpretation of new practice information, (ii) Generation of new practice knowledge, (iii) Exchange of the new practice knowledge among peers and its validation, and (iv) Dissemination of the validated knowledge among practitioners in the LC. Each one of the above four activities, which constitutes a learning cycle, has a one-to-one correspondence with the KM activities in an organization. The learning cycle revolves around the issue of knowledge creation, which is a direct result of the social interaction between tacit and explicit knowledge. This is illustrated in Figure 2 where the socio-technical approach adopted by a LC for knowledge creation is shown. Quite often, organizations work on the premise that information is the same as knowledge and propose various information technology (IT) based solutions for KM. However, information (raw data, reports, etc.) by itself does not contribute to the learning process. Information at best can be utilized as an input for an interrogation process in which people as represented by the practitioner community in the LC discover and evolve new knowledge pertinent to their work practices leveraging the support of various computer based tools (design & manufacturing tools, databases, expert systems) and methodologies (e.g. QFD, FMEA, etc.).

The framework for implementing the learning cycle should be developed bearing in mind the following facets of learning:

- Practitioners learn depending on the extent of the feedback process (feedback for questions raised and solutions provided) nurtured within the LC.
- Learning in groups involves exploring open-ended questions and exchange of tacit knowledge through interactions between group members.
- Learning involves mental models [5] - Mental models are surfaced, recognized, debated, modified and shared between practitioners. Mental models assist in enabling discussions and communication between practitioners in a LC and form the basis for learning.
- Learning from a LC perspective involves 'Double Loop' or 'Top-down and Bottom-up' learning. In other words, practitioners with various levels of expertise learn by exchanging knowledge in a LC.

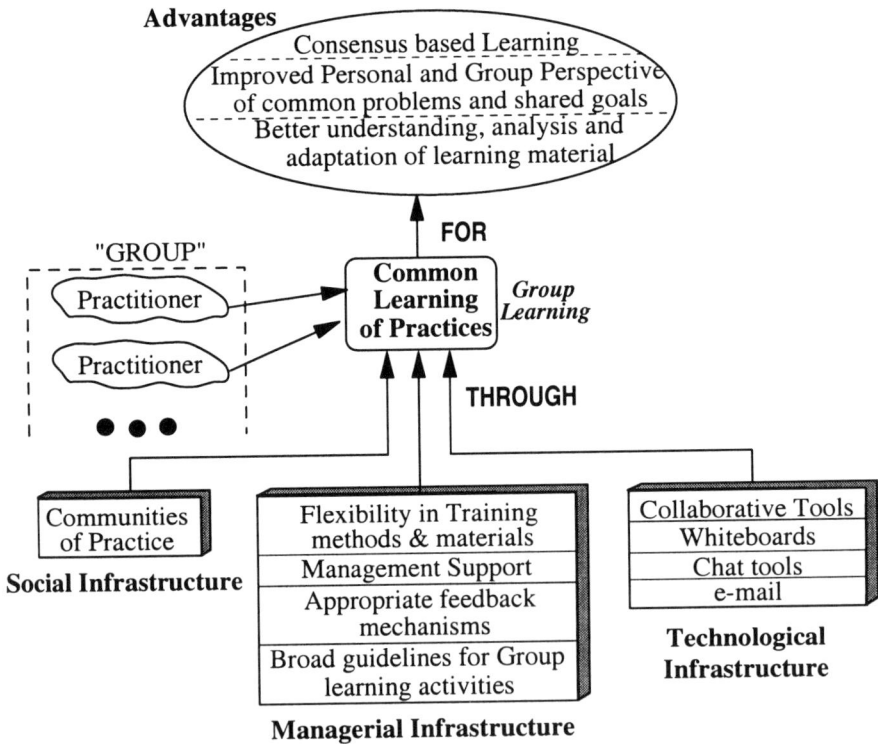

Figure 3: The concept of Group Based Learning

Traditionally, organizations confined learning activities at the individual and group level to classroom training type initiatives. Furthermore, these activities were fairly disconnected from the actual work of product development. However, learning is a continuous process and involves both 'training' and 'learning on the job' as evident from the learning cycle. In the past, objectives for learning initiatives were limited in scope to improving the technical competency of the individuals. At present, these objectives have been heaved up to achieve improvements in the complete product development process. Such improvements cannot be achieved by training alone, but would involve in situ learning that is situated in everyday work of product development as well as distributed peer-to-peer learning, rather than just instructor-to-student learning. This leads to the further realization that learning cannot be performed in isolation by individuals, but would have to be conducted in a group setting leveraging the intellectual resources of the community (i.e. Learning Community). Group Based Learning (GBL) (Figure 3) provides one such model framework for imparting 'group wide' learning activities. It is the GBL framework with its three infrastructure components (namely social, management and technology) that we would like to propose as the framework for conducting KM activities in organizations involving sharing of engineering practice knowledge. Some of the advantages offered by GBL as reflected in internal programs conducted at various organizations consist of the following:

- Improvements in understanding, analysis and adaptation of internal technical learning material (e.g. training manuals, new processes, guidelines, etc.).
- Improved personal and group perspective of common problems and shared goals (system and multi-functional design issues).
- Promotion of consensus based learning. This would improve the consensus among the practitioners regarding the need, implementation strategy and policy for incorporating PD process activities within their functional tasks.
- Promotes sharing of 'best design practices' and design data as well as acquisition of new problem solving and design skills.

A current challenge encountered in various organizations is to find better ways to provide appropriate infrastructure to support a framework for group based learning across geographical and process boundaries. Various issues associated with the requirements for the GBL infrastructure is presented in the next section.

3. Infrastructure for Group Based Learning: Knowledge Management Framework

Current frameworks for KM are very much IT-centric in terms of their infrastructure [6][7]. Infrastructure for GBL is more than just technology [2][4][6], as learning involves a *social* process of interaction among the practitioners in an organizational environment, which is *conducive* (requires management support) for such interactions across organizational, geographical and functional *barriers* (requires technology support for bridging the barriers).

Social infrastructure is dependent on aspects of LC/CoP such as apprenticeship, interpersonal interaction, group membership, and identity for support. The same factors which lead to the formation of LC such as social affinity among practitioners caused as a result of interpersonal interaction (working and training together) as well as the personality traits of practitioners (competence, style, etc.) provide the core of the social infrastructure support for GBL. The LC provides a forum through which practitioners with varying levels of expertise can interact with one another, thus promoting the various aspects of KM.

Managerial infrastructure involves providing broad management support for the social infrastructure, namely the LC. This would involve providing support to the following activities in the organization:

- Identifying and facilitating the formation of Learning communities by encouraging practitioner's to participate in LC related activities.
- Providing official recognition for the roles and responsibilities of different individuals within the LC community - Support mentoring efforts.
- Providing a charter for incorporating LC activities within the organization: (i) promotion of flexible training methods, (ii) resource allocation for GBL, (iii) promote practitioners to provide feedback regarding PD activities through LC forums, and (iv) guidelines for activities of practitioners in the LC.

GBL would also require technological infrastructure support in the form of collaborative tools such as whiteboards, chat tools, e-mail/multimedia utilities and other web-based applications in order to facilitate learning and interaction within the LC. Features of web-based technologies such as ubiquity across various information access platforms, widespread availability and affordability make it a focal point for providing support to the LC.

A major task at hand in various organizations is to draw appropriate levels of synergism between the above three types of infrastructure. Imbalances in participatory levels of practitioners, management support or the absence of appropriate technology would hinder the exchange of practice knowledge, particularly among those engineering practices (e.g. QFD, Robust design, Manufacturing processes – Machining, Injection molding, etc.) which are performed across various functional, geographical and organizational domains [6]. Cognizant of the roles played by LC in GBL, various organizations [6][7] have attempted to develop a 'virtual network' of 'individual learning community' initiatives, using widely available information technology tools. The information infrastructure of this 'virtual network' is called a Learning Network (LN) and can be supported using world-wide-web technology. Associated with the LN are the following issues which would influence the KM activities espoused by the LC: (i) Design, content and anticipated applications of the LN, (ii) Technology component – databases, user interfaces, collaborative tools and communication techniques for exchange and sharing of information between users, etc., and (iii) Social responsibilities of practitioners in the LN.

Learning Networks by themselves are not all encompassing in providing the support framework for collaboration in the LC for the following reasons: (i) Current modes for collaboration lack the ability to replicate face-to-face interaction (involving all of the 5 senses) and do not support various learning modes, (ii) Current collaborative tools only assist in establishment of weak social ties among the practitioners since they only support inter-exchange of information devoid of its underlying intent (i.e. knowledge), and (iii) Emphasis is placed on information access as opposed to knowledge creation. The above drawbacks for collaboration in prevalent LNs prompted us to introspect on what kinds of support might be required for a LC in the future. This leads us to believe that focus should be placed on the following aspects:

⇒ **Fostering Cooperation and Information transfer:** (i) ability of the practitioner to effectively communicate thoughts, insights and recommendations for achieving the objectives of the LC, and (ii) effective creation of content in the appropriate form for the appropriate subject and its effective communication within the LC.

⇒ **Facilitation of Learning:**
  • Information acquisition: Capability to acquire information from a variety of resources using pre-defined mental models which in turn requires an effective representation scheme for the mental model as well as associated search tools.
  • Information distribution: Exchange of the acquired information through the use of mental models and its effective expression using data dependent content creation facilities.
  • Information interpretation: Interpretation of information based on pre-defined mental models and identifying the appropriate knowledge requiring the services of: (i) information filters that can extract the knowledge content from the underlying data, and (ii) information presenters that can present the derived knowledge in an appropriate information format (graphical, data, rules, etc.).
  • Knowledge induction: Encoding the identified knowledge in an appropriate format (e.g. guidelines, expert system, database, etc.) within the organization.

Analyzing the above two issues leads us to the following additional requirements as desired of practitioners within the LC: (i) Mental models for engineering practices, (ii) Better methods for acquiring and interpreting information, and distributing as well as expressing the derived knowledge, and (iii) Better interaction from the underlying knowledge architecture. Consolidation of the above requirements further directs us to the notion of incorporating 'Mental models' and 'Active documents' within the LC framework for promoting GBL.

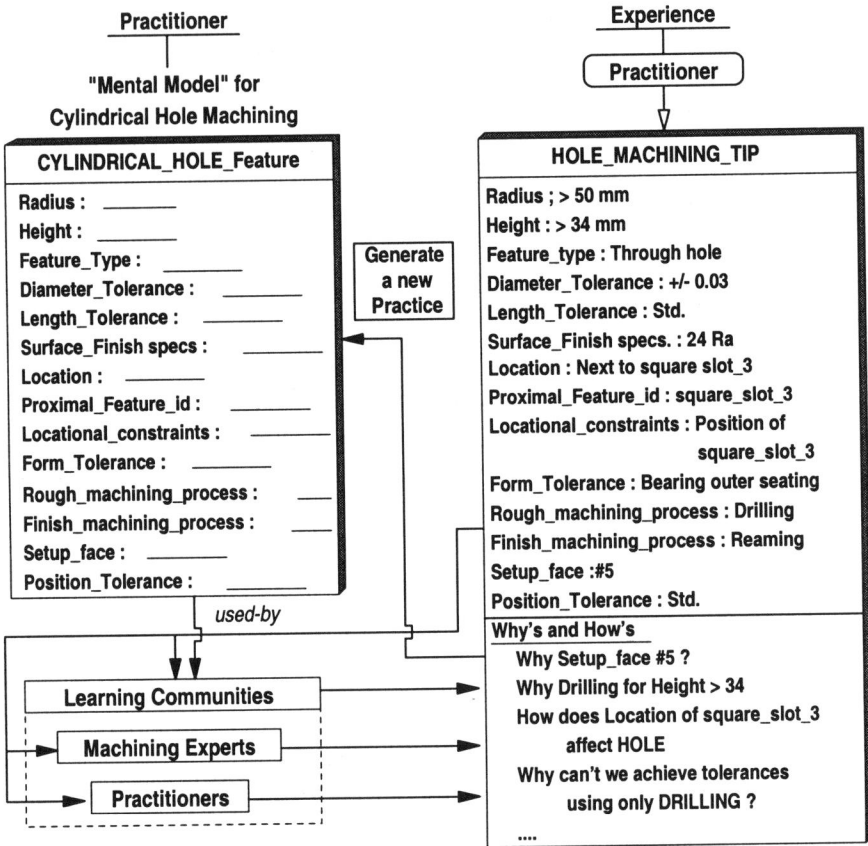

Figure 4: Engineering 'Mental Models' - A Knowledge representation scheme for acquiring, ameliorating and encoding engineering practices

Mental modeling involves, "developing the capacity to reflect on our internal pictures of the world to see how they shape our thinking and actions" [5]. Mental models of engineering practices are strongly reflected in the thought processes of practitioners and their actions within an organization. These have consequences for the conduct of product design and development activities. A sample illustration of an engineering mental model in the domain of machining practices is shown in Figure 4. The mental model for cylindrical hole machining (Figure 4) describes attributes which characterize various aspects of machining a cylindrical hole feature and can be easily related by practitioners who perform similar activities in an organization. In other words, these attributes reflect the common professional language parlance of the practitioners. Mental models are particularly useful for the conduct of KM activities in view of the following observations:

- They provide the practitioner with the ability to conduct interactions at the individual and group level through the provision of models for information management based on practice area and interaction with other organizational processes and practices. Furthermore, these models provide a medium for communication of practice information and knowledge.

- Using mental models, practitioners can specify practice information (based on individual experience), interrogate it using the resources (people and technology) of the LC, and generate new knowledge in the domain of the specific practice. This aspect is best illustrated in Figure 4, where a practitioner provides a 'HOLE_MACHINING_TIP' based on experience to the LC associated with machining related activities. The information presented to the LC is utilized by other participating individuals to ameliorate knowledge as observed from this individual experience to existing machining practices in the LC. The structure of the mental model is such that it assists in the generation and usage of new knowledge by facilitating the study and analysis of various cause-and-effect relations. Problem solving techniques such as QFD, PFMEA, etc. can be used by the practitioners to ask the *Why's and How's* to relate observed experiences to knowledge of new practices.
- The mental model has similar functional responsibility as that of a product model, although it is used for the task of knowledge creation in the domain of engineering practices.
- Mental models are the basis for sharable practice knowledge, which reflects itself as sharable knowledge resources when used by multiple practitioners in the LC.

'Active Documents' complement the role of mental models and provide the IT mechanism through which these models can be effectively utilized within the LC. It is a common fact that documents are the means and resources through which product design and development activities are conducted, communicated and documented within an organization. Additionally, documents contain knowledge pertinent to various engineering practices that are often encoded as design and process guidelines. However, engineering practices and their interactions cannot be effectively captured in 'traditional' paper documents due to differences between the expression and conduct of these practices as well as their inability to express all interactions through a single media. The inadequacy of traditional documents led to the manifestation of requirement for 'Active Documents'. With the emergence of the WWW, the development of active documents that can be effectively utilized for GBL in LC has become a possibility. These documents provide support for: (i) Multiple styles for expressing, acquiring and interrogating knowledge, (ii) Context specific rendering of relevant information, (iii) Support diverse working and learning styles of individuals, and (iv) Access to a wide variety of design and manufacturing services on and through the WWW. With reference to Figure 5, an 'Active Document' would be accessed by practitioners of the LC at their workplace. The document would consist of both information and pertinent practice knowledge encoded in it. Moreover, the document would provide a single gateway for accessing and handling multimedia information. Implementation specific details would involve developing embeddable 'objects' in HTML documents using Java/ActiveX/XML based technology that would dynamically control the document content. These embeddable objects would have built-in hooks (middle-level application logic) to databases for data access, expert systems for rules and drawing inferences, CAD/CAM systems for accessing and performing engineering product design functions, other dedicated resources such as parallel computers for computing intensive tasks as well as a variety of tools which provide interactive content. Moreover, these objects can be customized based on user preferences and would be capable of filtering information and presenting it in multiple formats. At present, numerous applications have been developed based on this concept as illustrated in reference [8].

In addition to the above infrastructure for GBL, a framework for performing GBL activities within the PD process would have to be developed. An outline for one such framework is provided in the next section.

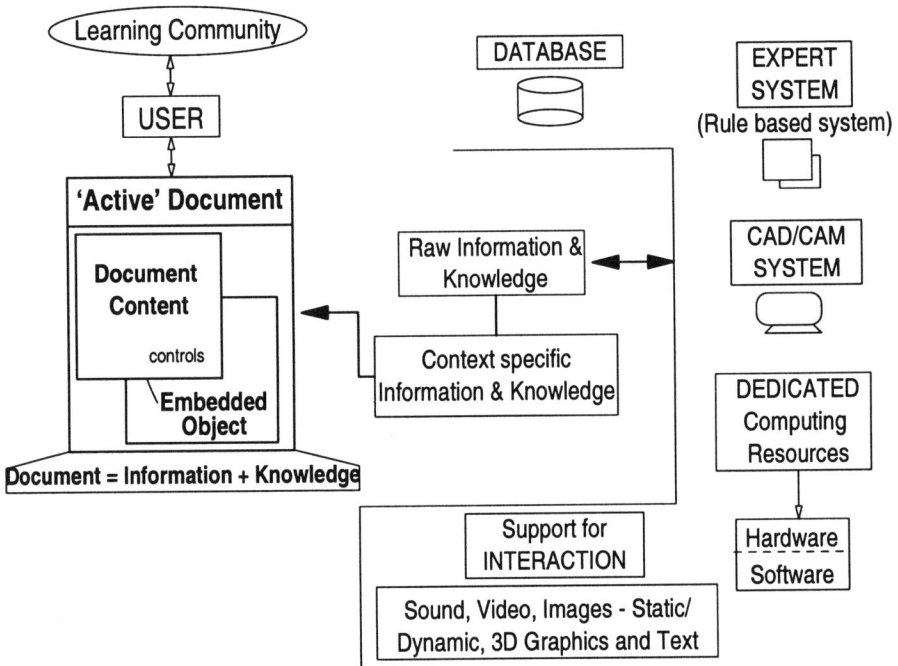

Figure 5: "Active Document" - Organization and Structure

4. Community Based Design: A Framework for Group Based Learning in the Product Development Process

The objective of any product development activity involves the creation of a product with a certain amount of value to a customer. Three elements which influence the outcome of a product development activity are: (i) The customer who suggests new requirements for a desired product and is the ultimate user of the product (Voice of Customer - VOC), (ii) The tasks which constitute the product development activity and assist in the realization of the desired product, and (iii) The practitioners involved in the product development activity who provide the intellectual framework for the translation of VOC requirements into a complete product specification.

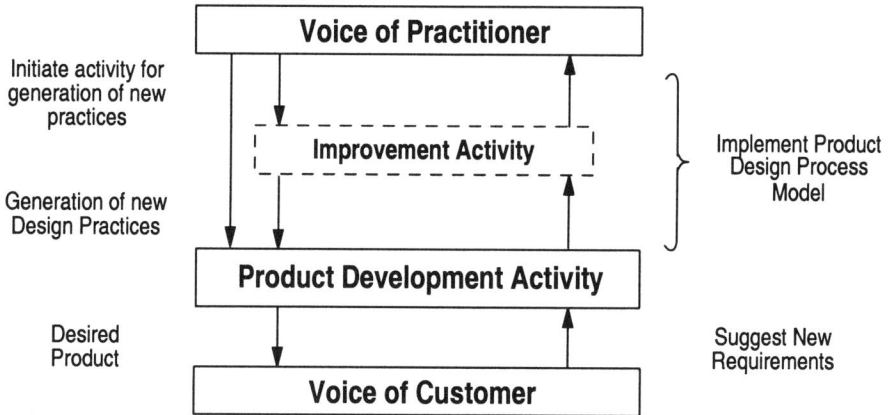

Figure 6: A model for Community Based Design

For sustained excellence in product development, a fourth element called as "Improvement Activity" is required [6]. The "Improvement Activity" is mainly concerned with the task of improving PD capabilities with practitioners as the internal customer of improvements. Voice of the Practitioner (VOP) holds the similar level of importance as VOC is for products with regard to the "Improvement Activity" (Figure 6). The importance of VOC and new PD techniques has been recognized in many organizations and these have been successfully incorporated in their PD activity [1]. However, these organizations have failed to properly address and incorporate the VOP into their PD activities resulting in less than optimal performance of the organization as a whole entity. The VOP is important because practitioners can recognize both problems and opportunities with existing PD activities, and are in the best position to implement any improvements in the domain of their functional activities. The incorporation of "Improvement Activity" in the product design process leads to the notion of Community Based Design (CBD) [6]. CBD differs from existing PD process models in that it fosters both process and practice improvements through the collective learning experience of practitioners in the learning community which manifests itself as VOP. As discussed earlier in Section 2, KM activities cannot be carried out in isolation from day-to-day PD activities of the practitioners. Furthermore, KM initiatives would need to be finely enmeshed within the PD process if its benefits are to be accrued on a continuous basis (a core objective of KM). CBD provides the appropriate "organizational framework" in which KM activities can be carried out by incorporating VOP at appropriate phases in the PD process. A view of some scenarios involving KM activities in LC within the realm of the PD process is examined in the next section.

5. Knowledge Management in Learning Communities: Some Scenarios

Knowledge Management activities initiated by learning communities would not only involve implementing the learning cycle but also supporting and promoting the following modes of learning: (i) Skill refinement through learning by practice/simulation, (ii) Experience based learning, (iii) Learning by feedback (through peer review and mentor-mentee relationships), (iv) Learning from examples (induction), (v) Discovery of new knowledge by learning through inferences (patterns, hypothesis and test), and (vi) Learning through similarity recognition (analogy).

Bearing in mind the above learning modes, a snapshot of how KM activities can be carried out in a LC (detailed case studies are described in references [6][8]) is presented through the following scenarios: (i) Representation and usage of tacit information, (ii) Knowledge generation through social roles and responsibilities, (iii) Active documents for knowledge creation, (iv) Interactive refinement of engineering design information and (v) Learning through simulation.

Representation and Usage of Tacit Information: KM frameworks are centered on the notion of converting tacit information to explicit knowledge. An essential step in the conversion process involves the representation and usage of tacit information possessed by practitioners in LC. This KM activity is challenging because of the personal and context-specific nature of tacit information, which makes it hard to formalize and encode within the framework of mental models used in LC. We will examine one such scenario involving the setup of a LN for the QFD design practice (Figure 7) [6]. This LN strives to serve as a forum for the exchange of QFD practice information. The objective is to improve the position of the practitioner in the 'Knowledge-Function-Organization' box such that the individual attains a high level of practice knowledge as a result of KM activities in the QFD LC. The QFD LN is moderated by a group of QFD practice experts (PE) with direct support from management to overseeing and nurturing various activities in the LC. The LN consists of information repositories of training manuals and past QFD project reports as well as a QFD practice database, which represents a mental model of the QFD practice information for assistance in GBL. The mental model is built on the relationship information between different parameters at various levels in the HOQ matrix and consists of the following two models: (i) Parameter-DataModel – Details of parameters such as nature of influence on design, interaction with other parameters in the design and the underlying model (heuristic or practice) governing the parameter, and (ii) Relationship-DataModel – Details of relationships between parameters and their dependencies, observations from practice, and the underlying model (practice or

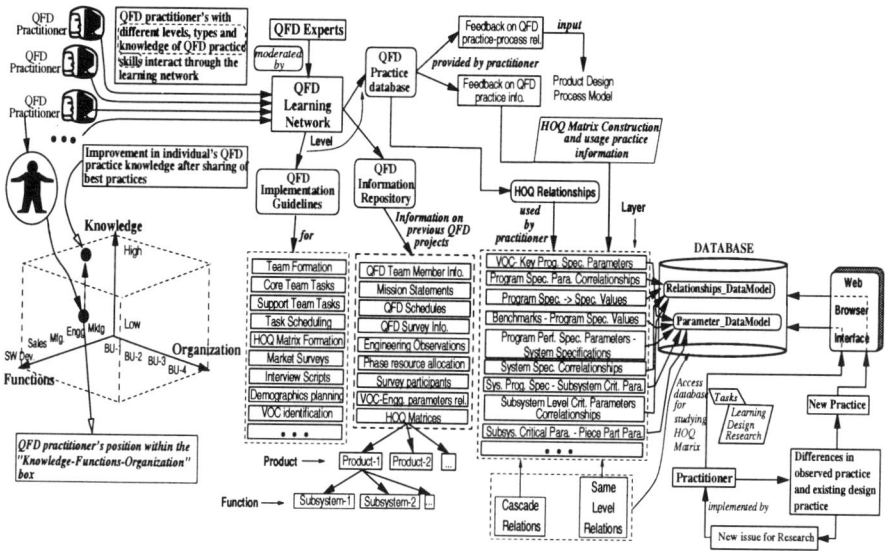

Figure 7: A model of the prototype QFD Engineering Practice Learning Network illustrating the HOQ relationships and its application for Group Based Learning

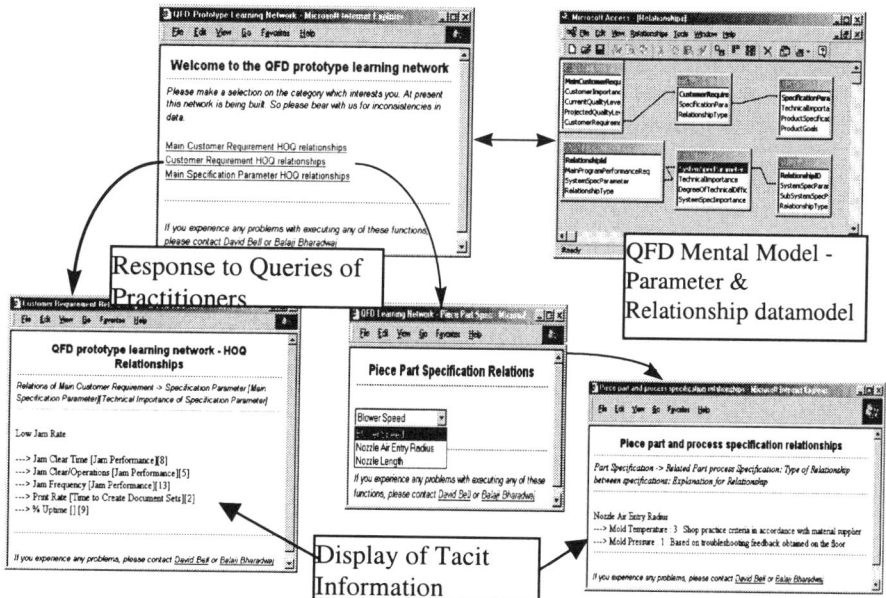

Figure 8: Representation and display of tacit information within the QFD Engineering Practice Learning Network

theory) governing these relationships. The above two data models provide the appropriate mechanism for encoding the tacit information pertinent to QFD practices and can be implemented within any commercial database (in this case MS-Access) as database tables (Figures 7 and 8). The above two data models are mutually dependent on each other and would have to be used simultaneously in order to make an useful interpretation of the HOQ Matrix. The information contained within the data models can be leveraged for a variety of tasks such as learning, and regular design activities through suitable web-based interfaces in the following manner: (i) dependency relations between parameters and the nature of their relationships, (ii) co-relationship information between parameters, and (iii) explanation generation for details of certain parameter details and relationships. An illustrative application of the data models for learning activities is shown in Figure 8, where tacit information pertinent to a dependency relationship between a customer requirement (e.g. Low Jam Rate) and different specification parameters (such as Jam Clear Time, % Uptime, etc.) for a photocopier design is displayed on a web browser along with the technical importance rating of the parameter. Additionally, tacit information concerning relationships between part and process specifications is also displayed in Figure 8.

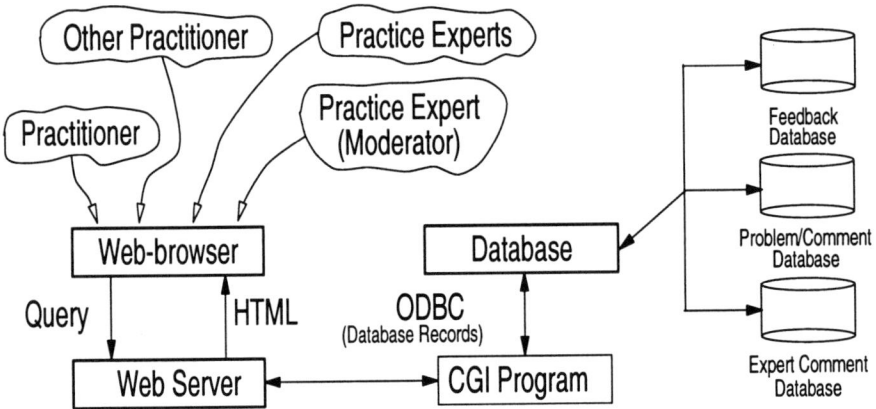

Figure 9: Framework for knowledge generation through the roles and responsibilities of various individuals in a Learning Network

Knowledge generation through social roles and responsibilities: The creation of new knowledge is a socio-technical process as shown in Figure 2 and involves inter-exchange of information as facilitated by the social roles and responsibilities of the practitioners in the LC. The societal responsibilities of practitioners towards other members of the LC would help ensure their participation in the various phases of the learning cycle. This is essential as information exchange involves a give-and-take (reciprocal rewards) relationship, which is very much dictated by the social actions of the individuals. This aspect of KM involving social roles would be examined in the context of the LN for the QFD practice (Figure 7). In addition to providing a placeholder for storage of tacit information, the QFD LN provides a mechanism through which practitioners at varying levels of expertise (novice to expert) can interact (virtually through collaborative means) with each other in the QFD LC. This interaction is made possible through the 'QFD Practice database' that in addition to storing various levels of tacit information also contains feedback mechanisms for QFD practice and QFD practice-product development process model relationships. The QFD practice database mirrors a virtual laboratory in which practitioners can look-up for existing information and acquire knowledge while at the same time provide feedback to the LC based on their individual experiences. Other practitioners ('peer-to-peer review') as well as practice experts (PE) ('mentoring') can in turn provide feedback based on their experiences or might explore them further. New knowledge pertinent to the practice area is created as a result of the feedback process. Sustaining the practitioner feedback process is critical to the task of creating new knowledge and would involve support from the social infrastructure of the LC. In the QFD LN, a practitioner might provide feedback on either the product design process model or on specific practice information. This feedback would be based on the practitioner's observations or experience and might result in creation of new practice knowledge or a new issue for research in the LC. At this juncture, the practitioner can post the relevant observation on the notice board (refers to web board), which is accessible to other members of the QFD LC in a format that is best suited for describing the relevant practice information. A HTML based form along with various features such as file uploads can be utilized for performing this task (Figure 9). The system would automatically generate an initial response providing a special tracking number with which the status/responses of a particular observation can be tracked. This information is stored in the Feedback database. Other practitioners and PE can review and respond to the practice information provided by practitioners on the LN using appropriate utilities. These responses would be stored in the Problem/Comment database and validated by a designated PE who has the authority and the responsibility to evaluate submitted practices, provide response to practice related problems/feedback. The validated practices and feedback responses to practitioners would be stored in the Expert Comment database as well as posted on the web board of the LC.

Active documents for knowledge creation: Active documents provide a focal point for promoting various learning modes by drawing on their rich interactive presentation style and knowledge content. These documents by themselves do not provide any automatic means for creating knowledge. They only provide the necessary tools to the practitioners for effective creation of new knowledge in the LC. We will examine the model framework of the collaborative product development system described in reference [8] in this context. This framework assists in the virtual collocation of PD teams and design services. Although different in end-user functionality, this model framework provides an indication of how an active document should be structured. This active document consists of 'shared' webpages, which provide common user access to the following (structure of the document is shown in Figure 10): (i) Geometric and product models of component design, and (ii) Different product design services. The end users would interact with each other through 'shared' VRML models that provide for complete bi-directional connectivity between the presentation layer (CAD model in VRML format) and the underlying knowledge architecture (product design model). The 'shared' webpage and VRML models simulate the notion of 'interactive' design sessions and have the ability to capture all user interactions which take place in these sessions during the evolution of the new product design. This is very much similar to the creation of new product design/engineering knowledge. This ability of the active document to capture every form of user interaction and assist in knowledge creation is particularly relevant in the context of GBL.

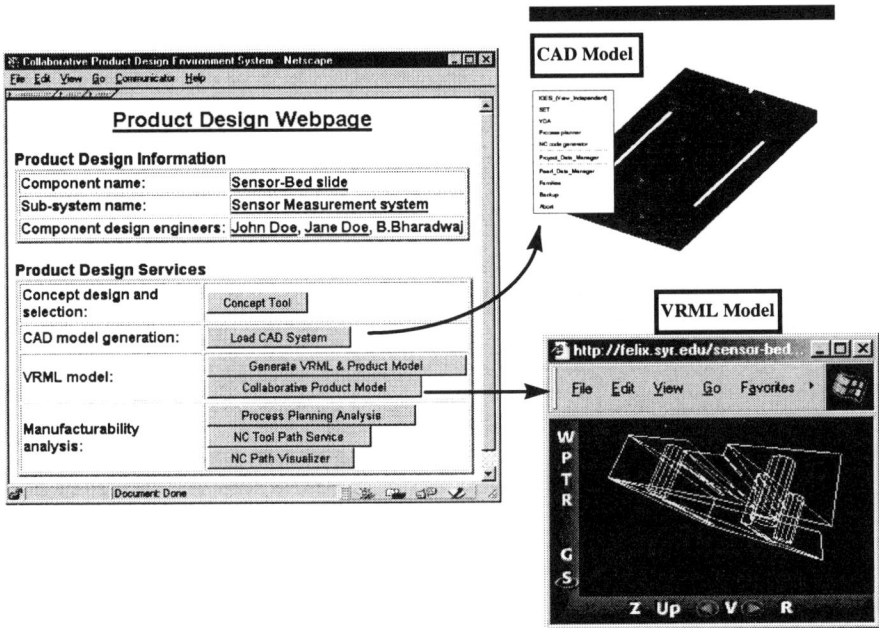

Figure 10: A model framework for an "Active Document" - Sharable Product Design Webpage and VRML model

Interactive refinement of engineering design information: Learning is situated in everyday work of product development and engineering involving a variety of learning modes (discovery, analogy, induction, etc.). Moreover, engineering activities involve group-oriented tasks involving interaction among various individuals and groups. One often gets to hear from an individual who participates in engineering activities that they have something to learn from their everyday tasks. Organizations soon realized the potential of recording such observations (individual/group experiences) in the form of documents such as 'Lessons learnt/Best Practices' or 'Design guidelines'. These documents are meant for the benefit of engineers in the future so that they can *learn* from the wisdom of their predecessors. Additionally, interactivity in the creation and usage of engineering design information promotes shared (group wide) understanding of PD knowledge and practices. GBL also strives to achieve the similar objectives for practitioners within a LC. The model framework of the collaborative PD system described earlier as well as in reference [8] would be examined in this context.

A key objective of the model framework is to provide an environment for the visual presentation, interactive refinement and specification of product design information on VRML based sharable CAD models on a WWW (Intranet/Internet) based PD platform along with web integrated product data management system support (illustrated in Figure 10). The architecture of this framework promotes: (i) Shared understanding of PD design data, and (ii) Visual interaction with geometric as well as PD design information for the engineer in a PD team. The virtual design and prototyping environment provided by this framework simulates the interactive atmosphere prevalent in a design session involving engineers across a "drawing board" or around a "physical prototype part". The individual would have real time view of all PD data and the associated engineering knowledge (customer requirements, design specifications and manufacturability information). The unique advantages offered by this framework from the perspective of GBL are the following: (i) Automatic updates of CAD geometric models with design revisions eliminating the ambiguity between CAD models of different engineers, and (ii) Reduced communication overheads for PD data through real time and simultaneous visual display, manipulation and specification of CAD models.

Learning through Simulation: Skill refinement through learning by simulation is a powerful technique for stimulating the thought process through observable actions. The ability to imbibe the learning material is influenced by the individual capabilities of the practitioners. Although individualistic, this technique can be applied for GBL by utilizing appropriate computer based simulation techniques wherever possible. This is particularly relevant in present day organizational environments where LC are spread across the globe and practitioners prefer to take self paced courses as opposed to classroom lectures. We examine this learning scenario in the context of process plan generation and NC code visualization for machining operations. The services of a computer based process planner and NC tool path generator of the collaborative PD system described earlier (additional details provided in reference [8]) was utilized for this purpose. The simulation features consisted of (shown in Figure 11): (i) Interactive examination of various machining operations, (ii) Identification of relationship between NC tool path and machinability characteristics, and (iii) Support for various "what-if" type analysis. We observed that practitioners (in this case, students in engineering course) were enthusiastic learners as long as the simulations were presented in the context of what they would observe in day-to-day practice. We believe that, richer levels of user interaction would help further improve the capability of individuals to learn.

6. Networks of Learning Communities in a Corporate Scenario: Implementation Guidelines, Benefits and Performance Evaluation Metrics

In this section, a roadmap for implementation of the proposed framework along with the benefits to be accrued as well as the metrics for evaluating the effectiveness of learning communities is discussed. A phased plan for implementation of a framework for sustaining networks of learning communities would consist of the following phases:

- **Phase 1**: Formation of a core group (practitioners and PE) providing support to formation of the LC for envisioning KM objectives and goals for a particular engineering practice. This is a crucial phase, which relies on the support of the social framework for jump-starting the formation of the LC.
- **Phase 2**: Identification of appropriate content providers (individuals, PD groups, PE's and outside resources) and the content (e.g. mental models, training materials, etc.) for the KM initiative. This phase requires extensive support of senior management in the form of following: (i) development of operational procedures and time schedules for PD teams to incorporate CBD principles, and (ii) commitment of appropriate support infrastructure (support personnel, computer software, etc.) for the LC.
- **Phase 3**: Development of computer based models and tools for: (i) creation, visual presentation and interactive refinement of information, (ii) knowledge acquisition, encoding and usage of tacit information, and (iii) support knowledge generation through the social roles and responsibilities of the practitioners in the LC. This phase would also require extensive support from the IT department of the organization.
- **Phase 4**: Sustainment of Learning Community – KM activities are part of continuous improvement initiatives and hence do not have a concluding phase. On the contrary, this phase is very critical from the perspective of sustaining and evolving KM activities within the LC. The current vision envisages the involvement of the practitioners in this phase. This activity further symbolizes the spirit of the Community Based Design philosophy.

Knowledge Management initiatives in various organizations across a wide spectrum of business activities involving networks of learning communities are suggestive of some of the following observable benefits: (i) Enhanced collaboration resulting in faster execution of projects (Sandia National Labs, Silicon Graphics), (ii) Reduced project/application development costs (Ernst & Young, Anderson Consulting), (iii) Faster distribution and access of information for employees (Eli Lilly, Rockwell, 3M) and (iv) Exchange and distribution of best practices (Olivetti R&D Labs, National Semiconductor, Xerox). The above listed benefits further strengthen the importance and significance of learning communities within the overall realm of the business and PD process.

Organizations often employ various metrics for evaluating the performance of reengineering initiatives for studying and forecasting their effectiveness. Knowledge management is a reengineering effort with its own measures. These measures can be categorized as quantitative as well as qualitative and have been summarized in Table 1.

Table 1: Listing of Qualitative and Quantitative Metrics

Quantitative measures
1. Time/Cost reductions in executing a process as a result of suggestions provided by the practitioner community,
2. Measure of Things Gone Wrong (TGW) per project: (a) Reduction in TGW of a similar nature, (b) Discovery and resolution of new TGW and (c) Communication of TGW to other project teams,
3. Throughput of process – Effectiveness (time/cost/quality) in accomplishing the process as a result of the collective knowledge of the practitioners,
4. Number of Tips/Suggestions in the knowledge repository pertaining to a particular practice area,
5. Mistake proofing – Reduction in effort (time and cost) required for removing and validating mistakes eliminated as a result of earlier detection within the practitioner community,
6. Number of decisions bucks – Number of requests to higher levels in the organizational hierarchy for problem resolution,
7. Number of individuals contributing and utilizing the knowledge repository – This measure would help determine if people are having the right incentives to contribute and use the collective knowledge of the individuals.

Qualitative measures
1. Effectiveness in change management – How can practitioners organize their process to accommodate for changes in project management and execution,
2. Process redesign – How practitioners can design the process they are collectively responsible for with enough flexibility for redesign without affecting performance.
3. Intellectual asset – How much and of what importance is the knowledge provided by the practitioner of use to the organization

The above listed metrics are by no means all encompassing in the scope of the business activities performed by various organizations and can at best provide generic guidelines. A point to note is that the metrics should not be used as the yardstick for supporting the networks of learning communities. This is because of the fact that learning within the practitioner community is an ongoing process and any shortcomings or accomplishments need to be shared within the learning community. Shortcomings or accomplishments highlight the deficiencies or positive effects of the supporting framework.

7. Conclusion

The role of learning communities in facilitating knowledge management activities among engineering practitioners in the context of industrial organizations was examined in this chapter. A model framework encompassing the appropriate forms of social, management and technological infrastructure for supporting and sustaining the knowledge creation activities within a learning community was proposed to facilitate learning at both the individual and group level among the practitioners within these communities. The organization and evolution of learning communities suggest the following important issues to consider from the perspective of knowledge management and product development:

- Knowledge management activities cannot be performed in isolation from product development activities.
- The focus of knowledge management initiatives should not be centered on the creation of information repositories. The thrust should be rather placed on: (i) creation of new knowledge, (ii) emergence of new practices/processes as well as improvements in current tasks, and (iii) educating other members of the learning community about these new practices employing various IT based solutions.
- The Voice of the Practitioner (VOP) has similar importance to the product development improvement activity as Voice of the Customer has to the product development process itself. KM activities have an important role in promoting improvements in the product development process through the VOP.
- Community Based Design can help remove bottlenecks and provide the appropriate framework for organizations to reorganize around the process centric framework of product development.

Current published research [2][4][6] has already highlighted the importance of learning communities in knowledge management activities. It should be noted that the 'knowledge factory' encompassing these learning communities could be only sustained by promoting the appropriate synergism among the practitioners, corporate management and the available technology.

Acknowledgment: The author would like to thank Dr. David Bell from Xerox Palo Alto Research Center and Prof. Utpal Roy from Syracuse University for many useful suggestions and discussions.

References

[1] Hammer, M., and Champy, J., 1993, *Reengineering the Corporation: A Manifesto For Business Revolution.* Harper Collins, New York.
[2] O'Day, V. L., Bobrow, D. G., and Shirley, M., 1996, "The Social-Technical Design Circle," Proc. of the ACM Conference on CSCW.

[3] Lave, J., and Wenger, E., 1991, *Situated Learning: Legitimate peripheral participation.* Cambridge University Press, NY.

[4] Brown, J. S. and Duguid, P., 1991, "Organizational Learning and Communities-of-practice: Toward a unified view of working, learning, and innovation," Organization Science, Vol. 2., No. 1.

[5] Senge, P., 1990, *The Fifth Discipline: The Art and Practice of the Learning Organization. Doubleday Currency,* New York.

[6] Bharadwaj, B. and Bell, D. G., 1997, "A Framework for Sharing of Engineering Practices Through Networks of Learning Communities," Advances in Concurrent Engineering CE97, Technomic Publications.

[7] Hansen, M., Nohria, N., and Tierney, T., 1999, "What's Your Strategy for Managing Knowledge?" Harvard Business Review 77, No. 2, pp. 106-116.

[8] Roy, U., et al., 1997, "Product Development in a Collaborative Design Environment," International Journal of Concurrent Engineering, Research and Application, Vol. 5, No. 4, pp. 347-366.

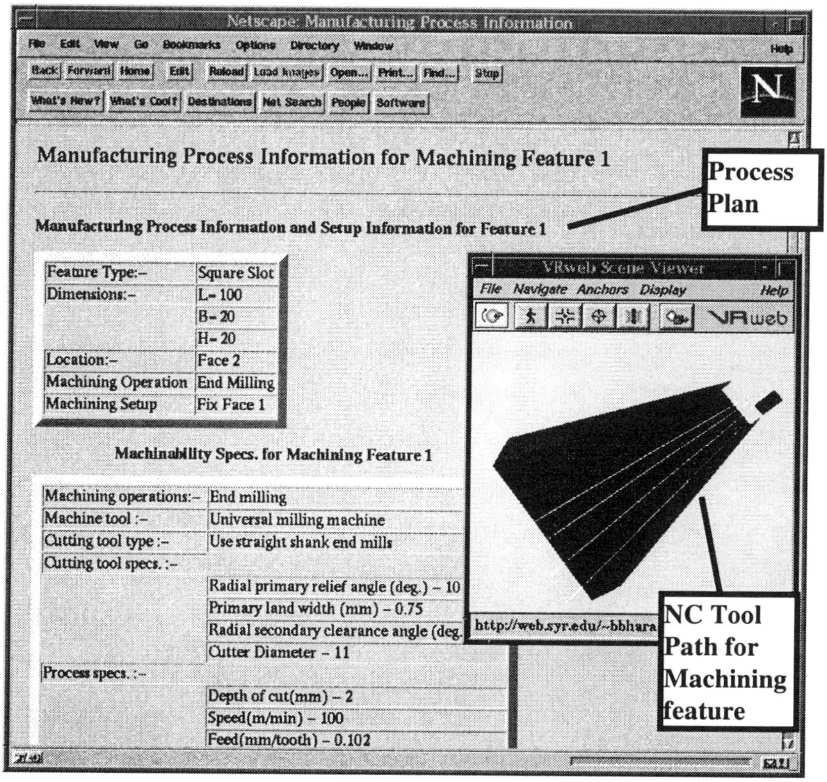

Figure 11: Web based on-line simulation of various manufacturability evaluation criteria -
Promotes better understanding of production issues

Managing Engineering Knowledge Federations

Fatma Mili

Oakland University, Rochester MI 48309-4478, mili@oakland.edu

Abstract: The knowledge relevant to the design of engineering systems comes from different sources, is owned by different bodies, and is evolving over time. In this paper, we present a framework where the different sources of knowledge form a federation of independent knowledge bases. Design support systems access the knowledge in these complementary sources and integrate it into a coherent set. We motivate the need for the monitoring of knowledge base properties that capture the quality and coherence of the evolving and distributed knowledge. We discuss the technical issues related to the integration of the different sources in light of the properties defined.

1. Introduction

The knowledge related to the design of any engineering artifact is typically generated from different sources of expertise and sources of regulation. The design of a mechanical system, such as an engine for example, is bound by functionality constraints defined by mechanical engineers, casting constraints defined by die casting experts, assembly constraints defined by assembly experts, and environmental constraints defined by federal and professional agencies, to name just a few. Design decisions need to be made in light of knowledge from all of these sources. Such knowledge must be collected, organized, and activated to bear on the design decisions. The multifaceted nature of this knowledge, and the facts that it is distributed and evolving pose a unique set of challenges.

Early knowledge based systems such as MYCIN [8], PROSPECTOR [17], and XCON [4] were process-oriented with a tight integration between the content and representation of the knowledge and the process for which it is used. Further, these systems have been successful thanks to the painstaking efforts of teams of pioneer researchers who single handedly elicited, encoded, tested, and maintained large collections of domain knowledge. These systems provided the initial proof of concept, but their success could not be readily reproduced in the less uniform and less controlled environment of most organizations. General concepts, tools, and methodologies were needed to support the tasks of knowledge elicitation and management, and to promote the reusability, manageability, and maintainability of the resultant knowledge bases.

A number of converging ideas have since emerged with the goal to make knowledge-based systems viable outside of research laboratories. There is now a consensus on the benefits of raising the level of abstraction of knowledge bases, of separating factual knowledge from processing knowledge, and of developing tools and methods for knowledge acquisition, validation, and maintenance

[7,11,13,14,21,22,25,27,32]. The focus of these approaches remains, nevertheless, on developing knowledge bases from scratch, and on creating knowledge bases that are dedicated to a specific purpose, even if they are more modular and more easily reusable than first generation systems. The more recent knowledge management movement led to a shift in focus from creating new knowledge repositories to using existing knowledge sources. It recognizes corporate knowledge as a valuable resource that needs to be reckoned with, cultivated, and exploited [5, 12,19, 28, 31.]

In the context of engineering design, the majority of the knowledge is already available. Rules and regulations are generated and documented by a variety of stakeholders who keep them up to date to reflect continuous scientific findings, technological innovations, and changes in standards and regulations. In this paper we consider the technical issues related to using these existing knowledge sources. We focus in particular on the issues of integration of these different sources representing different perspectives, and updated at different times with different frequencies. The paper is organized as follows. In Section 2, we discuss the general approach used. In Section 3, we motivate the need for schema invariants used to safeguard the quality of the knowledge. In Section 4, we discuss technical issues relating to the integration of these resources in a way that preserves the invariants defined. We summarize and conclude in Section 5.

2. General Approach

2.1 Federation of knowledge sources

As discussed in the introduction, the knowledge relevant to the design of engineering artifacts is typically widely documented and maintained by different departments and stakeholders. It can be tempting to synthesize all of these sources into a single monolithic integrated knowledge base that is dedicated to the task of engineering design support. Creating a design specific knowledge base would provide sufficient freedom to formalize, encode and organize the knowledge in a way that serves best the process being supported. This approach has major drawbacks, though. In addition to recreating the wheel, it makes the timely update and maintenance of the knowledge base a challenging task, to say the least. We choose an approach that balances between the two competing goals of efficiency of the knowledge base and modularity and liveliness of the knowledge by using the concept of federation [5].

The collection of independent knowledge sources forms a loosely bound system, called a knowledge federation. Stakeholders maintain ownership of their knowledge base and full responsibility of its contents. The knowledge in these sources remains use-independent and is naturally distributed along the lines of expertise or stakes. These different sources participate in the federation by providing access to some selected subset of their contents. The federation can be used by multiple systems. The knowledge needed by the design support system is not recreated, but extracted automatically from the existing sources. The design support system has its extraction component that accesses the various sources and conceptually integrates them to fit its needs. This knowledge filter creates a customized view that can be computed on a need basis or stored and updated incrementally. The architecture is illustrated in Figure 1.

The figure shows the different knowledge sources to the left. Each of the sources has its own local knowledge base that it makes available to the outside world. The different local knowledge bases are accessed by the knowledge filter. The filter is used to *conceptually* integrate the sources, extracting knowledge of interest, combining it, and synthesizing it for the design system.

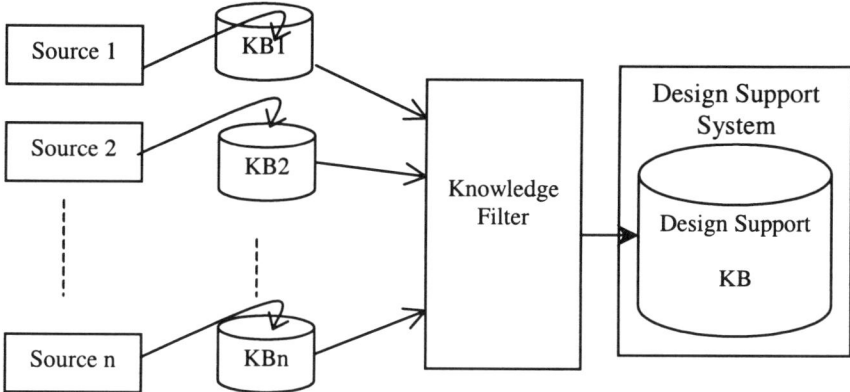

Figure 1: Knowledge based federation

This framework is similar to the one used in HKBMS [30], with the difference notably that HKBMS stores knowledge in the form of production rules.

2.2 Knowledge model

The data model most commonly used in federated databases is the object-oriented model. The object model is favored because its expressive power encompasses that of other models and because other models can be easily translated into it. The same reasoning applies for knowledge bases, although the information is of a different nature as is described in [24]. We briefly summarize the knowledge model used here.

Because the different knowledge bases belong to different stakeholders, they do not necessarily use a common model. Yet, to participate in the federation of knowledge bases, a view of the shared knowledge must be expressed in some common global model. It is the global model that we describe here.

Given a domain of expertise or area of interest, we see the knowledge maintained as a set of requirements on the objects to be designed. For example, the mechanical engineering knowledge base can be thought of as a set of specifications of the mechanical properties of the engines, pistons, radiators, and so forth. The specification information is represented by sets of constraints associated with object classes. In sum, each knowledge base is a set of classes encapsulating sets of constraints representing the knowledge base's owner's perspective. The different classes are related via pre-defined associations. The key associations used are the traditional subclass/superclass relationship as well as other relationships of assembly, spatial containment, and part-feature. We briefly describe each association. More details can be found in [24].

Subclass/Superclass association. This association is used primarily to define classes in a modular way. A class Engine for example is used to house constraints that apply to all engines. Specific requirements to diesel engines, or turbo engines, on the other hand, are defined in their respective subclasses. Subclasses inherit all attributes, methods, and constraints of their superclasses.

Assembly-part association. This association captures the fact that a system is assembled from parts. This relationship creates a dependency between the different parts of the same system and between the parts and the whole.

Spatial containment. This association is a "packaging" association. It captures the common occurrence whereby parts are packaged together within the same container without necessarily implying any dependency other than the sharing of a common space. For example, the relationship between the under-hood and the various parts that it houses is a spatial containment relationship.

Part-feature association. In an effort to reduce redundancy, we abstract the concept of feature and use separate classes to describe them and define their constraints. Features are generally abstract concepts (e.g. opening, depression) that have no existence outside their context (e.g. door, hood), yet they have their own attributes and constraints that are independent of the context in which they occur.

With a number of independent knowledge bases defined, the major issue is that of integrating the knowledge between them. Intuitively, the contents of two knowledge bases together is the sum of their individual contents. In practice, computing the conceptual sum of two knowledge bases is far from being trivial as it involves identifying gaps to be filled, ambiguities to be resolved, and inconsistencies to be addressed. In the next two sections we discuss issues related to the integration process.

3. Knowledge Base Invariants

3.1 Related research

The quality of the information stored in a database or a knowledge base is of utmost importance. Ruggles [29] identifies four dimensions of knowledge quality: *timeliness, comprehensiveness, accuracy, and cognitive authority*. These qualities are, at least in part, non-intrinsic qualities; they are defined in reference to external concepts. As a result, they cannot be verified internally. Generally, weaker intrinsic qualities have been defined and monitored within databases and knowledge bases.

Two classes of invariants have typically been of concern: model (intention) invariants and data (extension) invariants. In relational database systems, the model invariants are the normal forms of the relations; the data invariants are the entity integrity and the referential integrity constraints. The richer is the semantics of a data or knowledge model, the more critical is the need for quality measures and the more opportunity there is for defining them. Semantic and object-oriented database systems have been the target of research on defining and preserving model consistency [1,9,15]. We mention some examples here.

For the Object-oriented database system O_2 [2], two classes of schema consistency have been defined: Structural consistency, and behavioral consistency. Structural consistency is defined purely in terms of the inheritance hierarchy. A schema is said to be structurally consistent if the three following conditions are met.

1. The inheritance graph defined by the schema is a directed acyclic graph.

2. Attributes and methods' types are inherited as is or specialized (subclassing is consistent with subtyping).

3. Multiple inheritance raises no ambiguity (multiply inherited attributes have a common origin).

A schema is said to be behaviorally consistent if the run-time behavior of the system is "correct". The O_2 data definition language includes a set of primitive schema update operators. Whenever such operator is executed, the Schema Consistency Checker verifies that the update does not violate the consistency of the schema. The structural consistency invariants of O_2 are very similar to the invariants defined for the Orion system [3]. In Orion, the invariants are also centered on the semantic and structural consistency of the inheritance graph. The Orion system has a set of schema change operations and a set of rules that define the corrective action to be taken in response to these changes. The set of change operators is shown to be complete, and the set of rules is guaranteed to preserve the schema invariants.

The focus of the O_2 and Orion systems is on the attributes and methods of the classes, and on the inheritance hierarchy. Because we are mainly interested in the knowledge captured by the class constraints, we mention two object-oriented systems that support some form of class constraints. The Chimera system [10] is an active object-oriented database system. Chimera's classes have active rules attached to them. Ceri *et al* define the qualities of a database model from the active rules point of view (termination, confluence, etc.) [10]. They define a methodology to analyze a database model and to evolve it in such a way to preserve these qualities. Ode [18] is another object oriented database system that supports the notions of integrity constraints and business rules. Classes are associated with constraints and rules. Constraints are inherited by subclasses along with attributes and methods. The developers of Ode have devoted some attention to the issues of data integrity and behavioral integrity. They define a comprehensive set of rules for the preservation of the integrity of the data and the proper propagation of reactive behavior [20].

The formulation of schema invariants and the monitoring of their preservation take a special significance in the context of interest where the knowledge is distributed and continuously evolving. Because the overall system is a federation, there is no centralized oversight over consistency, and there is no guarantee of coordination. Because of the life span of these knowledge bases (long term investments) and the typical turnover of employees, even within a single system, there is no built-in continuity. Different people are likely to be updating the same source at different times. All of this emphasizes the critical need for some system support of consistency or at least system support for detection of lack of consistency.

3.2 Knowledge base invariants

The set of invariants defined for such systems as O_2, Orion, and Chimera are invaluable in unearthing fatal errors in schemas such as cyclic inheritance chains and conflict between two superclasses of the same class. Similar invariants for the relational model would be requirements on the atomicity of domains, and uniqueness of relation names and attribute names for example. Violating such invariants is fatal, but meeting them does not necessarily indicate a good database design. We are interested in building on the basic invariants and defining a set of properties that would capture more qualitative measures of consistency. Inconsistency across classes is generally a result of some form of redundancy. In the traditional relational database model, a set of normal forms has been defined to indicate the absence of specific types of redundancies, and thus the absence of resulting inconsistencies. We use the same model to define four normal forms.

Whereas with the relational model, higher normal forms represent stronger conditions, it is not the case here. Therefore, except for the first normal form, we will not number the normal forms, we name them instead. The first normal form, as is the case in the relational model, is in fact a pre-requisite, not directly related to redundancy.

Definition 1 A knowledge base is in *first normal form* if every property of interest is individually identified.

This normal form identifies the need for the comprehensive and complete documentation of all knowledge of interest. The intended use defines what knowledge must be documented. For example, if we need to be able to identify all constraints from a specific source, or all constraints based on a given assumption, then the source and that assumptions of a constraints must be explicitly recorded and adequately labeled. The comprehensive and explicit documentation of constraints enable many uses such as: identify constraints affected by a technological change, differentiate between constraints having the same formula but different rationale, and so forth. All of these considerations are important for integration and adaptation.

Definition 2 A knowledge base is in *generalization normal form* if every constraint is associated with the most general class to which it applies.

The underlying guiding principle here is that every constraint should appear only once, attached to only one class, and inherited and propagated automatically to all the contexts where it applies. For example, a constraint associated with the class RectangularOpening and with the class RoundOpening should, instead, be placed with the common parent, viz. Opening. This eliminates redundancy across siblings and redundancy between parents and children.

Definition 3 A knowledge base is in *smallest context normal form* if every constraint is associated with the smallest context to which it applies.

This is another embodiment of the unique attachment of constraints. The class model must be designed in such a way that every constraint finds a unique class to which it can be attached. For example, all constraints relating to the shape, size, and whatnot of window openings should be placed with the class WindowOpening rather than repeated with every occurrence of this feature and associated with the wall in which it appears. This eliminates repetition within a same class with multiple occurrences of a feature, as well as repetition across classes sharing the same feature.

Definition 4 A knowledge base is in *encompassing context normal form* if every constraint is associated with a class that includes all of the attributes being constrained.

One of the difficulties of associating constraints to classes is that some constraints restrict associations, and thus, involve more than a single class [20]. In our model, associations are defined by classes. These classes house the associative constraints and are responsible for the coordination between the participant classes. This is always possible in the context of design where the associations of interest are structural in nature.

4. Knowledge Integration

Integrating knowledge from different sources is in some sense an additive process. In this section, we first define an ordering relationship that captures the intuitive idea that a model captures more information than some other model. We then use this ordering to formulate the process of integration. A more detailed account of the information presented in this section can be found in [23].

4.1 More defined ordering relation

We define classes by the quadruplet $<N,A,M,K>$ where N is the class name, and A, M, and K are the sets of attributes, methods, and constraints respectively. A schema is defined by its set of classes C and the set of inheritance, assembly, and part-feature relationships on C. A schema with an empty set of classes is an empty schema. We will restrict our discussion here to the inheritance relationship. Given

U, a universe of possible schemas, we define an ordering relationship on this universe.

Definition 5: We define the relation *more-defined* on the universe *U* of schemas as follows:
- *Universal lower bound:* Any schema is more defined than the empty schema.
- *Class ordering:* Given a schema S consisting of a single class C=<N,A,M,K>, and given a schema S' consisting of a single class C'=<N',A',M',K'> such that N'=N, A'⊆ A, M'⊆ M, and K'⊆ K, schema S is more defined than schema S'. In other words, a class C with the same name as class C' but with more attributes, methods, and constraints, is considered to be more defined than C'.
- *More classes:* Given two disjoint schemas S_1 and S_2 (no common classes), the schema S=$S_1 \cup S_2$ is more defined than each of S_1 and S_2.
- *More associations:* Given a schema S, two classes C and C' in S if the association subclass <C,C'> does not conflict with the inheritance graph, then the schema S' defined by adding <C,C'> to the inheritance graph is more defined than S.

The above definition of the ordering relationship accounts only for the inheritance association. It can be easily expanded to include other associations [23].

4.2 Integration

The integration of different knowledge bases consists of executing the following steps:

1. Identify syntactic and semantic conflicts between the different sources.

2. Reconcile the different sources to resolve the conflicts.

3. Create the integrated knowledge base.

The two first steps of identifying and resolving conflicts will not be addressed here. Most of the issues are identical to those faced with multi-databases [16,26]. We focus instead on the third step keeping in mind the potential inconsistencies.

We define integration as follows:

Definition 6 Given two schemas S_1 and S_2, integrating S_1 and S_2 consists of building a schema S --if it exists -- such that:
- S is more defined than S_1 and more defined than S_2.
- S is the smallest such schema.

In other words, when it exists, S is the least upper bound (lub) of S_1 and S_2. When the two schemas are not consistent, they admit no least upper bound. When the two schemas are consistent, the lub is unique as can be easily shown. We show below very simple examples of schema integration and rely on the reader's intuitive understanding of the concept of least upper bound.

Disjoint schemas: Consider the schema S1 consisting of a single class, Door, and the schema S_2 consisting of the single class, Wheel. The schema S, union of S_1 and S2, is the least upper bound of S1 and S2.

Complementary schemas: Consider the schema S_1 consisting of a single class, Door with stamping constraints, and the schema S_2 consisting of the same class Door, but with the styling constraints. The least upper bound of S1 and S2 is the schema S, consisting of the class Door with constraints from both sources.

Sibling classes: Consider the schema S1 consisting of the class FrontDoor, and the schema S_2 consisting of the class RearDoor. The least upper bound of S1 and S2 is the schema S consisting of the two classes FrontDoor and RearDoor.

In the last example, it is clear that the least upper bound, S, of S_1 and S_2 is most likely not in normal form. If there are any constraints common to the FrontDoor and the RearDoor, these constraints are repeated in both classes. This violates the generalization normal form. To correct the situation, we need to add a class Door to capture the common attributes, methods, and constraints between the FrontDoor and the RearDoor classes. This example illustrates the fact that the lub of normal form schemas is not necessarily in normal form; it may contain some redundancy between the classes. We amend the integration process to account for normal forms. We integrate the schemas in such a way that if the participant schemas are in normal form, the resultant schema is also in normal form.

Definition 7 Given two schemas S_1 and S_2 in normal form, the normal integration of S_1 and S_2 consists of building a schema S --if it exists-- such that S meets the following conditions:

- S is more defined than lub (S_1, S_2).
- S is in normal form.
- S is the smallest such a schema.

The additional condition of normal form means that, part of the integration process, classes and associations can be changed or added as needed to meet the normalization criteria. In the case of the door example, the normalized integration would produce a three classes schema with Door superclass of the FrontDoor and RearDoor. In practice, the process of identifying the legitimacy and logical adequacy of adding a common parent, a common child, or any other transformation cannot be fully automated. The process is bound to be heuristic. The system must be able to detect the possible need for transformations, possibly suggest transformations, and let the users decide. In [23], we define a set of heuristics used by a system for the detection of possible normal form violations and the generation of possible corrective actions. This approach is in line with other approaches that acknowledge that schemas do not contain sufficient information to allow the automatic detection and correction of semantic conflicts and semantic overlap [26].

5. Summary, Conclusion

In this paper, we have identified the need to build infrastructures that allow us to make use of existing resources relevant to the design of engineering artifacts. We have advocated the use of federations of knowledge bases with a global object oriented model centered on constraints. We have justified this approach by the need for modularity, knowledge independence, and autonomy between the different bodies setting standards and legislating products and processes. For this approach to be viable, we need a formal mechanism for assessing the quality of a knowledge base, and operators for ``adding'' together the contents of different knowledge bases. The two aspects are not independent since whatever qualities are defined, they need to be preserved by the addition operators.

In this paper we have mostly introduced the issues and outlined some solutions. There is an obvious gap in the area of knowledge bases and object oriented models that have not been filled yet. More work in this domain will be most beneficial.

References:

[1] S. Abiteboul. Restructuring hierarchical database objects. *Theoretical Computer Science*, 62, 1988.

[2] F. Bancilhon, C. Delobel, and P. Kanellakis. *Building an Object-Oriented Database System: The story of O2*. Morgan Kaufmann Publisher,1992.

[3] J. Banerjee, W. Kim, H.J. Kim and H.F. Korth. Semantics and implementation of schema evolution in object-oriented databases. *In Proceedings of the ACM SIGMOD Conference*,1987.

[4] V.E. Barker and D.E. O'Connor. Expert systems for configuration at digital: Xcon and beyond. *Communications of the ACM*, 32(3), March 1889.

[5] Frank Blackler, Michael Reed, and Alan Whitaker. Editorial introduction: Knowledge workers and contemporary organizations. *Journal of Management Studies*, 30(6):852-862,November 1993.

[6] O.A. Boukhres, et al. The integration of databases systems. In O.A. Boukhres and A.K. Elmagarmid, editors. *Object-Oriented Multidatabase Systems*, pages 37-56. Prentice Hall, 1996.

[7] D.C. Brown and B. Chandrasekaran. Design problem solving: Knowledge structures and control strategies. In *Research Notes in Artificial Intelligence*. Pitman, 1989.

[8] Bruce G. Buchanan and Edward H. Shortliffe. Rule-Based Expert Systems: *The MYCIN Experiments of the Stanford Heuristic Programming Project*. Addison Wesley publishing company, 1985.

[9] S. Castano,V. De Antonellis, M. G. Fugini and B. Percini. Conceptual schema analysis: Techeniques and applications. *ACM TODS*, 23(3):286-332,September 1998.

[10] S. Ceri snd P. Fraternelli. *Designing Database applications with Objects and Rules: The IDEA methodology*. Addison-Wesley, 1997.

[11] B. Chandrasekaran. Design problem solving: A task analysis. *AI Magazine*, 40(11):59-71,1990.

[12] Chun Wei Choo. *The Knowing Organization*. Oxford University Press, 1998.

[13] W.J. Clancey. The epistemology of a rule based system –a framework for explanation. *Artificial Intelligence*, 20:215-251,1981.

[14] H. Eriksson, Y. Shahar, S.W. Tu, A.R. Puerta and M.A. Musen. Task modeling with reusable problem-solving methods. *Artificial Intelligence*, 79(2):293-326,1995.

[15] A. Formica, H.D. Groger, and M. Missokof. An efficient method for checking object-oriented database schema correctness. *ACM TODS*, 23(3):333-369, September 1998.

[16] M. Garcia-Solaco, F. Saltor, and M. Castellanos. Semantic heterogeneity in multidatabase systems. In O. A. Boukhres and A.K. Elmagarmid, editors, *Object-Oriented Multidatabase Systems*, pages 129-202. Prentice Hall, 1996.

[17] J. Gasching. Application of the prospector system to geological exploration. In J.E. Hayes, D. Michie, and Y-H Pao, editors, *Machine Intelligence*. Wiley, 1982.

[18] N.H. Gehani, H.V. Jagadish and O. Shmueli. Event specification in an active object-oriented database. In *Proceedings of the ACM-SIGMOD 1992 International Conference of Management of Data*, 1992.

[19] William E. Halal, editor. *The infinite resource: Creating and leading the knowledge enterprise*. Jossey-Bass Publishers, 1998.

[20] H.V. Jagadish and Xiaolei Qian. Integrity maintenance in an object-oriented database. In *Proceedings of the 18th VLDB*, August 1992.

[21] Sandra Marcus, editor. *Automating Knowledge Acquisition for Expert Systems*. Kluwer Academic Publishers, 1998.

[22] J. McDermott. Preliminary steps toward a taxonomy of problem solving methods. In Sandra Marcus, editor, *Automating Knowledge Acquisition for Expert Systems*, pages 225-256. Kluwer Academic Publishers, 1988.

[23] F. Mili. Knowledge integration in federated knowledge bases. Technical report. Oakland University, Rochester MI48309-4478.

[24] F. Mili, K. Narayanan, and D. VanDenBossche. Domain knowledge in engineering design: Nature, representation and use. In Rajkumar Roy, editor, *Industrial Knowledge Management- A Micro Level Approach*. Springer-Verlag London limited, 2000.

[25] M. A. Musen, et al. Protégé-II:computer support for development of intelligent systems from libraries of components. *In Proceedings of MEDINFO'95, Eighth World Congress on Medical Informatics*, pages 766-770,1995.

[26] Sh. Navathe and A. Savasere. A schema integration facility using object-oriented model. In O.A. Boukhres and A.K. Elmagarmid, editors, *Object-Oriented Multidatabase System*, pages 105-128. Prentice Hall, 1996.

[27] A. Newell. The knowledge level. *Artificial Intelligence*, 18:87-128,1982.

[28] Deniel E. O'Leary. Enterprise knowledge management. *IEEE Computer*, 31(3):54-61,March 1998.

[29] R. Ruggles. Knowledge tools: Using technology to manage knowledge better. Technical support, Ernest & Young, LLP Center for business innovations, April 1997.

[30] S. Y.W. Su, A. Doshi and L. Su. Hkbms: An integrated heterogeneous knowledge base management system. In O.A. Boukhres and A.K. Elmagarmid, editors, *Object-Oriented Multidatabase Systems*, pages 589-620. Prentice Hall, 1996.

[31] Gerald H. Taylor. Knowledge companies. In William E. Halal, editor, *The infinite resource*, pages 97-110. Jossey-Bass Publishers, 1998.

[32] B.J. Wielinga, A. Th. Schreiber, and J.A. Breuker. KADS: A modeling approach to knowledge engineering. *Knowledge Acquisition*, 1(4), 1992.

Languages for Knowledge Capture and their Use in Creation of Smart Models

Biren Prasad[1]

[1]Unigraphics Solutions, Knowledge-based Engineering (KBE) PBUs , CERA Institute, P.O. Box 3882, Tustin, CA 92782, Tel: (714) 952-5562, Fax: (714) 505-0663, Email: <prasadb@ugsolutions.com>

Abstract: Languages are means of capturing the knowledge for the design and development of a product. Smart Models are the results of such knowledge capture. The author, first describes how languages for knowledge capture have evolved over a thirty year time period. Author through literature search finds such languages to fall into three major classes: (a) Geometry-based language (b) Constraint-based language (c) Knowledge-based language. The paper then describes the differences and similarities of these languages that can be employed to capture life-cycle intent. The second part of the paper describes how such languages are being used in creation of smart models. A smart model is a reusable conceptualization of an application domain. The models contain the knowledge (attributes, rules or relations) of the application domains forming the basis for future problem solving. The paper also describes two popular ways of formulating a problem that leads to such smart models: (1) Constraint-based programming (2) Knowledge-based programming. Through analysis of existing practices, new development and trends, the paper then discusses some "new emerging directions in the use of languages for the knowledge capture". Finally, the benefits of knowledge capture and creation of smart models over conventional models are discussed.

1. Introduction

Except in a few rare cases, products are now so complex that it is extremely difficult to correctly "capture" their life-cycle intent right the first time no matter what C4 (CAD/CAM/CAE/CIM) tools, productivity gadgets or automation widgets are used. Traditionally, CAD tools are primarily used for activities that occur at the end of the design process. Such usage of CAD tools, for instance, during detailing geometry of an artifact, is in generating a production drawing, or in documenting geometry in a digitized form (See Figure 1). CAM systems are conventionally used to program machining or cutting instructions on the NC machines for a part whose mock-up design, clay or plaster prototype may already exist. CAE systems are used to check the integrity of the designed artifact (such as structural analysis for stress, thermal, etc.), when most of the critical design decisions have already been made.

Studies have revealed that 75% of the eventual cost of a product is determined before any full-scale development or a CAD tool usage actually begins [1].

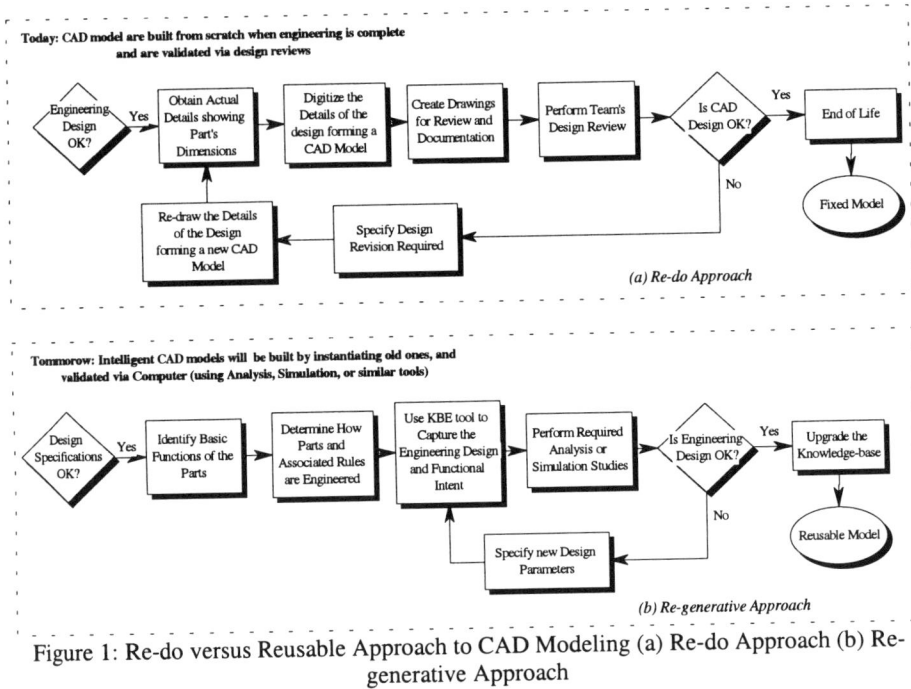

Figure 1: Re-do versus Reusable Approach to CAD Modeling (a) Re-do Approach (b) Re-generative Approach

Most C4 tools in use today are not really *"capture"* tools. The need for *"change"* after a design model is initially *"built"* is all but inevitable. Today, CAD models are built from scratch only when engineering activity is complete, and are validated via a series of design reviews. Design work-groups typically document the design through CAD software only after the completion of major engineering processes and after resolving all of the pressing engineering issues. A work-group captures the geometry in a static form, such as lines and surfaces. Static representation is actually a documentation that tells a designer what the final design looks like but not how it has come to be. If changes are required in the design, a new CAD model is recreated (see Figure 1) using some types of computer-aided "re-do" or "back-tracking" methods. Such CAD methods of activating change or modification (e.g., a redo or a backtracking) can be extremely time-consuming and costly being that late in the life-cycle process. In such static representations of geometry, configuration changes cannot be handled easily, particularly when parts and dimensions are linked. In addition to the actual process that led to the final design, most of the useful lessons learned along the way are also lost. In the absence of the latter, such efforts have resulted in loss of configuration control, proliferation of changes to fix the errors caused by other changes, and sometimes-ambiguous designs. Hence, in recent days, during a PD^3 process, emphasis is often placed on the methods used for capturing the life-cycle intent with ease of modifications in mind. The power of a "capture" tool comes from the methods used in capturing the design intent initially so that the anticipated changes can be made easily and quickly later if needed. By capturing a "design intent" as opposed to a "static geometry," configuration changes could be made and controlled more effectively using the power of the computer than through the traditional CAD attributes (such as line and surfaces). *"Life-cycle capture"* refers to the definition of a physical object and its environment in some generic form [2]. *"Life-cycle intent"* means representing a life-cycle capture in a form that can be modified and iterated until all the life-cycle specifications for the product are fully satisfied. *"Design-capture"* likewise refers to the design definitions of the physical objects and its surroundings. *"Design-intent"* means representing the *"design capture"* in a form (such as a parametric or a variational scheme) that can be iterated. Design in this case means one of the life-cycle functions (see Figure 4.2 of Volume I [3]). In the future, CAD models will be reusable. The new models will be built by instantiating the old ones and validating them via computer (using simulation, analysis, sensitivity, optimization, etc., see Figure 1). Such models will have some level of intelligence built into them [4].

2. Languages for Life-cycle Capture

Languages are means of capturing the knowledge for the design and development of a product. Models are the results of such knowledge capture. The primary goal of knowledge-capture formalism is to provide a means of defining ontology. Ontology is a set of basic attributes and relations comprising the vocabulary of the product realization domain as well as rules for combining the attributes and relations. Engineering Analysis Language (EAL), for example, provides a means of creating analysis or design models as run streams. Later, they form the basis for iterative analysis and design [5]. ICAD/IDL, on the other hand, captures the knowledge about the process of designing and developing a product [6][7]. There are three types of languages that can be employed to capture life-cycle intent:

a) Geometry-based language
b) Constraint-based language
c) Knowledge-based language

Figure 2 shows an evolution of languages for capturing knowledge over a thirty year time period. These are C4 (CAD/CAM/CIM/CAE) specification languages for product engineers or designers to define configurations of parts and assemblies. They are not computer languages for software programmers (such as C, or C++). During this thirty-year period, there had been a tremendous innovation.

- The *first generation* of C4 languages, first introduced during 1960, only dealt with 2-D drafting and 2-D wire-frame design.
- The *second generation* of C4 languages dealt with surfaces and 3-D solids.

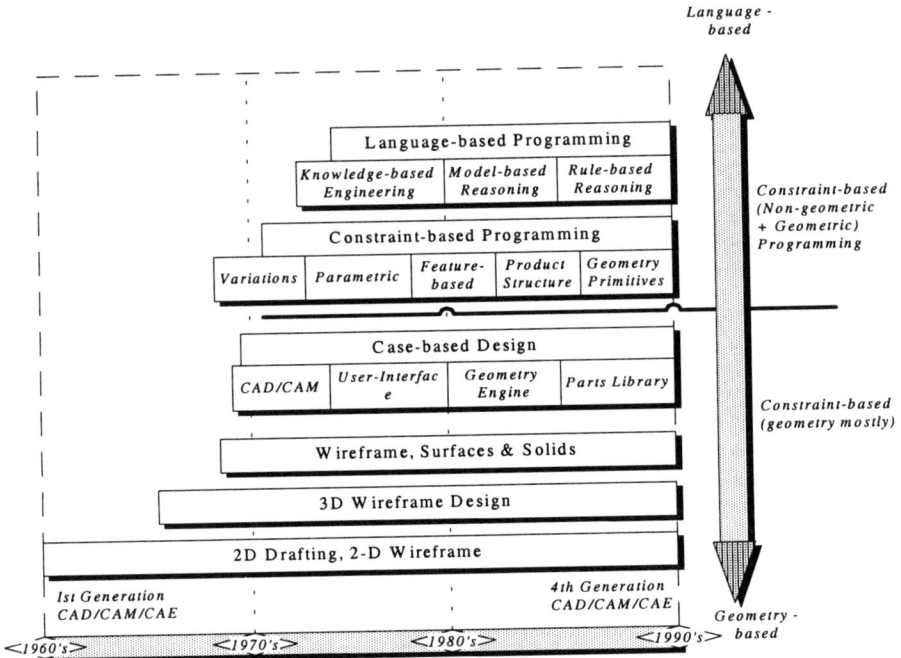

Figure 2: Evolution of Languages for Knowledge Capture

- The *third generation* of C4 language was constraint-based but mostly dealt with geometry. Examples include case-based design, parametric scheme, variational scheme, etc. During the period from 1980 onwards, there was a history of developments in making C4 codes more user friendly, use of a solid-based geometry engine (CSG versus B-Rep) and introduction of part library concept. There was also a flurry of activities in the use of techniques, such as parametric schemes, variational schemes, featured-based concepts for creating product structure and for defining geometric primitives [4].
- The *fourth generation of languages*. Today is the age of fourth generation C4 languages, which is quite different from the past. Fourth generation of languages are knowledge-based techniques giving CE design work-groups the ability to capture both geometric and non-geometric information.

Languages for Life-cycle Capture = $\cup$ [Geometry-based language, Constraint-based language, Knowledge-based language]

3029

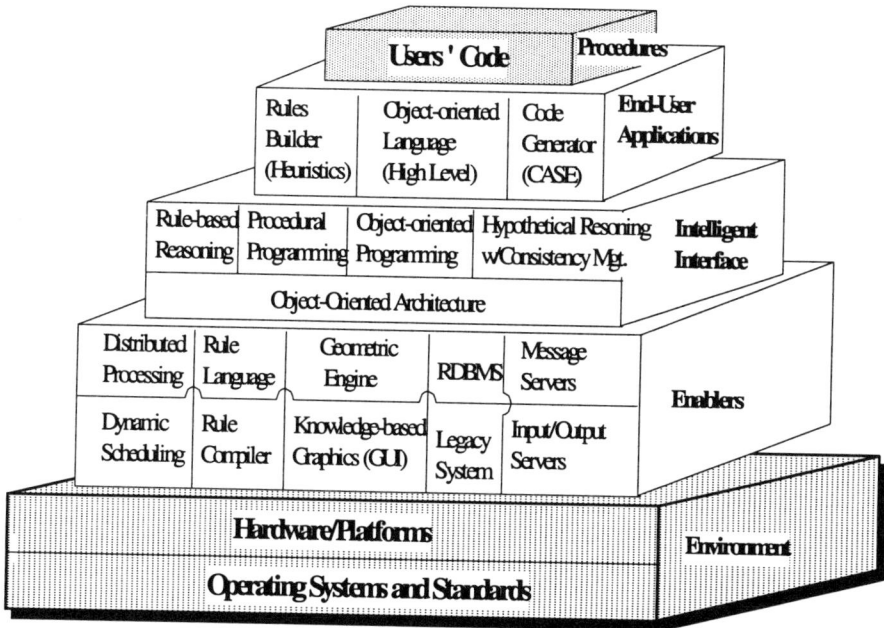

Figure 3: A Computational Architecture for a typical Knowledge-based System

3. Knowledge-based Systems (KBS)

Knowledge-based Systems (KBS) are software programs designed to capture and apply domain-specific knowledge and expertise in order to facilitate solutions of problems. Languages can be used as means to build KBS. Knowledge-based Engineering (KBE) deals with processing of knowledge. There are many ways to capture knowledge to control its processing. KBE is a process of implementing knowledge-based systems in which domain-specific knowledge about a part or a process is stored along with other attributes (geometry, form features, etc.). A computational architecture for a typical knowledge-based system is shown in Figure 3. It consists of five layers, each layer supporting the others.

- **Environment:** The *first layer* is an environment, which provides a foundation for the rest of the layers. A typical environment consists of a slew of operating systems, standards, and compute platforms (workstations, hardware, etc.).
- **Enablers:** The *second layer* consists of core enablers. Some of the key enablers included at this layer are: distributed processing, dynamic scheduling, real time rule language, rule compiler, knowledge-based graphics, GUI, geometry engine, relational database management system (RDBMS), legacy system, input/output servers, message servers, etc.

- *Intelligent Interface*: The ***third layer*** adds intelligence to the enabling tools (second layer) and gives a programming interface to build the end-user applications. Most knowledge-based engineering tools encompass five critical technologies within an object-oriented architecture:

  (a) Rule-based reasoning
  (b) Procedural programming
  (c) Object-oriented programming
  (d) Hypothetical reasoning with consistency management
  (e) Case-based reasoning.

- ***End-user applications:*** The ***fourth layer*** is made out of end-user applications. It consists of a high level object-oriented language, a rule builder, and a code generator similar to a CASE tool.
- *Procedures:* The ***topmost layer*** embodies the procedures for the users' code.

3.1 Geometry-based Language

In the past, knowledge about products was mainly present in the form of geometry. Now 3-D solid geometry, as opposed to surfaces, wire-frames and other forms of geometry, is being used more often. Most traditional languages are geometry-based. They capture the attributes of solid primitives including lines and curves of a modeled object and their relationships to each other. Some high-end languages also capture information about the space inhabited by an object or about its enclosure (for example, Constructive Solid Geometry—solids). Some modelers develop complex solids by adding an extension to the traditional Boolean (join, intersect, and subtract) operations. For example, a combined solid can be driven by a 2-D sketch. As illustrated in I-DEAS Master series [8], the sketches can also be driven by geometric elements of other solids. Most solid modelers, however, fail to draw on knowledge about what the object is, its relationship to other objects or components, or its life-cycle aspects. Constraint-based CAD programs speed the design-change process by controlling and constraining object relationships based on dimensions (size, orientations, etc.), positioning, or geometrical inputs. However, such programs still focus on the geometrical aspects of the product development, not the knowledge about its life-cycle manufacture.

3.2 Constraint-based Language

Constraint-based language provides facilities for defining constraints. Most constraint-based languages provide means of incorporating arithmetic, logical functions, and mathematical expressions within a procedure. Such constraints may have a simple, linear algebraic relationship between entities to control shape (e.g., the length of line A is twice the length of line B) or geometry. The examples of geometric relationships include horizontal and vertical leveling, parallelism, perpendicularity, tangency, concentricity, coincidence, etc. Some constraints provide means to define and solve a system of linking equations that constitute a set of necessary design constraints and bounds. Finite element analysis and sensitivity analysis are some of the options that are generally considered an integral part of a constraint-based language. Such languages encompass command structures, symbol substitution, user-written macros, control branching, matrix analysis functions, engineering data base manager and user interface to integrate complex multi-disciplinary analysis, design, and pre- and post-processing work tasks.

Finite element systems usually consist of:

(a) A set of preprocessors through which a team defines finite element meshes, applied loads, constraints, etc.,
(b) A central program that primary performs numerical computation
(c) A set of post-processors for displaying the results.

They do not provide instantiation needs. An issue often encountered in a conventionally structured program is how can CE work-groups go beyond what the FEA programs provide. If it is necessary to perform functions that are not explicit capabilities of a program, the only recourse available to the teams is a very expensive and time-consuming one. The product developer has to write a new module (in a conventional language, such as FORTRAN, C or C $^{++}$) that operates on an output data file produced by the finite element analysis program. The constraint-based language largely eliminates this difficulty, making it very easy for teams to integrate complex and highly specialized analysis and design tasks, and to create specialized input formats and output displays. Another class of constraint-based languages that use AI techniques is based on solving a constraint satisfaction problem (CSP). Formally a CSP is defined as follows [9]:

Given a set of n variables each with an associated domain and a set of constraining relations each involving a subset of the variables, find an n-tuple that is an instantiation of the n variables satisfying the relations.

In the CSP approach, most of the efforts are in the area of solving a constraint satisfaction problem automatically. Most design problems, on the other hand, are open-ended problems. They are evolutionary in nature requiring a series of frequent model updates and user interactions, such as what is encountered during a loop and track methodology (discussed in section 9, [3]). How to manage such team interactions in the CSP approach has been the topic of research in the AI community for some time.

3.3 Knowledge-based Language

In a knowledge-based engineering (KBE) system, work-group members use a design language to build a smart model of the product. Formalism for defining smart models is a knowledge-based representation paradigm for describing the life-cycle domain knowledge. KBE languages go well beyond parametric, variational, or feature-based geometry capture mode to a knowledge-based life cycle capture mode. A design work-group does not just design parts. Work-groups design products -- a collection of functional parts placed in an assembly to form a finished (that is a functional) product. A KBE language provides ways to capture geometry and non-geometric attributes, and to write the rules that describe the process to create the assembly.

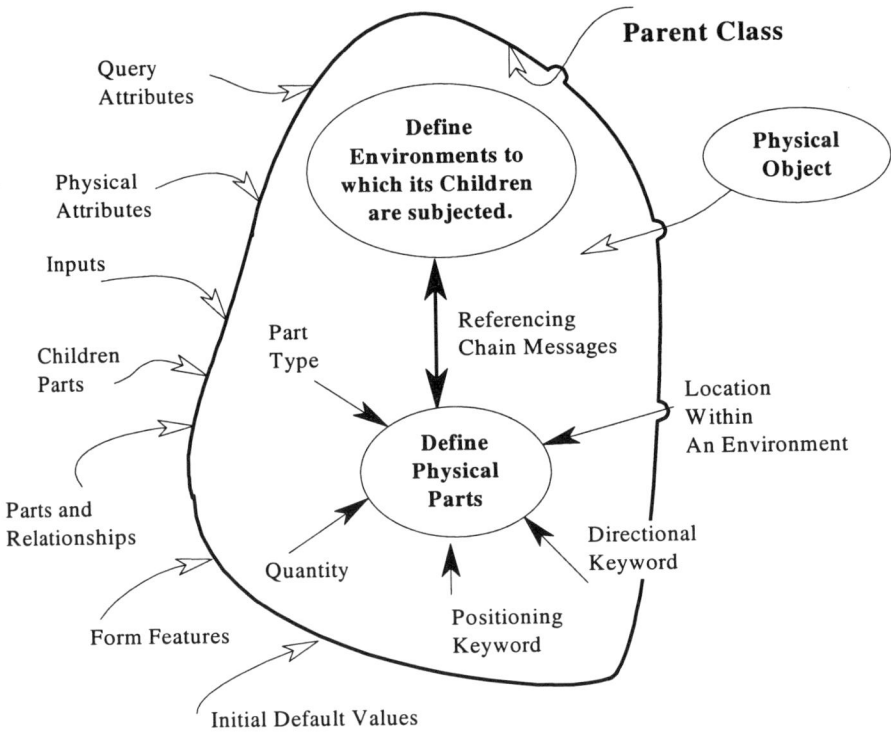

Figure 4 Definition of the "physical" Object

Such rules might include design stresses, resultant volume, or other   parts' positioning, mating and orientation rules. These rules form a part of an intelligent planning procedure derived from a domain specific knowledge. The use of intelligent planning procedure in KBE replaces an exhaustive enumeration of all feasible assembly plans that would have been needed otherwise. Most of the present KBE languages use object-oriented techniques. Unlike constraint programming languages, which define procedures for the manipulation of objects and entities, knowledge-based languages define classes of "objects," and their characteristics and behaviors that possess built-in manipulation capabilities. KBE languages capture the totality of the functions and relationships between model elements.

The following are some of the characteristics of a knowledge-based language:

- *Object symbols or Attributes*: In a KBE language, object symbols or attributes are the backbone of the system. Attributes describe object geometry, overall physical parts, its environments, location of the parts within that environment, and any other characteristics that are required. Figure 4 shows a definition of a physical object and a few examples of some associated attributes. Some attributes that are fixed are defined as constant-attributes. Variable attributes are design specifications whose values change. Inputs and children of an object are considered as variable attributes. Most of the physical attributes are fixed attributes. KBE languages allow a team member to define attributes in any order but they are internally recognized as "keywords." Directional and positioning keywords are keyword examples shown in Figure 4. Keywords enable "demand-driven-operations" to take place [6][7].

- *"Demand-driven" Operations*: In this mode the system determines the "order" and the "necessity" to evaluate an attribute. If a value for an attribute is demanded for the first time, the system computes the value and remembers it. In subsequent operations of the same attribute, when its value is required, the system returns the "cached" value instead of recomputing it. This method of evaluation, called "demand-driven evaluation" [6], is considered an important property for recalculating the value each time it is demanded. This type of operation relieves the programmer from assigning the order in which to evaluate the attributes. This makes programming in KBE languages significantly easier than in other languages.

- *Frame Structures with Rules*: Frames are object-oriented structures that allow for the storage of attribute information as object hierarchies. Frame representation is, thus, convenient for the storage of geometric dimensional and quantitative knowledge. Rules are used to implement the procedural expressions. The combination of an object-oriented frame structure with rules results in an adequate framework for capturing life-cycle manufacture knowledge.

- *Symbolic logic*: Symbolic logic is an underlying logic theory used in KBE to let knowledge engineers represent and manipulate the various types of knowledge required in CE. Symbolic logics are composed of object symbols (attributes), predicates, frame structures with rules, classes and instances, and kind-of inheritances. Simple logic statements can be connected using logical connectives to form compound logic statements. The set of such logic statements -- simple or compound -- is commonly called the logic theory. A full accounting for how objects and relations in the real world map to the logic symbols forms an "interpretation" of this logic theory [10].

- *Classes and Instances*: Most KBE languages allow definition of classes and instances. Classes are generic descriptions of objects, and "instances" are specific outcomes of an object-class. An object is a software packet that contains a set of related data and procedures. An object's procedures are called its methods. Objects communicate by sending messages to other objects requesting that they perform one of their methods. Object is an occurrence, or instance of a class. KBE languages often provide tools such as browser to represent objects and review instances, both graphically and non-graphically.

- *Kind-of inheritance*: The language allows definition of a "new class" from an "old-class" where the "new class" is derived from the "old class" with some "same but except" characteristics. A new class is said to inherit a portion of definitions from an existing class. Users only define the "except" changes, for example, "square" is a "kind of inheritance" from a "rectangle" object class. Inheritance allows the developer to define generalized behavior classes that can be used by multiple, slightly different subclasses. It also allows existing classes to be extended and modified without changing the source code. This is accomplished by overriding methods at the subclass levels. It supports the creation of object models by allowing object designers or programmers to specify class hierarchies through selection of methods. The resulting object is maintained in a storage-independent form.

- *Generic Parts*: A generic part is an object-oriented structure that includes engineering rules, methods, attributes, and references to the children of sub-parts. The KBE language provides options for specifying the parts' attributes as variable attributes with no initial values specified. Other attributes are defined as a function of the variable attributes. Generic parts receive their attribute-values by means of "inputs" at run-time. The concept is useful since the generic parts' family can be replicated or instantiated at run time merely by specifying the required inputs for each part throughout an assembly. The generic parts can encapsulate other sub-parts or contain positioning or assembly information.
- *Referencing-chain*: Referencing chain is a useful concept to access an object or an attribute of a tree from any other place in the tree. It is often used to define dependencies that exist or are desired between children. An access is permitted by identifying a path that leads to the object whose attribute descriptions are required. In the definition of a physical object, shown in Figure 4, a referencing chain is shown connecting an "environment definition" with "physical parts" definitions. The concept is useful since attributes or parts can be retrieved by passing messages without actually replicating the source code, reasoning or logic behind the definition of the parts or the attributes.

Methods of Capturing Life-cycle Intent = $\cup$ [*Attribute definition, "Demand-driven" Operations, Frame Structures with Rules, Classes and Instances, Kind-of inheritance, Generic Parts, Referencing-chain*]

Depending upon the available library of primitive parts, some languages are easier than others are.

4. Creation of Smart or Intelligent Models

An important component of a smart or an intelligent model is the ability to define geometry in terms of parameters and constraints. Constraints are rules about dimensions, geometric relationships or algebraic relationships. A smart model is a reusable conceptualization of an application domain. The models contain the knowledge (attributes, rules or relations) of the application domains forming the basis for future problem solving. Rules define how design entities behave, for example, whether a hole-feature is through or blind. In the case of a blind-feature, if the part becomes thicker and the cylinder is not long enough, the hole will become a blind hole. Unlike blind-hole feature, a through-hole feature understands the rule that cylinder must pass completely through the part and will do so no matter how the part changes. There are two ways of formulating a problem that leads to smart models:

a)   Constraint-based programming
b)   Knowledge-based programming

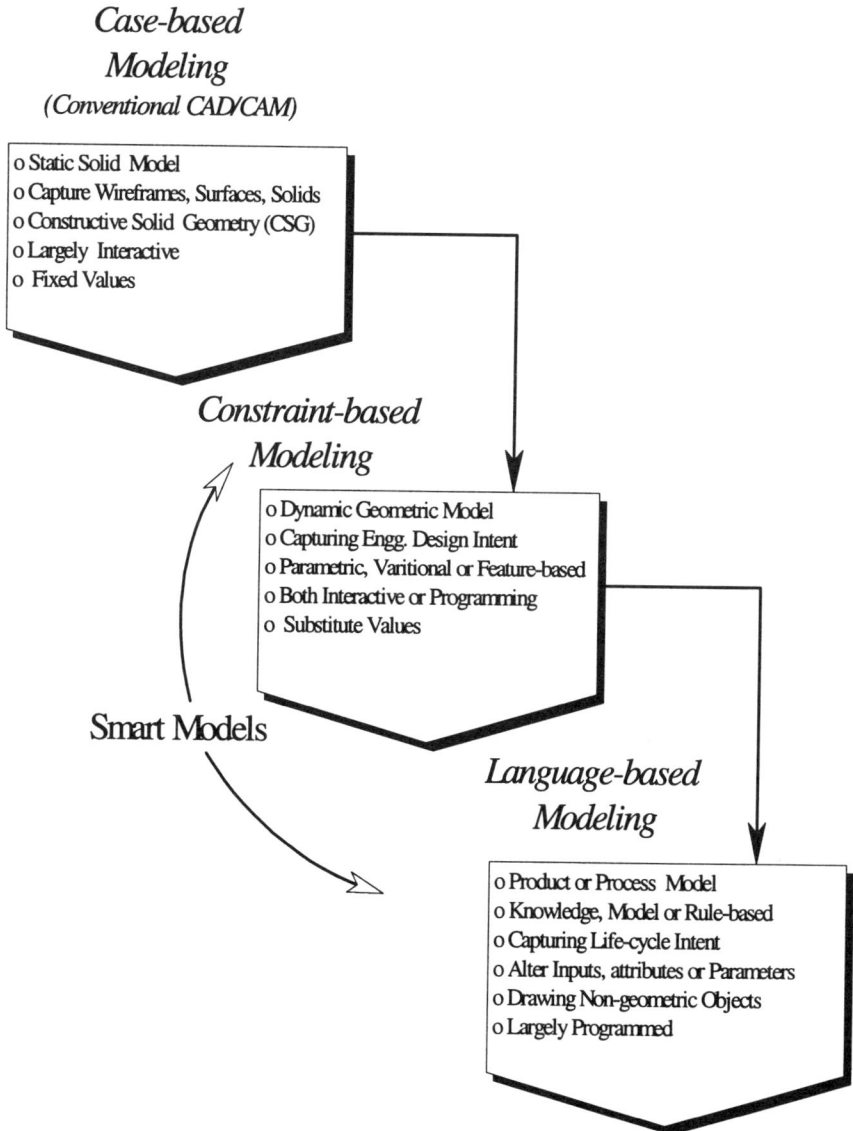

Case-based Modeling
(Conventional CAD/CAM)

o Static Solid Model
o Capture Wireframes, Surfaces, Solids
o Constructive Solid Geometry (CSG)
o Largely Interactive
o Fixed Values

Constraint-based Modeling

o Dynamic Geometric Model
o Capturing Engg. Design Intent
o Parametric, Varitional or Feature-based
o Both Interactive or Programming
o Substitute Values

Smart Models

Language-based Modeling

o Product or Process Model
o Knowledge, Model or Rule-based
o Capturing Life-cycle Intent
o Alter Inputs, attributes or Parameters
o Drawing Non-geometric Objects
o Largely Programmed

Figure 5: Properties of Conventional and Smart Models

	Representation of Parameters and Constraints	Type of Relationships between Parameters & Constraints	Method of Solving the Constraint Satisfaction Problem
Conventional Models	Non Interpretative (Fixed Dimensions)	Fixed Attributes: Points, Lines and Surfaces	Computational Geometry B-Splines, NURBS
	Interpretative (Variable Dimensions)	Variable Attributes: Points, Lines & Surfaces	Computational Geometry B-Splines, NURBS, Linear Algebra
Constraint-based Programming (Smart Models)	Parametric/Symbolic	Explicit/Algebraic	Equation Solver, Variational Geometry/Analysis
	Features/Forms	Explicit/Algorithmic	Linear/Non-Linear Programming, Optimization
	Mixed	Explicit/Algebraic + Algorithmic	Equation Solver/Linear Algebra, Multi-Criterian Optimizatiion, Optimal Remodeling
Knowledge-based Programming (Smart Models)	Rule-based Reasoning	Implicit/Heuristics	Inference Engine (Object-Oriented Programming – OOP)
	Model-based Reasoning	Implicit/Heuristics	Inference Engine (OOP)
	Case-based Reasoning	Implicit/Heuristics	Inference Engine (OOP)

Figure 6: Comparison between the Conventional and Smart Modeling Approaches.

Constraint-based modeling (CBM) or programming yields constraint-based models. Knowledge-based programming results in knowledge-based models. The major differences between the two model types (constraint-based modeling and language-based modeling) in contrast to the conventional modeling (traditional CAD/CAM system) are shown in Figure 5. In conventional modeling, the geometry is captured using a static representation of wire-frames, surfaces or solids, i.e., the geometry is captured in digitized (fixed value) form. The modeling process is largely interactive. In constraint-based modeling, since the mechanism of geometry capture is through parametric, variational or feature-based techniques, each model represents an instantiated (or dynamic) geometry. Thus, by setting new values to the CBM attributes, several instances of the geometry can be obtained. Knowledge-based modeling (KBM) is similar to constraint-based when it comes to capturing the geometry. However, because of its abilities to capture non-geometric information and to associate rules with attributes, KBM is also suited for capturing life-cycle intent. Other contrasting features of CBM and KBM are listed in Figure 5 and further explained in the following:

4.1 Constraint-based programming

Constraint-based Programming (CBP) is a concept of formulating a problem in terms of "constraints," which may be a part of a product definition, a process definition, or an environment for the problem definition. No distinction is made between types of constraints or their sources. A spreadsheet program is a simple example of a constraint-based programming. Here equations representing constraint relationships are input to cells in a spreadsheet program. There is a close resemblance between design rationale (DR) [11] and spreadsheets. Equations that are entered into cells of a spreadsheet are analogous to "capturing" DRs, and computed cell values in the spreadsheet program are analogous to "identifications" of DRs. The cells themselves constitute the "knowledge" of CBP. If smart models are thought of as a series of spreadsheets for a concurrent team, "programming" in CBP is analogous to specifying the relationships between the cells of a spreadsheet.

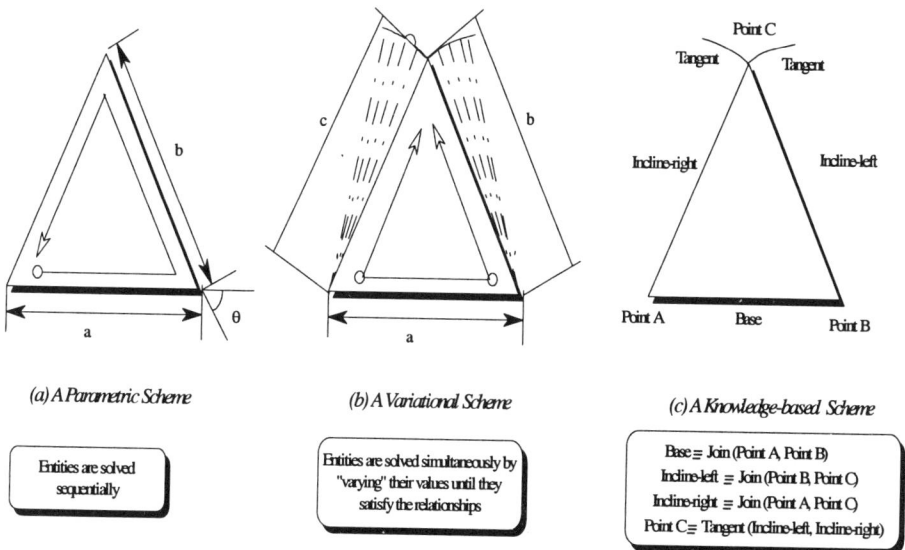

(a) A Parametric Scheme   (b) A Variational Scheme   (c) A Knowledge-based Scheme

Entities are solved sequentially

Entities are solved simultaneously by "varying" their values until they satisfy the relationships

Base ≡ Join (Point A, Point B)
Incline-left ≡ Join (Point B, Point C)
Incline-right ≡ Join (Point A, Point C)
Point C ≡ Tangent (Incline-left, Incline-right)

Figure 7: Difference between Parametric, Variational and Knowledge-based Schemes

The following are some typical constraint parameters that can be employed during constraint-based programming:

⇒ Design specifications
⇒ Design criteria
⇒ Subjective qualifications

⇒ Design constraints
⇒ Manufacturing constraints and tolerances
⇒ Material properties
⇒ Geometry
⇒ Sectional properties
⇒ Configuration and topology
⇒ Heuristics or rules
⇒ Historical data
⇒ Performance requirements
⇒ Test specifications and data

Figure 6 shows a three-way comparison between a set of key characteristics of smart models (created using constraint-based and knowledge-based programming approaches) and conventional models. Three key categories employed for comparison are: "types of representations", "types of relationships between parameters and constraints", and the method of solving the constraint satisfaction problem. They are shown in Figure 6 as columns of a matrix. The modeling categories (conventional and smart models) are listed as rows. A cell of the matrix shows the differences in the approaches used in each modeling category. In the conventional models, the geometry compatibility (such as line and arc constraints) and consistency issues are resolved through computational geometry, linear algebra, B-spline and NURBS techniques. In CBP, product design problem is defined in terms of constraints and the inter-relationships that exist between them (see Figure 7). Constraints may be a part of

• Product definition, such as geometry, materials, size, etc.
• Process constraints such as assembly constraints, tolerances, fits/clearances, etc.
• Product environment such as loads, performance, test results, etc.

In Constraint-based modeling, the constraints are of explicit/algorithmic or algebraic types. They are resolved through a set of linear and nonlinear programming, optimization and optimal remodeling techniques (see Figure 7). Such methods help develop a set of individualized criteria for life-cycle design improvement (more than what it generally appears to be the case). For example, on the surface it would appear that parametric generation of parts would not be a significant improvement over the "fixed-dimension" (static geometry) approach. In both cases, initial geometry definitions have to be captured and shared with other users of the information. The real advantages come if there is a large amount of change processing to be done to the original design. In a traditional CAD environment, this could be very time consuming and cumbersome. In knowledge-based modeling, in addition to the types of relationships specified for CBP, the constraint set also contains implicit/heuristics type of rules. Inference engines (backward and forward chaining) or constraint propagation techniques are commonly used in conjunction with object-oriented programming to resolve the imposed constraints (see Figure 7). The knowledge-based programming approach is discussed in Section 4.2 in greater depth.

Figure 8 shows a sample set of parameters for a typical CBP environment. The types of environment depend on the descriptions of the functional intent behind the product or the process that are modeled. The key parameters surrounding a constraint-based smart model are:

⇒ Geometrical, sectional and configuration variables
⇒ Manufacturing engineering design criteria and heuristic rules
⇒ Performance requirements, cost, efficiency data, and customer satisfaction
⇒ Design specifications or constraints
⇒ Manufacturing tolerances or constraints
⇒ Material selection or qualification
⇒ Library of parts, CAD data
⇒ Historical or subjective qualifications

In constraint-based programming, algorithmic tools such as analysis, simulation, generic modeling, sensitivity, and optimization are often used as an integral part of the design process. The original set of parameters is grouped into three categories: design variables, v_i, performance functions, f_i, and design constraints, c_i. The dependencies between performance and design constraints with respect to design variables are controlled through sensitivity analysis:

$$\text{Sensitivity} = \frac{\partial f_i}{\partial v_i}$$

Additional discussion of the problem formulation can be found in Section 4.8 of Chapter 4 [12].

4.2 Knowledge-based Programming

Knowledge-based Programming is another way of creating a smart model as shown in Figure 9. In order to develop an integrated view of PD^3, which is rich and comprehensive, it is necessary to include a variety of knowledge sources and representations. Knowledge-based representations deal with *explicit* knowledge, *implicit* knowledge and *derived* knowledge.

• *Explicit Knowledge*: Statements of explicit knowledge are available in product or process domains as retrievable information like CAD data, procedures, industrial practice, computer programs, theory, etc. They can be found both within and outside of a company. The explicit knowledge can be present as a set of engineering attributes, rules, relations or requirements. Inside knowledge deals with observations and experiences of the concurrent work-groups. Outside knowledge sources include papers, journals, books, and other product design and standard literature.

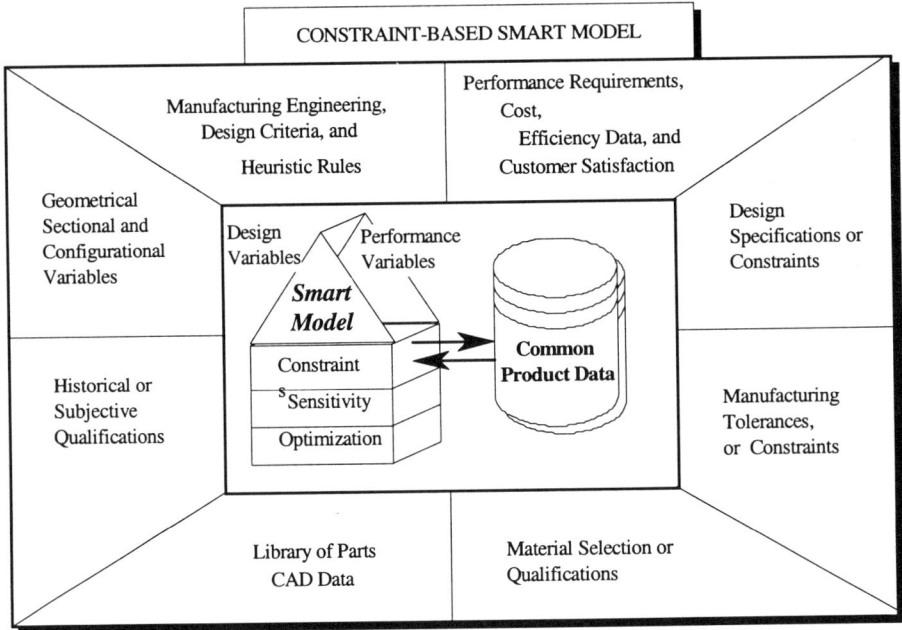

Figure 8: A Constraint-based Smart Model with Key Parameters

- **Implicit Knowledge**: Statements of implicit knowledge are mainly available as process details such as memory of past designs, personal experience, intuition, myth, what worked, what did not, etc. Difficulties arise when such processes are obscure, e.g., intuitive or creative. Implicit knowledge includes skills and abilities of the work-groups towards the application tasks and problem solving methods. Implicit knowledge that is found outside the work-group circles is mostly in case studies and discussion dialogues. Difficulties arise when such implicit knowledge has not been articulated in a form that allows ease of use and transfer.
- **Derived Knowledge**: Statements of derived knowledge are those that are discovered only by running external programs, such as analyses, simulation, etc. Derived knowledge is like extra- or interpolation of the current domain for which explicit knowledge is missing or incomplete. Undiscovered knowledge has been the driving force of most research and development organizations.

Knowledge-based programming software offers three basic benefits: capture of engineering knowledge, quick alterations of the product within its acceptable gyration, and facilitation of concurrent engineering.

Examples

o *Geometry definitions*
o *Product decompositions*
o *Engineering rules*
o *Mfg. constraints*
o *Parts library*

SMART **MODEL**

Implicit Knowledge

(Personal experience, intuition, common sense, application tasks, etc.)

Explicit Knowledge

(Papers, books, journals, theory, procedures, engineering attributes rules, relations, requirements, etc.)

Desired Knowledge

(Analysis, simulation, search, sensitivity, optimization, etc.)

Inputs

Geometric and
Non-Geometric
Attributes

Outputs

o Engineering Spec.
o Geometric Constraints

o 3D CAD Model
o Mfg. Reports
o Cost Analysis
  Reports

P R O D U C T   **P R O C E S S**

o Methodolgy
o Standards
o Compliance with
  Design Codes

o Procedures
o Company Policy/Regulations
o Practices

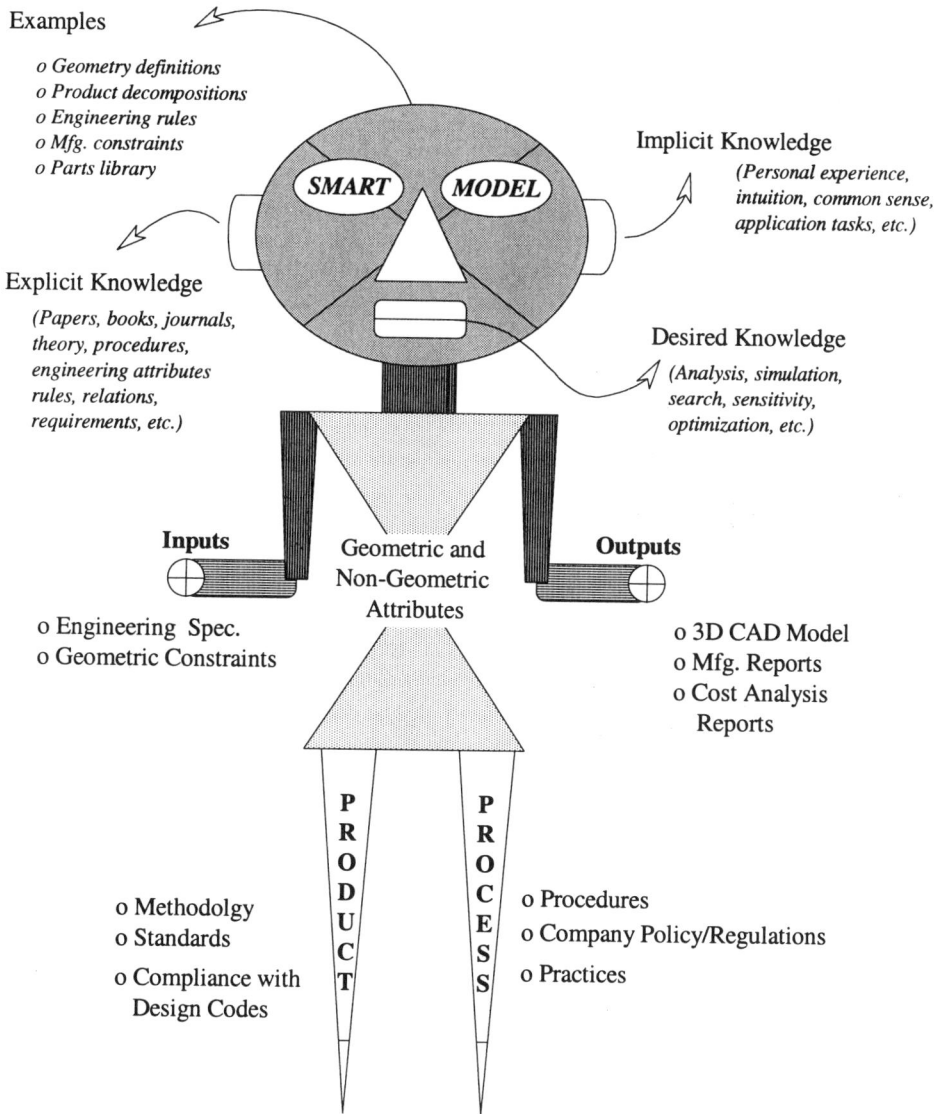

Figure 9: Salient Features of a Knowledge-based Smart Model

- The *first* strategic benefit is due to KBE system capturing engineering knowledge electronically. This capture allows companies to leverage scarce engineering expertise and to build on the knowledge acquired slowly over time.
- *Second*, the system permits design variations to be generated rapidly by quickly changing a long list of inputs while maintaining model integrity.

Products designed on the KBE system can practically design their own tooling. The system also enables designs to rough out their own macro process plans automatically by drawing on knowledge for similar GT designs.

- **Third**, KBE systems have been shown to enable concurrent engineering. Design, tooling, and process planning all benefit by working from a common integrated smart model that is able to represent, retrieve, and integrate engineering knowledge from many different sources and disciplines. Knowledge-based engineering reduces two of the most common problems that arise with team-oriented CE: boredom and time.

⇒ **Boredom:** Boredom crops into most traditional processes as part and parcel of their detail. Work-group members do not find it attractive to check hundreds of common drudgery details that occur every-time a new design is obtained -- from checking its specifications to tolerances—as part of a PD^3 cycle. The idea is to capture those design and manufacturing issues which impact the design of "most products, most of the time." This action is justified based on *8020 rule*. This is commonly called the 80:20 heuristic or Pareto's law of distribution of costs [13]. In a typical situation, 80% of the assignments are routine or detail works and only 20% are creative (Figure 10). Pareto's law states that while only 20% of the possible issues are creative, they generally consume about 80% of work-group resources. The 80% of the assignments that are routine do not seem to cause any significant product's problem or consume as much resources.

⇒ **Shortage of time and resources:** The second problem is shortage of 7Ts, and resources. Many concurrent team members are not able to find enough time to devote on actual design process due to their heavy occupation with other time demanding chores such as staff-meeting, E-mail notes, management briefings, technical-walk-throughs, design reviews, etc.

KBE reduces boredom by attending to the 3Ps details in ways that reflect design procedures, standards, company policies, and compliance of design and manufacturing codes and regulations. By packaging the life-cycle behaviors into a smart model, KBE improves productivity and automation. The 80% of routine tasks is reduced to a 20% level (see Figure 10). The CE work-groups spend more time adding functional value to the design (in the saved time) rather than repeating engineering calculations for each case or recreating existing design items. If the traditional cycle-time were reduced from an initial total of 30 months to 20 months using smart models, the following calculation applies (see Table 1).

The work-group is able to concentrate more on satisfying the creative tasks up to a maximum of 80% of the reduced cycle time. They are freed from worrying about meeting the drudgery details (routine tasks) that previously took up 80% of the total cycle time in the traditional method. Thus, even after spending 40% of the work-group engineers' time more on creative tasks, there was a surplus of two months using smart models compared to the time it took following traditional method.

Time Saved = Total Cycle-time - Reduced Cycle-time

Traditional Method   Smart Model

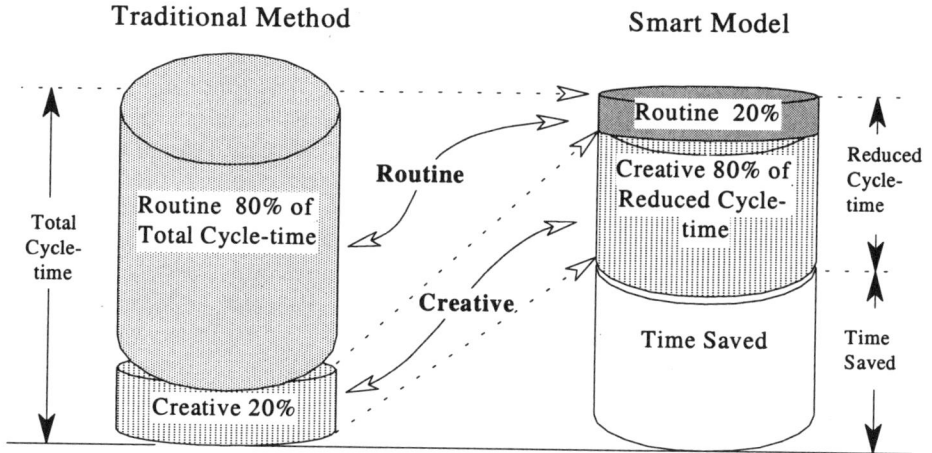

Time Saved = Total Cycle-time - Reduced Cycle-time

Figure 10: Key Benefits of Knowledge-based Engineering

Table 1: Comparison of Savings

Actions or tasks	Traditional Methods	Smart Models	Remarks
Total time taken to finish the tasks	30 months	20 months	Assumptions
Ratio of time spent doing routine/creative tasks in months (percentage)	24/6 months (80%/20%)	4/16 months (20%/80%)	Following the definitions of smart and traditional models
If 40% more time is spent doing creative tasks then number of extra months needed	12 months (40%)	8 months (40%)	This shows how you can do more with less.
Difference in time compared to traditional Method	12 Months (deficit)	2 months (surplus)	

5. Discussion and Future Trends

Depending upon the frequency and need for creating modified designs, it is desirable to weigh the benefits. Alternatives are either to apply an additional effort and the time required to develop the smart models or to carry on the design in a traditional way. Knowledge-based development of design may not be worth the effort if each design is a unique design and significant changes to the product over its life cycle are not expected. However, this perspective changes quickly if one attempts to view the worth with respect to the overall company performance. Techniques like parametric, feature-based and knowledge-based models all facilitate Concurrent Engineering and Collaborative Engineering. Some techniques offer better capabilities than others do. For example, parametric or feature-based techniques can change the geometry of the design very quickly. However, in doing so, design work-groups no longer have the assurance of knowing if all, or indeed, any of the non-geometric (e.g., engineering and manufacturing rules) have been violated. Through Knowledge-based engineering, one can capture, besides parametric geometry, the engineering and manufacturing rules for geometric modifications. When the specifications demand a new geometric design, the corresponding rules are automatically engaged to meet the engineering and manufacturing requirements and to achieve the best possible compromise. An interesting aspect of this approach is that one can capture and build trial processes (such as levels of analysis iterations, sensitivity, optimization, etc.) into a TPM in order to establish the best design. The effect of these trials is to automatically run thousands of analysis iterations in the background before the final design is selected. All of this can be transparent to the work-group members, who simply want to feed in specifications and are interested in reviewing the outcome that works. Most of the smart techniques provide some form of an electronic control over the design process which can afford a dramatic reduction in change processing such as design revisions, design changes, etc. The term automatic generation is construed as computer-aided generation of outputs such as design renderings, process plans, bill-of-materials, numerical control instructions, software prototyping, machining, replacement parts, product illustrations, etc. If an electronic capture of design intent is extended to include automatic or partially automatic design generation, the worth of electronic capture of design increases substantially. Furthermore, if such generation is done in 3-D solids that do not require assembly information in downstream processing, electronic capture becomes almost an irrefutable requirement of the product definition process.

In recent years use of smart models are increasing. More and more product development teams are capturing product and process-related knowledge. While their usage is increasing there is this need to make the captured knowledge in smart models used (leveraged) more widely across an enterprise. To facilitate such "knowledge reuse" product development teams are following standard techniques and tools for building such smart models. Thus, creation process for building Smart models is becoming more and more structured. The four main elements that make a smart model more structured are:

1. *Objects, independent problem domains and tasks:* Objects and the system are partitioned into a number of relatively independent loosely coupled problem domains so that the decomposed tasks are of manageable size. Each sub-problem can be solved somewhat independently from other domains and each can be performed in parallel. This is discussed in Chapter 4 -- "Product Development Methodology [12]."

2. *Design Rationale:* Irrespective of the modeler used for creating smart models (knowledge-based modeler or constraint-based modeler), the modeler splits the data into *rule-sheets* and *frame-sheets*. The rule-sheets contain the equations, and the variables are automatically placed in the frame-sheets. Additional sheets can be defined as needed, such as data settings, constraints-sheet and tables where results are tabulated, and plot-sheets which define how results are plotted. This gives the design problem a symbolic structure rather than a numerical processing structure. The other advantages of using design rationales are [11]:

   ⇒ Definition of objectives and constraints are in terms of problem parameters, design rules, or intent with possibly incomplete data.
   ⇒ DRs establish a set of relationships (explicit or implicit) between parameters and constraints.
   ⇒ DRs capture the informal heuristics, or chains of reasoning, rather than a set of well-defined algorithms.

3. *Method of solving the constraint based problem*: This is discussed at greater length in Chapters 4.1 and 4.2 of Concurrent Engineering Fundamental Book [12].

4. *Databases, Technical memory or knowledge-bases*: Databases, technical memory or knowledge-bases derived from a surrogate product can serve as a basis for developing smart models and conducting strategic studies [13]. When actual product data is not available, a surrogate object can take the place of a technical memory. Later, when actual data becomes available, the information in technical memory is replaced on a modular basis and the information model can be updated dynamically.

6. Concluding Remarks

Knowledge-based programming (KBP) provides an environment for storing explicit and implicit knowledge, as well as for capturing derived knowledge. When these sets of knowledge are combined into a total product model (TPM), it can generate designs, tooling, or process plans automatically. Unlike traditional computer-assisted drafting (e.g., a typical CAD) programs that capture geometric information only, knowledge-based programming captures the complete intent behind the design of a product — "HOWs and WHYs," in addition to the "WHATs" of the design. Besides design intent there are other knowledge (such as materials, design for X-ability, 3Ps, process rules) that must be captured. Knowledge-based Engineering (KBE) is an implementation paradigm in which complete knowledge about an object (such as a part) is stored along with its geometry. Later when the part is instantiated, the captured knowledge is utilized to verify the manufacturability, processiability and other X-abilities concerns of the part. One important aspect of knowledge-based engineering is the ability to generate quickly many sets of consistent designs instead of just capturing a single idea in a digitized CAD form that cannot be easily changed. Knowledge-based programming technology encourages development of a "generic" smart model that synthesizes -- what is needed in many life-cycle instances in complete detail. This is a most flexible way of creating many instances of a model, each being a consistent interpretation of the captured design intent. The interpretation is the result of acting on rules captured through the smart models by feeding in the specific inputs at the time of the request.

The current trend in smart models creation is to:

(a) Use some structured process for capturing the knowledge content and
(b) Store those rules and knowledge outside the smart models in some neutral object-oriented databases.

This way others could access them in more places, if they need them, during the product development across an enterprise.

References

[1] Nevins, J.L. and D.E. Whitney, eds.,1989, *Concurrent Design of Products and Processes*, New York: McGraw-Hill Publishing, Inc.
[2] Kulkarni, H.T., B. Prasad, and J.F. Emerson, 1981, "Generic Modeling Procedure for Complex Component Design", *SAE Paper 811320*, Proceedings of the Fourth International Conference on Vehicle Structural Mechanics, Detroit, Warrendale, PA: SAE.
[3] Prasad, B., 1996, Concurrent Engineering Fundamentals: Integrated Product and Process Organization, Volume 1, Upper Saddle River, NJ: Prentice Hall PTR, Inc.
[4] Finger, S., M.S. Fox, D. Navinchandra, F.B. Prinz, and J.R. Rinderle, 1988, "Design Fusion: A product life-cycle View for Engineering Designs," Second *IFIP WG 5.2 Workshop on Intelligent CAD*, University of Cambridge, Cambridge, UK, (September 19-22, 1988).
[5] Whetstone, W.D., 1980, "EISI-EAL:Engineering Analysis Language", Proceedings of the Second Conference on Computing in Civil Engineering, New York: American Society of Civil Engineering (ASCE), pp. 276-285.
[6] Rosenfeld, L.W., 1989, "Using Knowledge-based Engineering," *Production*, Nov., pp. 74-76.
[7] ICAD, 1995, *understanding the ICAD Systems*, Internal Report, Concentra Corporation, Burlington, MA.
[8] IDEAS Master Series, 1994, Structural Dynamics Research Corporation, SDRC IDEAS Master Series Solid Modeler, Internal Report, Version 6, 1994.
[9] Kumar V., 1992, "Algorithms for Constraint-Satisfaction Problems: A Survey", *AI Magazine*, Volume 13, No. 1.

[10] Winston, pH., 1992, *Artificial Intelligence (3rd Edition),* New York: Addison-Wesley Publishing Company.

[11] Klein, M., 1993, "Capturing Design Rationale in Concurrent Engineering Teams," *IEEE Computer*, Jan. 1993, pp. 39-47.

[12] Prasad, B., 1997, Concurrent Engineering Fundamentals: Integrated Product Development, Volume 2, Upper Saddle River, NJ: Prentice Hall PTR, Inc.

[13] Nielsen, E.H., J.R. Dixon, and G.E. Zinsmeister, 1991, "Capturing and Using Designer Intent in a Design-With-Features System," Proceedings, 1991 ASME Design Technical Conferences, 2nd International Conference on Design Theory and Methodology, Miami, FL.

Ontologies for Knowledge Management

Michael Uschold[1] and Robert Jasper[2]

[1]The Boeing Company, P.O. Box 3707, m/s 7L-40, Seattle, WA 98124-2207, USA
[2]The Boeing Company, P.O. Box 3707, m/s 7L-20, Seattle, WA 98124-2207, USA

Abstract: We introduce the idea of an ontology, and show how an ontology can be used as an important knowledge management tool for improving the sharing and reuse of knowledge assets. We introduce and describe four major categories of ontology applications: neutral authoring, ontology as specification, common access to information, and ontology-based search. For each, we highlight the role of the ontology, and indicate how it helps improve the sharing and reuse of knowledge. We conclude by discussing some limitations of the current state of the art, and what we expect in the future.

1. Introduction

The goal of this chapter is to give readers a practical understanding of what an ontology is and how ontologies may be applied to knowledge management. We begin by highlighting important activities in the lifecycle of knowledge assets from initial creation of knowledge assets, all the way to their eventual application and reuse. We point out that a major barrier to effective knowledge management throughout the lifecycle is the *lack of shared understanding and agreement about what terms to use and what they mean*, in a given domain. Ontologies are specifically designed to overcome this barrier by providing an explicit conceptualization of a domain, using a controlled vocabulary of terms, each with an agreed meaning. We describe technical approaches and scenarios for applying ontologies. For each approach, we highlight which part(s) of the knowledge asset lifecycle it is applicable to and the specific benefits of using the ontology.

2. Knowledge Asset Lifecycle

The Gartner Group created a KM 'Process Framework' that identified 5 key KM activities: create, capture, organize, access and use. We elaborate on this original framework, and present it as key activities in the lifecycle of a knowledge asset.

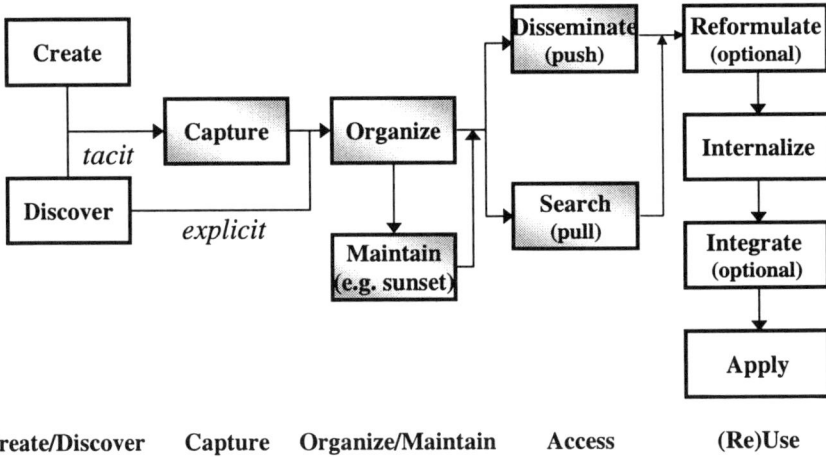

| Create/Discover | Capture | Organize/Maintain | Access | (Re)Use |

Figure 1: Activities in the knowledge asset lifecycle (KAL)

Create or Discover: A knowledge asset comes into existance or an existing knowledge asset is found somewhere in the organization. An asset may be tacit, in someone's head, or in an explicit form.

Capture: Tacit knowledge is identified and converted into explicit knowledge.

Organize & Maintain: A knowledge asset is placed in context, classified and indexed for the purpose of being rendered accessible for later use. A procedure is put in place, for maintaining the assets; e.g sunsetting those assets which are no longer useful.

Access: Knowledge is made available to knowledge workers who may actively search for a knowledge asset, or it could be disseminated using push technology.

(Re)Use: The knowledge asset is put to use. This require reformulating the knowledge so that it is appropriate for the current context. This must be done by someone who is familiar with the knowledge asset. It must then be internalized, (i.e., clearly understood) by the person who is going to use it – which may entail training. The asset may require integration with any other assets. Finally, it is applied to the current task at hand, typically resulting in a computer application.

3. Why Ontologies and What are They?

Barriers to exploiting knowledge assets–Exploiting knowledge assets depends on the ability of people, organizations and software systems to communicate between and among themselves. However, due to different needs and contexts, there can be varying viewpoints and assumptions regarding what is essentially the same subject matter. This can be due merely to the use of different terms for the same concepts, or the same term can be used to refer to different concepts. In general, the lack of a shared understanding leads to poor communication between people and their organizations.

Ontologies to the rescue–The way to overcome these barriers is to reduce or eliminate conceptual and terminological confusion and come to a *shared understanding*. This understanding is expressed as a common vocabulary, with an agreed meaning for the terms; it is called an `ontology'. The ontology functions as a unifying framework for the different viewpoints and serves as the basis for achieving a wide range of benefits.

What is an ontology?–The word `ontology' has long used by philosophers to refer to the study of existence. The goal is to identify and characterize all the kinds of entities that make up the world. In the nineties, this term came into common usage in the artificial intelligence community. Instead of referring to the *study* of what exists, it refers to the *output* of the study, in the form of an explicit conceptual model. An ontology is the result of careful analysis identifying all the important entities and relationships in a chosen domain, typically, a restricted subject area. Analogously, the word `methodology' used to refer to the study of methods in general. Now it is commonly used to refer to a specific method. See [1,2] for further introductory material on ontologies.

What does an ontology look like?–In the artificial intelligence community, an ontology is generally presumed to consist of set of terms with formal axioms that define each term's meaning. More recently, the word has been much more widely used to include sets of terms which may not have explicit definitions nor carefully defined relationships among the terms. A good example of this is the hierarchically presented Yahoo! subject classification, which has been referred to as an ontology. For the purposes of this paper, we will adopt a broad meaning of the term `ontology', characterizing it as follows: *An ontology may take a variety of forms, but necessarily includes a **vocabulary of terms**, and some indication of what the terms **mean**.*

There are many different ways that an ontology may be represented, depending on where it lies on a `semantic richness' continuum. We use this phrase to denote a rough measure of 1) the extent to which the semantics of the representation language used to express the meaning of terms is explicit and unambiguous and 2) how much meaning is specified. There are both costs and benefits of increasing the semantic richness of an ontology. For some applications, a `lightweight ontology' is most appropriate. Others applications require richer and more formal semantics.

At one extreme of the semantic richness continuum, there are vocabularies only, with little or no specification of meaning. This can still be an important step in reaching agreement, and in achieving KM benefits. However, there is no explicit agreement about what the terms mean. A classification taxonomy, is one step beyond this, in that there is an explicit generalization/specialization relationship which reduces possible ambiguity on meaning of terms. The Yahoo! system of categories contains much taxonomic information, and is presented hierarchically, but it also contains many other kinds of relationships. For example, the connection between two categories, variously means: part-of, generalization/specialization, similar-subject, to name a few. There is no

552

indication of which is which. A thesaurus for cataloging is another common example of a set of agreed terms, with some explicitly defined relationships which serve to characterize their meaning. Moving further along the semantic richness continuum, there are a variety of more expressive languages, including frame-based (object-oriented) ones [3] description logics [4,5] on to more general formal logics.

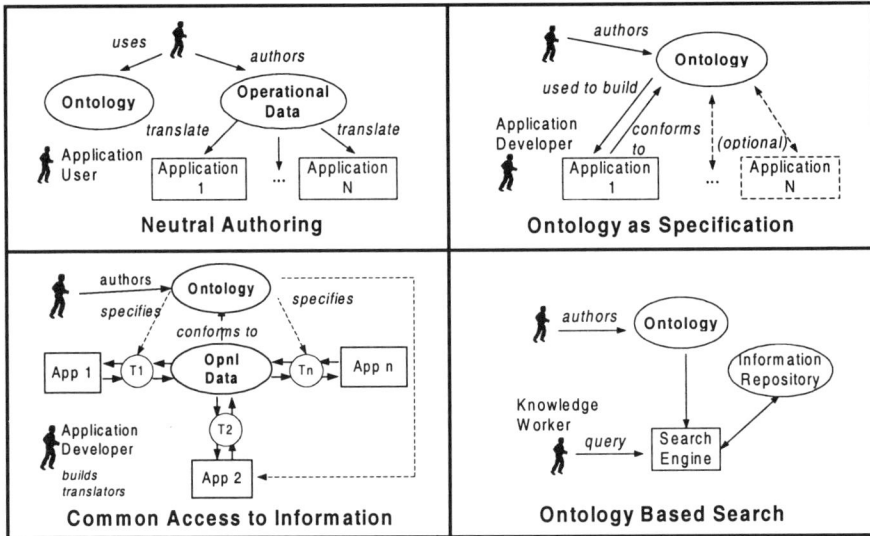

Figure 2: Approaches for applying ontologies

4. Approaches for Applying Ontologies

In this section, we introduce four main categories for ontology applications, emphasizing the specific role of the ontology (Figure 2). We describe each of these, indicate the main benfits, and note which of the KM activities these benefits accrue during. This section is based on [6], where further details may be found.

Neutral Authoring–An information artifact is authored in a single language, and is converted into a different form for use in multiple target systems. Benefits of this approach include knowledge reuse, improved maintainability and long term knowledge retention. This approach is relevant during the *creation*, *capture*, *organize & maintain*, and *(re)use* phases of the knowledge asset lifecycle (KAL).

Ontology as Specification–An ontology of a given domain is created and used as a basis for specification and development of some document or software. Benefits of this approach include documentation, maintenance, reliability and knowledge (re)use. This approach is relevant during the *creation*, *capture*, *organize & maintain*, and *(re)use* phases of the KAL.

Common Access to Information–Information is required by one or more persons or computer applications, but is expressed using unfamiliar vocabulary, or in an inaccessible format. The ontology helps render the information intelligible by providing a shared understanding of the terms, or by mapping between sets of terms. Benefits of this approach include inter-operability, and more effective use and reuse of knowledge resources. This approach is relevant during the *organize & maintain, access,* and *(re)use* phases of the KAL.

Ontology-Based Search–An ontology is used for locating information resources from a repository. The main benefit is more efficient access which leads to more effective use and reuse of knowledge resources. This approach is relevant during the *organize & maintain, access,* and *(re)use* phases of the knowledge asset lifecycle.

4.1 Neutral authoring

The goal of neutral authoring is to reduce costs of maintenance and to improve sharing and reuse of knowledge assets which are required to be delivered in multiple target formats. Knowledge assets are authored using a neutral format and then translated into the desired target format(s) as needed.

To implement this approach, one must 1) design an appropriately expressive neutral format and 2) build one-way translators from the neutral format into each target format. Designing a neutral format entails identifying the underlying conceptual models of all the target formats, and merging them into a standard conceptual model, or ontology. *This is a challenging and time-consuming task.* The ontology defines the agreed vocabulary which is the foundation for defining the neutral authoring format. It is also used as the basis for translation into target formats. Formal, semantically rich ontologies make it easier to build accurate translators.

4.1.1 Examples

In a large manufacturing organization, part designs may need to be delivered in different formats to different sub-contractors because they use different application software.

Figure 2 shows neutral authoring of operational data which is based on an ontology. A variation of this scenario is for the neutral authoring of ontologies themselves. An ontology is then translated into various target knowledge representation languages, for use in multiple knowledge based systems. See [7] for a detailed description of this approach, and [8] for a case study.

4.1.2 Benefits, tradeoffs and maturity

Benefits–Neutral authoring helps in the following activities in the KAL:
Creation, Capture: Adopting a neutral authoring format as a standard in an organization can result in reduced costs in learning and using the authoring tool because there is a greater pool of expertise in a single format.

Organize & Maintain: Maintenance costs are reduced because the major expense incurred by updating and keeping different versions in synch is eliminated. If a neutral format is designed to capture information at a higher conceptual level than the vendor-executable formats, then this can increase understandability and also contribute to reduced maintenance costs. Together, these benefits contribute to cheaper and more effective long-term retention of important knowledge assets.

(Re)Use: Increased understandability also makes it easier for knowledge assets to be reformulated and internalized so they can be adapted to a new application.

Tradeoffs–The large upfront cost of designing the neutral format must be amortized by subsequent savings due to easier maintenance and the value of retaining knowledge assets which otherwise may be lost.

Typically, not all target formats can express the same range of concepts. If the neutral format is very expressive, then something will be lost when translating to less expressive target formats. This can result in authors choosing to use only the restricted subset of the neutral format which is known to translate well into the desired target format. We call this *translator bias.* If different authors aim their creations at different target formats, then this undermines the benefits of neutrality—re-introducing problems with maintenance and long-term knowledge retention. Translator bias is reduced if the neutral format is less expressive, but this comes at the cost of no longer being able to use advanced language features found in target applications. In summary, designing the `best' neutral format, entails choosing the right balance between degree of translator bias that is acceptable versus the need for advanced features in some of the target formats. See [9] for further analysis of these tradeoffs.

Maturity–Totally automated translation of ontologies into operational targets has been difficult and typically relies on translation of idiomatic expressions [10] . For case studies and analysis of some of these problems see [6,9,11,12,13]. Common practice in industry is to build point to point translators when the need arises.

4.2 Ontology as specification

Ontologies may be used to improve on the development and deployment of software applications–an important kind of knowledge asset. An ontology is created that models the application domain and provides a vocabulary for specifying the requirements for one or more target applications. The richer the ontology is in expressing meaning, the less the potential for ambiguity in creating requirements. The ontology guides the development of the software; the software conforms to the ontology. The benefits of this approach include documentation, maintenance, reliability and knowledge (re)use.

Structurally, this scenario is similar to neutral authoring (Figure 2). In both cases, the ontology plays an important role in the development of the application. There are two important differences. First, there are benefits that apply if there is only one application (e.g. documentation and maintenance). Second, the role of the ontology is to guide the development of the target application(s); there is no translated artifact which becomes part of the target application.

4.2.1 Examples

KADS/CML–An ontology author creates an ontology using the conceptual modeling language (CML) from the KADS methodology [14]. The application developer uses this ontology as part of the requirements specification when developing the target knowledge based system (e.g., for diagnosing faults).

In this example, documentation is improved because there is an explicit representation of the ontology that the software is based on. Better documentation always makes maintenance easier. Reuse is achieved if the ontology is used for different applications in the same domain.

Information Modeling–Data modeling languages such as EXPRESS [15] and IDEF1X [16] are used to create data base schemas, which can be thought of as ontologies. In the case of EXPRESS, the data base schema specifies the physical file format for a target implementation. More generally, information and system modeling languages like UML can also be used to create ontologies which serve to specify target software. A special case of information modeling is object-oriented analysis and design [17]. The analysis phase of this approach consists, in part of building an ontology (classes and attributes) which provides the vocabulary for specifying the software requirements.

There are many other examples of this use of ontologies, see [6] for a few more.

4.2.2 Benefits, tradeoffs and maturity

Benefits – Using an ontology as a specification has benefits in the following areas in the knowledge asset life cycle:

Creation, Capture: By building an ontology first, important tacit knowledge may become explicit, possibly for the first time. Thus, the ontology is itself a knowledge asset.

Organize & Maintain: The ontology could be used as a way to classify the application which helps in the organization step. Making an explicit link between an ontology and a target application, increases understandability of the application, which in turn reduces maintenance costs.

(Re)Use: Increased understandability of the application also makes it easier to reuse in a new context. Furthermore, the ontology itself can become a reusable asset, e.g. by supporting other applications in the same domain.

Tradeoffs – The major tradeoff for using an ontology as a specification, is the up-front time it takes to build the ontology. See [8] for a compelling case arguing that it is worth the effort.

Maturity – The KADS methodology is widely used for building knowledge based systems. Typically, however there is no formal connection between the ontology and the developed software.

For Express, IDEF1X, UML and various other such languages, there are commercial tools for automatically generating parts of the target software from the specifications. There are also various tools for translating between these languages (e.g. Express and UML).

4.3 Common access to information

This approach uses ontologies to enable multiple target applications (or humans) to have access to heterogeneous sources of information which is otherwise unintelligible. The lack of intelligibility derives both from 1) different formats and 2) different terminology [even if the format is the same]. The main benefits of this approach include inter-operability, and knowledge reuse.

The first step towards achieving interoperability is to identify the local ontologies for each target application. These specify the meaning of models encoded in the corresponding formats and are necessary before any software can be developed which automatically converts between formats. We will look at two different scenarios in this category.

Common Access via Shared Ontology – In this case, an ontology is created which is a combination of all the local ontologies for each target format. This ontology is used as the basis for a neutral interchange format; thus it must be agreed on by all target applications. Knowledge assets are converted to the required formats by a two step translation process, first into the interchange format, and from there into the desired target format (Figure 2). This requires two-way translation between each target format and the neutral format. The process of designing the neutral interchange format is much the same as for designing a neutral authoring format.

Common Access via Mapped Ontologies – In this case, there is no explicit shared ontology; instead, each application has it's own local ontology. For each pair of applications wishing to exchange data, the ontology authors cooperate to create and agree on a set of mapping rules that define what each term in one ontology means in the other ontology. This mapping is used to generate a mediator, which maps operational data expressed in the terminology of one ontology into operational data expressed using terms in the other ontology. A mediator uses these rules at runtime so that applications can access each other's data.

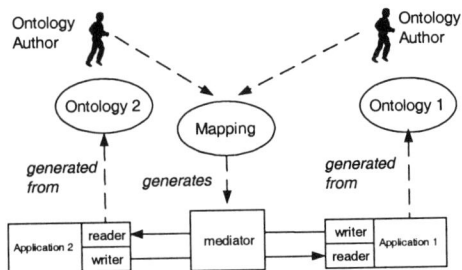

Figure 3: Common access via mapped ontologies

4.3.1 Examples

Shared Ontology–EcoCyc [18] is a commercial product that uses a shared ontology to enable access to heterogeneous databases in the field of molecular biology. The ontology is a conceptual schema (i.e. ontology) that is an integration of the conceptual schemas for each of the separate databases.

Mapped Ontologies–A developer of an application (e.g., electrical power suppliers) wants to share data with another application (e.g., schematic viewer). Each application has its own ontology created in EXPRESS. The developers agree on a mapping (e.g., represented in EXPRESS-X), which relates terms in the power supply application with electrical schematics. The mapping is used to generate a mediator that maps those portions of the electrical power supply data into schematic data.

4.3.2 Benefits, tradeoffs and maturity

Benefits–Using an ontology to achieve common access to information has benefits in the following areas in the knowledge asset life cycle:
Access: Knowledge assets that would otherwise be inaccessible because they are in the wrong format, are made available for use. However, the ontology does not directly assist in helping to find them; for that, see section 4.4.
(Re)Use: The main benefit of using an ontology in this way is to make it possible for systems to make use of a wider range of knowledge assets, through interoperability. Reaching agreement on terms can also facilitating greater communication between humans, by alleviating problems with specialized jargon.

Tradeoffs–Having a neutral format is one approach to handling the problem of requiring multiple target formats. An alternative is to use existing vendor formats for authoring, and to write point to point translators to other target formats. One way to do this, is by using explicit mapping rules and mediators, as described above. In either case, translation is still required. Point to point translators tend to be more robust and tailored to the individual target formats. This is traded off against the increased cost of developing and maintaining many different translators between many combinations of target formats. Every time a single format changes, then several translators may need to be updated, instead of only needing to change one, if a neutral format is used.

Maturity–The degree of maturity of the technologies described above depends a lot on the context and the particular application. In some cases, there is mature commercial technology, (e.g. EXPRESS [15], EcoCyc [18]) in other cases, there are mainly research prototypes (e.g. PSL [19]). Fully automated translation of ontologies is beyond the current state of the art. Problems such as translator bias, are inherent, and will not go away, as long as different formats have different levels of expressive power. See [6,9,11,12,13] for further discussion of these issues.

4.4 Ontology-based search

In this scenario, an ontology author creates an ontology that is used as a basis for structuring and indexing an information repository. Knowledge workers identify concepts that they are interested in. The search engine uses these concepts to locate desired resources from a repository. The motivation is to improve precision and/or recall and reduce the amount of time spent searching.

Supporting technologies include ontology browsers, search engines, automatic tagging tools, automatic classification of documents, natural language processing, meta-data languages (e.g., XML), natural language ontologies, large general-purpose knowledge bases and thesauri, and knowledge representation and inference systems.

4.4.1 Examples

Yahoo!–Here, the ontology is used as a conceptual structure to help users understand the information repository, and also as an index into the repository. This requires that all items in the repository are linked to items from the index. An ontology author creates an ontology which is used to limit the scope of the search. A specialized search engine uses these terms to locate relevant documents in a repository.

Knowledge-Based Discovery Tool [20]–The user enters words related to the concept of interest (e.g. tank). The browser consults a large WordNet-based ontology, identifies any potential ambiguities, and asks the user to identify the correct concept (e.g. a storage tank, or an army tank). The browser re-formulates and improves the query by automatically. For example, if the desired concept is storage tank, then responses which contain words strongly associated with `army' are excluded.

These examples illustrate two different roles that an ontology may play to assist search; they may be used separately, or in concert. First, an ontology may be used as *a basis for semantically structuring and organizing the information repository*. This structure may be used in at least two distinct ways: a) as a conceptual framework to help the user think about the information repository and formulate queries (e.g. Yahoo!) and b) as a vocabulary for meta-data used for indexing and/or tagging the documents in the repository (this does not require user interaction).

The second role an ontology may play is *to assist in query formulation*; here also, the ontology can be used in at least two different ways, depending on the involvement of the user. It can be used to a) drive the user interface for creating and refining queries, for example by disambiguating –or– b) perform inference to improve the query (not requiring user interaction). See [21,22] for additional examples of using ontologies for search.

4.4.2 Benefits

Ontology-based search has benefits in the following areas in the knowledge asset lifecycle:

Organize: The ontology plays a key role in structuring and organizing the repository of knowledge assets, so that they may be effectively indexed and classified.

Access: The ontology is used as the basis for browsing the repository, and in query (re)formulation.

5. An Ontology Application of the Future

At this point, we turn our attention to an important knowledge management problem: *enabling integration of diverse information resources on the web.* We motivate the problem, indicate the limitations of current technologies, and outline an approach for automated support, highlighting the role of ontologies.

5.1 Motivation

The ease of publishing web content has produced explosive growth in the quantity of knowledge available in corporate intranets and the Internet as a whole. Currently, most web publishing focuses on human understanding and does not provide the kinds of semantic markup required to adequately automate common activities. The most common and widely recognized of these activities is automating search. Many research projects have focused on improving automated search using language analysis, statistical techniques, and semantic tags.

To ensure more effective use of knowledge assets, future applications will also need to assist in other activities, such as reformulation, integration and human internalization of knowledge (Figure 1). Computers can play an important role in automating these activities. In part, we can enable automation by meeting the computer halfway—providing markup to increase computer understanding of web-based content. Currently, the near-ubiquitous use of HTML and lack of semantic markup has made it difficult for software programs (e.g., agents) to assist humans in these activities.

5.2 Current approaches

A significant amount of knowledge management research has focused on locating and searching web content. Ontologies using various degrees of semantic richness, are fundamental to several techniques including: structured frameworks for navigation (e.g., Yahoo), concept-based search [23] , and semantic tags (e.g., Ontobroker [24], SHOE [25]).The key activities of reformulation, integration and human internalization of knowledge are not well supported by automation.

Users must often reformulate content or integrate it with other sources based on the context of a specific problem. In simple cases, this reformulation occurs implicitly (in people's heads) until they reach cognitive overload. More experienced users open multiple browser windows on different sources to handle slightly more complex integration problems. Moderate to complex problems may require that users explicitly copy, integrate, and reformulate information to internalize enough content to solve a problem.

Software developers have become adept at building web front-ends to legacy databases and applications. HTML is the predominant mechanism for encoding the underlying knowledge. While HTML is adequate for *presenting* content for human users, it does not provide built-in mechanisms for machines to *identify* and *reason* about semantic content. Identification of semantic content is necessary for content reformulation and construction of a consistent view (i.e., with consistent semantics).

XML provides a mechanism for programmers to develop their own so-called semantic tags, typically defined in a DTD. The common usage of the word *semantic*, here is somewhat misleading. Programmers can agree amongst themselves the meaning of the tags, but the DTD provides no means of publicly and formally declaring their meaning. Programs using these XML documents must embed the semantics of the tags at compile time to understand their meaning or make useful inferences.

Researchers have developed languages further along the 'semantic richness' continuum, (e.g., SHOE, Ontobroker). These languages provide mechanisms for representing semantic content in a way that is explicit, public and available for programs to exploit. A simple example in which semantic markup can be used is location of documents containing specific tags. A user fills out a form identifying specific values for various semantic tags. A search engine uses these tags and the underlying document to return a set of documents consistent with the user query. Semantic markup used this way, helps in the important step of accessing knowledge assets. Figure 4 shows the relationship between various markup languages. The problem remains: how to enable agents to assist in reformulation and integration of information necessary for solving complex problems?

	Presentation	Content	Content Semantics	Supports Inference
HTML	Yes	implicit, no access	no	no
XML/XSL	Yes	explicit, accessible	implicit, inaccessible for reasoning	no
SHOE, Ontobroker	Yes	explicit, accessible	explicit, accessible for reasoning	yes

Figure 4: Comparing web markup languages

5.3 A sample integration problem

Suppose you start a new job in the purchasing department of a large international aerospace company. Part of your job involves purchasing *jauges* for the 797 aircraft. You've been told there are numerous suppliers of *jauges* around the world, each using different terminology to describe their products. Using existing web technology, you search the web and identify 3 *jauges*, all from France! As the international *jauges* market is small, all of the sites are still written in French. A closer inspection of the sites, with foreign dictionary in hand, reveals that *jauges* are dipsticks. Using an online translation service, you are able to locate dipstick manufactures in Germany, Britain, Brazil, Iceland, and Moldova. Each manufacture website uses HTML to present their products in a different form using different terminology. Furthermore, searching of the websites requires filling in various forms using the local terminology. Solving your purchasing problem requires that you select, reformulate (e.g., translate) and integrate data from each of the web sites. Integrating heterogeneous content provides another way in which knowledge workers can exploit ontologies.

Reformulation and integration today—Selecting, reformulating, and integrating dipstick data using standard HTML would be difficult, error prone, and brittle. Human users, with an understanding of the language, are able (sometimes) to make good guesses about meaning using context, formatting, and experience as guides. Because HTML does not provide a standard mechanism for identifying semantic content, programs are unable to interpret the data reliably. While researchers have developed wrappers [26] to extract semantic content, slight changes in website structure or formatting ultimately make this approach brittle.

Reformulation and integration tomorrow—XML provides a means of tagging information content that is independent of the way it is presented. The current XML mechanisms (DTDs and XML Schema) provide little for making the semantics external and explicit. Users of XML tagged data must agree on the meaning of the tags prior to their use. Applications can only make sense of tags they are "aware of" (i.e., their programmers have built this knowledge in). Changes to the underlying schema (i.e., adding, removing, or changing tags) is likely to cause these programs difficulty. In essence, all information about the nature of the document content must be compiled into the application before it's used. As inherent meaning of the tags is implicit and internal in the applications, changes to the meaning can occur without any agreement or external notification.

Reformulation and integration the day after tomorrow (ontologies)—The ontological approaches, such as in SHOE and Ontobroker, provide a means of describing semantics that is pubic, and external to the applications which use it. Applications can read the public, explicit ontology to gain an understanding of meaning of the data, without hard coding of implicit semantics. A simple example would be ontological specification of units-of-measure. Suppose that British dipstick manufacturers uses inches, and the rest of the manufacturers use centimeters as a means of measuring dipstick length. Normally, applications would have to contain some built-in logic for distinguishing different manufacturers, which units they were using, and how to convert the data. Ideally, an ontology would contain enough information about different metrics in such a way that the application could automatically convert between these data types without having any pre-built understanding of inches and centimeters. Similar ontological specifications could be used to normalize data between different units and measures across the manufacturers.

6. Summary

In this chapter, we have argued that ontologies play an important role in achieving a wide range of knowledge management benefits. We described key activities in the life cycle of a knowledge asset, and noted some barriers to exploiting knowledge assets. We showed that many of these barriers arise from a lack of a common vocabulary and common usage of terms, and can thus be addressed by using ontologies. We described four major approaches to applying ontologies, indicating benefits, tradeoffs and the level of maturity of the relevant technologies. Finally, we argued that ontologies will play a key role in solving a problem of growing importance: *enabling integration of diverse information resources on the web.*

References

[1] M. Uschold and M. Gruninger. Ontologies: Principles, methods and applications. *Knowledge Engineering Review*, 11(2), 1996

[2] M. Uschold. (editor) Knowledge level modelling: Concepts and terminology. *Knowledge Engineering Review*, 13(1), 1998.

[3] M. Kifer, G. Lausen, and J. Wu. Logical foundations of object-oriented and frame-based languages. *Journal of the ACM*, 42, 1995.

[4] R. Brachman, D. McGuinness, P. Patel-Schneider, et al. Living with classic: When and how to use a kl-one-like language. In J. Sowa, editor, *Principles of Semantic Networks*, pages 401-456. Morgan Kaufman, 1991.

[5] C Welty and J. Jenkins. Formal ontology for subject. *J. Knowledge and Data Engineering*, 31(2):155-182, 1999.

[6] R. Jasper and M. Uschold. A framework for understanding and classifying ontology applications. In *Proceedings of the 12th Workshop on Knowledge Acquisition for Knowledge-Based Systems Workshop*, Banff, Alberta, Canada, October 1999. Department of CS, University of Calgary.

[7] R. Fikes, A. Farquhar, and J. Rice. Tools for assembling modular ontologies in ontolingua. In *Proceedings of AAAI-97*, pages 436-441, 1997.

[8] G. van Heijst, Schreiber A.Th., and B.J. Wielinga. Using explicit ontologies in kbs development. *Journal of Human Computer Studies*, 46(2/3):183-292, 1997.

[9] M. Uschold, R. Jasper, and P. Clark. Three approaches for knowledge sharing: A comparative analysis. In *Proceedings of the 12th Workshop on Knowledge Acquisition for Knowledge-Based Systems Workshop*, pages 4-13-1 to 4-13-13, Banff, Alberta, Canada, October 1999. Department of Computer Science, University of Calgary.

[10] T. Gruber. A translation approach to portable ontology specifications. *Knowledge Acquisition*, 5(2):199-220, 1993.

[11] W. Grosso, J. Gennari, R. Fergeson, et al. When knowledge models collide (how it happens and what to do). In *Proc. of the 11th Workshop on Knowledge Acquisition, Modeling and Management*, Banff, Alberta, Canada, April 1998.

[12] A. Valente, T. Russ, R. MacGregor, et al. Building and (re)using an ontology of air campaign planning. *IEEE Intelligent Systems*, 14(1):27-36, January/February 1999.

[13] M. Uschold, M. Healy, K. Willamson, et al. Ontology reuse and application. In N. Guarino, editor, *Formal Ontology in Information Systems*, pages 179-192, Trento, Italy, June 1998.

[14] G. Schreiber, B. Wielinga, H. Akkermans, et al. *CML: The CommonKADS Conceptual Modelling Language*. Kads-ii project deliverable, University of Amsterdam and others, 1994.

[15] D. Schenck and P. Wilson. *Information Modeling the EXPRESS Way*. Oxford University Press, 1994.

[16] T.A. Bruce. *Designing Quality Dababases with IDEF1X Information* Models. Dorset House Publishing, 1992.

[17] C. Larman. *Applying Uml and Patterns : An Introduction to Object-Oriented Analysis and Design*. Prentice Hall, 1997.

[18] P.D. Karp, M. Riley, S.M. Paley, et al. Ecocyc: encyclopedia of e.coli genes and metabolism. *Nucleic Acids Res.*, 24:32-40, 1996.

[19] C. Schlenoff, M. Ciocoiu, D. Libes, et al. Process specification language: Results of the first pilot implementation. *In Proceedings of the International Mechanical Engineering Congress and Exposition*, Nashville, TN, 1999.

[20] E. Eilerts, K. Lossau, and C. York. Knowledge base discovery tool. In *Proceedings of AAAI-99*, 1999.

[21] N. Guarino, C. Masolo, and G. Vetere. Ontoseek: Using large linguistic ontologies for accessing on-line yellow pages and product catalogs. *IEEE Intelligent Systems*, 14(3):70-80, May/June 1999.

[22] D. McGuinness. Ontological issues for knowledge-enhanced search. In N. Guarino, editor, *Formal Ontology in Information Systems*, pages 302-316, Trento, Italy, June 98.

[23] P. Clark, J. Thompson, H. Holmbeck, et al. *Exploiting a thesaurus-based semantic net for knowledge-based search*. Submitted to Conference: Innovative Applications of AI, 2000.

[24] S. Decker, M. Erdmann, D. Fensel, et al. Ontobroker: Ontology based access to distributed and semi-structured information. In R. Meersman et al., editor, *Semantic Issues in Multimedia Systems*. Kluwer, 1999.

[25] S.. Luke, L. Spector, D. Rager, et al. Ontology-based web agents. *In First International Conference on Autonomous Agents* (AA '97), 1997.

[26] I. Muslea, S. Minton, and C.A. Knoblock. A hierarchical approach to wrapper induction. In *Proceedings of the Third International Conference on Autonomous Agents*, Seattle, WA, 1999.

Chapter 7

Resource Guide

Resource Guide

The Resource Guide is developed to provide the practitioners and researchers with a selected list of Internet resources, journals, conferences and research groups in the area of Micro Knowledge Management. Most of the information is based on searches on the Internet, and the site addresses are valid at the time of this chapter development.

Resource Sites on the Internet

- Knowledge Based Systems:
  http://www.emsl.pnl.gov:2080/proj/neuron/kbs/homepage2.html
- Expert Systems:
  http://www.pcai.com/pcai/New_Home_Page/ai_info/expert_systems.html
- Engineering and Management of Knowledge: http://www.commonkads.uva.nl/
- Knowledge Representation: http://www.cs.man.ac.uk/~franconi/kr.html
- Knowledge Management and the Internet: http://www.eknowledgecenter.com/
- Machine Learning Network: http://www.mlnet.org/
- Machine Learning Resources:
  http://www.aic.nrl.navy.mil/~aha/research/machine-learning.html
- Ontology projects: http://www.cs.utexas.edu/users/mfkb/related.html
- Ontology Resource Page: http://wings.buffalo.edu/philosophy/ontology/
- Data Mining, Web Mining and Knowledge Discovery:
  http://www.kdnuggets.com/
- Artificial Intelligence: http://spinoza.tau.ac.il/hci/dep/philos/ai/links.html
- Knowledge Management Network: http://ceres.cibit.nl/index.html
- Knowledge Management Reference Site:
  http://cis.kaist.ac.kr/research/kmsite.htm
- Knowledge Management Resources: http://www.brint.com/km/

Journals

- Data Mining and Knowledge Discovery, Kluwer Academic Publishers,
  http://www.wkap.nl/journalhome.htm/1384-5810
- International Journal of Uncertainty, Fuzziness and Knowledge-Based Systems, World Scientific Publishing,
  http://www.wspc.com.sg/journals/ijufks/ijufks.html
- Machine Learning, Kluwer Academic Publishers,
  http://www.wkap.nl/journalhome.htm/0885-6125
- Data and Knowledge Engineering, Elsevier Science,
  http://www.elsevier.nl/inca/publications/store/5/0/5/6/0/8/
- IEEE Transactions on Knowledge and Data Engineering, IEEE Computer Society, http://www.computer.org/tkde/

- International Journal of Software Engineering & Knowledge Engineering, World Scientific, http://www.ksi.edu/ijsk.html
- The Knowledge Engineering Review, Cambridge University Press, http://www.cup.cam.ac.uk/Journals/JNLSCAT/ker/ker.html
- Artificial Intelligence for Engineering Design, Analysis and Manufacturing, Cambridge University Press, http://www.cup.cam.ac.uk/Journals/JNLSCAT/aie/aie.html
- Intelligent Data Analysis, IOS Press, http://www.iospress.nl/html/1088467X.html
- International Journal of Human-Computer Studies/Knowledge Acquisition, Academic Press, http://www.academicpress.com/ijhcs
- Knowledge and Information Systems - An International Journal, Springer, http://link.springer.de/link/service/journals/10115/index.htm
- Journal of Knowledge Management Practice, TLA Inc., http://www.tlainc.com/jkmp.htm
- Journal of Knowledge Management, MCB University Press, http://www.mcb.co.uk/cgi-bin/journal1/jkm

Conferences

- PAKeM2000, London (UK), http://www.practical-applications.co.uk/PAKeM2000/
- DaWaK 2000, London (UK), http://www.dexa.org/
- PAKM2000, Basel (Switzerland), http://research.swisslife.ch/pakm2000
- CIKM2000, Washington (USA), http://www.cs.umbc.edu/cikm/index.html
- KDD2000, Boston (USA), http://www.acm.org/sigkdd/kdd2000
- KBCS2000, Mumbai (India), http://www.ncst.ernet.in/kbcs2000/
- ES2000, Cambridge (UK), http://www.bcs-sges.org/es2000/
- PAKDD2000, Kyoto (Japan), http://www.kecl.ntt.co.jp/icl/about/ave/PAKDD00/
- ICML2000, San Francisco (USA), http://www-csli.stanford.edu/icml2k/
- EKAW'2000, French Riviera (France), http://www.inria.fr/acacia/ekaw2000
- ECAI2000, Berlin (Germany), http://www.ecai2000.hu-berlin.de/
- KMWorld2000, California (USA), http://www.kmworld.com/00/
- Communities of Practice 2000, California (USA), http://www.iir-ny.com/conference.cfm?EventID=MW129&

Magazines

- Knowledge Management, Learned Information Europe Ltd., http://www.knowledge-management.co.uk/kbase/index.asp
- KM World Online, http://www.kmworld.com/
- The International Knowledge Management Newsletter, Management Trends International, http://www.mjm.co.uk/knowledge/index.html

Research Groups

- Knowledge Acquisition, Nottingham (UK),
  http://www.psychology.nottingham.ac.uk/research/ai/themes/ka/
- Data Mining Research Lab, Arizona (USA), http://insight.cse.nau.edu/
- Institute for Knowledge and Language Engineering, Magdeburg (Germany),
  http://www-iik.cs.uni-magdeburg.de/iik-eng-neu.html
- Knowledge Based Systems, Urbana Champaign (USA),
  http://www-kbs.ai.uiuc.edu/
- Knowledge Modelling, Milton Keynes (UK),
  http://kmi.open.ac.uk/knowledge-modelling/
- Knowledge Science Institute, Calgary (Canada),
  http://ksi.cpsc.ucalgary.ca/KSI/KSI.html
- Knowledge Engineering Methods and Languages, Amsterdam (The
  Netherlands), ftp://swi.psy.uva.nl/pub/keml/keml.html
- Advanced Knowledge Technologies, Aberdeen (UK),
  http://www.csd.abdn.ac.uk/research/aktgroup.html
- Knowledge Engineering, Northern Ireland (UK), http://nikel.infj.ulst.ac.uk/
- Micro Knowledge Management, Cranfield (UK),
  http://www.cranfield.ac.uk/sims/cim
- Software Engineering and Knowledge Engineering, Madrid (Spain),
  http://zorro.ls.fi.upm.es/udis/i_d/isic_e.html
- Artificial Intelligence, Edinburgh (UK), http://www.dai.ed.ac.uk/
- Applied Artificial Intelligence, Washington (USA),
  http://www.aic.nrl.navy.mil/
- Expert Systems and Knowledge-Based Systems, NRC (Canada),
  http://ai.iit.nrc.ca/subjects/Expert.html
- Experimental Knowledge Systems Laboratory, Massachusetts (USA),
  http://eksl-www.cs.umass.edu/eksl.html
- Automated Reasoning Systems, Trento (Italy), http://sra.itc.it/index.epl
- Knowledge Systems and Ontology, Osaka (Japan),
  http://www.ei.sanken.osaka-u.ac.jp/
- Ontology, Padova (Italy),
  http://www.ladseb.pd.cnr.it/infor/Ontology/ontology.html
- Machine Intelligence, Calcutta (India), http://www.isical.ac.in/~miu/
- Knowledge Management, Warwick (UK),
  http://bprc.warwick.ac.uk/Kmweb.html
- Knowledge Management, Arizona (USA),
  http://www.cmi.arizona.edu/research/Kno_mgmt/
- Knowledge Management, Los Alamos (USA), http://km.lanl.gov/research.htm